Selected Topics in Structronics and Mechatronic Systems

SERIES ON STABILITY, VIBRATION AND CONTROL OF SYSTEMS

Founder and Editor: Ardéshir Guran
Co-Editors: A. Belyaev, C. Christov, G. Stavroulakis
& W. B. Zimmerman

About the Series

Rapid developments in system dynamics and control, areas related to many other topics in applied mathematics, call for comprehensive presentations of current topics. This series contains textbooks, monographs, treatises, conference proceedings and a collection of thematically organized research or pedagogical articles addressing dynamical systems and control.

The material is ideal for a general scientific and engineering readership, and is also mathematically precise enough to be a useful reference for research specialists in mechanics and control, nonlinear dynamics, and in applied mathematics and physics.

Selected Volumes in Series B

Proceedings of the First International Congress on Dynamics and Control of Systems, Chateau Laurier, Ottawa, Canada, 5–7 August 1999
Editors: A. Guran, S. Biswas, L. Cacetta, C. Robach, K. Teo, and T. Vincent

Selected Volumes in Series A

Vol. 1   Stability Theory of Elastic Rods
Author: T. Atanackovic

Vol. 2   Stability of Gyroscopic Systems
Authors: A. Guran, A. Bajaj, Y. Ishida, G. D'Eleuterio, N. Perkins, and C. Pierre

Vol. 3   Vibration Analysis of Plates by the Superposition Method
Author: Daniel J. Gorman

Vol. 4   Asymptotic Methods in Buckling Theory of Elastic Shells
Authors: P. E. Tovstik and A. L. Smirinov

Vol. 5   Generalized Point Models in Structural Mechanics
Author: I. V. Andronov

Vol. 6   Mathematical Problems of Control Theory: An Introduction
Author: G. A. Leonov

Vol. 7   Analytical and Numerical Methods for Wave Propagation in Fluid Media
Author: K. Murawski

SERIES ON STABILITY, VIBRATION AND CONTROL OF SYSTEMS

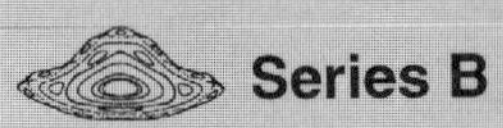 **Series B**   **Volume 3**

Founder and Editor: **Ardéshir Guran**

Co-Editors: **A. Belyaev, C. Christov, G. Stavroulakis & W. B. Zimmerman**

Selected Topics in Structronics and Mechatronic Systems

Editors

Alexander Belyaev

State Technical University of St. Petersburg, Russia

Ardéshir Guran

Institute for Structronics, Canada

Published by

World Scientific Publishing Co. Pte. Ltd.

5 Toh Tuck Link, Singapore 596224

USA office: Suite 202, 1060 Main Street, River Edge, NJ 07661

UK office: 57 Shelton Street, Covent Garden, London WC2H 9HE

British Library Cataloguing-in-Publication Data
A catalogue record for this book is available from the British Library.

SELECTED TOPICS ON STRUCTRONICS & MECHATRONICS SYSTEM

ISBN 981-238-083-3

Printed in Singapore.

STABILITY, VIBRATION AND CONTROL OF SYSTEMS

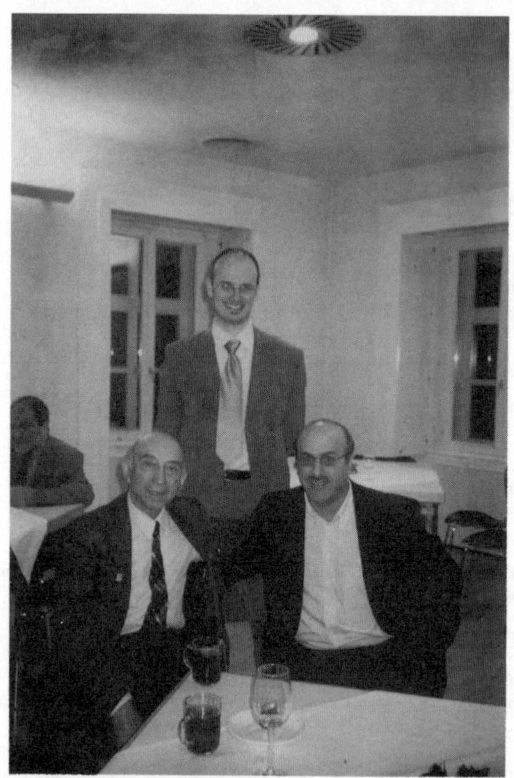

Prof. A. Lotfi Zadeh (University of California, Berkeley) and Prof. Ardeshir Guran (Institute of Structronics, Canada) during reception following the ceremony of the award of Doctorate Honoris Causa bestowed to Prof. Lotfi Zadeh, Johannes Kepler University, Linz, February 2003. (Photo by Dr. Christina Leitner)

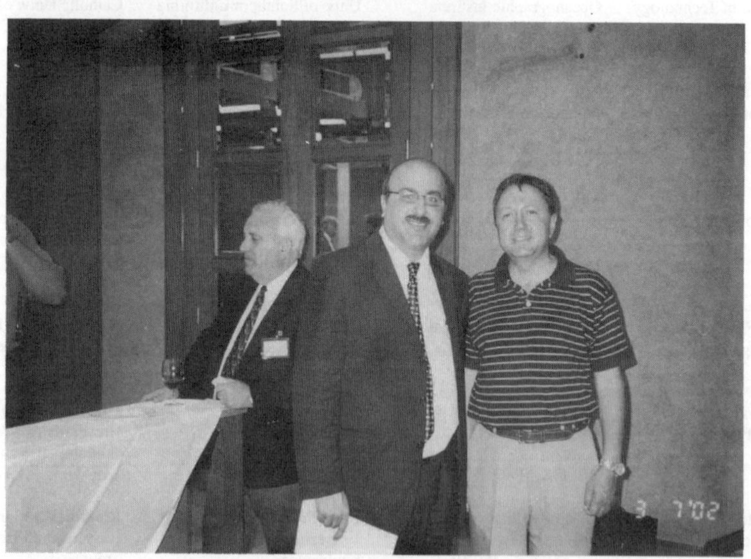

Prof. Ardeshir Guran (Institute of Structronics, Canada) and Prof. Bahram Ravani (University of California, Davis) during a reception sponsored by mayor of the city of Linz in honor of participants of the First International Congress on Mechatronics, Johannes Kepler University, July 3–6, 2002. (Photo by Professor Herbert Überall)

Preface

Structronics and Mechatronics are frequently reported as a combination of mechanics, structural engineering, electronics and computer science. The main applications of this new science is on controlled mechanical devices and structural systems that may lead the engineers to the statement that there is nothing new behind, simply because authomatic control and computing has been within engineering tools since long time ago. Thus, there are obstacles in this reunion process. We believe Structronics or Mechatronics is a philosophy and belongs to all of the above mentioned branches but to none of them exclusively.

Both authors of this book have been invited as Professor at Johannes Kepler University in Austria where they taught undergraduate as well as graduate courses in Mechatronics Programme. The main objective of this volume is to present recent developments in this field.

Alexander Belyaev   Ardeshir Guran

Linz, Austria
February 2002

Some of the participants of the first symposium on Structronics and Mechatronic Systems are celebrating the success of the meeting in professor Belyaev's house. (Photo by Mrs. Olga Belyaev)

In clockwise direction: Prof. Miha Boltezar, Prof. Ardeshir Guran, Prof. Alexander Belyaev, Prof. Hartmut Bremer, Mrs. Bremer, Mr. J. Gerstmayr, Mr. F. Fuchshumer, and Mr. R. Novak. (Photo by Mr. J. Mikota)

Contents

Chapter 3: Some Aspects of Washing Complex Non-Linear Dynamics   83
Miha Boltežar

Chapter 4: Analysis and Nonlinear Control of Hydraulic Systems in Rolling Mills   121
Rainer M. Novak

Chapter 5: Mathematical Modelling and Nonlinear Control of a Temper Rolling Mill   175

Stefan Fuchshumer, Kurt Schlacher and Andreas Kugi

Chapter 6: Combining Continuous and Discrete Energy Approaches to High Frequency Dynamics of Structures   221

Alexander K. Belyaev

Chapter 7: Computational Methods for Elasto-Plastic Multibody Systems   269

Johannes Gerstmayr

Chapter 9: Ionic Polymer-Conductor Composites (IPCC) as Biomimetic Sensors, Actuators and Artificial Muscles

Mohsen Shahinpoor and Ardeshir Guran

ON THE USE OF NONHOLONOMIC VARIABLES IN ROBOTICS

HARTMUT BREMER

Dept. Mechatronics, Chair of Robotics
Johannes Kepler Universität Linz
Altenbergerstr. 69, A-4040 Linz/Austria
E-mail: bremer@mechatronik.uni-linz.ac.at

Robots generally represent Multi Body Systems with apparently simple topology (mostly a topological chain). Design elements are typically motor-gear-arm units or, in case of stiff gears, single arms (rigid bodies) with may be looked at as subsystems. The obviously easiest access to subsystem dynamics is the use of cartesian velocities (a first set of nonholonomic variables). When combining the subsystems to the model under consideration, these are constrained by the actual design requirements leading to a second set of velocities, denoted as minimal velocities according to the minimal (constrained) state. These are nonholonomic if nonholonomic constraints arise but they may also be nonholonomic for the holonomic case. A first look on the available methods leads to the use of the Projection Equations for simplest procedure. Next, specifying reference frames yields a recursive scheme for kinematics as well for kinetics which avoids the total mass matrix inversion. Finally, one may ask why to inspect the constraints at all instead of to confine oneself on minimal state space considerations. The answer is simply that constraints may become active or remain passive which typically arises in mobile robots (friction at the contacts with the environment in wheeled robots, closing and opening contacts along with impact and friction in legged locomotion). Here, correponding motion equations lead directly to an adequate formulation of the Gauss Principle. Having these backgrounds in mind, the main question to be investigated in this contribution is wether the use of nonholonomic variables offers advantages even in the holonomic case.

1 Introduction

Robots generally consist of "Motor-Gear-Arm" units (rigid and/or flexible). A typical representation is the stationary industrial robot where the combination of units leads to (a kind of generalized) n-fold pendulum ("topological chain"). In contrary to stationary ones mobile robots are characterized by its basis moving w.r.t to the inertial frame. Mobility is hereby achieved either by wheels (or chains) or by legs. Obviously, any kind of wheeled motion (without sliding) needs to fulfill nonholonomic constraints, while in legged locomotion at least one leg will contact the ground during motion, leading to holonomic constraints because of the temporarily "fixed" contact point (but changing the number of degree of freedom during motion). However, things are not so different as it seems: Every Motor-Gear-Arm unit (more general: every "subsystem") undergoes, if treated separately, a "guided motion" (motion of e.g.

body-fixed reference frame) with superimposed relative motion with no difference wether corresponding velocities are nonholonomically constrained or not. Clearly, in case of nonholonomic constraints non-integrable "minimal velocities" are unevoidable. However, these may also be used for holonomic systems. Its advantage seems clear: Using for the free (but guided) motion velocities, these need not be specified at this stage of investigation. Thus, one has already to deal with nonholonomic *variables*. They can, on the other hand, be specified (e.g. using Euler angles or parameters), but with an enormous effort. And, furthermore, one has to ask the question: The free motion will lateron be constrained when all the subsystems are joined together. Do we have, in that case, to expect nonholonomic *constraints*? Finally, if we have to, are these (or the holonomic ones) permanent, or may they disappear once more under certain circumstances? To give an example: The typical case of nonholonomic constraints is the wheeled robot travelling on a rough surface, i.e. the wheels are not sliding perpendicular to their path. However, when passing into a slippery domain, this kind of constraint gets lost. In order to be aware of such possibly active or passive constraints it is obviously the easiest way to in advance describe the situation using adequate nonholonomic variables.

2   Choice of Procedure

The main procedures in dynamics are listed in Figure 1 at the end of this section (with no claim for completeness, of course). All those methods which use $\mathbf{q}, \dot{\mathbf{q}}$ only (minimal coordinates $\in \mathbb{R}^f$ and its time derivatives) are restricted to the holonomic case, while those which use $\dot{\mathbf{s}}$ instead of $\dot{\mathbf{q}}$ for minimal velocities may be used either for holonomic or nonholonomic systems. The reader who is not so much interested in the backgrounds may therefore start with the next section (or with Figure 1, resp.) although we felt it essential to clearly elaborate and define what we are speaking about in the following. It should, however, be emphasized that the aim of this chapter lies purely in applications.

2.1   *Constraints and Minimal Velocities*

The notation "minimal coordinates" is used in the following in Lagrange's sense ("le plus petit nombre des variables indetérminées" [25] or "variables en moindre nombre" [26]) which seems clearer than the often used term "generalized coordinates" which has (probably) been introduced by Horace Lamb [28]. The term "minimal velocity" is then selfexplaining.

The most general access to minimal velocities is obtained by inspection of the constraints. Let the system under consideration consist of n rigid bodies the (cartesian) positions of which are characterized by variables $\mathbf{z} \in \mathbb{R}^{6n}$ (translation and orientation). These are constrained by $\boldsymbol{\Phi}(\mathbf{z}) = \mathbf{0} \in \mathbb{R}^m$. Looking for a minimal representation $\mathbf{s} \in \mathbb{R}^f, f = 6n - m$, requires $\mathbf{z} = \mathbf{z}(\mathbf{s})$. Thus, differentiating both the variables $\mathbf{z}$ and the constraint w.r.t. time followed by partial differentiation of the latter w.r.t. $\dot{\mathbf{s}}$ yields a representation of $\dot{\mathbf{z}}$ in terms of $\dot{\mathbf{s}}$, see Table 1. However, due to the time differentiation, the result is a velocity $\dot{\mathbf{s}}$ which needs by no means to be integrable. ($\mathbf{s}$ then represents "quasi-coordinates", a highly successful slang expression, obviously introduced by Whittaker[33]). But, because the constraints themselves are (up to here) purely holonomic, there exists also an integrable subset $\{\dot{\mathbf{q}}\} \in \{\dot{\mathbf{s}}\} : \int \dot{\mathbf{q}} dt \rightarrow \mathbf{q}$. These are chosen for positional description, leading to a (regular linear) combination of the $\dot{\mathbf{q}}$'s for velocity representation.

Table 1:   Constraints, Minimal Velocities: The Holonomic Case (g=f)

$$\boldsymbol{\Phi}(\mathbf{z}) = \mathbf{0} \in^m : \quad \left(\frac{\partial \boldsymbol{\Phi}}{\partial \mathbf{z}}\right) \dot{\mathbf{z}} = \mathbf{0} : \quad \left(\frac{\partial \boldsymbol{\Phi}}{\partial \mathbf{z}}\right)\left(\frac{\partial \dot{\mathbf{z}}}{\partial \dot{\mathbf{s}}}\right) = \mathbf{0} \;\Rightarrow\quad \left(\frac{\partial \dot{\mathbf{z}}}{\partial \dot{\mathbf{s}}}\right) \quad (1)$$

$$\mathbf{z} = \mathbf{z}(\mathbf{s}) : \quad \dot{\mathbf{z}} = \left(\frac{\partial \mathbf{z}}{\partial \mathbf{s}}\right)\dot{\mathbf{s}} : \quad \left(\frac{\partial \mathbf{z}}{\partial \mathbf{s}}\right) = \left(\frac{\partial \dot{\mathbf{z}}}{\partial \dot{\mathbf{s}}}\right) \;\Rightarrow\; \dot{\mathbf{z}} = \left(\frac{\partial \dot{\mathbf{z}}}{\partial \dot{\mathbf{s}}}\right)\dot{\mathbf{s}} \quad (2)$$

$$\{\dot{\mathbf{q}}\} \in \{\dot{\mathbf{s}}\} \qquad\qquad\qquad\qquad\qquad \Rightarrow \dot{\mathbf{z}} = \left(\frac{\partial \dot{\mathbf{z}}}{\partial \dot{\mathbf{q}}}\right)\dot{\mathbf{q}} \quad (3)$$

$$(3) \text{ in } (2)^{-1}: \quad \Rightarrow \dot{\mathbf{s}} = \left(\frac{\partial \dot{\mathbf{s}}}{\partial \dot{\mathbf{z}}}\right)\left(\frac{\partial \dot{\mathbf{z}}}{\partial \dot{\mathbf{q}}}\right)\dot{\mathbf{q}} = \left(\frac{\partial \dot{\mathbf{s}}}{\partial \dot{\mathbf{q}}}\right)\dot{\mathbf{q}} = \mathbf{H}(\mathbf{q})\dot{\mathbf{q}} \in \mathbb{R}^g \quad (4)$$

Remark. The differentiation rules are not unique in literature, sometimes even pretty mixed up. We use here the following concept: Let $s = \mathbf{a}^T\mathbf{x}$ be a scalar with variable $\mathbf{x}$ and constant $\mathbf{a}$. Differentiation w.r.t $\mathbf{x}$ then yields $(\partial s/\partial \mathbf{x}) = \mathbf{a}^T$ (a *row* vector while $\mathbf{a}$ represents a column). Consequently, one obtains for $(\partial \mathbf{r}(\mathbf{x})/\partial \mathbf{x})$ the k-th *row* as $(\partial r_k/\partial \mathbf{x})$. This is in agreement with e.g. $\dot{\mathbf{r}} = (\partial \mathbf{r}/\partial \mathbf{x})\dot{\mathbf{x}}$ ("row times column - rule"). Clearly, one has to state that the "symbolic" notation here has drawbacks when compared to tensor notation. On the other hand, because in the present context no higher order tensors arise (with exceptions, e.g. differentiation of a matrix w.r.t. a vector) the notation itself is much simpler.

Note that, if the constraints are properly posed, then $(\partial \mathbf{z}/\partial \mathbf{s})$ has maximum rank and the inverse $(\partial \mathbf{s}/\partial \mathbf{z})$ always exists. (It should be emphasized that regularity and invertibily refers to the *differential* (or: functional matrix) in order not to be confused with such thoughtless arguments like "how are partial derivatives $(\partial \mathbf{s}/\partial \mathbf{z})$ possible if $\mathbf{s}$ is not defined?"[Anonymous (1985)])

One has to state, however, that from the engineering point of view, the $\mathbf{q}$'s are, in general, prechosen from modeling assumptions. This is much more efficient than to let the computer numerically choose $\mathbf{q}$ from given $\boldsymbol{\Phi} = \mathbf{0}$ as Roberson/San Diego once proposed. The same argument holds for the velocities $\dot{\mathbf{s}}$: the result of Table 1 trivially shows that, if $\mathbf{q}$ defines the positional state of the system in view of the holonomic constraints, then $\dot{\mathbf{q}}$ may be used for the velocity state *as well as* any linear but regular combination of these. Of course, "an inconvenient choice (of $\mathbf{H}(\mathbf{q})$) leads to even more complicated relations than those ones obtained by the Lagrangian approach". However, concluding [Anonymous (1986)] that "the result (i.e using $\dot{\mathbf{s}}$ for efficient procedure) is, therefore, wrong" is ridiculous. In order to avoid such misleading comments one should ask, in the mechanical sense, for some kind of natural approach instead of any (mathematical) arbitraryness in the choice (or calculation) of state variables. One insight is given with a look on the classical results on nonholonomic systems. Nonholonomic relations $\dot{\boldsymbol{\Psi}}(\mathbf{q}, \dot{\mathbf{q}}) = \mathbf{0} \in \mathbb{R}^k$ constrain the $\dot{\mathbf{q}}$ *additionally* to the honolomic ones. The (global) position is not affected, the $\mathbf{q}$'s are therefore already defined with Table 1. For the treatment of nonholonomic constraints, a short look on the relations given in Table 2 ("sometimes referred to as Helmholtz's auxiliary equation" [19]) will be helpful.

Table 2:   The "Helmholtz Auxiliary Equation"

$$\boldsymbol{\Psi}(\mathbf{q}) = \mathbf{0}$$

$$\dot{\boldsymbol{\Psi}}(\mathbf{q}, \dot{\mathbf{q}}) = \left(\frac{\partial \boldsymbol{\Psi}}{\partial \mathbf{q}}\right)\dot{\mathbf{q}} = \mathbf{0} \qquad \Rightarrow \left(\frac{\partial \boldsymbol{\Psi}}{\partial \mathbf{q}}\right) = \left(\frac{\partial \dot{\boldsymbol{\Psi}}}{\partial \dot{\mathbf{q}}}\right)$$

$$\ddot{\boldsymbol{\Psi}}(\mathbf{q}, \dot{\mathbf{q}}, \ddot{\mathbf{q}}) = \left(\frac{\partial \dot{\boldsymbol{\Psi}}}{\partial \mathbf{q}}\right)\dot{\mathbf{q}} + \left(\frac{\partial \dot{\boldsymbol{\Psi}}}{\partial \dot{\mathbf{q}}}\right)\ddot{\mathbf{q}} = \mathbf{0} \quad \Rightarrow \left(\frac{\partial \ddot{\boldsymbol{\Psi}}}{\partial \ddot{\mathbf{q}}}\right) = \left(\frac{\partial \dot{\boldsymbol{\Psi}}}{\partial \dot{\mathbf{q}}}\right)$$

$$\ldots\ldots$$

$$\Rightarrow \left(\frac{\partial \boldsymbol{\Psi}}{\partial \mathbf{q}}\right) = \left(\frac{\partial \dot{\boldsymbol{\Psi}}}{\partial \dot{\mathbf{q}}}\right) = \left(\frac{\partial \ddot{\boldsymbol{\Psi}}}{\partial \ddot{\mathbf{q}}}\right) = \cdots$$

(The first line in Table 2 denotes a "quasi-constraint" in the same slangful sense as the "quasi-coordinate"). Because of $(\partial \boldsymbol{\Psi}/\partial \mathbf{q}) = (\partial \dot{\boldsymbol{\Psi}}/\partial \dot{\mathbf{q}})$, the nonholonomic constraints are treated the same way as in Table 1 (replace $(\partial\boldsymbol{\Phi}/\partial\mathbf{z})$ by $(\partial\dot{\boldsymbol{\Psi}}/\partial\dot{\mathbf{q}})$ and $\mathbf{z}$ by $\mathbf{q}$).

Table 3:   Constraints, Minimal Velocities: The Nonholonomic Case (g=f-k)

$$\left(\frac{\partial\dot{\boldsymbol{\Psi}}}{\partial\dot{\mathbf{q}}}\right)\dot{\mathbf{q}} = \mathbf{0}: \quad \left(\frac{\partial\dot{\boldsymbol{\Psi}}}{\partial\dot{\mathbf{q}}}\right)\left(\frac{\partial\dot{\mathbf{q}}}{\partial\dot{\mathbf{s}}}\right) = \mathbf{0} \quad \Rightarrow \dot{\mathbf{s}} = \left(\frac{\partial\dot{\mathbf{s}}}{\partial\dot{\mathbf{q}}}\right)\dot{\mathbf{q}} = \mathbf{H}(\mathbf{q})\dot{\mathbf{q}} \in \mathbb{R}^g$$

Remark. We retain the dots in formulating the nonholonomic constraints for the same reason as we did for $\dot{\mathbf{s}}$: All these are expressions on the velocity level. The usefulness of this notation lies in the comparability of holonomic/nonholonomic terms and can easily be seen in the above treatment of minimal velocities: The solutions $\{\dot{\mathbf{s}}\}$ (see Table 1) may be integrable or not, thus the dot represents a total differential or not.

Remark. For $\dot{\boldsymbol{\Psi}}$ being nonlinear w.r.t. $\dot{\mathbf{q}}$ one may replace $(\partial\dot{\boldsymbol{\Psi}}/\partial\dot{\mathbf{q}})$ by $(\partial\ddot{\boldsymbol{\Psi}}/\partial\ddot{\mathbf{q}})$ in Table 3, see Table 2. However, do nonlinear nonholomic constraints exist? Let us assume that: first, the position $\mathbf{z} \in \mathbb{R}^{6n}$ is constrained with $\boldsymbol{\Phi}(\mathbf{z}) = 0 \in \mathbb{R}^m$, leading to minimal coordinates $\mathbf{q} \in \mathbb{R}^{f=6n-m}$. Then, clearly, the volocity $\dot{\mathbf{z}}$ and the acceleration $\ddot{\mathbf{z}}$ (and any higher derivative) is automaticly also constrained (the holonomic case). Second, the velocity may *additionally* be constrained with $\dot{\boldsymbol{\Psi}}(\mathbf{q},\dot{\mathbf{q}}) = 0 \in \mathbb{R}^k$ (not being integrable, of course, otherwise one would have additional holonomic constraints) leading to minimal velocities $\dot{\mathbf{s}} \in \mathbb{R}^{g=6n-m-k}$. Third, the acceleration may *additionally* be constrained with $\ddot{\boldsymbol{\Theta}}(\mathbf{q},\dot{\mathbf{q}},\ddot{\mathbf{q}}) = 0 \in \mathbb{R}^r$ (not being integrable, of course, otherwise one would have additional nonholonomic constraints) leading to minimal accelerations $\ddot{\mathbf{s}} \in \mathbb{R}^{h=6n-m-k-r}$. However, this third assumption does not make much sense because of the basic axioms (momentums are already linear w.r.t. accelerations). Therefore, operating on the acceleration level for nonlinear nonholonomic constraints is nothing but the same "trick" as for honolomic ones: Differentiation reveals linear terms which can then elementarily be treated. But the way back is always possible (integrability of differentiated constraints is of course trivially assured). Therefore, the frequent statement that nonlinear nonholonomic constraints do not exist because corresponding acceleration constraints would contradict the basic "natural laws" does not hold (consequently, not having the integrability in mind, holonomic constraints would also contradict the axioms). The only statement one can extract here is that "nonlinear nonholonomic constraints do not seem to exist in daily experience"(Hamel) [19], although for control purpuses velocities and even accelerations may be computed anyhow with no restriction at all.

Next, if we introduce with Hamel[14] a set of "variables"

$$\dot{\mathbf{s}}^T = (\dot{\mathbf{s}}^T_{dep.}\ \ \dot{\mathbf{s}}^T_{indep.}) = \underbrace{\left[\dot{\mathbf{\Psi}}_1 = 0 \cdots \dot{\mathbf{\Psi}}_k = 0 \right.}_{\in \mathbb{R}^k} \mid \underbrace{\left. \dot{s}_1 \cdots \dot{s}_g \right]}_{\in \mathbb{R}^{g=f-k}}, \overset{def}{=} \mathbf{f}(\mathbf{q}, \dot{\mathbf{q}}) \in \mathbb{R}^f \quad (1)$$

(hence denoted as "Hamel's variables") then the $\dot{\mathbf{q}} = (\partial\dot{\mathbf{q}}/\partial\dot{\mathbf{s}})\dot{\mathbf{s}} \in \mathbb{R}^f$ satisfy the nonholonomic constraints $\dot{\mathbf{\Psi}}(\mathbf{q},\dot{\mathbf{q}}) = \mathbf{0} \in \mathbb{R}^k$ *identically*[32]. From the engineering point of view, Eq.(1) consists of f (translational, rotational) velocities, where, in the presence of nonholonomic constraints, some of them are "forbidden". These are, of course, in advance known from the modeling assumptions. Thus, one "natural" choice of velocity variables is given with a coordinate representation in a reference frame taking the "forbidden" and "allowed" velocities directly into account.

(**Remark:** mathematicians, in general, do not like this kind of argumentation because of loosing, in their opinion, generality. However, we should not forget the sequence of foregoing: The first step is modeling. The second is its mathematical description. So why forget the first step while discussing the second? This kind of "generality" is, from the engineering point of view, shadow-boxing.)

Clearly, using Eq.(1) for description, $\mathbf{f}(\mathbf{q}, \dot{\mathbf{q}})$ may be used as a whole which enables to cancel the dependent componentents at the end of calculation. However, one has to be cautious: The motion equations one would abtain for $\dot{s}_i = 0, i = 1\cdots k$ are not (at least not directly) fulfilled. The reason is: looking at the corresponding virtual work in the sense $[\cdots]\delta s_i = 0$, the δs_i are arbitrary for the independent part but zero for the dependent part (as direct consequence of $\dot{\mathbf{s}}_{dep} = \mathbf{0}$). In view of structurally variant systems where one has to deal with closed and open constraints simultaneously, this situation is not very satisfactory. On the other hand, one may easily introduce Lagrange parameters to obtain arbitrary δs_i for the dependent part as well and thus free the dynamics from the corresponding constraints[16]. Generally, one has for the (generalized) constraint forces e.g. $(\partial\dot{\mathbf{\Psi}}_i/\partial\dot{\mathbf{s}})^T\lambda_i$ where $(\partial\dot{\mathbf{\Psi}}_i/\partial\dot{\mathbf{s}})$ indicates its directions. Evidently, using $\dot{\mathbf{\Psi}}$ itself for coordinate, then $(\partial\dot{\mathbf{\Psi}}_i/\partial\dot{\mathbf{\Psi}})^T$ reduces to the i-th unit vector. One has then just to add λ_i to the corresponding equation: $[\cdots - \lambda_i]\delta s_i = 0, i = 1\cdots k$.

Remark. In order not to trivialize Hamel's "Principle of the relaxation of the constraints" one should emphasize that the above denotes a very simple case. What Hamel did in his famous paper was of course much deeper: Starting with the rigid body, formulating the conditions of rigidity along with the corresponding Lagrange parameters, then freeing the system from the constraints where the constraint forces (Lagrange parameters) convert into impressed forces, he obtains immediately the (symmetric) stress-strain relations of continuum mechanics.

2.2 On Virtual Displacements and Variations

Although already mentioned in the context, "virtual" terms have not yet been defined up to here. These have caused a lot of confusion during the centuries. It may be that Lagrange[27] himself initiated these misleadings with his "I have to emphasize ... that I introduced a new characteristic δ. By this, δZ shall express a differential of Z which is not the same as dZ but which is built by the same rules." 44 years after Lagrange, Poinsot published a paper (reprinted in his textbook[35]) where he stated that the "virtuals" leave something obscure in one's mind and he therefore replaced the virtual displacements by the actual velocities, calling that a "new principle". This is completely wrong (see Lagrange: same rules, but *different* things), although contemporary authors call this "a better foundation, where Lagrange's principle appears as a simple corollar"[36]. Other contemporarians share Poinsot's opinion: "virtual displacements ... are the closest thing in dynamics to black magic", they are "ill-defined, nebulous, hence objectionable" (Kane[37], followed by his disciples and colleagues: "too vague for practical use" (Levinson), "esoteric quantities" (Angeles)). This kind of *non*understandment (not misunderstandment) can easily be rejected by trivial interpretation of Lagrange's concept: A mechanical system the position of which is restricted to $\mathbf{\Phi}(\mathbf{z}) = \mathbf{0}$ undergoes (generalized) constraint forces which are perpendicular to this (hyper-) plane and force the (considered reference point of the) system to remain in that plane. Thus, any arbitrary but "allowed" motion direction $\delta\mathbf{z}$ is element from the tangent plane $((\partial\mathbf{\Phi}/\partial\mathbf{z})\delta\mathbf{z} = 0 \Rightarrow \delta\mathbf{z})$ *except* the singular solution $(\partial\mathbf{\Phi}/\partial\mathbf{z})d\mathbf{z} = 0 \Rightarrow d\mathbf{z}$ (see Poinsot's error, see Lagrange: same rules, but *different* things). The reason is simple: using $d\mathbf{z}$, the scalar product projects motion into the unconstrained minimal space, eliminating all nonworking forces at once, not only the constraint forces (as desired) but also the Coriolis reactions.

The question arises, how to calculate "allowed" $\delta\mathbf{z}$ (or any other $\delta(\ )$). To start with, let us first investigate $\delta\mathbf{q}$, which is not restricted by any constraint ($\mathbf{q}$ represents minimal coordinates). The easiest procedure is then that one from variational calculus

$$\mathbf{q} = \mathbf{q}^* + \delta\mathbf{q} = \mathbf{q}^* + \epsilon\eta, \quad \eta \mid_{t_o}^{t_1} = 0 \tag{2}$$

where $\mathbf{q}^*$ denotes the true solution (from boundary conditions). Hence, one

8   *H. Bremer*

can define the "deviations" in the sense of a Taylor series expansion w.r.t. the variational parameter ϵ yielding the calculation rules for the δ-process, e.g.

$$\mathbf{r} = \mathbf{r}(\mathbf{q}, t) \Rightarrow \delta\mathbf{r} = \left[\frac{d\mathbf{r}}{d\epsilon}\right]_{\epsilon=0} \epsilon = \left[\frac{\partial\mathbf{r}}{\partial\mathbf{q}}\frac{\partial\mathbf{q}}{\partial\epsilon}\right]_{\epsilon=0} \epsilon + \left[\frac{\partial\mathbf{r}}{\partial t}\frac{\partial t}{\partial\epsilon}\right]_{\epsilon=0} \epsilon = \frac{\partial\mathbf{r}}{\partial\mathbf{q}}\delta\mathbf{q} \qquad (3)$$

thus,

$$\mathbf{\Phi} = \mathbf{\Phi}(\mathbf{r}, t) = 0 \Rightarrow \delta\mathbf{\Phi} = \left[\frac{d\mathbf{\Phi}}{d\epsilon}\right]_{\epsilon=0} \epsilon = \left(\frac{\partial\mathbf{\Phi}}{\partial\mathbf{r}}\right)\left(\frac{\partial\mathbf{r}}{\partial\mathbf{q}}\right)\delta\mathbf{q} = \left(\frac{\partial\mathbf{\Phi}}{\partial\mathbf{r}}\right)\delta\mathbf{r} = 0 \quad (4)$$

(see above: (in case of skleronomicity) same rule but different things).

Remarks. 1) The Taylor expansion is "complete": due to the linear approach, no higher oder terms arise. That means that $\mid \delta\mathbf{q} \mid$ has by no means to be assumed "infinitesimal".

2) This is in agreement with variational calculus where, looking for an (unknown) optimal solution J_{opt} in terms of a Taylor series w.r.t. the optimum $(J = J_{opt} + (\partial J/\partial\mathbf{q})_{opt}\delta\mathbf{q} + \cdots)$, the $\delta\mathbf{q}$ has to be kept arbitrary in magnitude because the (unknown) J_{opt} may be far away from J. *However*, this is a totally different question. The only solution required here is the tangent vector $\delta\mathbf{r}$ according to Eq.(4).

3) In optimization, using Eq.(2), the problem is shifted to the question what class of functions η have to enter consideration, e.g. in common words: if not every club member is known, one can never decide who of them is the tallest one (P.Funk) [8]. *However*, this question does not arise in the present at all, because not one special but *any* ("allowed") tangent vector from e.g. Eq.(4) may be used for calculation.

(The often stated "$\delta\mathbf{r}$ is in agreement with the constraints, it is infinitesimal small and happens with infinite speed" is wrong and/or not understandable, hence objectionable.)

Conclusion. Except the definition of virtual displacements, the (sometimes so-called) differential principles of dynamics have nothing in common with the usual calculus of variation. In fact, adding nonholonomic constraints to the (so-called) Hamilton's Principle by means of Lagrangian multipliers leads to wrong results when treated with the rules of variational calculus (see Hamel [17], Papstavridis [31]) although Hamilton's Principle is not at all restricted to holonomic systems as often but erroneously stated. The only task here is to derive suitable orthogonality relations (Eq.(4)) with the aim to eliminate constraints (or, if need be, to introduce Lagrange parameters).

Remark on "higher variations" (Jourdain)[24], (Gauss)[13]. Looking at the nonlinear nonholomic constraints, or, as mentioned above, its time derivative, some authors believe on the necessity of "higher order variations". However, applying Lagrange's variation to $\dot{\Phi}$ or on $\ddot{\Phi}$ for instance, one has

$$\delta\dot{\Phi} = \left(\frac{\partial\dot{\Phi}}{\partial\mathbf{r}}\right)\delta\mathbf{r} + \left(\frac{\partial\dot{\Phi}}{\partial\dot{\mathbf{r}}}\right)\delta\dot{\mathbf{r}} = \mathbf{0},$$

$$\text{or}\quad \delta\ddot{\Phi} = \left(\frac{\partial\ddot{\Phi}}{\partial\mathbf{r}}\right)\delta\mathbf{r} + \left(\frac{\partial\ddot{\Phi}}{\partial\dot{\mathbf{r}}}\right)\delta\dot{\mathbf{r}} + \left(\frac{\partial\ddot{\Phi}}{\partial\ddot{\mathbf{r}}}\right)\delta\ddot{\mathbf{r}} = \mathbf{0} \tag{5}$$

Now, Jourdain requires $\delta\mathbf{r} = \mathbf{0}$ (commonly denoted δ'), Gauss requires $\delta\mathbf{r} = \mathbf{0}$ and $\delta\dot{\mathbf{r}} = \mathbf{0}$ (commonly δ'') *because the required orthogonality relation can only be achieved with these "artificial" assumptions.* These kinds of "higher variations" are already accomplished with the "Helmholtz auxiliary equation" (Table 2): for Eq.(4) one obtains

$$\frac{\partial\Phi}{\partial\mathbf{r}}\delta\mathbf{r} = \frac{\partial\dot{\Phi}}{\partial\dot{\mathbf{r}}}\delta\mathbf{r} = \frac{\partial\ddot{\Phi}}{\partial\ddot{\mathbf{r}}}\delta\mathbf{r} = \mathbf{0}. \tag{6}$$

This is a *homogeneous* equation the solution of which is the tangent vector $\delta\mathbf{r}$. Replacing $\delta\mathbf{r}$ with $\delta'\dot{\mathbf{r}}$ or $\delta''\ddot{\mathbf{r}}$ (or anything else) means just to give the same thing another name[5].

These considerations coincide with e.g. Hamels textbook[19], pp. 496: Of course, dealing with nonlinear nonholonomic constraints $\dot{\mathbf{s}} = \mathbf{f}(\mathbf{q},\dot{\mathbf{q}}) \Leftrightarrow \dot{\mathbf{q}} = \mathbf{F}(\mathbf{q},\dot{\mathbf{s}})$, it would not make sense to write for instance $\delta\mathbf{q} = \mathbf{F}(\mathbf{q},\delta\mathbf{s})$. "Therefore, Lagrange's Principle fails at first but Gauss' Principle leads further". We don't agree with this for the moment, because with the above statements these principles should be identical. However, some lines below one reads: "we can bring the above expressions into the Lagrangian form if we define the virtual displacements according to $\delta\mathbf{s} = (\partial\mathbf{f}/\partial\dot{\mathbf{q}})\delta\mathbf{q}$ or, vice versa, $\delta\mathbf{q} = (\partial\mathbf{F}/\partial\dot{\mathbf{s}})\delta\mathbf{s}$". This is in total agreement with the above. It should, however, be emphasized that these considerations do not affect Gauss' principle of least constraints which is not the same as the corresponding "differential principle" and which will be used later.

2.3 The Transitivity Equation

Using the $\delta\mathbf{q}$ in the sense of Eq.(2) answers the (vital! see below) question: Is the d–operation (the total differentiation) and the δ–operation interchangeable? First, looking at $\mathbf{r} = \mathbf{r}(\mathbf{q})$, one obtains by elementary calculation

$$d\delta r_i - \delta dr_i = \sum_{k,j} \left(\frac{\partial^2 r_i}{\partial q_j \partial q_k} - \frac{\partial^2 r_i}{\partial q_k \partial q_j} \right) dq_k \delta q_j + \sum_j \frac{\partial r_i}{\partial q_j} \left(d\delta q_j - \delta dq_j \right) \quad (7)$$

where

$$d\delta \mathbf{q} - \delta d\mathbf{q} = \mathbf{0} \tag{8}$$

if $\delta \mathbf{q}$ is used in the sense of Eq.(2). Then, Eq.(7) reduces to the integrability condition which is obviously fulfilled if $\mathbf{r}$ represents the (objective, inertial) position of a considered mass element dm. This is simply because "it is impossible that an element vanishes at a certain location in order to simultaneously appear at another one" (Hamel) [19], thus

$$d\delta \mathbf{r} - \delta d\mathbf{r} = \mathbf{0} \tag{9}$$

However, for $\dot{\mathbf{s}}$ (nonintegrable) we are left with the "transitivity equation"

$$d\delta \mathbf{s} - \delta d\mathbf{s} \neq \mathbf{0} \tag{10}$$

Conclusion. Once more Poinsot: for d and δ being the same, the transitivity equation $d\delta(\ ) - \delta d(\ )$ always vanishes which is obviously false. In order to calculate the transitivity equation one has to chose a rule for the δ's. *This is not at all selfevident.* Applying the transitivity equation to the (inertial, objective) position vector $\mathbf{r}$ (of considered mass element), i.e. $d\delta(\mathbf{r}) - \delta d(\mathbf{r})$, for instance, yields zero only if one assumes $\delta \mathbf{q}$ in the sense of e.g. variational calculus. Here, Lur'e [29] for instance, was not correct with his statement "avec l'égalité évidente" of $d\delta(\mathbf{r}) - \delta d(\mathbf{r}) = \mathbf{0}$ because this kind of "selfevidence" already implies the use of the $\delta \mathbf{q}$'s in the above sense and automaticly eliminates other procedures (like the Lie group approach for instance which requires, in the sense of Hamel [14], the interchangeability w.r.t. $\mathbf{s}$ instead of $\mathbf{q}$).

With $\dot{\mathbf{s}}$ from Eq.(1) one obtains, having the "Helmholtz" equation in mind,

$$\delta \mathbf{s} = \left(\frac{\partial \mathbf{f}}{\partial \dot{\mathbf{q}}} \right) \delta \mathbf{q} \;\Rightarrow\; \frac{d}{dt} \delta \mathbf{s} = \frac{d}{dt} \frac{\partial \mathbf{f}}{\partial \dot{\mathbf{q}}} \delta \mathbf{q} + \frac{\partial \mathbf{f}}{\partial \dot{\mathbf{q}}} \frac{d}{dt} \delta \mathbf{q}$$
$$\wedge \; \delta \frac{d\mathbf{s}}{dt} = \frac{\partial \mathbf{f}}{\partial \mathbf{q}} \delta \mathbf{q} + \frac{\partial \mathbf{f}}{\partial \dot{\mathbf{q}}} \delta \frac{d}{dt} \mathbf{q} \tag{11}$$

yielding with Eq.(8)

$$\left(\frac{d\delta \mathbf{s} - \delta d\mathbf{s}}{dt} \right) = \left[\frac{d}{dt} \frac{\partial \mathbf{f}}{\partial \dot{\mathbf{q}}} - \frac{\partial \mathbf{f}}{\partial \mathbf{q}} \right] \delta \mathbf{q} \tag{12}$$

where $\delta\mathbf{q}$ has to be expressed in terms of $\delta\mathbf{s}$ which is always possible since $\dot{\mathbf{s}} = \mathbf{f}(\mathbf{q}, \dot{\mathbf{q}})$ is always invertible: $\dot{\mathbf{q}} = \mathbf{F}(\mathbf{q}, \dot{\mathbf{s}})$, yielding (once more with Helmholtz) $\delta\mathbf{q} = (\partial\mathbf{F}/\partial\dot{\mathbf{s}})\delta\mathbf{s}$. For the linear case we obtain for the m-th component $\dot{s}_m = f_m(\mathbf{q}, \dot{\mathbf{q}}) = [(\partial s_m/\partial\dot{\mathbf{q}})\dot{\mathbf{q}}]$

$$\sum_j \left[\frac{d}{dt}\frac{\partial f_m}{\partial\dot{q}_j} - \frac{\partial f_m}{\partial q_j} \right] \delta q_j = \sum_{k,j} \left(\frac{\partial^2 s_m}{\partial q_j \partial q_k} - \frac{\partial^2 s_m}{\partial q_k \partial q_j} \right) \dot{q}_k \delta q_j \qquad (13)$$

(compare to Eq.(7) for $r_i \rightarrow s_m$ along with Eq.(8)). If one inserts $\mathbf{q} = \mathbf{q}(\mathbf{s})$ explicitly then the above expression can be freed from the δs_m leading to the (linear) Hamel coefficients. Hamel concedes that "its use is not always practicable" and prefers in his book to directly carry out the $d\delta - \delta d$- process and from there to conclude on those coefficients. However, his foregoing, applied to several examples in his book, is a bit "tricky" so that we prefer a direct calculation of Eq.(13) coming out with

$$\sum_{k,j} \left(\frac{\partial^2 s_m}{\partial q_j \partial q_k} - \frac{\partial^2 s_m}{\partial q_k \partial q_j} \right) \dot{q}_k \delta q_j =$$
$$\delta\mathbf{s}^T \left(\frac{\partial\mathbf{q}}{\partial\mathbf{s}}\right)^T \left\{ \left[\frac{\partial}{\partial\mathbf{q}}\left(\frac{\partial s_m}{\partial\mathbf{q}}\right)^T \right] - \left[\frac{\partial}{\partial\mathbf{q}}\left(\frac{\partial s_m}{\partial\mathbf{q}}\right)^T \right]^T \right\} \left(\frac{\partial\mathbf{q}}{\partial\mathbf{s}}\right) \dot{\mathbf{s}} \qquad (14)$$

Here, $(\partial\mathbf{q}/\partial\mathbf{s})$ is the (generalized) inverse of $(\partial\mathbf{s}/\partial\mathbf{q})$.

We are leaving now nonlinear nonholonomic relations at this point with our view directed to robot applications. Here, nonlinear constraints do not arise (do they exist at all?). The above mentioned control constraints (where accelerations and/or velocities might be linearly or nonlinearly combined) will also not arise because, following the usual philosophy, control inputs always belong to the impressed (working) forces/torques. Here, we are, in the first step, interested in the plant in order to later determine control.

2.4   Dynamical Procedures

With the $\delta \mathbf{r}$ defined one may proceed with Lagrange's Principle [25] (..."which up to the time beeing has been confused with d'Alembert's Principle" (Heun) [21])

$$\int\limits_{(S)} (dm\ddot{\mathbf{r}} - d\mathbf{f}^e)^T \, \delta\mathbf{r} = 0 \tag{15}$$

(symmetric stress tensor assumed) which can be reformulated with $\mathbf{r} = \mathbf{r}(\mathbf{s})$

$$\frac{d}{dt} \int \frac{\partial}{\partial \dot{\mathbf{r}}} \left(\frac{1}{2}\dot{\mathbf{r}}^T dm\dot{\mathbf{r}} \right) \frac{\partial \mathbf{r}}{\partial \mathbf{s}} \delta\mathbf{s} - \int \left[\frac{\partial}{\partial \dot{\mathbf{r}}} \left(\frac{1}{2}\dot{\mathbf{r}}^T dm\dot{\mathbf{r}} \right) \right] \frac{d}{dt}\delta\mathbf{r} - \delta W^e = 0 \tag{16}$$

(δW^e: virtual work of impressed forces). With $d\delta\mathbf{r} = \delta d\mathbf{r}$ according to Eq.(9) and $(\partial\mathbf{r}/\partial\mathbf{s}) = (\partial\dot{\mathbf{r}}/\partial\dot{\mathbf{s}})$ one obtains the "central equation"

$$\frac{d}{dt} \left[\left(\frac{\partial T}{\partial \dot{\mathbf{s}}} \right) \delta\mathbf{s} \right] - \delta T - \delta W^e = 0 \tag{17}$$

(discussed e.g. in (Bremer) [3]) where T: kinetic energy. Integrating Eq.(17) over a fixed time domain yields the so-called Hamilton Principle and shows its range of applicability (will not be used here). On the other hand, elementary calculation yields (the variational form of) Hamel's equations

$$\left[\frac{d}{dt} \left(\frac{\partial T}{\partial \dot{\mathbf{s}}} \right) - \left(\frac{\partial T}{\partial \mathbf{s}} \right) - \mathbf{Q}^T \right] \delta\mathbf{s} + \left(\frac{\partial T}{\partial \dot{\mathbf{s}}} \right) \left(\frac{d\delta\mathbf{s} - \delta d\mathbf{s}}{dt} \right) = 0 \tag{18}$$

where $T = T(\mathbf{q}, \dot{\mathbf{s}})$. Here, $(\partial T/\partial\mathbf{s})$ is formal notation for $(\partial T/\partial\mathbf{q})(\partial\mathbf{q}/\partial\mathbf{s})$.

From Eq.(18) one obtains immediately the Lagrange equations (of second kind) using $\dot{\mathbf{q}}$ for minimal velocity which clearly excludes any nonholonomicity (not only for the chosen variables but also concerning constraints). However, one may use $\dot{\mathbf{q}}$ only for the first step and then express the $\delta\mathbf{q}$ in terms of nonholonomic variables, either for introducing new variables (congruence transformation) or for simultaneously introducing nonholonomic constraints (Maggi) [30]. One should, however, have in mind that the application of the Lagrange equations is, for general mechanical systems, already tedious. The more is Maggi's procedure (although later used, but with a slightly changed aim). Obviously, the direct access to nonholonic systems is given with Eq.(18) along with the corresponding "transitivity equation".

With the aim of (robot) applications, we are going to search for simple procedures. Two examples may give insight and lead to that way.

Example 1 First, let us investigate the plane motion of a wheeled robot as depicted in Figure 4 neglecting the SCARA and concentrating on the car body (i.e. index 1 in the sketch, here suppressed). Minimal coordinates are $\mathbf{q} = [x, \; y, \; \gamma]^T$, where (x, y) denotes the reference point O as seen from the inertial frame and γ is the orientation angle. The vector of minimal velocities $\dot{\mathbf{s}}$ contains the cartesian velocities towards ahead and aside (in car body fixed components), and the angular velocity $\dot{\gamma}$. We have then the following situation:

$$\dot{\mathbf{s}} = \left(\frac{\partial \mathbf{s}}{\partial \mathbf{q}} \right) \dot{\mathbf{q}} = \mathbf{A}_\gamma \dot{\mathbf{q}} \Leftrightarrow \dot{\mathbf{q}} = \left(\frac{\partial \mathbf{q}}{\partial \mathbf{s}} \right) \dot{\mathbf{s}} = \mathbf{A}_\gamma^T \dot{\mathbf{s}}$$

$$\text{where } \mathbf{A}_\gamma = \begin{bmatrix} \cos\gamma & \sin\gamma & 0 \\ -\sin\gamma & \cos\gamma & 0 \\ 0 & 0 & 1 \end{bmatrix} \tag{19}$$

In order to calculate Eq.(14), one may for instance consider the first component $\dot{s}_1$, leading to

$$\left[\frac{\partial}{\partial \mathbf{q}} \left(\frac{\partial s_1}{\partial \mathbf{q}} \right)^T \right] = \frac{\partial}{\partial \mathbf{q}} \begin{bmatrix} \cos\gamma \\ \sin\gamma \\ 0 \end{bmatrix} = \begin{bmatrix} 0 & 0 & -\sin\gamma \\ 0 & 0 & \cos\gamma \\ 0 & 0 & 0 \end{bmatrix} \tag{20}$$

hence

$$\frac{d\delta s_1 - \delta d s_1}{dt} =$$

$$\delta \mathbf{s}^T \mathbf{A}_\gamma \begin{bmatrix} 0 & 0 & -\sin\gamma \\ 0 & 0 & \cos\gamma \\ \sin\gamma & -\cos\gamma & 0 \end{bmatrix} \mathbf{A}_\gamma^T \dot{\mathbf{s}} = \delta \mathbf{s}^T \begin{bmatrix} 0 & 0 & 0 \\ 0 & 0 & 1 \\ 0 & -1 & 0 \end{bmatrix} \dot{\mathbf{s}} \tag{21}$$

Example 2 The preceeding example is an almost simple one due to the othonormality of $\mathbf{A}_\gamma$. Let us now consider the rotations of a gyro using Euler angles, i.e. $\mathbf{q} = [\psi, \; \vartheta, \; \psi]^T$, leading to the well known relations

$$\dot{\mathbf{s}} = \begin{bmatrix} \sin\vartheta \sin\varphi & \cos\varphi & 0 \\ \sin\vartheta \cos\varphi & -\sin\varphi & 0 \\ \cos\vartheta & 0 & 1 \end{bmatrix} \dot{\mathbf{q}} = \left(\frac{\partial \mathbf{s}}{\partial \mathbf{q}} \right) \dot{\mathbf{q}}$$

$$\dot{\mathbf{q}} = \begin{bmatrix} \frac{\sin\varphi}{\sin\vartheta} & \frac{\cos\varphi}{\sin\vartheta} & 0 \\ \cos\varphi & -\sin\varphi & 0 \\ -\cot\vartheta \sin\varphi & -\cot\vartheta \cos\varphi & 1 \end{bmatrix} \dot{\mathbf{s}} = \left(\frac{\partial \mathbf{q}}{\partial \mathbf{s}} \right) \dot{\mathbf{s}} \tag{22}$$

Once more taking $m = 1$ into account one has

$$\frac{\partial}{\partial \mathbf{q}}\begin{bmatrix} \sin\vartheta\sin\varphi \\ \cos\varphi \\ 0 \end{bmatrix} = \begin{bmatrix} 0 & \cos\vartheta\cos\varphi & \sin\vartheta\cos\varphi \\ 0 & 0 & -\sin\varphi \\ 0 & 0 & 0 \end{bmatrix} \tag{23}$$

yielding

$$\frac{d\delta s_1 - \delta ds_1}{dt} =$$

$$\delta\mathbf{s}^T\left(\frac{\partial\mathbf{q}}{\partial\mathbf{s}}\right)^T\begin{bmatrix} 0 & \cos\vartheta\cos\varphi & \sin\vartheta\cos\varphi \\ -\cos\vartheta\cos\varphi & 0 & -\sin\varphi \\ -\sin\vartheta\cos\varphi & \sin\varphi & 0 \end{bmatrix}\left(\frac{\partial\mathbf{q}}{\partial\mathbf{s}}\right)\dot{\mathbf{s}} =$$

$$\delta\mathbf{s}^T\begin{bmatrix} 0 & 0 & 0 \\ 0 & 0 & 1 \\ 0 & -1 & 0 \end{bmatrix}\dot{\mathbf{s}} \tag{24}$$

Remarks. 1) The first example contains (or may contain) a nonholonimc constraint ($\dot{s}_2 = 0$). Note, however, that with Eq.(1), the procedure includes the constraints looking at them as (dependent) variables. Thus, the calculation refers to IR^3.

2) Both the examples result in the same: the Hamel coefficients are zero or plus/minus one. We are not going to discuss wether the coefficients have to be constant all the time (they will surely be not in case of nonlinear nonholonomic constraints). But we can already conclude that they are constant in case of freely moving rigid bodies.

3) The latter has, for the moment, nothing in common with (nonholonomic) constraints. However, the velocities of a freely moving body represent already nonholonomic variables. This leads to the concept of a) considering first the unconstrained bodies b) and then insert constraints in the sense of Maggi to c) making benefit of the above conclusion (constant Hamel coefficients) *and* d) simultaneously calculate ($\partial T/\partial\dot{\mathbf{s}}$) explicitly in order to allow direct insertion of constraints (which is, as well known, not allowed for T itself because of the required partial derivatives).

2.5   Analytic Approach vs. Synthetical Approach

As will come out later, the sketched concept leads from analytical dynamics to the so-called synthetical approach(es). As usual within the scientific community there is no agreement how understand these notations. It is therefore essential to define these, e.g. in the same sense as from historical development (synthesis: descriptive geometry (Newton's Principia); analysis: analytical geometry (Lagrange's Méchanique Analytique (note spelling))). Synthesis means (literally) to combine together, while analysis means (literally) to extract things from a whole. Considering dynamics, the "whole" is (in a generalized sense, if we include e.g. the Gibbs-Appell-function) the energy, and the aim here is to extract (analyze) motion equations (generalized force balance), while synthesis means to put single parts (forces/torques) together to obtain motion equations.

Let us first consider a freely moving rigid body the kinetic energy T of which reads

$$2T = (\mathbf{v}\ \boldsymbol{\omega})^T \begin{bmatrix} m\mathbf{E} & \mathbf{0} \\ \mathbf{0} & \mathbf{J} \end{bmatrix} \begin{pmatrix} \mathbf{v} \\ \boldsymbol{\omega} \end{pmatrix} \tag{25}$$

($\mathbf{v}, \boldsymbol{\omega}$: translational and angular mass center velocity, m: mass). In body-fixed coordinates, the tensor of moments of inertia $\mathbf{J}$ is constant. Then, obviously, $\dot{\mathbf{s}}^T = [\dot{\mathbf{s}}_1^T\ \dot{\mathbf{s}}_2^T] = [\mathbf{v}^T\ \boldsymbol{\omega}^T]$ is predestinated for the use as minimal velocity. Next, in order to make benefit from Eq.(9) (inertial frame required) and, simultaneously, from the fact that every mass element dm has a constant distance $\mathbf{r}_p$ from the body's mass center when using a body-fixed frame for representation, the absolute velocity of dm may be calculated via

$$\dot{\mathbf{r}} = \mathbf{A} \begin{bmatrix} \mathbf{E} & \widetilde{\mathbf{r}}_p^T \end{bmatrix} \dot{\mathbf{s}} \tag{26}$$

where: $\dot{\mathbf{s}}$: velocities in body fixed representation, $\mathbf{r}_p$: distance vector of dm in body fixed representation (thus being constant), $\widetilde{(\ )} \in \mathbb{R}^{3,3}$: skew symmetrix tensor providing the vector product, $\mathbf{A}$: transformation matrix from body-fixed to inertial frame. From Eq.(26) we obtain immediately the corresponding δ-expression (with the above defined calculation rules) enabling to calculate $(d\delta\mathbf{r} - \delta d\mathbf{r}) = \mathbf{0}$. Premultiplying with $\mathbf{A}^T$ and having the basic properties $\mathbf{A}^T\dot{\mathbf{A}} = \widetilde{\boldsymbol{\omega}} = \widetilde{\dot{\mathbf{s}}}_2 \Rightarrow \mathbf{A}^T\delta\mathbf{A} = \widetilde{\delta\mathbf{s}}_2$ in mind, we are eventually left with

$$\mathbf{A}^T \left(\frac{d\delta\mathbf{r} - \delta d\mathbf{r}}{dt} \right) = \begin{bmatrix} \mathbf{E} & \widetilde{\mathbf{r}}_p^T \end{bmatrix} \left\{ \begin{pmatrix} \widetilde{\boldsymbol{\omega}} & \widetilde{\mathbf{v}} \\ \mathbf{0} & \widetilde{\boldsymbol{\omega}} \end{pmatrix} \delta\mathbf{s} + \left(\frac{d\delta\mathbf{s} - \delta d\mathbf{s}}{dt} \right) \right\} = \mathbf{0}\ \forall\ \mathbf{r}_p \tag{27}$$

Hence, for every rigid body i under consideration,

$$\left(\frac{d\delta\mathbf{s} - \delta d\mathbf{s}}{dt} \right)_i = \begin{bmatrix} \widetilde{\boldsymbol{\omega}}_i^T & \widetilde{\mathbf{v}}_i^T \\ \mathbf{0} & \widetilde{\boldsymbol{\omega}}_i^T \end{bmatrix} \delta\mathbf{s}_i;\ \ \delta\mathbf{s}_i = \left(\frac{\partial\dot{\mathbf{s}}_i}{\partial\dot{\mathbf{s}}} \right)\delta\mathbf{s} = \begin{pmatrix} \dfrac{\partial\mathbf{v}_i}{\partial\dot{\mathbf{s}}} \\ \dfrac{\partial\boldsymbol{\omega}_i}{\partial\dot{\mathbf{s}}} \end{pmatrix} \delta\mathbf{s} \tag{28}$$

($\dot{\mathbf{s}}_i \in \mathbb{R}^6$, $\mathbf{s} = \mathbf{s}(\mathbf{s}_i) \in \mathbb{R}^g$. Obviously, the corresponding Hamel coefficients are zero or plus/minus one). The Hamel equation (18) yields along with Eq.(25) and Eq.(28)

$$\delta\mathbf{s}^T \sum_{i=1}^{n}\left\{\left(\frac{\partial\mathbf{v}}{\partial\dot{\mathbf{s}}}\right)^T(\dot{\mathbf{p}}+\widetilde{\omega}\mathbf{p}-\mathbf{f}^e)+\left(\frac{\partial\omega}{\partial\dot{\mathbf{s}}}\right)^T(\dot{\mathbf{L}}+\widetilde{\omega}\mathbf{L}-\mathbf{M}^e)\right\}_i =0, \quad \begin{matrix} \mathbf{p} = m\mathbf{v} \\ \mathbf{L} = \mathbf{J}\omega \end{matrix}$$

$$(29)$$

where, from derivation, $\mathbf{v}, \omega$ and the linear and angular momentum vectors $\mathbf{p}$, $\mathbf{L}$ are represented in a body-fixed frame. Of course, the result does not depend on any chosen frame (Eq.(29) is a scalar product). However, one has to be careful with the notations. Considering an arbitrary (orthonormal) reference frame R with $\mathbf{A}_{RB}$ (transformation from body-fixed frame B into reference frame R) one has, with index I for inertial frame and with left index for coordinate representation

$$\sum_{i=1}^{n}\left\{\left(\frac{\partial_B\mathbf{v}}{\partial\dot{\mathbf{s}}}\right)^T\mathbf{A}_{BR}\,\mathbf{A}_{RB}({}_B\dot{\mathbf{p}}+{}_B\widetilde{\omega}_B\mathbf{p}-{}_B\mathbf{f}^e)+\cdots =\right.$$

$$\sum_{i=1}^{n}\left\{\left(\frac{\partial_R\mathbf{v}}{\partial\dot{\mathbf{s}}}\right)^T\mathbf{A}_{RB}\left[\mathbf{A}_{BI}\frac{d}{dt}\left(\mathbf{A}_{IBB}\mathbf{p}\right)-{}_B\mathbf{f}^e)\right]+\cdots =\right.$$

$$\sum_{i=1}^{n}\left\{\left(\frac{\partial_R\mathbf{v}}{\partial\dot{\mathbf{s}}}\right)^T\mathbf{A}_{RB}\left[\mathbf{A}_{BI}\frac{d}{dt}\left(\mathbf{A}_{IRR}\mathbf{p}\right)-{}_B\mathbf{f}^e)\right]+\cdots =\right.$$

$$\sum_{i=1}^{n}\left\{\left(\frac{\partial_R\mathbf{v}}{\partial\dot{\mathbf{s}}}\right)^T({}_R\dot{\mathbf{p}}+{}_R\widetilde{\omega}_{IRR}\mathbf{p}-{}_R\mathbf{f}^e)+\cdots\right.$$

$$(30)$$

This leads to a representation in an arbitrary but orthonormal reference frame R

$$\sum_{i=1}^{n}\left\{\left(\frac{\partial\mathbf{v}_c}{\partial\dot{\mathbf{s}}}\right)^T(\dot{\mathbf{p}}+\widetilde{\omega}_{IR}\mathbf{p}-\mathbf{f}^e)+\left(\frac{\partial\omega_c}{\partial\dot{\mathbf{s}}}\right)^T(\dot{\mathbf{L}}+\widetilde{\omega}_{IR}\mathbf{L}-\mathbf{M}^e)\right\}_i =0, \quad \begin{matrix} \mathbf{p} = m\mathbf{v}_c \\ \mathbf{L} = \mathbf{J}^c\omega_c \end{matrix}$$

$$(31)$$

where, in order to distiguish the ω's, the index c (for mass center) has been introduced. However, the question how to chose $\dot{\mathbf{s}}$ has not yet been answered. **Interpretation.** Eq.(31) contains the (linear and angular) momentum theorems

$$(\dot{\mathbf{p}}+\widetilde{\omega}_{IR}\mathbf{p}-\mathbf{f}^e-\mathbf{f}^{\text{constr.}}) = \mathbf{0}$$

$$(\dot{\mathbf{L}}+\widetilde{\omega}_{IR}\mathbf{L}-\mathbf{M}^e-\mathbf{M}^{\text{constr.}}) = \mathbf{0}$$

$$(32)$$

with free eligeable reference frame, e.g. inertial frame: $\widetilde{\omega}_{IR} = \widetilde{\omega}_{II} \equiv \mathbf{0}$ (mostly used for linear momentum, see (Euler)[6]), or body-fixed frame: $\widetilde{\omega}_{IR} = \widetilde{\omega}_{IB} = \widetilde{\omega}_c$ (mostly used for momentum of momentum, see (Euler)[7]). However, in many applications, e.g. rotor dynamics, the use of a guidance frame ("Führungssystem") R is advantageous, see example 3.

The constraint forces and torques are eliminated in Eq.(31) by means of functional matrices. This means, in other words, motion is projected into the unconstrained space which is characterized by $\dot{s}$. Eq.(31) shall therefore be called *The Projection Equation*. With this interpretation, Eq.(31) is the typical representative of the synthetical method the aim of which is to synthesize motion equations by force/torque relations (simultaneously, of course, eliminating constraint forces). It should, therefore, not astonish that the huge amount of "synthetical procedures", mainly elaborated within the last four decades, can eventually and without any exception be interpreted in the sense of Eq.(31).

On the other hand, premultiplying once more with $\delta\mathbf{s}^T$, one has the dimension of a (virtual) energy and, recalling the derivation, Eq.(31) is an intepretation of the central equation (17) which is a typical representative of the analytical approach. However, Eq.(17) is a rewriting of Lagrange's principle (15) where the virtual displacements may be separated from the integral

$$\left\{ \int\limits_{(S)} \left[(dm\ddot{\mathbf{r}} - d\mathbf{f}^e)^T \frac{\partial \mathbf{r}}{\partial \mathbf{s}} \right] \right\} \delta\mathbf{s} = \mathbf{0} \tag{33}$$

Does Eq.(33) belong to the synthetical or to the analytical approach? The momentum theorem arises already, but the physical dimension is that one of an energy. Leaving the arbitrary virtual displacements aside the dimension is that one of a (generalized) force. So what? We believe that this kind of discussion is somewhat useless (although things should be called anyhow). The only (but important) statement one can extract here is that, for the different procedures, one always deals with the same (classical) mechanics, looking at it from different view points. According to the actual aim of investigation, one should focus that view point to the corresponding procedure which offers minimum effort.

The main procedures are listed in Figure 1.

Legend: T: kinetic energy, V: potential, H: Hamilton-function, G: Gibbs-Appell-function, $\mathbf{Q}$: generalized force, $\mathbf{p}$: (generalized) momentum, $\mathbf{q}$: minimal coordinates, $\dot{\mathbf{s}}$: minimal (nonholonomic) velocities, δW: virtual work, $\mathbf{v}, \boldsymbol{\omega}$: translational, angular velocity (cartesian), $\mathbf{p}, \mathbf{L}$: momentum, momentum of momentum (referring to mass center), $\mathbf{f}, \mathbf{M}$: force, torque. Indices: e: impressed, I: inertial frame, C: bodyfixed frame at the mass center , R: reference frame

Historical remark. The present context mainly concentrates on Georg Hamel (1877-1954) because, in our opinion, his contributions offer the easiest and deepest access to the problems under consideration. However, we do not want, by no means, to insult other authors in the field. Hamel himself, as a question of fairness, did never ignore other contributions as can be seen troughout all his contributions.

Historically, Hamel's research had been initiated by the situation of (scientific) mechanics around the turn of the 20th century. In that period, although characterized by the euphorical foundations of *technical* universities in Europe, *technical* mechanics found itself in a severely desolate state due to the influence of their practicioners who denied the role of kinetics for practical problems. It is not that these people had been wrong – in a statical sense – like most of today's robot practitioners are either: As long as everything moves slowly, kinetics do not play a significant role. But these mecanicians had lost, almost arrogantly, the roots to their own science, thus consequently being helpless when faced with fast moving steam engines. These short sighted practicioners had been the reason for the absurdity that practical engineering failed due to the loss of practice itself. In this critical situation Karl Heun (1859-1929) came up with his famous (about 120 page-) paper "on the kinetic problems in scientific engineering" [20]. By the influence of the famous mathematician Felix Klein (1849-1925), Karl Heun became professor in Karlsruhe where Georg Hamel was his assistant from 1903 till 1905. During that period, Hamel published his habilitation thesis [14].

For those readers who are interested in more background and historical development we can therefore recommend Hamel's contributions themselves, as a contemporarian of that time, or, with the viewpoint from now backwards, the thoughtful contributions of Papastavridis [33].

$$(1673)\ 1847 \qquad T + V = H$$

$$(1879)\ 1899 \qquad \left(\frac{\partial G}{\partial \ddot{\mathbf{s}}}\right)^T = \mathbf{Q}$$

$$(1828) \qquad \int_{t_o}^{t_1} (\delta T + \delta V)\,dt = 0$$

$$1835 \qquad \dot{\mathbf{p}} = -\left(\frac{\partial H}{\partial \mathbf{q}}\right)^T \;;\; \dot{\mathbf{q}} = \left(\frac{\partial H}{\partial \mathbf{p}}\right)^T$$

$$1780 \qquad \left[\frac{d}{dt}\frac{\partial T}{\partial \dot{\mathbf{q}}} - \frac{\partial T}{\partial \mathbf{q}} - \mathbf{Q}^T\right] = 0$$

$$1903 \qquad \left[\frac{d}{dt}\frac{\partial T}{\partial \dot{\mathbf{q}}} - \frac{\partial T}{\partial \mathbf{q}} - \mathbf{Q}^T\right]\left(\frac{\partial \dot{\mathbf{q}}}{\partial \dot{\mathbf{s}}}\right) = 0$$

$$1904 \qquad \left[\frac{d}{dt}\frac{\partial T}{\partial \dot{\mathbf{s}}} - \frac{\partial T}{\partial \mathbf{s}} - \mathbf{Q}^T\right]\delta\mathbf{s} + \frac{\partial T}{\partial \dot{\mathbf{s}}}\left(\frac{d\delta\mathbf{s} - \delta d\mathbf{s}}{dt}\right) = 0$$

$$\begin{array}{l}\textit{CENTRAL}\\ \textit{EQUATION}\end{array} \qquad \frac{d}{dt}\left[\left(\frac{\partial T}{\partial \dot{\mathbf{s}}}\right)\delta\mathbf{s}\right] - \delta T - \delta W = 0$$

$$\sum_{i=1}^{n}\left\{\left(\frac{\partial \mathbf{v}_{IC}}{\partial \dot{\mathbf{s}}}\right)^T(\dot{\mathbf{p}} + \widetilde{\boldsymbol{\omega}}_{IR}\mathbf{p} - \mathbf{f}^e) + \left(\frac{\partial \boldsymbol{\omega}_{IC}}{\partial \dot{\mathbf{s}}}\right)^T(\dot{\mathbf{L}} + \widetilde{\boldsymbol{\omega}}_{IR}\mathbf{L} - \mathbf{M}^e)\right\}_i = 0$$

THE TRIANGLE OF METHODS

(1673: Huygens "Horologium Oscillatorum" (conservative)), 1847: Helmholtz "Über die Erhaltung der Kraft" (general), (1879: Gibbs "On the Fundamental Formulae of Dynamics" (holonomic)), 1899: Appell "Sur une forme générale des équations de la dynamique" (general), 1828: Hamilton "Theory of Systems of Rays", 1835: Hamilton "Second Essay on a General Method in Dynamics", 1780: Lagrange "Théorie de la libration de la lune", 1903: Maggi "Stereodinamica", 1904: Hamel "Die Lagrange-Eulerschen Gleichungen der Mechanik" (compare also also Boltzmann, Chapligyn, Rumyantsev, Volterra, Woronetz)

Fig 1. Main Procedures in Dynamics

3   Choice of Reference Frame

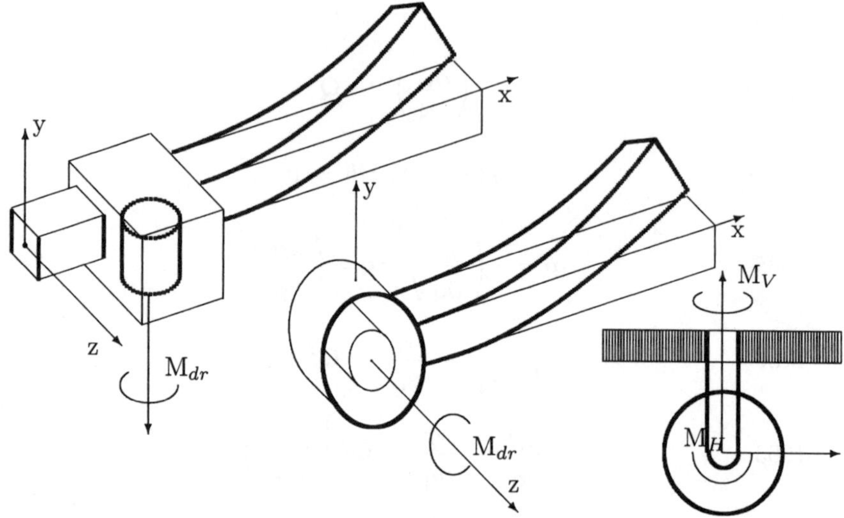

Figure 2   Modeling Elements (Subsystems): Prismatic Joint, Revolute Joint, Wheel

In order to come out with a representation according to Eq.(1) it is useful to separate the "guidance" (which is purely depending on the predecessor's motion) from the superimposed relative motion. This is achieved with the choice of a reference frame at the body fixed "hinge point". One obtains thus a constant tensor of moments of inertia $\mathbf{J}$ along with, for elastic bodies, the usual notation of elastic deflections being function of the undeformed state. However, elastic bodies will not be persued here.

3.1   Element Matrices

Using the above references the mass center velocities in Eq.(31) have to be rewritten in terms of hingepoint velocities

$$
\begin{pmatrix} \mathbf{v}_c \\ \boldsymbol{\omega}_c \end{pmatrix}_i = \underbrace{\begin{bmatrix} \mathbf{E} & \tilde{\mathbf{r}}_c^T \\ \mathbf{0} & \mathbf{E} \end{bmatrix}_i}_{[(\partial \mathbf{v}_c/\partial \dot{\mathbf{y}})\,(\partial \boldsymbol{\omega}_c/\partial \dot{\mathbf{y}})]_i} \underbrace{\begin{pmatrix} \mathbf{v}_{oi} \\ \boldsymbol{\omega}_{oi} \end{pmatrix}}_{\dot{\mathbf{y}}_i}
\tag{34}
$$

(o: inertial frame, i: reference frame origin of i-th body) leading to a representation in element matrices

$$\sum_{i=1}^{n} \left(\frac{\partial \dot{\mathbf{y}}_i}{\partial \dot{\mathbf{s}}} \right)^T \left[\overline{\mathbf{M}}_i \ddot{\mathbf{y}}_i + \overline{\mathbf{G}}_i \dot{\mathbf{y}}_i - \mathbf{Q}_i \right] = 0 \quad \text{where}$$

$$\overline{\mathbf{M}}_i = \begin{bmatrix} m\mathbf{E} & m\widetilde{\mathbf{r}}_c^T \\ m\widetilde{\mathbf{r}}_c & \mathbf{J} \end{bmatrix}_i, \quad \overline{\mathbf{G}}_i = \begin{bmatrix} m\widetilde{\boldsymbol{\omega}}_o & -[m\widetilde{\boldsymbol{\omega}}_o\widetilde{\mathbf{r}}_c] \\ [m\widetilde{\boldsymbol{\omega}}_o\widetilde{\mathbf{r}}_c]^T & \widetilde{\boldsymbol{\omega}}_o\mathbf{J} \end{bmatrix}_i \qquad (35)$$

($\mathbf{E}$: unit matrix, $\mathbf{r}_c$: mass center distance, $\mathbf{J} = \mathbf{J}^c + m\widetilde{\mathbf{r}}_c\widetilde{\mathbf{r}}_c^T$, $[\boldsymbol{\omega}_o]_i = \boldsymbol{\omega}_{oi}$)

The advantage of nonholonomic variables even for holonomic systems lies obviously mainly in the field of rotations. Let us therefore have a short look on a single gyro for motivation.

Example 3: Gyroscope. The reference frame is body-fixed and the mass center is used for "hinge point" moving with (x y 0). Kinematics are calculated with Kardan angles (rotation sequence $z \to y \to x$ (related angles $\Omega t, \beta, \alpha$), all deviations from the desired position are assumed small.

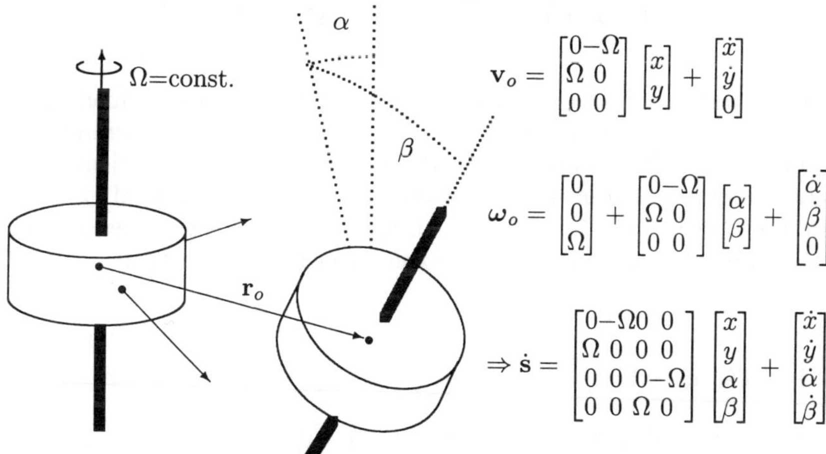

Figure 3   Gyroscope Using Kardan-Angles for Kinematics

Here, one obtains suitable minimal velocities by direct inspection of the kinematics in Figure 3

$$\dot{\mathbf{y}} = \begin{pmatrix} \mathbf{o} \\ \boldsymbol{\Omega} \end{pmatrix} + \left(\frac{\partial \dot{\mathbf{y}}}{\partial \dot{\mathbf{s}}} \right) \dot{\mathbf{s}} =$$

$$\begin{pmatrix} \mathbf{o} \\ \boldsymbol{\Omega} \end{pmatrix} + \begin{bmatrix} 1 & 0 & 0 & 0 \\ 0 & 1 & 0 & 0 \\ 0 & 0 & 0 & 0 \\ 0 & 0 & 1 & 0 \\ 0 & 0 & 0 & 1 \\ 0 & 0 & 0 & 0 \end{bmatrix} \left\{ \begin{bmatrix} 0 & -\Omega & 0 & 0 \\ \Omega & 0 & 0 & 0 \\ 0 & 0 & 0 & -\Omega \\ 0 & 0 & \Omega & 0 \end{bmatrix} \begin{pmatrix} x \\ y \\ \alpha \\ \beta \end{pmatrix} + \begin{pmatrix} \dot{x} \\ \dot{y} \\ \dot{\alpha} \\ \dot{\beta} \end{pmatrix} \right\}$$

$$(36)$$

where $\mathbf{o} = (0\ 0\ 0)^T, \boldsymbol{\Omega} = \Omega \mathbf{e}_3 = \Omega(0\ 0\ 1)^T$. (The reason why $\dot{\mathbf{s}}$ here is composed by two terms is due to the linearization process: The Taylor expansion, with $\dot{\mathbf{s}}_o \equiv 0, \mathbf{q}_o \equiv 0, \mathbf{q}, \dot{\mathbf{q}}, \dot{\mathbf{s}}$ small, yields $\dot{\mathbf{s}} = [\partial(\mathbf{H}(\mathbf{q})\dot{\mathbf{q}})/\partial\mathbf{q}]_o \mathbf{q} + [\partial(\mathbf{H}(\mathbf{q})\dot{\mathbf{q}})/\partial\dot{\mathbf{q}}]_o \dot{\mathbf{q}})$

The matrices from Eq.(35) reduce to

$$\overline{\mathbf{M}} = \begin{bmatrix} m\mathbf{E} & \mathbf{0} \\ \mathbf{0} & \mathbf{J} \end{bmatrix}, \quad \overline{\mathbf{G}} = \begin{bmatrix} m\widetilde{\boldsymbol{\Omega}} & \mathbf{0} \\ \mathbf{0} & \widetilde{\boldsymbol{\Omega}}\mathbf{J} - (\mathbf{J}\boldsymbol{\Omega})^{\sim} \end{bmatrix} \tag{37}$$

with $\mathbf{J} = diag\{A, A, C\}$. Using the corresponding equations for modelling a wheel along with its suspension, for instance, needs a transformation into a car fixed frame (see next example below). If the gyro is considered as element of a planetary gear (e.g. (Bremer)[4] for nonstationary Ω), as often used in robotic design, then, also, one is interested in an arbitrarily rotating reference frame. It is therefore advantageous to apply a congruence transformation

$$\dot{\mathbf{s}} \to \mathbf{T}\dot{\mathbf{s}}, \quad \mathbf{q} \to \mathbf{T}\mathbf{q}$$

$$\mathbf{T}^T\overline{\mathbf{M}}\mathbf{T}\ddot{\mathbf{s}} + \left[\mathbf{T}^T\overline{\mathbf{G}}\mathbf{T} + \mathbf{T}^T\overline{\mathbf{M}}\dot{\mathbf{T}} \right]\dot{\mathbf{s}} - \mathbf{T}^T\mathbf{Q} = 0 \tag{38}$$

with $\mathbf{T} = \begin{bmatrix} \overline{\mathbf{T}} & \mathbf{0} \\ \mathbf{0} & \overline{\mathbf{T}} \end{bmatrix}$, $\overline{\mathbf{T}} = \begin{bmatrix} \cos\overline{\gamma} & \sin\overline{\gamma} \\ -\sin\overline{\gamma} & \cos\overline{\gamma} \end{bmatrix}$, $\dot{\overline{\gamma}} = (\Omega - \Omega_R)$

yielding

$$\dot{\mathbf{s}} = \begin{bmatrix} 0 & -\Omega_R & 0 & 0 \\ \Omega_R & 0 & 0 & 0 \\ 0 & 0 & 0 & -\Omega_R \\ 0 & 0 & \Omega_R & 0 \end{bmatrix} \begin{bmatrix} x \\ y \\ \alpha \\ \beta \end{bmatrix} + \begin{bmatrix} \dot{x} \\ \dot{y} \\ \dot{\alpha} \\ \dot{\beta} \end{bmatrix}$$

$$\ddot{\mathbf{s}} + \left[\begin{array}{cc|cc} 0 & -\Omega_R & 0 & 0 \\ \Omega_R & 0 & 0 & 0 \\ \hline 0 & 0 & 0 & \dfrac{C}{A}\Omega - \Omega_R \\ 0 & 0 & -\dfrac{C}{A}\Omega + \Omega_R & 0 \end{array} \right] \dot{\mathbf{s}} - \begin{pmatrix} \dfrac{1}{m}\begin{pmatrix} f_x \\ f_y \end{pmatrix} \\ \dfrac{1}{A}\begin{pmatrix} M_x \\ M_y \end{pmatrix} \end{pmatrix} = \begin{pmatrix} 0 \\ 0 \\ 0 \\ 0 \end{pmatrix}$$

$$(39)$$

where Ω_R is the rotation of chosen reference frame. Obviously, this notation is much simpler than the usual one commonly used in rotor dynamics.

3.2 Recursive Kinematics

The (cartesian) variables $\dot{\mathbf{y}}_i$ may now be separated into its parts from the predecessor (p) and the superimposed relative motion (i), where $\mathbf{r}_{pi}$ denotes the location of i-th reference frame w.r.t. its predecessor:

$$\dot{\mathbf{y}}_i = \begin{bmatrix} \mathbf{v}_{oi} \\ \boldsymbol{\omega}_{oi} \end{bmatrix}, \dot{\mathbf{y}}_p = \begin{bmatrix} \mathbf{v}_{op} \\ \boldsymbol{\omega}_{op} \end{bmatrix}, \dot{\mathbf{y}}_{ri} = \begin{bmatrix} \mathbf{v}_{pi} \\ \boldsymbol{\omega}_{pi} \end{bmatrix}$$

$$\Rightarrow \dot{\mathbf{y}}_i = \mathbf{T}_{ip}\dot{\mathbf{y}}_p + \dot{\mathbf{y}}_{ri}, \quad \mathbf{T}_{ip} = \begin{bmatrix} \mathbf{A}_{ip} & \mathbf{A}_{ip}\tilde{\mathbf{r}}_{pi}^T \\ \mathbf{0} & \mathbf{A}_{ip} \end{bmatrix}, \mathbf{T}_{ip}^{-1} = \begin{bmatrix} \mathbf{A}_{ip}^T & \tilde{\mathbf{r}}_{pi}\mathbf{A}_{ip}^T \\ \mathbf{0} & \mathbf{A}_{ip}^T \end{bmatrix}, \quad (40)$$

The accelerations are then $\ddot{\mathbf{y}}_i = \mathbf{T}_{ip}\ddot{\mathbf{y}}_p - \widetilde{\boldsymbol{\Omega}}_{pi}\mathbf{T}_{ip}\dot{\mathbf{y}}_p + \ddot{\mathbf{y}}_{ri}$ where $\widetilde{\boldsymbol{\Omega}}_{pi} = blockdiag\{\widetilde{\boldsymbol{\omega}}_{pi}, \widetilde{\boldsymbol{\omega}}_{pi}\}$. The minimal velocities $\dot{\mathbf{s}}$ are composed of $\dot{\mathbf{s}}_i$ of i-th body. These are elements from $\dot{\mathbf{y}}_{ri}$. Thus, $\dot{\mathbf{y}}_{ri} = \mathbf{F}_i\dot{\mathbf{s}}_i$.

Example 4 In case of plane motions, with the mass center located on the x-axis, Eqs.(35) and (40) reduce to

$$\overline{\mathbf{M}} = \begin{bmatrix} m & 0 & 0 \\ 0 & m & mc \\ 0 & mc & J \end{bmatrix}, \overline{\mathbf{G}} = \begin{bmatrix} 0 & -m & -mc \\ m & 0 & 0 \\ mc & 0 & 0 \end{bmatrix} \omega_o \qquad (41)$$

$$\mathbf{T}_{ip} = \begin{bmatrix} \cos\gamma_i & \sin\gamma_i & x_{pi}\sin\gamma_i - y_{pi}\cos\gamma_i \\ -\sin\gamma_i & \cos\gamma_i & x_{pi}\cos\gamma_i + y_{pi}\sin\gamma_i \\ 0 & 0 & 1 \end{bmatrix}, \mathbf{T}_{ip}^{-1} = \begin{bmatrix} \cos\gamma_i & -\sin\gamma_i & y_{pi} \\ \sin\gamma_i & \cos\gamma_i & -x_{pi} \\ 0 & 0 & 1 \end{bmatrix}$$

$$(42)$$

along with $\widetilde{\boldsymbol{\Omega}}_{pi} \in \mathbb{R}^{6,6} \to \dot{\gamma}_i \widetilde{\mathbf{e}}_3 \in \mathbb{R}^{3,3}, \boldsymbol{\omega}_{oi} = [\omega_{op} + \dot{\gamma}_i]\mathbf{e}_3, \dot{\mathbf{y}}_i = [v_{xi}\, v_{yi}\, \omega_{oi}]$.

Let for the example in Figure 4 be nr.1: car body, nrs.2 and 3 the SCARA, nrs.4,5,6 right, left and front wheel. One obtains then with $p(i)$ (predecessor of body i) the "topology matrix" (Huston)[22] (which already contains every information from graph theory, e.g. (Wittenburg)[41], which is needed for the kinematics of Multi Body Systems). Especially, its columns represent the "path" the numbering scheme of which is essential for the sequence of summations in Eq.(35).

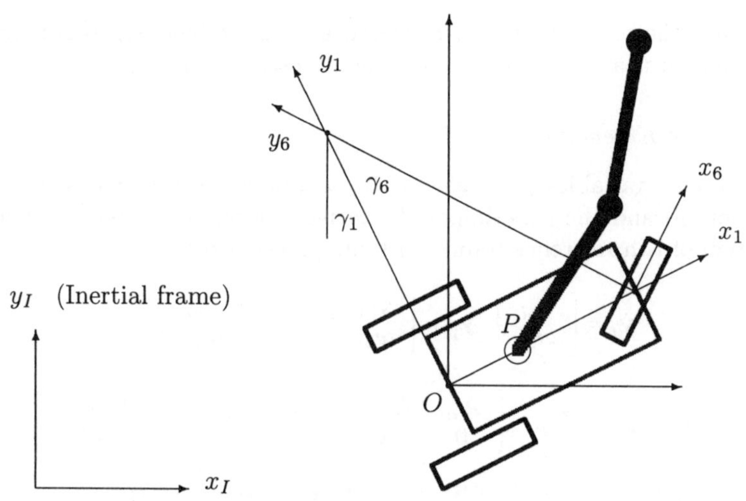

Figure 4　Plane Motion of a Wheeled Robot

Table 4:　The Topology Matrix

i=	1	2	3	4	5	6
p(i)=	0	1	2	1	1	1
p[p(i)]=	0	0	1	0	0	0
p{p[p(i)]}=	0	0	0	0	0	0

Let furthermore be $x_{12} = b, x_{23} = L, y_{14} = -d, y_{15} = d, x_{16} = a$ (all other x_{pi}, y_{pi} zero). One may then write down the cartesian velocities with the aid of Eq.(40) following the index scheme from Table 4. The velocities may contain possible constraints, characterized with an upper index pc ("possibly constrained"). We have then the following situation

$$\dot{\mathbf{y}}_1^{pc} = \dot{\mathbf{y}}_{r1} \qquad \text{chassis}$$

$$\left.\begin{aligned}\dot{\mathbf{y}}_2 &= \mathbf{T}_{21}\,\dot{\mathbf{y}}_1^{pc} + \dot{\mathbf{y}}_{r2} \\ \dot{\mathbf{y}}_3 &= \mathbf{T}_{32}\,\dot{\mathbf{y}}_2 + \dot{\mathbf{y}}_{r3}\end{aligned}\right\} \text{SCARA}$$

$$\left.\begin{aligned}\dot{\mathbf{y}}_4 &= \mathbf{T}_{41}\,\dot{\mathbf{y}}_1^{pc} \\ \dot{\mathbf{y}}_5 &= \mathbf{T}_{51}\,\dot{\mathbf{y}}_1^{pc}\end{aligned}\right\} \text{rear wheels}$$

$$\dot{\mathbf{y}}_6^{pc} = \mathbf{T}_{61}\,\dot{\mathbf{y}}_1^{pc} + \dot{\mathbf{y}}_{r6} \qquad \text{front wheel}$$

$$(43)$$

Remark. The rolling condition is not of interest for the moment; wether the wheels roll or slip does not enter the above kinematics. Generally, one would leave the plane motion conditions with the wheel angular velocity. However, this case can trivially be treated in the following way: The velocity at the wheel conctact is $(v_{x4} - R\dot{\beta}_4)$ for the right rear wheel for instance. (R: wheel radius, β: wheel rotation about its y-axis). Considering pure rolling, this relation represents a constraint

$$\dot{\psi}_4 = \begin{bmatrix} 1 & -R \end{bmatrix}\begin{pmatrix} v_{x4} \\ \dot{\beta}_4 \end{pmatrix} = 0 \tag{44}$$

If one inserts the constraint along with the corresponding Lagrangian parameter λ and frees the system from the constraint (Hamel), then the λ converts into an impressed force f_{x4}^e which is obviously the actual wheel resistive force. One obtains thus, with $[\partial\dot{\psi}_4/\partial(v_{x4}, \dot{\beta}_4)] = \begin{bmatrix} 1 & -R \end{bmatrix}$, the drive equation

$$J_{y4}\ddot{\beta}_4 + R\,f_{x4}^e = M_{\text{drive},4} \tag{45}$$

and the impressed force for subsystem 4 (right wheel) is $\mathbf{Q}_4 = (f_{x4}^e,\ 0,\ 0)^T$. The physical relations at the contact point will then decide on the actual behaviour (friction or slip). Eq.(43) is, however, not influenced by these considerations.

Because the rear wheels do not contibute, one may rewrite Eq.(35)

$$\sum_{i=1}^{6} \left(\frac{\partial\dot{\mathbf{y}}_i}{\partial\dot{\mathbf{s}}}\right)^T \left[\overline{\mathbf{M}}_i\ddot{\mathbf{y}}_i + \overline{\mathbf{G}}_i\dot{\mathbf{y}}_i - \mathbf{Q}_i\right] =$$

$$\sum_{\substack{i=1 \\ i \neq 4,5}}^{6} \left(\frac{\partial \dot{\mathbf{y}}_i}{\partial \dot{\mathbf{s}}} \right)^T \left[\overline{\mathbf{M}}_i \ddot{\mathbf{y}}_i + \overline{\mathbf{G}}_i \dot{\mathbf{y}}_i - \mathbf{Q}_i \right] = 0 \quad \text{where}$$

$$\overline{\mathbf{M}}_1 \rightarrow \overline{\mathbf{M}}_1 + \mathbf{T}_{41}^T \overline{\mathbf{M}}_4 \mathbf{T}_{41} + \mathbf{T}_{51}^T \overline{\mathbf{M}}_5 \mathbf{T}_{51}$$

$$\overline{\mathbf{G}}_1 \rightarrow \overline{\mathbf{G}}_1 + \mathbf{T}_{41}^T \overline{\mathbf{G}}_4 \mathbf{T}_{41} + \mathbf{T}_{51}^T \overline{\mathbf{G}}_5 \mathbf{T}_{51}$$

$$\overline{\mathbf{Q}}_1 \rightarrow \overline{\mathbf{Q}}_1 + \mathbf{T}_{41}^T \overline{\mathbf{Q}}_4 + \mathbf{T}_{51}^T \overline{\mathbf{Q}}_5 \tag{46}$$

since $\mathbf{T}_{41}, \mathbf{T}_{51}$ are constant. One is thus left with the independent $\dot{\mathbf{y}}_i$, and one has to chose $\dot{\mathbf{s}}$. This can be done in the following way:

If in Eq.(43) at any stage new possible contraints arise, then the previous velocities are expressed by the new ones, and simultaneously new variables have to enter calculation in order to take the new situation into account. This is the case here for $\dot{\mathbf{y}}_6^{pc}$. Denoting the possible constraint herein by v_{y6} (that means possibly (hopefully) no front wheel sliding) one obtains

$$\dot{\mathbf{y}}_6 = \left(\begin{array}{c} v_{x6} \\ v_{y6} \\ \omega_{o6} \end{array} \right) = \left[\begin{array}{ccc} \cos\gamma_6 & \sin\gamma_6 & a\sin\gamma_6 \\ -\sin\gamma_6 & \cos\gamma_6 & a\cos\gamma_6 \\ 0 & 0 & 1 \end{array} \right] \left(\begin{array}{c} v_{x1} \\ v_{y1} \\ \omega_{o1} \end{array} \right) + \left(\begin{array}{c} 0 \\ 0 \\ \dot{\gamma}_6 \end{array} \right) \right\} \begin{array}{c} \text{front} \\ \text{wheel} \end{array} \tag{47}$$

Resolving for $\dot{\mathbf{y}}_1^{pc}$ by means of Eq.(42) yields

$$\left(\begin{array}{c} v_{x1} \\ v_{y1} \\ \dot{\gamma}_1 \end{array} \right) = \left[\begin{array}{ccc} \cos\gamma_6 & -\sin\gamma_6 & 0 \\ \sin\gamma_6 & \cos\gamma_6 & -a \\ 0 & 0 & 1 \end{array} \right] \left(\begin{array}{c} v_{x6} \\ v_{y6} \\ \dot{\gamma}_1 \end{array} \right) \tag{48}$$

Replacing $\dot{\mathbf{y}}_1^{pc}$ in Eq.(47) with Eq.(48) does not make sense because: if both constraints will never be active – although correct, why should one? If, on the other hand, two constraints may become active ($v_{y1} = 0, v_{y6} = 0$), then, inserting the above relation, one takes care of $v_{y6} = 0$ but neglects $v_{y1} = 0$, or vice versa if one leaves everything as it is. The aim is therefore to introduce new variables such that with these, replacing the old $\dot{\mathbf{y}}_1^{pc}$, both constraints may become active simultaneously. That means that both (possibly active) constraints have to enter the term one is looking for. This is easily achieved: resolve the possible constraint in Eq.(48), i.e. $(0\ 1\ 0)\dot{\mathbf{y}}_1^{pc}$, for one of the remaining free variables:

$$\cancel{v_{y1}} = v_{x6}\sin\gamma_6 + \cancel{v_{y6}}\cos\gamma_6 - a\dot\gamma_1 :$$

either

$$\begin{pmatrix} v_{x6} \\ \cancel{v_{y6}} \\ \dot\gamma_1 \end{pmatrix} = \frac{1}{\sin\gamma_6} \begin{bmatrix} 1 & -\cos\gamma_6 & a \\ 0 & \sin\gamma_6 & 0 \\ 0 & 0 & \sin\gamma_6 \end{bmatrix} \begin{pmatrix} \cancel{v_{y1}} \\ \cancel{v_{y6}} \\ \dot\gamma_1 \end{pmatrix}$$

$$\text{in (48)} \Rightarrow \begin{pmatrix} v_{x1} \\ \cancel{v_{y1}} \\ \dot\gamma_1 \end{pmatrix} = \frac{1}{\sin\gamma_6} \begin{bmatrix} \cos\gamma_6 & -1 & 0 \\ \sin\gamma_6 & 0 & 0 \\ 0 & 0 & \sin\gamma_6 \end{bmatrix} \begin{pmatrix} \cancel{v_{y1}} \\ \cancel{v_{y6}} \\ \dot\gamma_1 \end{pmatrix} = \mathbf{T}\begin{pmatrix} \cancel{v_{y1}} \\ \cancel{v_{y6}} \\ \dot\gamma_1 \end{pmatrix} \tag{49}$$

or

$$\begin{pmatrix} v_{x6} \\ \cancel{v_{y6}} \\ \dot\gamma_1 \end{pmatrix} = \begin{bmatrix} 0 & 0 & 1 \\ 0 & 1 & 0 \\ -\frac{1}{a} & \frac{1}{a}\cos\gamma_6 & \frac{1}{a}\sin\gamma_6 \end{bmatrix} \begin{pmatrix} \cancel{v_{y1}} \\ \cancel{v_{y6}} \\ v_{x6} \end{pmatrix} \tag{50}$$

$$\text{in (48)} \Rightarrow \begin{pmatrix} v_{x1} \\ \cancel{v_{y1}} \\ \dot\gamma_1 \end{pmatrix} = \begin{bmatrix} 0 & -\sin\gamma_6 & \cos\gamma_6 \\ 1 & 0 & 0 \\ -\frac{1}{a} & \frac{1}{a}\cos\gamma_6 & \frac{1}{a}\sin\gamma_6 \end{bmatrix} \begin{pmatrix} \cancel{v_{y1}} \\ \cancel{v_{y6}} \\ v_{x6} \end{pmatrix} = \mathbf{T}\begin{pmatrix} \cancel{v_{y1}} \\ \cancel{v_{y6}} \\ v_{x1} \end{pmatrix}$$

Although arbitrary in principle, the first choice contains singularities (for vanishing steering angle γ_6) while the second does not.

The minimal velocities are composed of $\dot s_i, i = 1,2,3,6$. Collecting the variables yields the situation according to Eq.(1),

$$\dot{\mathbf{s}} = \begin{bmatrix} \cancel{v_{y1}} & \cancel{v_{y6}} & v_{x6} & \dot\gamma_2 & \dot\gamma_3 & \dot\gamma_6 \end{bmatrix}^T \tag{51}$$

The corresponding time derivative of minimal coordinates $\mathbf{q}$ according to Table 1 are obtained the from the fact that the path $(x,\ y,\ \gamma_1)$ is given within the inertial plane (in robotics referred to as "world coordinates")

$$\begin{pmatrix} v_{x1} \\ v_{y1} \\ \dot\gamma \end{pmatrix} = \begin{bmatrix} \cos\gamma_i & \sin\gamma_i & 0 \\ -\sin\gamma_i & \cos\gamma_i & 0 \\ 0 & 0 & 1 \end{bmatrix} \begin{pmatrix} \dot x \\ \dot y \\ \dot\gamma \end{pmatrix} = \mathbf{A}_\gamma \begin{pmatrix} \dot x \\ \dot y \\ \dot\gamma \end{pmatrix} \tag{52}$$

Along with the remaining variables $(\dot\gamma_2\ \dot\gamma_3\ \dot\gamma_6)$ and with Eq.(50) one has

$$\dot{\mathbf{q}} = \begin{bmatrix} \mathbf{A}_\gamma^T \mathbf{T} & | & \mathbf{E} \end{bmatrix} \dot{\mathbf{s}} = \mathbf{H}(\mathbf{q})^+\dot{\mathbf{s}} : \tag{53}$$

$$
\begin{bmatrix} \dot{x} \\ \dot{y} \\ \dot{\gamma}_1 \\ \dot{\gamma}_2 \\ \dot{\gamma}_3 \\ \dot{\gamma}_6 \end{bmatrix} = \begin{bmatrix} -\sin\gamma_1 & -\cos\gamma_1\sin\gamma_6 & \cos\gamma_1\cos\gamma_6 & 1 & 0 & 0 \\ \cos\gamma_1 & -\sin\gamma_1\sin\gamma_6 & \sin\gamma_1\cos\gamma_6 & 0 & 1 & 0 \\ -\frac{1}{a} & \frac{1}{a}\cos\gamma_6 & \frac{1}{a}\sin\gamma_6 & 0 & 0 & 1 \end{bmatrix} \begin{bmatrix} v_{y1} \\ v_{y6} \\ v_{x6} \\ \dot{\gamma}_2 \\ \dot{\gamma}_3 \\ \dot{\gamma}_6 \end{bmatrix}
\tag{54}
$$

($[\ ]^{+}$: generalized inverse).

The last step for the motion equations according to Eq.(35) is determination of the Jacobians $(\partial\dot{\mathbf{y}}_i/\partial\dot{\mathbf{s}})$. Because the basis (body nr. 1) does not undergo a "guidance motion" its velocity is characterized by $\dot{\mathbf{y}}_1 = \dot{\mathbf{y}}_{r1}$ leading with Eq.(40) to

$$
\begin{bmatrix} \dot{\mathbf{y}}_1 \\ \dot{\mathbf{y}}_2 \\ \dot{\mathbf{y}}_3 \\ \vdots \\ \vdots \\ \dot{\mathbf{y}}_n \end{bmatrix} = \begin{bmatrix} \mathbf{E} & & & & & \\ \mathbf{T}_{21} & \mathbf{E} & & & & \\ \mathbf{T}_{31} & \mathbf{T}_{32} & \mathbf{E} & & & \\ \vdots & \vdots & & \ddots & & \\ \vdots & \vdots & & & \mathbf{E} & \\ \mathbf{T}_{n1} & \cdots & \cdots & & \mathbf{T}_{n,n-1} & \mathbf{E} \end{bmatrix} \begin{bmatrix} \dot{\mathbf{y}}_{r1} \\ \dot{\mathbf{y}}_{r2} \\ \dot{\mathbf{y}}_{r3} \\ \vdots \\ \vdots \\ \dot{\mathbf{y}}_{rn} \end{bmatrix}
\tag{55}
$$

where $\mathbf{T}_{n,j} = \mathbf{T}_{n,p(n)} \times \mathbf{T}_{p(n),p[p(n)]} \times \cdots \mathbf{T}_{s(j),j}$ ($s(j)$: successor of j and $p(n)$: predecessor of n (compare Table 4).

Because the minimal velocities are compontents of the relative cartesian velocities, $\mathbf{y}_{ri} = \mathbf{F}_i\dot{\mathbf{s}}_i$, one obtains the Jacobians

$$
\frac{\partial}{\partial\dot{\mathbf{s}}} \begin{bmatrix} \dot{\mathbf{y}}_1 \\ \dot{\mathbf{y}}_2 \\ \dot{\mathbf{y}}_3 \\ \vdots \\ \vdots \\ \dot{\mathbf{y}}_n \end{bmatrix} = \begin{bmatrix} \mathbf{F}_1 & & & & & \\ \mathbf{T}_{21}\mathbf{F}_1 & \mathbf{F}_2 & & & & \\ \mathbf{T}_{31}\mathbf{F}_1 & \mathbf{T}_{32}\mathbf{F}_2 & \mathbf{F}_3 & & & \\ \vdots & \vdots & & \ddots & & \\ \vdots & \vdots & & & \mathbf{F}_{n-1} & \\ \mathbf{T}_{n1}\mathbf{F}_1 & \cdots & \cdots & & \mathbf{T}_{n,n-1}\mathbf{F}_{n-1} & \mathbf{F}_n \end{bmatrix}
\tag{56}
$$

yielding for Eq.(35) the general expression

$$
\mathbf{M}\ddot{\mathbf{s}} + \mathbf{G}\dot{\mathbf{s}} - \mathbf{Q} = \mathbf{0} \in \mathbb{R}^g
\tag{57}
$$

However, proceeding this way directly yields an $g \times g$ mass matrix which needs lateron inversion. This can be avoided using the recursive kinematics.

3.3 Recursive Kinetics

In order to avoid a total mass matrix inversion one can use the fact that the Jacobian (56) is upper triangular and therefore directly apply a Gauss elimination procedure. With Eq.(56), the last two (block-)rows in Eq.(35) (i=n, p=n-1) read

$$\begin{bmatrix} \mathbf{F}_p^T & \mathbf{F}_p^T \mathbf{T}_{ip}^T \\ 0 & \mathbf{F}_i^T \end{bmatrix} \begin{pmatrix} \overline{\mathbf{M}}_p \ddot{\mathbf{y}}_p + \overline{\mathbf{G}}_p \dot{\mathbf{y}}_p - \mathbf{Q}_p \\ \overline{\mathbf{M}}_i \ddot{\mathbf{y}}_i + \overline{\mathbf{G}}_i \dot{\mathbf{y}}_i - \mathbf{Q}_i \end{pmatrix} = \begin{pmatrix} 0 \\ 0 \end{pmatrix} \tag{58}$$

The last row can be resolved for $\ddot{\mathbf{s}}_i$ in terms of predecessor velocity and accelereration (see Eq.(40)):

$$\ddot{\mathbf{s}}_i = -\left[\mathbf{F}_i^T \overline{\mathbf{M}}_i \mathbf{F}_i \right]^{-1} \mathbf{F}_i^T \times$$

$$\left\{ \overline{\mathbf{M}}_i (\mathbf{T}_{ip} \ddot{\mathbf{y}}_p + \widetilde{\boldsymbol{\Omega}}_i^T \mathbf{T}_{ip} \dot{\mathbf{y}}_p + \dot{\mathbf{F}}_i \dot{\mathbf{s}}_i) + \overline{\mathbf{G}}_i (\mathbf{T}_{ip} \dot{\mathbf{y}}_p + \mathbf{F}_i \dot{\mathbf{s}}_i) - \mathbf{Q}_i \right\}. \tag{59}$$

Insertion into the forelast row then yields the same structure as the last row by definition of

$$\mathbf{F}_p^T \left[\overline{\mathbf{M}}_p \ddot{\mathbf{y}}_p + \overline{\mathbf{G}} \dot{\mathbf{y}}_p - \mathbf{Q}_p \right] = \mathbf{0} :$$

$$\mathbf{M}_i \overset{def}{=} [\mathbf{F}_i^T \overline{\mathbf{M}}_i \mathbf{F}_i], \quad \mathbf{J}_i \overset{def}{=} \left[\mathbf{E} - \overline{\mathbf{M}}_i (\mathbf{F}_i \mathbf{M}_i^{-1} \mathbf{F}_i^T) \right] \Rightarrow$$

$$\begin{aligned} \overline{\mathbf{M}}_p &:= \overline{\mathbf{M}}_p + \mathbf{T}_{ip}^T \mathbf{J}_i \overline{\mathbf{M}}_i \mathbf{T}_{ip} \\ \overline{\mathbf{G}}_p &:= \overline{\mathbf{G}}_p + \mathbf{T}_{ip}^T \mathbf{J}_i (\overline{\mathbf{G}}_i - \overline{\mathbf{M}}_i \widetilde{\boldsymbol{\Omega}}_{pi}) \mathbf{T}_{ip}, \\ \mathbf{Q}_p &:= \mathbf{Q}_p + \mathbf{T}_{ip}^T \mathbf{J}_i (\mathbf{Q}_i - \overline{\mathbf{M}}_i \dot{\mathbf{F}}_i \dot{\mathbf{s}}_i - \overline{\mathbf{G}}_i \mathbf{F}_i \dot{\mathbf{s}}_i) \end{aligned} \tag{60}$$

and can therefore be resolved for $\ddot{\mathbf{s}}_{i-1}$ in the same manner. Applying this scheme repeatedly till $p = 0$ one can solve $\ddot{\mathbf{s}}_1$ with Eq.(59) (where $\mathbf{y}_p \equiv \mathbf{0}$) and then go ahead for $\ddot{\mathbf{s}}_2$ to $\ddot{\mathbf{s}}_n$. This means, in total, a recursion in three steps: First: formulate the kinematics for the last body. All the needed quantities are known from initial conditions. Second: Formulate $\overline{\mathbf{M}}_p, \overline{\mathbf{G}}_p, \mathbf{Q}_p$ $p = n \cdots 1$ backwards till the basis is reached. Third: Calculate $\ddot{\mathbf{s}}_i, i = 1 \cdots n$. The benefit is obvious: the number of operations to calculate the total mass matrix is quadratic w.r.t. the number of d.o.f. and its inversion is at least cubic while the maximum dimension of necessary inversions here corresponds to the rank of $\mathbf{F}_i$, although the recursion has to be rerunned three times. Hence, for $n \geq 3$ the recursion scheme is preferable, for systems with nontrivial kinetic energy. (Note that the Gaussian scheme is not restricted to the single body dynamics as used here. Any row of Eq.(58) may represent an arbitrary subsystem the structure of which is the same as the single body equation.)

Example 5: Plane motion of a serial chain and the double pendulum. We define the x-axes connecting the hinge points (distance L_i) and the z-axes perpendicular to the plane of motion. The mass center is situated on the x-axis. One has then

$$\dot{\mathbf{y}}_1 = \dot{\gamma}_1 \mathbf{e}_3, \quad \mathbf{T}_{ip} = \begin{pmatrix} \cos\gamma_i & \sin\gamma_i & L_p\sin\gamma_i \\ -\sin\gamma_i & \cos\gamma_i & L_p\cos\gamma_i \\ 0 & 0 & 1 \end{pmatrix}, \quad \omega_{oi} = \sum_{j=1}^{i} \dot{\gamma}_j \qquad (61)$$

and all the functional matrices $\mathbf{F}_i$ reduce to the third unit vector $\mathbf{e}_3$. The solution scheme is therefore extremely simple. However, the algorithm is obviously advantageous for the use on the computer, not so much for demonstration by hand. We reduce therefore the chain to two contiguous bodies only (the double pendulum), or the non-moving wheeled robot ($v_i = 0, \gamma_6 = 0$) retaining the indices 2 and 3 for upper and fore arm angle of the SCARA, yielding

$$\overline{\mathbf{M}}_2 := \overline{\mathbf{M}}_2 + \mathbf{T}_{32}^T \begin{bmatrix} m_3 & 0 & 0 \\ 0 & m_3\left[1 - \dfrac{m_3 c_3^2}{J_3}\right] & 0 \\ 0 & 0 & 0 \end{bmatrix} \mathbf{T}_{32},$$

$$\overline{\mathbf{G}}_2 := \overline{\mathbf{G}}_2 + \mathbf{T}_{32}^T \begin{bmatrix} 0 & -m_3 c_3 \omega_{o2} & -m_3\left[\omega_{o2} + \dot{\gamma}_3\right] \\ m_3\left[1 - \dfrac{m_3 c_3^2}{J_3}\right]\omega_{o2} & 0 & 0 \\ 0 & 0 & 0 \end{bmatrix} \mathbf{T}_{32},$$

$$\mathbf{Q}_2 := \mathbf{Q}_2 + \mathbf{T}_{32}^T \begin{bmatrix} Q_{3x} - m_3 c_3 \omega_{o3} \dot{\gamma}_3 \\ Q_{3y} - \dfrac{m_3 c_3}{J_3} Q_{3z} \\ 0 \end{bmatrix}$$

$$(62)$$

Hence,

- Forward step: Kinematics from 2 to 3

$$\dot{\mathbf{y}}_2 = \mathbf{e}_3\,\dot{\gamma}_2, \quad \dot{\mathbf{y}}_3 = \begin{pmatrix} L\cos\gamma_2 \\ L\sin\gamma_2 \\ 1 \end{pmatrix} \dot{\gamma}_2 + \mathbf{e}_3\,\dot{\gamma}_3 \qquad (63)$$

- Backward step: Matrices from 3 to 2

$$\ddot{\gamma}_2 = - \left[\mathbf{e}_3^T \overline{\mathbf{M}}_2 \mathbf{e}_3\right]^{-1} \mathbf{e}_3^T \left\{ \overline{\mathbf{G}}_2 \mathbf{e}_3 \dot{\gamma}_2 - \mathbf{Q}_2 \right\} :$$

$$\mathbf{e}_3^T \overline{\mathbf{M}}_2 \mathbf{e}_3 = \left[J_2 + m_3 L_2^2 \left(1 - \frac{m_3 c_3^2}{J_3} \cos^2 \gamma_3 \right) \right]$$

$$\mathbf{e}_3^T \overline{\mathbf{G}}_2 \mathbf{e}_3 = - \left[m_3 c_3 L_2 (\dot{\gamma}_2 + \dot{\gamma}_3) \sin \gamma_3 - m_3 L_2^2 \left(\frac{m_3 c_3^2}{J_3} \right) \dot{\gamma}_2 \sin \gamma_3 \cos \gamma_3 \right]$$

$$\mathbf{e}_3^T \mathbf{Q} = Q_{2z} + Q_{3x} L_2 \sin \gamma_3 + \left[Q_{3y} - \left(\frac{m_3 c_3}{J_3} \right) Q_{3z} \right] L_2 \cos \gamma_3$$

$$- m_3 c_3 L_2 (\dot{\gamma}_2 + \dot{\gamma}_3) \dot{\gamma}_3 \sin \gamma_3, \quad \text{leading to, with } Q_i = 0 \text{ for simplicity,}$$

$$\ddot{\gamma}_2 = \frac{\left[m_3 L_2^2 \left(\frac{m_3 c_3^2}{J_3} \right) \dot{\gamma}_2^2 \sin \gamma_3 \cos \gamma_3 + m_3 L_2 c_3 (\dot{\gamma}_2 + \dot{\gamma}_3)^2 \sin \gamma_3 \right]}{\left[J_2 + m_3 L_2^2 \left(1 - \frac{m_3 c_3^2}{J_3} \cos^2 \gamma_3 \right) \right]} \tag{64}$$

- Forward step

$$\ddot{\gamma}_3 = - \left(\frac{m_3 c_3 L_2 \cos \gamma_3 + J_3}{J_3} \right) \ddot{\gamma}_2 - \frac{m_3 c_3 L_2 \dot{\gamma}_2^2 \cos \gamma_3}{J_3} \tag{65}$$

The result is easily proven with the well known equations of the double pendulum (relative angles):

$$\begin{bmatrix} J_2 + J_3 + m_3 L_2^2 + 2m_3 c_3 L \cos \gamma_3 & J_3 + m_3 c_3 L_2 \cos \gamma_3 \\ J_3 + m_3 c_3 L_2 \cos \gamma_3 & J_3 \end{bmatrix} \begin{bmatrix} \ddot{\gamma}_2 \\ \ddot{\gamma}_3 \end{bmatrix}$$
$$+ \begin{bmatrix} -(\dot{\gamma}_3^2 + 2\dot{\gamma}_2 \dot{\gamma}_3) \\ \dot{\gamma}_2^2 \end{bmatrix} m_2 s_2 L \sin \gamma_3 = \begin{bmatrix} 0 \\ 0 \end{bmatrix} \tag{66}$$

Solving the second line for $\ddot{\gamma}_3$ and insertion into the first line yields $\ddot{\gamma}_2$. Once $\ddot{\gamma}_2$ known, the second line yields $\ddot{\gamma}_3$.

The basic idea of decomposing the mass matrix in the sense of a Gaussian algorithm is not new. Brandl et. al[2] for example eliminated the the contraint forces step by step from the last body to the root coming out with comparable results. First attemptions in this field have already been done by Vereshagin [40]. There is, however, a big advantage in the present procedure: The use of nonholonomic variables leads to the well structurized element matrices (35), starting already at the velocity level.

4   Structurally Variant Systems

4.1   *Freeing from the Constraints*

One of the main advantages in using minimal velocities according to Eq.(1) is, that for the constraint case the corresponding velocities may directly be cancelled in Eq.(35). (Recall that this is not allowed for any of the so-called analytical procedures! There, one has still to calculate derivatives w.r.t. $\dot{s}$.) However, following Hamel's "concept of relaxation of the constraints", one may also use Eq.(1) for variables but introduce corresponding (constraint) forces instead of eliminating them, yielding for Eq.(57), along with $\ddot{\Phi} = (\partial\Phi/\partial s)\ddot{s} + [d(\partial\Phi/\partial s)/dt]\dot{s} = 0$

$$
\mathbf{M}\ddot{\mathbf{s}} + \mathbf{g} - \left(\frac{\partial\Phi}{\partial\mathbf{s}}\right)^T \lambda = 0, \quad \mathbf{g} = \mathbf{G}\dot{\mathbf{s}} - \mathbf{Q}
$$

$$
\Rightarrow \left(\frac{\partial\Phi}{\partial\mathbf{s}}\right)\left[-\ddot{\mathbf{s}} - \mathbf{M}^{-1}\mathbf{g} + \mathbf{M}^{-1}\left(\frac{\partial\Phi}{\partial\mathbf{s}}\right)^T\lambda\right] =
$$

$$
\underbrace{\left[\frac{d}{dt}\left(\frac{\partial\Phi}{\partial\mathbf{s}}\right)\dot{\mathbf{s}} - \left(\frac{\partial\Phi}{\partial\mathbf{s}}\right)\mathbf{M}^{-1}\mathbf{g}\right]}_{\mathbf{b}} + \underbrace{\left[\left(\frac{\partial\Phi}{\partial\mathbf{s}}\right)\mathbf{M}^{-1}\left(\frac{\partial\Phi}{\partial\mathbf{s}}\right)^T\right]}_{\mathbf{A}}\lambda = 0
$$

$$
: \mathbf{A}\lambda + \mathbf{b} = 0
$$

(67)

(see e.g. (Schiehlen) [39]. The first attempt is obviously due to Jacobi [23]: " one has twice to differentiate the constraints...". Note that we are consequently using the "Helmholtz auxiliary equation" and therefore not making difference in the nature of (holonomic or nonholonomic) constraints any more).

We release then, *secondly*, the constraints. This means that the λ convert into impressed (generalized) forces which "depend mainly on the afore restricted coordinates by forbidden directions" [16]. The (generalized) force directions are by this in any case known, but the λ's have then to be treated as work performing forces, i.e. the configuration space is widened up.

4.2   *Remark on the Choice of Minimal Velocities*

When freeing the system from the constraints, then minimal velocities according to Eq.(1) offer a clear cut advantage concerning the arising (generalized) force direction. This can easily be seen by inspection of Eq.(57): The chosen variables are such that $(\partial\Phi/\partial s)$ is simply an "invariance matrix" (consisting of 0 and 1 only). However, one has to pay the price for that in all the other arising terms, of course.

Example 6: Nonholonomic vehicle. Let for simplicity the SCARA in Figure 4 aside. Using the Gaussian recursive scheme for calculation, one has to replace the indices in Eq.(60) $2 \to 1$, $3 \to 6$, along with $c_6 = 0$ (front wheel) yielding

$$\overline{\mathbf{M}}_1 := \begin{bmatrix} m_1 & 0 & 0 \\ 0 & m_1 & m_1 c_1 \\ 0 & m_1 c_1 & J_1^o \end{bmatrix} + \begin{bmatrix} m_6 & 0 & 0 \\ 0 & m_6 & m_6 a \\ 0 & m_6 a & m_6 a^2 \end{bmatrix} = \begin{bmatrix} m & 0 & 0 \\ 0 & m & mc \\ 0 & mc & J^o \end{bmatrix} \tag{68}$$

$$\frac{\overline{\mathbf{G}}_1}{\dot{\gamma}_1} := \begin{bmatrix} 0 & -m_1 & -m_1 c_1 \\ m_1 & 0 & 0 \\ m_1 c_1 & 0 & 0 \end{bmatrix} + \begin{bmatrix} 0 & -m_6 & -m_6 a \\ m_6 & 0 & 0 \\ m_6 a & 0 & 0 \end{bmatrix} = \begin{bmatrix} 0 & -m & -mc \\ m & 0 & 0 \\ mc & 0 & 0 \end{bmatrix} \tag{69}$$

$$\mathbf{Q}_1 := \begin{bmatrix} 1 \\ 0 \\ 0 \end{bmatrix} f_x + \begin{bmatrix} 0 \\ 0 \\ 1 \end{bmatrix} M_z + \begin{bmatrix} 0 \\ 1 \\ 0 \end{bmatrix} \lambda_1 + \begin{bmatrix} -\sin\gamma_6 \\ \cos\gamma_6 \\ a\cos\gamma_6 \end{bmatrix} \lambda_6 \tag{70}$$

where: $m = m_1 + m_6$, $mc = m_1 c_1 + m_6 a$, $J^o = J_1^o + m_6 a^2$, f_x, M_z: driving force and torque (rear drive assumed).

Hamel's Variables

Choosing the minimal velocities according to Eq.(1) ("Hamel's variables") one obtains for the "backward step" along with $\mathbf{F}_1 = \mathbf{T}$ from Eq.(50)

$$\mathbf{F}_1^T \mathbf{Q}_1 = \begin{bmatrix} 0 \\ -\sin\gamma_6 \\ \cos\gamma_6 \end{bmatrix} f_x + \begin{bmatrix} -(1/a) \\ (\cos\gamma_6)/a \\ (\sin\gamma_6)/a \end{bmatrix} M_z + \begin{bmatrix} 1 \\ 0 \\ 0 \end{bmatrix} \lambda_1 + \begin{bmatrix} 0 \\ 1 \\ 0 \end{bmatrix} \lambda_6$$

$$\mathbf{F}_1^T \overline{\mathbf{M}}_1 \mathbf{F}_1 = \begin{bmatrix} m\left[1 - \frac{2c}{a}\right] - \frac{J^o}{a^2} & \left[m\frac{c}{a} - \frac{J^o}{a^2}\right]\cos\gamma_6 & \left[m\frac{c}{a} - \frac{J^o}{a^2}\right]\sin\gamma_6 \\ & m\sin^2\gamma_6 + \frac{J^o}{a^2}\cos^2\gamma_6 & -\left[m\frac{c}{2a} - \frac{J^o}{2a^2}\right]\sin 2\gamma_6 \\ & \text{symmetric} & m\cos^2\gamma_6 + \frac{J^o}{a^2}\sin^2\gamma_6 \end{bmatrix}$$

$$\mathbf{F}_1^T(\overline{\mathbf{M}}_1\dot{\mathbf{F}}_1 + \overline{\mathbf{G}}\mathbf{F}_1) =$$

$$\begin{bmatrix} 0 & +\left[\frac{J^o}{a^2}-m\frac{c}{a}\right]\dot{\gamma}_6\cos\gamma_6 & -\left[\frac{J^o}{a^2}-m\right]\dot{\gamma}_6\cos\gamma_6 \\[2mm] 0 & -\left[\frac{J^o}{a^2}-m\right]\dot{\gamma}_6\cos\gamma_6\sin\gamma_6 & \left[\frac{J^o}{a^2}\cos^2\gamma_6+m\sin^2\gamma_6\right]\dot{\gamma}_6 \\[2mm] 0 & -\left[\frac{J^o}{a^2}\sin^2\gamma_6+m\cos^2\gamma_6\right]\dot{\gamma}_6 & \left[\frac{J^o}{a^2}-m\right]\sin\gamma_6\cos\gamma_6\dot{\gamma}_6 \end{bmatrix} +$$

$$\begin{bmatrix} 0 & -m\left[1-\frac{c}{a}\right]\dot{\gamma}_1\sin\gamma_6 & +m\left[1-\frac{c}{a}\right]\dot{\gamma}_1\cos\gamma_6 \\[2mm] +m\left[1-\frac{c}{a}\right]\dot{\gamma}_1\sin\gamma_6 & 0 & +m\frac{c}{a}\dot{\gamma}_1 \\[2mm] -m\left[1-\frac{c}{a}\right]\dot{\gamma}_1\cos\gamma_6 & -m\left[\frac{c}{a}\right]\dot{\gamma}_1 & 0 \end{bmatrix} \tag{71}$$

Once $\ddot{\mathbf{s}}_1 = (\dot{v}_{y1},\; \dot{v}_{y6},\; \dot{v}_{x6})^T$ is known from Eq.(59) with $\dot{\mathbf{y}}_{ip} = \mathbf{0}$,

$$\ddot{\mathbf{s}}_1 = -\left[\mathbf{F}_1^T\overline{\mathbf{M}}_1\mathbf{F}_1\right]^{-1}\left[\mathbf{F}_1^T(\overline{\mathbf{M}}_1\dot{\mathbf{F}}_1 + \overline{\mathbf{G}}_1\mathbf{F}_1)\dot{\mathbf{s}}_1 - \mathbf{F}_1^T\mathbf{Q}_1\right], \tag{72}$$

one obtains the remaining $\dot{\mathbf{s}}_6 = \ddot{\gamma}_6$ from Eq.(59) with $\dot{\mathbf{y}}_{ip} = \dot{\mathbf{y}}_1 = \mathbf{F}_1\dot{\mathbf{s}}_1$ as

$$\ddot{\gamma}_6 = \left[\frac{1}{a}\;\; -\frac{1}{a}\cos\gamma_6\;\; -\frac{1}{a}\sin\gamma_6\right]\ddot{\mathbf{s}}_1 + \left[0\;\; \frac{1}{a}\sin\gamma_6\;\; -\frac{1}{a}\cos\gamma_6\right]\dot{\gamma}_6\dot{\mathbf{s}}_1 + \frac{M_6}{J_6} \tag{73}$$

(this result is easily proven: from Eq.(50) one has, for the constrained case, $\ddot{\gamma}_1 = d[(v_{x6}/a)\sin\gamma_6]/dt$ yielding for Eq.(73) $J_6(\ddot{\gamma}_6 + \ddot{\gamma}_1) = M_6$ where M_6: steering torque.) Insertion of the constraints (i.e. canceling the first two components of $\dot{\mathbf{s}}_1$) yields

$$\begin{bmatrix} \frac{J^o}{a^2}\sin^2\gamma_6+m\cos^2\gamma_6 & 0 \\[2mm] \frac{1}{a}\sin\gamma_6 & 1 \end{bmatrix}\begin{bmatrix} \dot{v}_{x6} \\[2mm] \ddot{\gamma}_6 \end{bmatrix} + \begin{bmatrix} \left(\frac{J^o}{a^2}-m\right)\sin\gamma_6\cos\gamma_6 & 0 \\[2mm] \frac{1}{a}\cos\gamma_6 & 1 \end{bmatrix}\dot{\gamma}_6\begin{bmatrix} v_{x6} \\[2mm] \dot{\gamma}_6 \end{bmatrix} =$$

$$\begin{bmatrix} \cos\gamma_6 & \frac{1}{a}\sin\gamma_6 & 0 \\[2mm] 0 & 0 & \frac{1}{J_6} \end{bmatrix}\begin{bmatrix} f_x \\[1mm] M_z \\[1mm] M_6 \end{bmatrix} \tag{74}$$

However, the calculation of the λ's according to Eq.(67) is, for this case, tedious because $\left[\mathbf{F}_1\overline{\mathbf{M}}_1\mathbf{F}_1\right]^{-1}$ is needed.

Original Variables

On the other hand, with the λ's explicitly treated as (generalized) forces, the corresponding configuration space is not more restricted. One may therefore choose as well the original $\dot{\mathbf{y}}_1 \in \mathbb{R}^3$ for minimal velocities, i.e. $\mathbf{F}_1 = \mathbf{E}$ (unit matrix). The backward step is then simply, with Eqs.(68), (69), (70),

$$\overline{\mathbf{M}}_1 \ddot{\mathbf{y}}_1 + \overline{\mathbf{G}}_1 \dot{\mathbf{y}}_1 - \begin{bmatrix} f_x \\ 0 \\ M_z \end{bmatrix} - \begin{bmatrix} 0 & \sin\gamma_6 \\ 1 & \cos\gamma_6 \\ 0 & a\cos\gamma_6 \end{bmatrix} \begin{bmatrix} \lambda_1 \\ \lambda_2 \end{bmatrix} = \mathbf{0} \tag{75}$$

which can easily be solved with

$$\overline{\mathbf{M}}^{-1} = \frac{1}{J^c} \begin{bmatrix} \frac{J^c}{m} & 0 & 0 \\ 0 & \frac{J^o}{m} & -c \\ 0 & -c & 1 \end{bmatrix} \tag{76}$$

where: $J^c = J^o - mc^2$. Along with the "forward step" one has eventually

$$\begin{bmatrix} \dot{v}_{x1} \\ \dot{v}_{y1} \\ \ddot{\gamma}_1 \\ \ddot{\gamma}_6 \end{bmatrix} = \begin{bmatrix} 0 & \dot{\gamma}_1 & c\dot{\gamma}_1 & 0 \\ -\dot{\gamma}_1 & 0 & 0 & 0 \\ 0 & 0 & 0 & 0 \\ 0 & 0 & 0 & 0 \end{bmatrix} \begin{bmatrix} v_{x1} \\ v_{y1} \\ \dot{\gamma}_1 \\ \dot{\gamma}_6 \end{bmatrix} + \begin{bmatrix} \frac{f_x}{m} \\ 0 \\ 0 \\ \frac{M_6}{J_6} \end{bmatrix} + \begin{bmatrix} 0 \\ -\frac{c}{J^c} M_z \\ \frac{1}{J^c} M_z \\ -\frac{1}{J^c} M_z \end{bmatrix} +$$

$$\frac{1}{J^c m} \begin{bmatrix} 0 & -J^c \sin\gamma_6 \\ J^o & (J^o - mac)\cos\gamma_6 \\ -mc & m(a-c)\cos\gamma_6 \\ mc & -m(a-c)\cos\gamma_6 \end{bmatrix} \begin{pmatrix} \lambda_1 \\ \lambda_2 \end{pmatrix} \tag{77}$$

yielding the λ's according to Eq.(67):

$$\left(\frac{\partial \mathbf{\Phi}}{\partial \mathbf{s}}\right) = \begin{bmatrix} 0 & 1 & 0 & 0 \\ -\sin\gamma_6 & \cos\gamma_6 & a\cos\gamma_6 & 0 \end{bmatrix} \Rightarrow$$

$$\mathbf{A} = \begin{bmatrix} \dfrac{J^o}{J^c} & \dfrac{J^o - mac}{J^c}\cos\gamma_6 \\ \dfrac{J^o - mac}{J^c}\cos\gamma_6 & 1 + \dfrac{m(a-c)^2}{J^c}\cos^2\gamma_6 \end{bmatrix}$$

$$\mathbf{b} = -\begin{pmatrix} \frac{mc}{J^c} M_z \\ f_x \sin\gamma_6 + m(c-a)\frac{M_z}{J^c}\cos\gamma_6 \end{pmatrix} -$$

$$m\begin{pmatrix} v_{x1}\dot{\gamma}_1 \\ v_{x1}(\dot{\gamma}_1 + \dot{\gamma}_6)\cos\gamma_6 + v_{y1}(\dot{\gamma}_1 + \dot{\gamma}_6)\sin\gamma_6 + (a\dot{\gamma}_6 + c\dot{\gamma}_1)\dot{\gamma}_1 \sin\gamma_6 \end{pmatrix} \tag{78}$$

However, elimination of constraints needs, in this case, more effort because now the constraint forces are "distributed" over the whole system. One obtains the solution by premultiplying Eq.(75) with an orthogonal complement to $(\partial\boldsymbol{\Phi}/\partial\mathbf{y}_1)^T\boldsymbol{\lambda}$ one of which is alreday known with Eq.(50): Insertion of the constraints yields $\dot{\mathbf{y}}_1 = (v_{x1}\ v_{y1}\ \dot{\gamma}_1)^T = [\cos\gamma_6\ 0\ (\sin\gamma_6)/a]^T v_{x6}$.
Thus, premultiplication with $(\partial\dot{\mathbf{y}}_1/\partial v_{x6})^T = [\cos\gamma_6\ 0\ (\sin\gamma_6)/a]$ eliminates the constraints and yields, of course, the same result (74).
Conclusion: Both choices may be used either for permanent constraints or for releasing the constraints. a) The choice of Hamel's variables $(\mathbf{F}_1 = \mathbf{T})$ is advantageous for the constraints explicitly inserted. The aim here is to reduce the equations to a minimal representation (Hamel's intention). The problem under consideration is then a mechanical system with permanent constraints. b) The use of Hamel's variables may not be advantageous (i.e. too restrictive) for the calculation of the (generalized) constraint forces. If one is interested in motion only, these do, of course, not play any role. From the numerical point of view it might sometimes be advantageous to calculate the system with a higher number of variables than the minimal space would require, along with the corresponding constraint forces[1]. (Note, however, that then part of the equations describe the same physical contents (although they may differ by a congruence transformation), thus leading to drift errors during numerical integration.) c) Considering Hamel's Principle of the relaxation of the constraints, Hamel's variables (although still tedious to handle) are unevoidable due to the physical background: Here, the constraints are à priori given (like e.g., as mentioned formerly, the rigidity conditions of a rigid body) and the question is: how do the corresponding (impressed) forces look like if the constraints relax (like, e.g., the stress tensor of a cable, a beam, a membrane etc., (Hamel)[16], (Hamel)[19]). d) The question, however, arises how to treat a system undergoing contraints with changing behaviour, being active during a certain time interval and then releasing once more, and vice versa. This case will be considered next.

4.3 Gauss' Principle of Minimal Constraints

No matter how mobility is achieved, the mobile robot may come into situations where the constraints are no longer valid. That is, in the considered example, when (one or more) wheels begin to slide aside or to slip at its contacts. The usual ("classical") way to proceed is, mostly, to generate as many equations of motion as situations can arise and to check these (numerically) for validity according to the actual state. However, this is neither satisfactory nor unproblematic because a solution may, in some cases, not exist, and the number of equations which has to be checked can grow enormously. To give an example: Every of the three wheels of the mobile robot may stick or slip or slide, leading already to $3^3 - 3 = 24$ equations which have to be checked. (Hereby, 3 conditions do not enter consideration because the rear wheels cannot slide independently). Or: Having a look on a two-legged two-armed walking

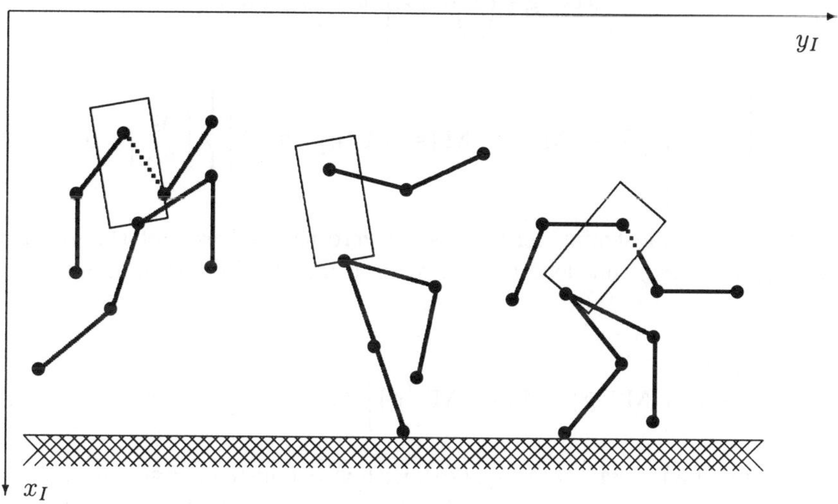

Figure 5   Legged Locomotion/Plane Motion

machine every arm or leg can contact and then either stick or slip in motion direction or against. From the modeling point of view, an in-plane motion for example is comparable to the wheeled robot from above, neglecting the wheels but adding some more "SCARA's" for legs and arms. Thus, the kinematics from Eq.(43) can directly be used because, kinematically, arms and legs are independent from each other as long as no contact with the environment occurs. The latter, however, changes the situation dramaticly. It leads, for an in-plane motion, with n extremities to n^4 possibilities, i.e. 256 equations for $n = 4$ which have to be checked. In case of six legs one has already more than

4000, in case of eight legs more than 65.000 corresponding equations. It is not efficient, or nearly impossible, to proceed this way. However, the problem can be solved in the following way: If additional constraints arise, Eq.(57) reads

$$\mathbf{g} = \mathbf{G}\dot{\mathbf{s}} - \mathbf{Q}: \quad \mathbf{M}\ddot{\mathbf{s}} + \mathbf{g} - \left(\frac{\partial \mathbf{\Phi}}{\partial \mathbf{s}}\right)^T \boldsymbol{\lambda} = 0, \quad \mathbf{\Phi}(\mathbf{s}) = \mathbf{0} \tag{79}$$

Shifting the constraint Jacobian to the acceleration level, i.e. ($[\partial \mathbf{\Phi}/\partial \mathbf{s}] = [\partial \ddot{\mathbf{\Phi}}/\partial \ddot{\mathbf{s}}]$), and differentiating the constraints themselves twice one may formulate some kind of "virtual work" by multiplying Eqs. (79) with $\delta \ddot{\mathbf{s}}$ and $\delta \boldsymbol{\lambda}$, resp., and summ up these equations, the result of which may be looked at in the sense of partial derivatives w.r.t. $\ddot{\mathbf{s}}$ and $\boldsymbol{\lambda}$:

$$\left[\mathbf{M}\ddot{\mathbf{s}} + \mathbf{g} - \left(\frac{\partial \ddot{\mathbf{\Phi}}}{\partial \ddot{\mathbf{s}}}\right)^T \boldsymbol{\lambda}\right]^T \delta \ddot{\mathbf{s}} + \left[\ddot{\mathbf{\Phi}}\right]^T \delta \boldsymbol{\lambda} =$$

$$\left\{\frac{\partial}{\partial \begin{bmatrix} \ddot{\mathbf{s}} \\ \boldsymbol{\lambda} \end{bmatrix}} \left[\frac{1}{2}\left(\ddot{\mathbf{s}} + \mathbf{M}^{-1}\mathbf{g}\right)^T \mathbf{M}\left(\ddot{\mathbf{s}} + \mathbf{M}^{-1}\mathbf{g}\right) - \ddot{\mathbf{\Phi}}^T \boldsymbol{\lambda}\right]\right\} \begin{bmatrix} \delta \ddot{\mathbf{s}} \\ \delta \boldsymbol{\lambda} \end{bmatrix} = 0. \tag{80}$$

The second constraint derivatives are zero themselves (not the gradients ($\partial \ddot{\mathbf{\Phi}}/\partial \ddot{\mathbf{s}}$), of course) thus leading to GAUSS' principle along with an additional constraint:

$$f = \left[\frac{1}{2}\left(\ddot{\mathbf{s}} + \mathbf{M}^{-1}\mathbf{g}\right)^T \mathbf{M}\left(\ddot{\mathbf{s}} + \mathbf{M}^{-1}\mathbf{g}\right)\right] \rightarrow \begin{array}{c} min \\ w.r.t.\ddot{\mathbf{s}} \end{array} \wedge \ddot{\mathbf{\Phi}} = 0 \tag{81}$$

Because of $\mathbf{M} = \mathbf{M}^T > 0$, Eqs.(79), with $\ddot{\mathbf{\Phi}} = 0$ instead of $\mathbf{\Phi} = 0$, represent necessary *and sufficient* conditions for minimizing f. Replacement of $\ddot{\mathbf{s}}$ in $\ddot{\mathbf{\Phi}}$ from motion equations (79) then yields

$$\ddot{\mathbf{\Phi}} = \underbrace{\left[\left(\frac{\partial \mathbf{\Phi}}{\partial \mathbf{s}}\right)\mathbf{M}^{-1}\left(\frac{\partial \mathbf{\Phi}}{\partial \mathbf{s}}\right)^T\right]}_{\mathbf{A}} \boldsymbol{\lambda} + \underbrace{\frac{d}{dt}\left(\frac{\partial \mathbf{\Phi}}{\partial \mathbf{s}}\right)\dot{\mathbf{s}} - \left(\frac{\partial \mathbf{\Phi}}{\partial \mathbf{s}}\right)\mathbf{M}^{-1}\mathbf{g}}_{\mathbf{b}} \tag{82}$$

for sufficient minimization condition (along with the equation of motion (79) itself) leading immediately to the solution – if exists. However, the above considerations only hold if all constraints are simultaneously active ($\ddot{\mathbf{\Phi}} \equiv 0$, see Eq.(67)). This does of course not yet contribute to the the multi-contact problem where the constraints may be active or not, at different locations and at different time. The solution here is to consider closed or opened contacts along with the exclusion that if a contact is closed then the constraint is zero and the contact force is not, and vice versa:

$$\ddot{\mathbf{\Phi}} \geq 0 \quad : \begin{cases} \ddot{\Phi}_i \neq 0, \lambda_i = 0 \ \ (\text{passive}) \\ \ddot{\Phi}_i = 0, \lambda_i \neq 0 \ \ (\text{active}) \end{cases}$$

$$\Rightarrow \ddot{\mathbf{\Phi}} = \mathbf{A}\boldsymbol{\lambda} + \mathbf{b}, \quad \ddot{\mathbf{\Phi}} \geq 0, \quad \boldsymbol{\lambda} \geq 0, \quad \boldsymbol{\lambda}^T \ddot{\mathbf{\Phi}} = 0 \tag{83}$$

(LAGRANGE-conditions → KUHN-TUCKER-conditions → Complementary Problem → generalized "simplex-tableau"-solution). Notice, however, that all these considerations go along with the second constraint derivatives. One has therefore simultaneously to guarantee fulfilment of the constraints themselves, (Glocker) [9], (Pfeiffer et al.) [34].

Remark. It has to be emphasized that any assumption on steadyness of the constraints has to be dropped when the the contact problem is treated. A deeper insight on the consequences concerning the basic principles is given by Glocker [10], [11]. The Helmholtz equation which leads to $\ddot{\mathbf{\Phi}} = \mathbf{0}$ instead of $\mathbf{\Phi} = \mathbf{0}$ refers to the gradient $(\partial\mathbf{\Phi}/\partial\mathbf{s}) = (\partial\ddot{\mathbf{\Phi}}/\partial\ddot{\mathbf{s}})$ which always (locally) exists without any additional assumption on $\ddot{\mathbf{\Phi}}$ itself, but $\mathbf{\Phi} = \mathbf{0}$ leads (more or less directly) to convex analysis as a powerful mathematical tool. However, the present discussion is not influenced by this.

Gauss' principle needs here the second constraint derivatives. An interpretation for this is obtained for instance with a look on the friction characteristics ("Stribeck-curve"): the force at a contact point is $| \lambda_T | \leq \mu_o \lambda_N$ (T: tangential, N: normal to to the contact surface, $\mu_o \lambda_N$: maximum of possible friction force) as long as the relative velocity $\dot{\mathbf{\Phi}}_T$ at the contact is zero. Obviously, the $\leq$-sign characterizes equilibrium: the actual friction reacts as a counterbalance to the resultant dynamical force as long as its value is less than the maximum possible friction force, leaving the point at rest. Once reaching the limit one has $\ddot{\mathbf{\Phi}}_T \neq 0 \Leftrightarrow \mu_o \lambda_N$ (*along with the either-or-condition* from Eq.(83)) which denotes transition from stiction to sliding (and vice versa. Note: μ_o, *not* μ). Thus, $\dot{\mathbf{\Phi}}_T = 0$ denotes a *possibly active* constraint (becoming active when $| \lambda_T | \geq \mu_o \lambda_N$, staying passive for equilibrium $| \lambda_T | < \mu_o \lambda_N$), where now $\ddot{\mathbf{\Phi}}$ $(= 0, < 0, > 0)$ answers the question wether stiction holds or not, and, if not, in which direction sliding will take place [9]. (For the formulation in the sense of Eq.(83), the friction characteristic has to be adjusted, e.g. by decomposition into the positive and negative parts of relative tangential acceleration, see e.g. Pfeiffer et al. [34] pp.75, Glocker [12]).

The (generalized) constraint force in Eq.(79) (looking at one single possible contact k for simplicity) may be rewritten

$$\left(\frac{\partial \dot{\mathbf{r}}_{cont}}{\partial \dot{\mathbf{y}}_k}\right)^T \left(\frac{\partial \boldsymbol{\Phi}}{\partial \mathbf{r}}\right)^T \boldsymbol{\lambda} \tag{84}$$

where: $\dot{\mathbf{r}}_{cont}$ (absolute) velocity of kth robot "point" which contacts (or may contact) the environment. The environment itself is characterized by a plane $\boldsymbol{\Phi}(\mathbf{r}) = 0$. Clearly, if the contact is closed, then $\mathbf{r}_{cont} = \mathbf{r}$ and $\boldsymbol{\lambda} \neq \mathbf{0}$. If, on the other hand, the contact is open, then $\boldsymbol{\lambda} = \mathbf{0}$, but the Jacobians in Eq.(84) are nevertheless well defined. *But* 1) the plane $\boldsymbol{\Phi}(\mathbf{r}) = \mathbf{0}$ denotes the "surrounding world" (therefore, naturally, given in "world coordinates") while the (possibly) contacting part of the robot is given in any adequate representation (the "configuraton space"). Following the above recursive scheme, for instance, one has already the corresponding velocities in body fixed representation (B), thus leading to

$$\dot{\mathbf{r}}_{cont} = {}_I\dot{\mathbf{r}}_{cont} = \mathbf{A}_{IB}\,{}_B\mathbf{v}_{cont} \Rightarrow$$

$$\left(\frac{\partial \boldsymbol{\Phi}}{\partial \mathbf{r}}\right)\left(\frac{\partial {}_I\mathbf{v}_{cont}}{\partial {}_B\mathbf{v}_{cont}}\right)\left(\frac{\partial {}_B\mathbf{v}_{cont}}{\partial \dot{\mathbf{y}}_k}\right) = \left[\left(\frac{\partial \boldsymbol{\Phi}}{\partial \mathbf{r}}\right)\mathbf{A}_{IB}\right]\left(\frac{\partial {}_B\mathbf{v}_{cont}}{\partial \dot{\mathbf{y}}_k}\right) = {}_B\mathbf{n}^T\mathbf{F}_{cont,k}$$
$$\tag{85}$$

Here, $\mathbf{n} = (\partial \boldsymbol{\Phi}/\partial \mathbf{r})$ is the normal vector of the contact plane in inertial representation, ${}_B\mathbf{n}$ is the normal vector transformed into the body-fixed frame B of the (contacting part of) the robot, $\mathbf{F}_{cont,k}$ is the corresponding functional matrix (Jacobian) at the (possible) contact point. Therefore, considering the transition from an arbitrary robot configuration to a contact situation, the corresponding constraint is always equivalent to a relative displacement (normal or tangential, see the wheeled robot example: the constraint force is always directed towards the "forbidden motion"). This statement would directly lead to the use of (in the most generalized sense) relative velocities w.r.t the constraints (i.e. Hamel's variables) for optimal choice. These are for instance the relative angle displacements between the toothing wheels in gear rattling problems. Or, in walking machines *as long as walking on a plane surface is considered* one may choose, with a look on Figure 5, the (possibly) contacting velocities of the "feet" (and "hands") in an inertial representation thus being always and directly aware of sudden contacts. The normal is, in this case, ${}_I\mathbf{n}^T = (-1,\ 0,\ 0)$ all the time. However, in more general cases (climbing a hill, for instance), $\mathbf{n}$ is not longer constant and a choice of normal and tangential directed contact point velocities for variables can ad hoc hardly be recommended – at least from the viewpoint of deriving the equations by hand. For the numerical treatment no experience exists for the time being. From the analytical viewpoint, i.e. analyzing the problem on the level of physical interpretation, the above considerations are, of course, helpful.

To conclude, Gauss' minimization approach is the most adequate procedure, or, may be, the only one to really succeed for the general case of multi contact impact and frictional problems. However, especially in the robotics area, one has to state that these considerations are not yet wide spread. Exception is, for instance, the research done at the Technical University of Munich, pars pro toto: the eight-legged walking maschine as investigated by Rossmann [38]. The reason might be twofold: 1) Going the "classical" (combinatorial) way, nobody would even try to calculate the corresponding $4^8 = 65.536$ equations, much less to check these for the one actual physically valid solution. Here, engineering feeling is asked, more than "theoretical completeness". For this purpose one has in mind that the robot is not totally helpless because of its control: Commonly, the open loop part achieves the desired gross motion where the superimposed closed loop guarantees a stable behaviour with respect to the control aim. However, there remains a certain portion of belief in this: Control does a good job as long its underlying assumptions are not violated. 2) These assumptions are, mainly, that the robot's feet which contact the ground will never slide. But is it allowed to restrict considerations always to that case? Obviously not. Now, "engineering feeling" usually argues that these basic assumptions are *mostly* fulfilled, and in those seldom situations they are not, one will switch to the corresponding sliding case... *This is nothing else* than the above "classical" procedure: How many possibilities do then exist (65.536 in case of the eight-legged robot!) and which one meets physical reality? This case may be seldom or not – if it occurs, there is no way to escape from the variety of arising possibilities.

One future aim in robotics is obviously, at least in the field of service robots, to survey the whole state (nothing new, of course) but then to develop strategies how to react when leaving the restricted state space the control is based on. Considering the two-legged-two-armed locomotion for example (Figure 5): If the gait control fails the robot may fall down. The robot has than, however, to react in such a manner that it prevents from damage and proceeds motion anyhow: shall it crawl on all fours or should it erect once more, or stop motion or what else? Here, Gauss' principle opens one view how to come around with these tasks.

The situation is now as follows: The robot is considered as "free" (i.e. without any contact, but including permanent constraints from the machine design itself). Its mathematical description using nonholonomic variables yields directly the gradients needed for the constraint forces, independently wether the constraints are active or not, Eq.(84). For numerical simulation one may advantageously use the outlined recursive algorithm. However, there arises, from the theoretical point of view, one problem: The algorithm yields $\ddot{\mathbf{s}}$ as a (vector) function at every time step without knowledge of the (inverse) mass matrix itself, while Eq.(83) needs the (inverse) mass matrix of the "free" (unconstrained) state. But, since the Jacobians $(\partial\mathbf{\Phi}/\partial\mathbf{s}_i)^T$ are known, one may use the Gaussian algorithm for the unconstrained case first and than stepwise add $\mathbf{Q}_i \rightarrow \mathbf{Q}_i + (\partial\mathbf{\Phi}_j/\partial\mathbf{s}_i)^T$,

$$
\begin{aligned}
\ddot{\mathbf{s}}_o &= -\mathbf{M}^{-1}\mathbf{g} \\
\ddot{\mathbf{s}}_1 &= -\mathbf{M}^{-1}\mathbf{g} + \mathbf{M}^{-1}\left[\frac{\partial\mathbf{\Phi}_1}{\partial\mathbf{s}}^T\right] \\
&\cdots \\
\ddot{\mathbf{s}}_m &= -\mathbf{M}^{-1}\mathbf{g} + \mathbf{M}^{-1}\left[\frac{\partial\mathbf{\Phi}_m}{\partial\mathbf{s}}^T\right]
\end{aligned}
\tag{86}
$$

to eventually come out with $\mathbf{M}^{-1}(\partial\mathbf{\Phi}/\partial\mathbf{s})^T$ by simple combinations of these partial results. The procedure requires then to rerun the algorithm $3 + 3m$ times (m: number of constraints). Once more, no experience on this exists for the time being.

5   Conclusions

The aim of the present contribution was to investigate wether the use of nonholonomic variables (in robotics) offers any advantage. To find an answer, one has obviously first to choose a (dynamical) procedure where the nonholonomic variables may be applied to. Having the robotics background in mind (holonomic and/or nonholonomic active or possibly active constraints) one obtains a first (and deep) inside with a look on the classical results in this field. Among these, Hamel's equations (nowadays mostly referred to as Hamel-Boltzmann-Equations) play an important role. They yield, along with Hamel's coordinates (velocities) which include the constraints as dependent components, already a straight-on procedure with (at least originally) the aim of minimal representation by eventually cancelling the dependent variables. Extended in the sense of "relaxation of the constraints" one has an à priori knowledge on the (generalized) constraint force directions.

The procedure belongs to the analytical methods and needs, for the derivation of the corresponding Hamel coefficients, the treatment of "virtual displacements" (or the corresponding calculation rules, respectively) which are, according to Kane, "the closest thing in dynamics to black magic, and they are the heart of Lagrange's approach" [37]. Although this citation is already about 15 years old and meanwhile a lot of statements on the so-called "Kane's dynamical equations" have been made, ranging from enthusiasm to indignation, the basic ignorance of Lagrange's fundamental concept which is stated with citations like the above once more comes up in our days (and will probably not have an end in future). Nowadays for instance philosophical-historical representatives have their say, citing great scientists as witnesses like Poinsot and Jacobi. However, Poinsot was scientificly false, and Jacobi, that brilliant mathematician, was misleaded in searching proofs for the mechanical principles (what a strange idea). It seems therefore necessary to first have a look on the basics because 1) it should be clear what one is speaking about, in order to (not at least) 2) take teaching serious (– what will our students and young scientists think when they become aware of such misleading statements?) and, from the practical point of view, to 3) build bridges between the different available methods thus to make sure that a special choice of procedure does not remain arbitrarily based on some kind of personal taste.

Concerning the choice of procedure, Hamel's equations contain some drawbacks. As in the case of all analytical procedures, the constraints cannot in advance be inserted because of the requirement of partial derivatives. Furthermore, the application of analytical procedures is cumbersome (as often and commonly already stated for Lagrange's equations of second kind for holonomic (nontrivial) high d.o.f. systems). Focusing on robot applications (or, more general, on Multi Body Systems) and leaving nonlinear nonholonomic constraints (w.r.t velocities) aside, it turns out that Hamel's coefficients the direct calculation of which is almost tedious are (mostly) constant.

Thus, although Hamel's equations may be considered *the classical access* to nonholonomic systems, they are not yet best suited for application. Application aspects also motivate the exclusion of nonlinear nonholonomic constraints which in nature do not seem to exist and which, artificially designed by control, will never enter consideration this way because of the fact that control forces will never belong to the contraints: they have to perform work, otherwise control would be useless. *However*, releasing the constraints could answer the question what kind of control is needed for special purposes. All in all and with due to care one may state that the analytical procedures have their advantages in theoretical analysis of the problem and physical interpretation, but not so much for hard core practical application, the first step of which is the derivation of motion equations.

Concerning the latter we are lead to the *Projection Equation* which benefits in advance from all the above considerations, e.g. constant Hamel coefficients, no partial derivatives required (– the functional matrices which may be looked at as partial derivatives are known in advance without any additional calculation, once minimal velocities are chosen). Here, two kinds of (nonholomic) velocities arise. The first ones, the cartesian velocities, are selfevident. Basicly, motion equations characterize the change in time of linear and angular momentum (along with elimination of constraint forces which do not contribute to motion). Thus, velocity variables for motion description simplify matters significantly. The second ones characterize the constrained (velocity) space, thus called (in comparison to Lagrange's notation) *minimal velocities* ("generalized speeds" in above cited Kane's contributions). The task is now twofold: Find suitable reference frames and suitable minimal velocities.

Multi Body Systems in general are topologically represented by a "tree", robots hereby often reduce to a "topological chain". It is therefore useful to define a "hingepoint"– fixed coordinate frame and use relative coordinates, leading immediately to a recursive scheme for kinematical description. The corresponding motion equations are, although highly nonlinear, very simple in structure and the overall functional matrix (Jacobian) results to be upper triangular. Thus, as kind of a by-product, one obtains also a simple recursive scheme for the kinetics which avoids the inversion of the total $g \times g$ mass matrix. Considering possible constraints the recursive kinematics show how to obtain Hamel's coordinates for the problem under discussion. One has then a first choice of minimal velocities which goes along with in advance known constraint force directions. However, as already mentioned, the main advantage in Hamel's coordinates lies in the treatment of constrained systems. When releasing the system from the constraints one has to pay the price for the simplicity of constraints by other calculations. Therefore, when the configuration space is enlarged (compared to the constrained case) the choice of Hamel's coordinates may be too restrictive. No further (especially numerical) experience exists on this for the time being. But the result confirms the general statement that an optimal choice of variables does not exist universally but depends on the actual problem under consideration. Here, the characteristics of the constraints themselves enter the problem formulation.

However, the question arises why to look at the contraint forces at all instead to confine investigations on the reduced constrained space. It might be for sizing the bearings, for instance, or for numerical aspects in integration routines – both aspects, however, being out of interest in this contribution. Here, the answer is simply that non-permanent constraints may arise like contact, impact and stick-slip which are typical phenomena in mobile robot applications. The usual way to proceed here is to determine as many motion equations as the dynamical possibilities require and to use the one valid according to the physical circumstances. However, one will be frighted by the number of arising equations. In case of legged locomotion one has, for n legs and m possibilities at each contact, already m^n equations. There is, for the general case, obviously no chance to succeed this way. However, the solution of this problem is immediately obtained with a closer look on the motion equations leading directly to the Gauss Principle of minimal constraints, with the second time derivative of the constraints to be zero as additional condition. One has then, from the correponding minimization procedure, the one physically consistent solution at once.

All these considerations go along with nonholonomic velocities (holonomically or nonholonomically constrained). The question wether the use of nonholonomic variables offers an advantage for applications has therefore to be answered as positive. Referring to the available methods, the use of the "nonholonomic procedures" which are often and mostly applied to nonholonomic systems (constraints) only, along with (of course) nonholonomic variables, represents the natural appoach to dynamics, also in the holonomic case. Closer inspection of the analytical procedures, taking some intermediate results of these in advance into account, leads to the *Projection Equation* which obviously fits best for application.

"There is nothing more practical than theory"

(after Ludwig Boltzmann)

References

1. Alishenas, T.: Zur numerischen Behandlung, Stabilisierung durch Projektion und Modellierung mechanischer Systeme mit Nebenbedingungen und Invarianten. *Ph.D. Thesis*, Königl. Techn. Hochschule Stockholm, 1992

2. Brandl,H., Johanni,R. Otter,M.: A Very Efficient Algorithm for the Simulation of Robots and Similar Multibody Systems without Inversion of the Mass Matrix. *Proc. IFAC Symp.*, Vienna (1986), 365-370

3. Bremer, H.: Über eine Zentralgleichung in der Dynamik. *Zeitschr. Angew. Math. u. Mech. (ZAMM)* **68** (1988), 307-311

4. Bremer, H.: Dynamical Aspects in Flexible Rotating Machinery. *J. Sound & Vibration* **152** (1) (1992), 39-55

5. Bremer, H.: Das Jourdainsche Prinzip. *Zeitschr. Angew. Math. u. Mech. (ZAMM)* **73** (1983), 184-197

6. Euler, L.: Découverte d'un nouveau principe de la mécanique. Berlin: *Akademieberichte* 1750

7. Euler, L.: Nova methodus motum corporum rigidorum determinandi. St. Petersburg: *Akademieberichte* 1775

8. Funk, P.: Variationsrechnung und ihre Anwendung in Physik und Technik. Berlin: Springer 1962

9. Glocker, C.: Dynamik von Starrkörpersystemen mit Reibung und Stössen. Ph.D. Thesis, Düsseldorf: *VDI-Fortschr.-Ber.* **18**, Nr. 182, 1995

10. Glocker, C.: The Principles of d'Alembert, Jourdain and Gauss in Nonsmooth Dynamics. Part I: Scleronomic Multibody Systems. *Zeitschr. Angew. Math. u. Mech. (ZAMM)* **78**, 1, (1998) 21-37

11. Glocker, C.: Discussion of d'Alembert's Principle for Nonsmooth Unilateral Constraints. *Zeitschr. Angew. Math. u. Mech. (ZAMM)* **79**, 1, (1999) 91-94

12. Glocker, C.: Decomposition of Scalar Force Interactions, in: *IUTAM Symposium on Unilateral Multibody Contacts* (Eds. F. Pfeiffer, C. Glocker) Kluwer Acad. Publ., 1999, 15-24

13. Gauss, C.F.: Über ein neues allgemeines Grundgesetz der Mechanik. *Journal f. Mathematik* (1829)

14. Hamel, G.: Die Lagrange-Eulerschen Gleichungen der Mechanik. *Zeitschr. Angew. Math. u. Phys.* **50** (1904), 1-57

15. Hamel, G.: Über die virtuellen Verschiebungen in der Mechanik. *Math. Annalen* **59** (1905) 426-434

16. Hamel, G.: Über ein Prinzip der Befreiung bei Lagrange. *Jahresbericht d. Deutschen Mathem.-Vereinigung* **25**, No. 1/3 (1917), 60-65

17. Hamel, G.: Das Hamiltonsche Prinzip bei nichtholonomen Systemen. *Math. Ann.* **111** (1935), 94-97

18. Hamel, G.: Nichtholonome Systeme höherer Art. *Sitzungsberichte der Berliner math. Gesellschaft* (1938), 41-48

19. Hamel, G.: Theoretische Mechanik, 1949. Reprint Berlin, Heidelberg, New York: Springer 1967

20. Heun, K.: Die kinetischen Probleme der wissenschaftlichen Technik. *Jahresbericht der DMVV* **9**, Part 2, (1900), 1-123

21. Heun, K.: Ansätze und allgemeine Methoden der Systemmechanik. *Enzyklop. d. math. Wiss.* **4** II, Leipzig and Berlin: Teubner 1914

22. Huston, R.L.: Useful Procedures in Multibody Dynamics, in: Bianchi/Schiehlen (Eds): *Dynamics of Multibody Systems*, Berlin, Heidelberg, New York: Springer 1985, 89-77

23. Jacobi, C.G.J.: Vorlesungen über Dynamik, Winter 1843/44. Berlin: Georg Reimer, 1866, p.55

24. Jourdain, P.E.B.: Note on an analogue on Gauss' principle of least constraints. *The quarterly Journal of pure and appl. math.* **40** (1909) 153-157

25. Lagrange, J.L.de: Récherches sur la libration de la lune. *Prix de l'academie royale de sciences de Paris* **IX**, 1764. Reprint Paris: Gauthier Villars 1873

26. Lagrange, J.L.de: Théorie de la libration de la lune. *Nouveaux mémoires de l'academie royale des sciences et des belles-lettres de Berlin,* année 1780. Reprint Paris: Gauthier Villars 1870

27. Lagrange, J.L.de: Essai d'une nouvelle méthode pour déterminer les maxima et minima des formes intégrales indéfinies, 1762. *Miscellanea Taurinensia* **II**, 173-195. Translated into German: Darmstadt: Wiss. Buchges. 1976

28. Lamb, Horace: Dynamics. Cambridge (England): Cambrigde University Press 1920

29. Lur'e, A.I.: Mécanique analytique. tome I, Louvain: Librairie universitaire 1968, p.22

30. Maggi, G.C.: Principii di Stereodinamica. Milano: Editore della Real Casa, 1903, 184-185

31. Papastavridis, J.G: Time-Integral Variational Principles for Nonlinear Nonholonomic Systems. *J .Appl. Mech.* **64**, (1997), 985 - 991

32. Papastavridis, J.G: A panoramic overview of the principles and equations of motion of advanced engineering dynamics. *Appl. Mech. Rev.* **4**, (1998), 239 - 265

33. Papastavridis, J.G: Analytical Mechanics. Oxford Univ Press (in print). To appear 2000

34. Pfeiffer,F., Glocker,C: Multibody Dynamics with Unilateral Constraints. New York etc: Wiley 1996

35. Poinsot, L.: Élemens de statique. Paris: Bachelier 1837, p.476

36. Pulte, H.: Jacobi's criticism of Lagrange. *Historia Mathematica* **25**, (1998), 154-184

37. Radetsky, P: The Man who Mastered Motion. *Science* (1986), 52-60

38. Rossmann, Th.: Eine Laufmaschine für Rohre. Ph.D. Thesis, Düsseldorf: *VDI-Fortschr.-Ber.* **8**, Nr. 732, 1998

39. Schiehlen, W.: Technische Dynamik. Stuttgart: Teubner 1986

40. Vereshagin, A.F.: Computer Simulation of the Dynamics of Complicated Mechanisms of Robot Manipulators, *Eng. Cybernetics* **6**, (1974), 65-70

41. Wittenburg, J.: Dynamics of Systems of Rigid Bodies, Stuttgart: Teubner (1977)

COMPENSATORS FOR THE ATTENUATION OF FLUID FLOW PULSATIONS IN HYDRAULIC SYSTEMS

JOSEF MIKOTA

Department of Foundations of Machine Design,
Johannes Kepler University of Linz,
Altenbergerstrasse 69, A-4040 Linz, Austria.
E-mail: josef.mikota@jku.at

Fluid borne noise is a problem in hydraulics. Most conventional devices for the attenuation of fluid flow pulsations are either too expensive, build too long or have poor attenuation characteristics. Furthermore, all conventional devices but multi-volume resonators fail to attenuate frequencies which are integer multiples of a base harmonic. In this publication several devices, such as multi-body mass-spring oscillators, multi-volume Helmholtz resonators, compact $\lambda/4$ side branch resonators, as well as systems based on various arrangements of plate/shell elements, will be investigated. Besides that, a novel concept for the frequency tuning of chain-structure mass-spring spring oscillators will be presented which places the natural frequencies $\omega_1 \ldots \omega_N$ at a base harmonic Ω_1 and $N-1$ integer multiples $\Omega_2 \ldots \Omega_N$. The usefulness of the presented concepts will be evaluated on a simulation study of two real world hydraulic systems (one low pressure system with $p_{Sys} = 40$ bar and one conventional system with $p_{Sys} = 200$ bar). Although the results appear promising, especially for multi DOF mass-spring compensators and $\lambda/4$ side-branch resonators, experimental work is required to confirm the results.

1 Introduction

Positive displacement pumps, as well as the utilisation of discontinuous control elements, create significant flow- and pressure pulsations in hydraulic circuits. Common measures to attenuate these unpleasant effects include accumulators, in particular featuring a *Pulse-Tone* design, $\lambda/4$ line silencers, $\lambda/4$ side branch resonators, *Helmholtz resonators* and novel multiple-volume resonators [6]. In this contribution several novel designs of oscillators, such as multi-body mass-spring oscillators, multi-volume Helmholtz resonators, compact $\lambda/4$ side branch resonators, as well as systems based on various arrangements of plate/shell elements will be investigated.

The investigation of $\lambda/4$ side branch resonators will cover the influence of wall flexibility, as well as the influence of additional flexible elements in the resonator chamber with respect to the natural frequencies of the coupled elastic/acoustic system.

Furthermore, a novel concept for the frequency tuning of multi degree-of-freedom mass-spring systems of a certain structure will be presented. It will be demonstrated that by exploiting this concept it is easily possible to place the natural frequencies of un-damped systems $\omega_1 \ldots \omega_N$ at a certain base harmonic Ω_1 and $N - 1$ integer multiples of it $\Omega_2 \ldots \Omega_N$. In order to evaluate the usefulness of this concept, the influence of damping with respect to system dynamics will be investigated. It will be shown that the presented concept works well for weakly damped systems.

In order to compare various oscillators, an illustrative example of two real-world hydraulic systems (one low pressure system with $p_{Sys} = 40$ bar and one conventional system with $p_{Sys} = 200$ bar) featuring compensators, such as multi degree-of-freedom mass-spring oscillators, compensators based on plate/shell elements and compact $\lambda/4$ side branch resonators will be discussed. Therefore criteria, such as the ability to compensate frequencies which are multiples of a base frequency, frequency tuning of the resonator, as well as compact and cheap design of the compensator will be utilised.

2   Sources of hydraulic noise

There are three main sources of noise in a hydraulic circuit: So-called *positive displacement pumps*, such as piston pumps, gear pumps etc. producing a non-steady flow stream in the first place, *positive displacement motors*, and the utilisation of discontinuous control elements, e.g. fast switching valves [5,12], in the hydraulic circuit. Some details concerning these elements will be discussed in the following paragraphs.

2.1   *Positive displacement pumps/motors*

By definition [4], positive displacement machines are characterised by a finite number of displacement elements, i.e. pistons in case of a piston pump/motor, teeth in gear pumps/motors etc. Hence, the flow stream originating at a pump is not constant over time, i.e. it is characterised by some harmonics repeating at the pump frequency and integer multiples of it. In addition to that, these periodic flow pulsations interact with other elements of the hydraulic circuit, such as piping, valves... or even notably with the load itself, and complex pressure waves are formed within the hydraulic circuit [3].

To the outside world, these fluctuations become notable as audible noise, vibrations of the pipe-work or fatigue problems of components.

Although the dynamics of positive displacement pumps is quite complicated (influence of port plate (or silencing) grooves, momentum of the fluid in the vicinity of the ports...), a very simple model for the flow generating mechanism of positive displacement pumps will be presented.

Neglecting the influence of the silencing grooves and the momentum of the fluid in the vicinity of the input/output ports, the flow generated by one pumping element may be written as

$$Q_i = \frac{Q_{spec}\,\pi\,n}{n_{Pist}}\,\sin(\omega\,t + \varphi_i)\,\sigma[\sin(\omega\,t + \varphi_i)] \tag{1a}$$

$$\text{where } \sigma(x) = \begin{cases} 1 & \text{if } x \geq 0 \\ 0 & \text{otherwise} \end{cases} \tag{1b}$$

$$\text{and } \varphi_i = \frac{2\,\pi}{n_{Pist}}(i - 1). \tag{1c}$$

Thus the total pump flow generated is

$$Q_{tot} = \sum_{i=1}^{n_{Pist}} Q_i. \tag{2}$$

Fig. 1 shows the flow characteristics of a 9 piston axial pump with $Q_{Nom} = 50$ l/min.

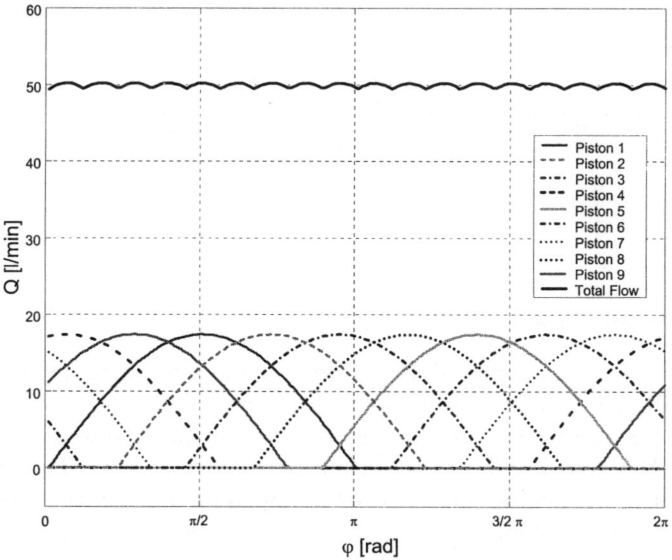

Figure 1. Flow characteristics of 9 piston pump

A spectral analysis of the pump flow (see Fig. 2 and [1]) reveals that the first harmonic of the pump flow is given as

$$f_1 = \begin{cases} \frac{2\,n_{Pist}\,n}{60} & \text{if } n_{Pist} \text{ is an odd number} \\ \frac{n_{Pist}\,n}{60} & \text{otherwise.} \end{cases} \tag{3}$$

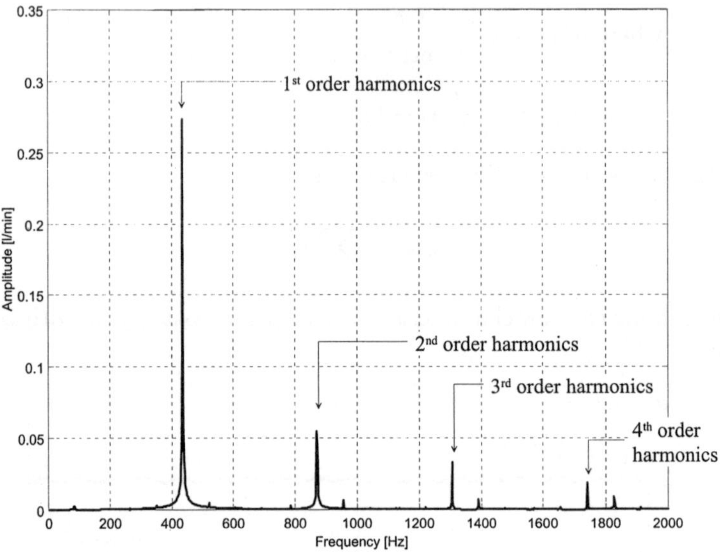

Figure 2. Frequency spectrum of flow characteristics of 9 piston pump

2.2   Switching valves

As explained in the papers [5,12], a fast (digital) switching valve in combination with a mechanical oscillator may be utilised to effectively adjust the pressure in hydraulic systems by periodically switching between a high pressure port P, a tank line T and an output port A (see Fig. 3). The pressure in the output line is hereby controlled by the pressure levels in the pressure and tank lines, the switching frequency f, the relative period of the pressure-on-phase and the relative period of the output-on-phase.

Although this method of pressure control is more energy efficient than conventional resistance control, there is of course the problem that, depending on the overall system dynamics, pressure transients are introduced in the hydraulic system due to the discontinuous nature of the switching process.

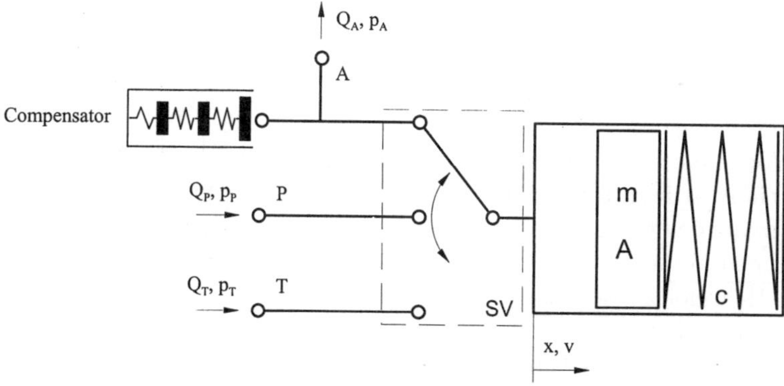

Figure 3. Schematics of resonance converter

3   Devices for the suppression of hydraulic noise

3.1   Conventional devices

As the reduction of flow pulsations, otherwise also commonly known as fluid borne noise, has been a topic in hydraulic Engineering for some time, numerous devices for the suppression of hydraulic noise exist.

As can be seen in Tab. 1, no conventional device but the multi volume resonator is capable of attenuating a base harmonic and integer multiples of it. Furthermore, the selection of compensators, such as accumulators and in-line noise suppressors, is mainly left to the experience of the engineer, some rules-of-thumb and is thus quite often a field for trial and error.

Table 1. Conventional noise suppressors

Accumulator	
⊕ simple device, readily available ⊖ low performance (especially at high frequencies) ⊖ frequency tuning by experimental means	Valve, Membrane, Casing, Connection block
Helmholtz resonator	
⊕ simple device ⊕ simple frequency tuning ⊖ tuned to attenuate 1 harmonic only	V, A, t, Pump, Load
In-line noise suppressor	
⊕ fairly simple device ⊖ expensive	Pump, Load
$\lambda/4$ Line silencer	
⊕ simple device ⊖ attenuation of odd order harmonics only ⊖ builds rather long for low frequencies	Pump, Load, $\frac{2n+1}{4}\lambda$, $n \in N_+$
$\lambda/4$ Side branch resonator	
⊕ simple device ⊖ attenuation of odd order harmonics only ⊖ builds rather long for low frequencies	$l=\lambda/4$, Pump, Load
Multiple volume resonator	
⊕ fairly simple device ⊕ may be tuned in such a way to attenuate integer multiples of a base harmonic ⊖ complex frequency tuning	OR Variable resonance type side branch resonator Mulit DOF type Helmholtz resonator

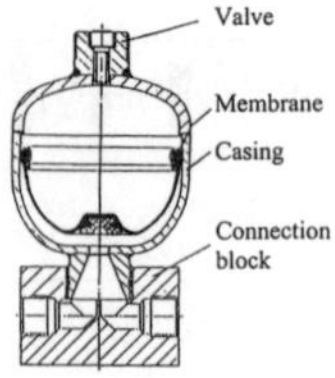

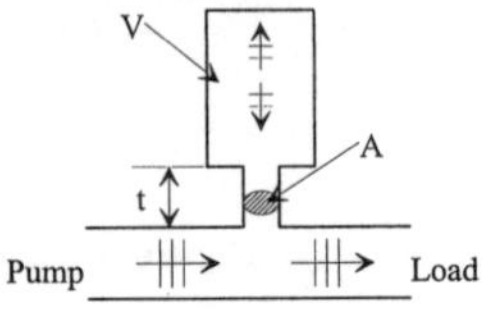

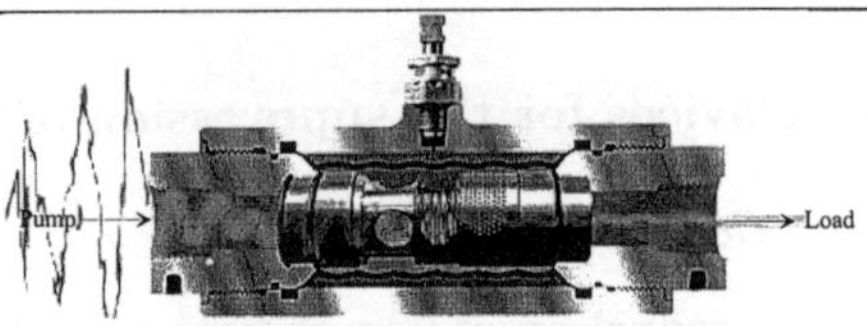

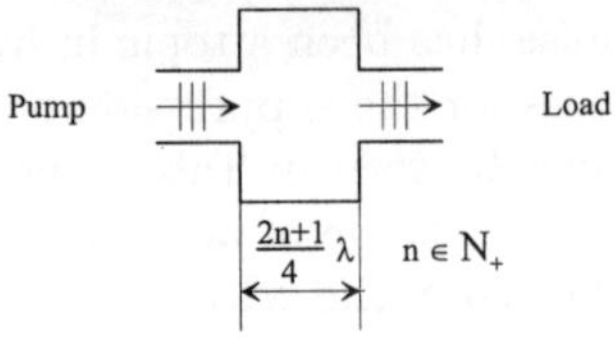

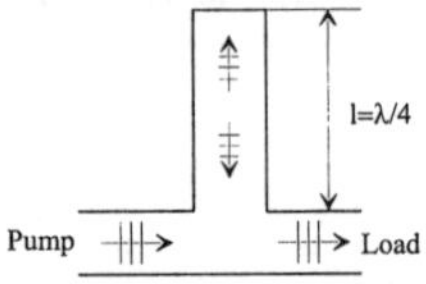

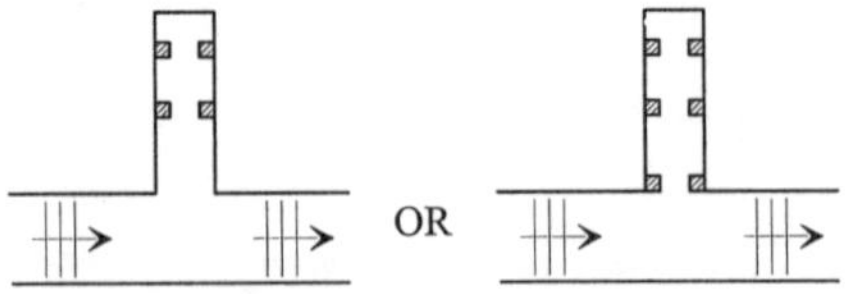

3.2 *Novel devices*

1. Common idea for all novel devices

   Although the concept of mechanical vibration compensation is commonly
   known in Engineering mechanics, a solution of the equations of motion for
   the simple case of an un-damped single degree-of-freedom (DOF) mass
   spring oscillator with mechanical compensation will be presented.

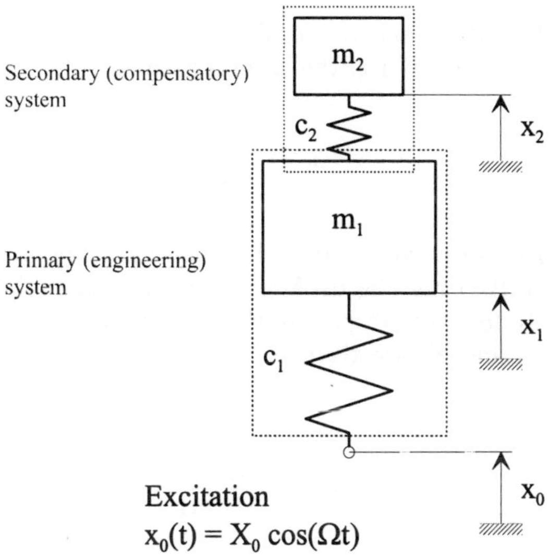

Figure 4. Schematics of Engineering and compensatory system

The equations of motion for this system may be written as [14, p. 449]

$$m_1\,\ddot{x}_1 = -c_1(x_1 - x_0) + c_2(x_2 - x_1) \tag{4a}$$
$$m_2\,\ddot{x}_2 = -c_2(x_2 - x_1). \tag{4b}$$

Using the functions

$$x_1(t) = X_1\,\cos(\Omega\,t) \tag{5a}$$
$$x_2(t) = X_2\,\cos(\Omega\,t) \tag{5b}$$

for the steady state solution of the equations of motion, the amplitudes

X_1 and X_2 of the masses m_1 and m_2 respectively are given as

$$X_1 = -\frac{(-m_2\,\Omega^2 + c_2)\,c_1\,X_0}{-m_1\Omega^4\,m_2 + m_1\,\Omega^2\,c_2 + c_1\,m_2\,\Omega^2 - c_1\,c_2 + c_2\,m_2\,\Omega^2} \tag{6a}$$

$$X_2 = -\frac{c_2\,c_1\,X_0}{-m_1\Omega^4\,m_2 + m_1\,\Omega^2\,c_2 + c_1\,m_2\,\Omega^2 - c_1\,c_2 + c_2\,m_2\,\Omega^2}. \tag{6b}$$

A closer look at the numerator of Eq. (6a) makes clear that a properly tuned secondary system, i. e. the natural frequency

$$\omega_C = \sqrt{\frac{c_2}{m_2}} \tag{7}$$

of the compensatory system placed at the harmonic of the excitation $x_0(t) = X_0\cos(\Omega t)$ may be used to effectively cancel the movement of the primary system with respect to the excitation $x_0(t)$. In that case the amplitudes X_1 and X_2 are given as

$$X_1 = 0 \tag{8a}$$

$$X_2 = -\frac{c_1\,X_0}{\Omega^2\,m_2}. \tag{8b}$$

Furthermore, the mechanical system Fig. 4 may be associated with its dual hydraulic system depicted in Fig. 5.

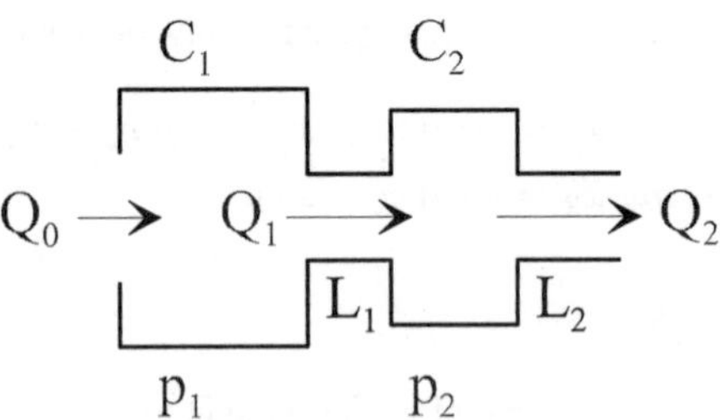

Figure 5. Dual hydraulic system to mechanical system Fig. 4

The equations describing the dynamics of the dual hydraulic system are

$$\dot{p}_1 = \frac{1}{C_1}(Q_0 - Q_1) \tag{9a}$$

$$\dot{p}_2 = \frac{1}{C_2}(Q_1 - Q_2) \tag{9b}$$

$$p_1 = L_1\,\dot{Q}_1 + p_2 \tag{9c}$$

$$p_2 = L_2\,\dot{Q}_2. \tag{9d}$$

Taking the derivatives of p_1 and p_2

$$\dot{p}_1 = L_1\,\ddot{Q}_1 + \dot{p}_2 \tag{10a}$$

$$\dot{p}_2 = L_2\,\ddot{Q}_2 \tag{10b}$$

and re-arranging the terms in the equations results in

$$\frac{1}{C_1}(Q_0 - Q_1) = L_1\ddot{Q}_1 + \frac{1}{C_2}(Q_1 - Q_2) \tag{11a}$$

$$\frac{1}{C_2}(Q_1 - Q_2) = L_2\ddot{Q}_2. \tag{11b}$$

The equations describing the mechanical system Eqs. (4) and the equations describing the hydraulic system Eqs. (11) may also be written as

Mechanical system

$$\begin{pmatrix} m_1 & 0 \\ 0 & m_2 \end{pmatrix}\begin{pmatrix} \ddot{x}_1 \\ \ddot{x}_2 \end{pmatrix} + \begin{pmatrix} c_1 + c_2 & -c_2 \\ -c_2 & c_2 \end{pmatrix}\begin{pmatrix} x_1 \\ x_2 \end{pmatrix} = \begin{pmatrix} c_1\,x_0 \\ 0 \end{pmatrix} \tag{12}$$

and

Hydraulic system

$$\begin{pmatrix} L_1 & 0 \\ 0 & L_2 \end{pmatrix}\begin{pmatrix} \ddot{Q}_1 \\ \ddot{Q}_2 \end{pmatrix} + \begin{pmatrix} \frac{1}{C_1} + \frac{1}{C_2} & -\frac{1}{C_2} \\ -\frac{1}{C_2} & \frac{1}{C_2} \end{pmatrix}\begin{pmatrix} Q_1 \\ Q_2 \end{pmatrix} = \begin{pmatrix} \frac{1}{C_1}\,Q_0 \\ 0 \end{pmatrix}. \tag{13}$$

As can be seen above, the equations describing the mechanical system may be led over into the equations describing the hydraulic system by using the following conventions

$$c \Rightarrow 1/C$$

$$m \Rightarrow L$$

$$x \Rightarrow Q.$$

Coming back to our original problem of reducing fluid borne noise in a hydraulic circuit, the excitation of the Engineering system $x_0(t)$ (in case of the mechanical system) and $Q_0(t)$ (in case of the hydraulic system), may be compensated by a properly tuned oscillator where the natural frequency of the oscillator ω_C is placed at the harmonic of the excitation Ω.

2. Mass spring oscillators

- **Single degree-of-freedom mass spring oscillator**

  Since a mass spring resonator for the compensation of fluid borne noise needs to be sealed from the hydraulic circuit and the dynamic behaviour of a single degree-of-freedom (DOF) mass spring system is commonly known, the attention of this paragraph is focused on the influence of damping.

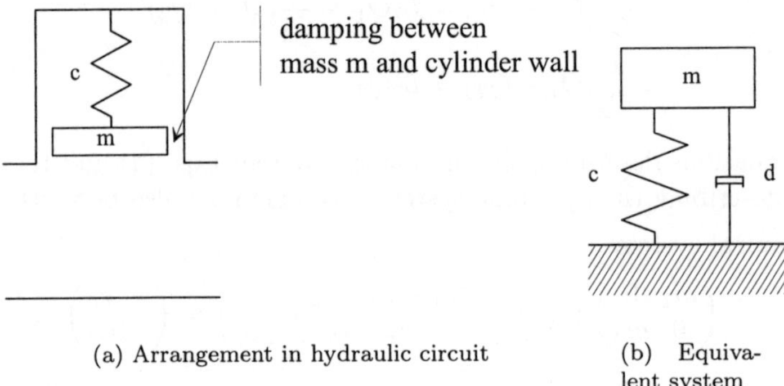

(a) Arrangement in hydraulic circuit   (b) Equivalent system

Figure 6. Single DOF system with damping

The equation of motion and the dimensionless damping ratio ζ for a (homogeneous) system depicted in Fig. 6 are given as

$$0 = m\,\ddot{x} + d\,\dot{x} + c\,x \tag{14}$$

$$\zeta = \frac{d}{2\sqrt{m\,c}}. \tag{15}$$

The different responses of a single DOF mass spring system with $m = 1$ kg, $\zeta \in \{0.01, 0.1, 1\}\ \frac{N}{m/s}$ and $c = 1$ N/m are depicted in Fig. 7.

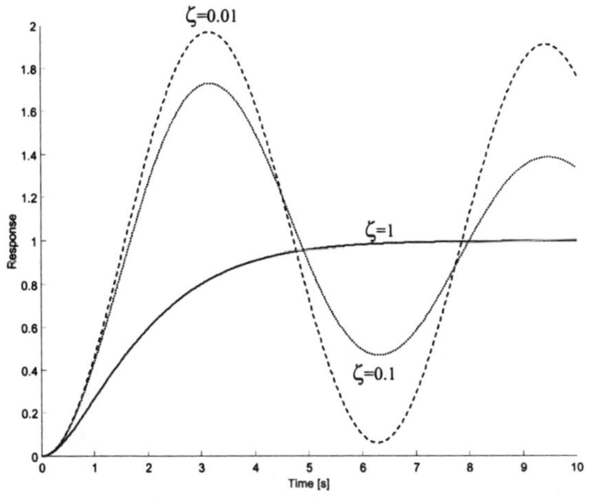

(a) Time response

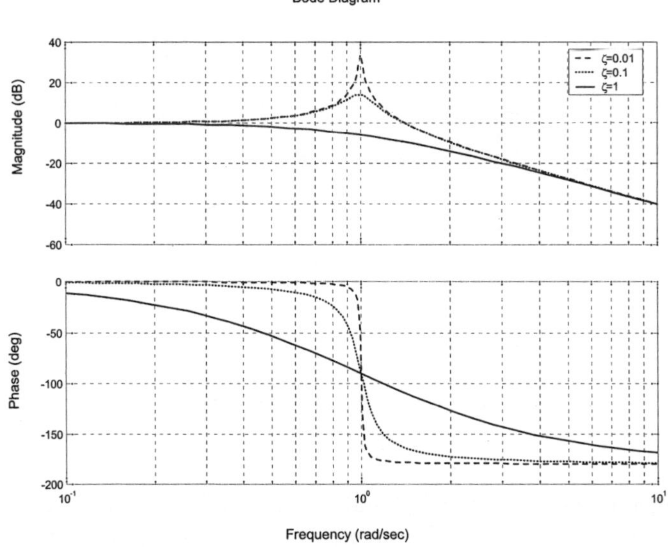

(b) Normalised frequency response

Figure 7. Characteristics of a single DOF mass spring oscillator with $m = 1$ kg, $\zeta \in \{0.01, 0.1, 1\}$ $\frac{N}{m/s}$ and $c = 1$ N/m

Although the time responses are considerably different for lightly damped systems ($\zeta \leq 0.1$), the change in terms of the resonance frequency is only of minor magnitude. Hence, the expression

$$\omega = \sqrt{\frac{c}{m}} \tag{16}$$

may be used for the natural frequency of lightly damped mass spring systems.

- Multi degree-of-freedom mass spring oscillator

In order to attenuate more than one harmonic of the pulsating flow stream, one may simply use several single DOF oscillators in a hydraulic circuit tuned to different harmonics (see Fig. 8), or utilise a multi DOF oscillator tuned to several harmonics of the flow pulsation [9].

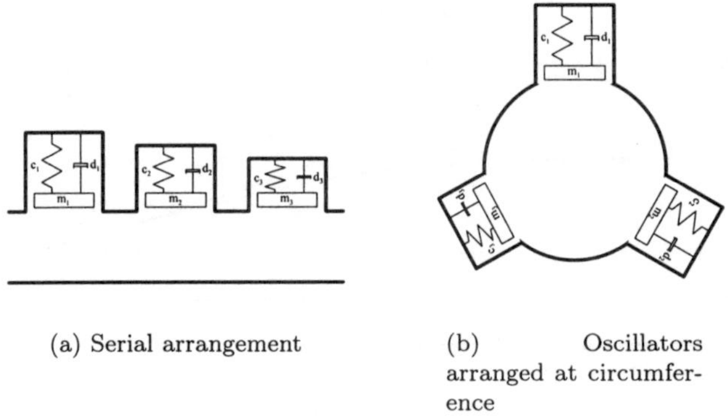

(a) Serial arrangement

(b) Oscillators arranged at circumference

Figure 8. Arrangement of several single DOF oscillators

Since a single device with multiple DOFs is sometimes of interest, a novel concept for the frequency tuning of multi DOF mass-spring oscillators (see Fig. 9), which places the natural frequencies $\omega_1 \ldots \omega_N$ at a certain base harmonic Ω_1 and $N - 1$ integer multiples of it $\Omega_2 \ldots \Omega_N$, will be presented [8].

The equations governing the dynamics of an N-body oscillator de-

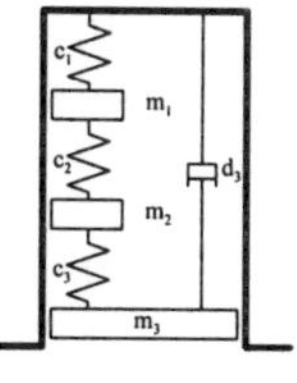

Figure 9. Multi DOF oscillator in hydraulic circuit

picted in Fig. 10 may be written as follows:-

$$m_1\,\ddot{x}_1 + c_1\,x_1 - c_2\,(x_2 - x_1) = 0 \quad (17\text{a})$$

$$m_2\,\ddot{x}_2 + c_2\,(x_2 - x_1) - c_3\,(x_3 - x_2) = 0 \quad (17\text{b})$$

$$\vdots$$

$$m_{N-1}\,\ddot{x}_{N-1} + c_{N-1}\,(x_{N-1} - x_{N-2}) - c_N\,(x_N - x_{N-1}) = 0 \quad (17\text{c})$$

$$m_N\,\ddot{x}_N + c_N\,(x_N - x_{N-1}) + d_N\,\dot{x}_N = F \quad (17\text{d})$$

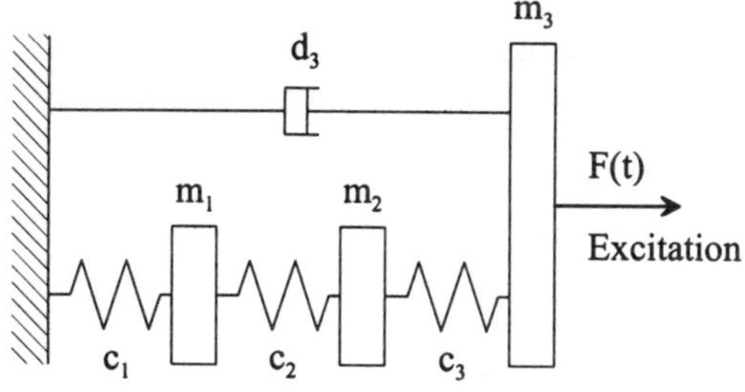

Figure 10. Structure of a third order system

In vectorial notation, above equations may also be written as

$$\mathbf{M}\,\ddot{\mathbf{x}} + \mathbf{D}\,\dot{\mathbf{x}} + \mathbf{C}\,\mathbf{x} = \mathbf{F} \tag{18}$$

where $\mathbf{x} = [x_1 \ldots x_N]^T$, $\mathbf{M}$ is the mass matrix, $\mathbf{D}$ is the damping matrix, $\mathbf{C}$ is the stiffness matrix and $\mathbf{F}$ is the force vector of the system.

Assuming an undamped system ($d_N = 0$), the natural frequencies of the oscillator may be calculated as the roots of the characteristic polynomial

$$|-\omega^2\,\mathbf{M} + \mathbf{C}| = 0. \tag{19}$$

A closer look at Eq. 19 reveals that the characteristic matrix $\left(-\omega^2\,\mathbf{M} + \mathbf{C}\right)$ has tri-diagonal structure and may be written as

$$\begin{pmatrix} -\omega^2\,m_1 + c_1 + c_2 & -c_2 & & & 0 \\ -c_2 & -\omega^2\,m_2 + c_2 + c_3 & -c_3 & & \\ & \ddots & \ddots & \ddots & \\ & & -c_{N-1} & -\omega^2\,m_{N-1} + c_{N-1} + c_N & -c_N \\ 0 & & & -c_N & -\omega^2\,m_N + c_N \end{pmatrix}.$$

By defining the first natural frequency of the oscillator as Ω and integer multiples of it as

$$\Omega_1 = \Omega \tag{20a}$$
$$\Omega_2 = 2\,\Omega \tag{20b}$$
$$\vdots$$
$$\Omega_N = N\,\Omega, \tag{20c}$$

the natural frequencies of an (un-damped) chain structure oscillator $\omega_1 \ldots \omega_N$ may be placed at $\Omega_1 \ldots \Omega_N$ simply by making the masses $m_1 \ldots m_N$ to

$$m_1 = m \tag{21a}$$
$$m_2 = m/2 \tag{21b}$$
$$\vdots$$
$$m_N = m/N \tag{21c}$$

and making the stiffnesses of the springs $c_1 \ldots c_N$ to

$$c_1 = N\,c \tag{22a}$$
$$c_2 = (N-1)\,c \tag{22b}$$
$$\vdots$$
$$c_N = c = \Omega^2\,m. \tag{22c}$$

To make this approach more plausible, the characteristic matrix [13,16] may then be written as

$$
m \begin{pmatrix}
-\omega^2+(2N-1)\Omega^2 & -(N-1)\Omega^2 & & & & 0 \\
-(N-1)\Omega^2 & -\frac{\omega^2}{2}+(2N-3)\Omega^2 & -(N-2)\Omega^2 & & & \\
& & \ddots & \ddots & \ddots & \\
& & & -2\Omega^2 & -\frac{\omega^2}{(N-1)}+3\Omega^2 & -\Omega^2 \\
0 & & & & -\Omega^2 & -\frac{\omega^2}{N}+\Omega^2
\end{pmatrix}.
\tag{23}
$$

Although a formal proof for arbitrary N cannot yet be presented, the characteristic polynomial has the form

$$
m^N \prod_{i=1}^{N} \left(\frac{-\omega^2}{i} + i\,\Omega^2 \right),
\tag{24}
$$

which places the natural frequencies ω of the oscillator exactly at $\Omega_1 \ldots \Omega_N$. Due to the fact that the resonance frequencies of a damped multi body oscillator are only marginally different from the natural frequencies in an un-damped case (see Fig. 11), the frequency tuning concept presented in this paragraph is also suitable for lightly damped oscillators ($\bar{\zeta} \leq 0.1$), where the dimensionless damping ratio $\bar{\zeta}$ is defined as

$$
\bar{\zeta} = \frac{d_N}{2\,m_N \sqrt{\frac{c_N}{m_N}}}.
\tag{25}
$$

3. Analogy between mechanical and acoustic systems

According to the derivations given in the previous paragraph of this section, the natural frequencies of a mechanical chain-structure oscillator may be placed at Ω_1 and $N-1$ integer multiples $\Omega_2 \ldots \Omega_N$. In this paragraph it will be shown that this concept of frequency tuning is not limited to mechanical systems but may also be applied to acoustic systems.

Since the general duality between mechanical, electrical and acoustic systems has been described in a number of excellent books (e. g. [15]), the analogy will be shown on a system with 2 DOFs.

The dynamic behaviour of the mechanical system depicted in Fig. 12 is described by

$$
m_1\ddot{x}_1 + c_1 x_1 - c_2(x_2 - x_1) = 0
\tag{26a}
$$

$$
m_2\ddot{x}_2 + c_2(x_2 - x_1) = F,
\tag{26b}
$$

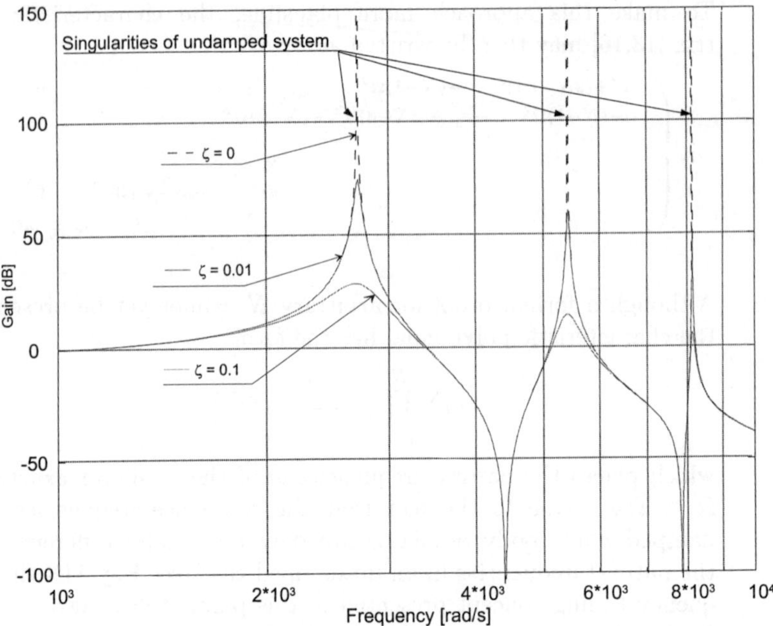

Figure 11. Normalised bode diagram of a 3rd order system featuring damping ratios $\bar{\zeta} \in \{0, 0.1, 0.01\}$

whereas the behaviour of the acoustic system is described by

$$p = L_2\dot{Q}_2 + p_2 \tag{27a}$$

$$\dot{p}_2 = \frac{1}{C_2}(Q_2 - Q_1) \tag{27b}$$

$$p_2 = L_1\dot{Q}_1 + p_1 \tag{27c}$$

$$\dot{p}_1 = \frac{1}{C_1}Q_1. \tag{27d}$$

By using different state variables, the equations describing the behaviour of the acoustic system may be also be written as

$$u = \dot{p} = L_2\ddot{Q}_2 + \dot{p}_2 = L_2\ddot{Q}_2 + \frac{1}{C_2}(Q_2 - Q_1) \tag{28a}$$

$$\frac{1}{C_2}(Q_2 - Q_1) = \dot{p}_2 = L_1\ddot{Q}_1 + \frac{1}{C_1}Q_1. \tag{28b}$$

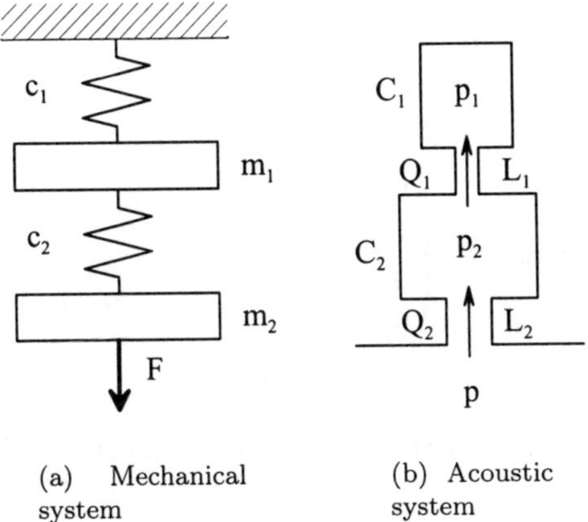

Figure 12. Mechanical and acoustic systems

In vectorial notation these equations are simply

Mechanical system

$$\mathbf{x_{mech}} = \begin{pmatrix} \dot{x}_1 & x_1 & \dot{x}_2 & x_2 \end{pmatrix}^T \tag{29a}$$

$$u_{mech} = F \tag{29b}$$

$$\dot{\mathbf{x}}_{mech} = \begin{pmatrix} 0 & -\frac{c_1+c_2}{m_1} & 0 & \frac{c_2}{m_1} \\ 1 & 0 & 0 & 0 \\ 0 & \frac{c_2}{m_2} & 0 & -\frac{c_2}{m_2} \\ 0 & 0 & 1 & 0 \end{pmatrix} \mathbf{x_{mech}} +$$

$$+ \begin{pmatrix} 0 & 0 & \frac{1}{m_2} & 0 \end{pmatrix}^T u_{mech} \tag{29c}$$

Acoustic system

$$\mathbf{x_{acou}} = \left(\dot{Q}_1 \; Q_1 \; \dot{Q}_2 \; Q_2\right)^T \tag{30a}$$

$$u_{acou} = \dot{p} \tag{30b}$$

$$\dot{\mathbf{x}}_{\mathbf{acou}} = \begin{pmatrix} 0 & -\frac{1/C_2+1/C_1}{L_1} & 0 & \frac{1/C_2}{L_1} \\ 1 & 0 & 0 & 0 \\ 0 & \frac{1/C_2}{L_2} & 0 & -\frac{1/C_2}{L_2} \\ 0 & 0 & 1 & 0 \end{pmatrix} \mathbf{x_{acou}} +$$

$$+ \left(0 \; 0 \; \tfrac{1}{L_2} \; 0\right)^T u_{acou}, \tag{30c}$$

where u_{mech} and u_{acou} are the inputs to the mechanical and acoustic system respectively.

Looking at these equations, it should be plausible that the natural frequencies of a multi DOF *Helmholtz resonator* may be placed at $\Omega_1 \ldots \Omega_N$, where

$$\Omega_1 = \Omega = \frac{1}{\sqrt{LC}} \tag{31a}$$

$$\Omega_2 = 2\,\Omega \tag{31b}$$

$$\vdots$$

$$\Omega_N = N\,\Omega \tag{31c}$$

by making the hydraulic inductivities $L_1 \ldots L_N$ to

$$L_1 = L \tag{32a}$$

$$L_2 = L/2 \tag{32b}$$

$$\vdots$$

$$L_N = L/N \tag{32c}$$

and making the hydraulic capacities $C_1 \ldots C_N$ to

$$C_1 = \frac{1}{N}\,C \tag{33a}$$

$$C_2 = \frac{1}{N-1}\,C \tag{33b}$$

$$\vdots$$

$$C_N = C. \tag{33c}$$

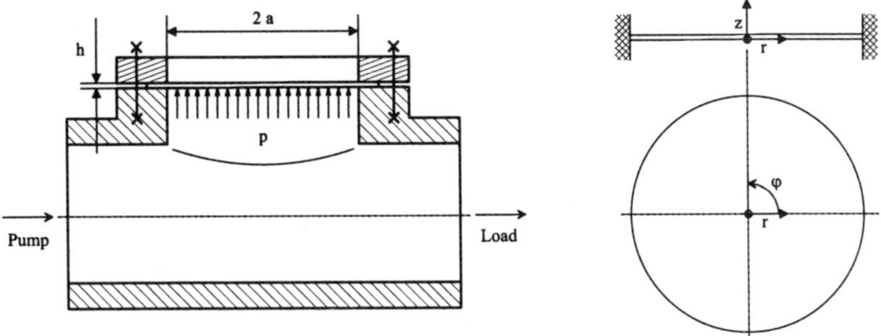

Figure 13. Principle of an oscillator based on a circular plate

4. Compensators based on plate/shell elements

In order to discuss the principle of compensators based on plate or shell elements, the simplest possible compensator of this kind (see Fig. 13) will be discussed in this section. This is a plate of homogeneous thickness h, clamped at the circumference and tuned in such a way to place the first natural frequency at the base harmonic of the pulsating flow stream. In addition to that, the stresses in the plate due to the maximum hydraulic pressure p_{Sys} must not exceed the maximum permissible stress σ_{max} of the material.

The first natural frequency Ω_{11} of a circular plate of constant thickness h being clamped at the circumference is given as [11]

$$\Omega_{11} = \frac{\lambda_{11}}{a^2}\sqrt{\frac{D}{\rho}} \qquad \text{where} \qquad \lambda_{11} \cong 10.216 \quad (34a)$$

and the flexural rigidity of the plate
$$D = \frac{E\,h^3}{12(1-\nu^2)}. \quad (34b)$$

The induced bending moments per unit length due to a constant pressure distribution at the bottom surface in radial and tangential direction, M_r and M_φ respectively, are given as

$$M_r = \frac{1}{16}\,p\,a^2[1+\nu-(3+\nu)\alpha^2] \qquad (35a)$$

$$M_\varphi = \frac{1}{16}\,p\,a^2[1+\nu-(1+3\,\nu)\alpha^2] \qquad (35b)$$

where $\alpha = r/a$.

At the circumference $r = a$, above equations simplify to

$$M_r = -\frac{1}{8}\, p\, a^2$$

$$M_\varphi = -\frac{\nu}{8}\, p\, a^2$$

and result in maximum stresses at the bottom/top layer of

$$\sigma_r = \frac{6\, M_r}{h^2} \quad \text{and} \quad \sigma_\varphi = \frac{6\, M_\varphi}{h^2}. \tag{37}$$

These stress components σ_r and σ_φ respectively may be combined to an equivalent stress σ_E according to the "von Mises" hypothesis

$$\sigma_E = \sqrt{\sigma_r^2 - \sigma_r\, \sigma_\varphi + \sigma_\varphi^2}. \tag{38}$$

5. $\lambda/4$ Side branch resonators

As known in the literature, a side branch resonator of length $\lambda/4$ may be used to compensate incoming pressure pulsations of a base harmonic $\Omega_1 = \Omega$ and odd multiples $\Omega_3 = 3\,\Omega$, $\Omega_5 = 5\,\Omega \ldots$ of it.

However, since the length of the device $l = \lambda/4$ directly determines the first filtering frequency f with

$$c_S = \sqrt{\frac{B}{\rho}} \quad \text{and} \quad \lambda = \frac{c_S}{f}, \tag{39}$$

long pipes are required for the attenuation of low frequency noise (typically $\lambda/4 \approx 1.1$ m for a resonance frequency of 300 Hz in hydraulic oil). In the following paragraphs, the influence of wall flexibility, as well as the influence of flexible elements in the resonator chamber will be discussed and how these effects may be exploited to build more compact resonators.

- Influence of wall flexibility

  According to the theory of thin walled tubes, the increase in volume of a tube due to an increase in pressure [1] is

$$\Delta V_{Tube} = V_0 \frac{\Delta p\, d}{E_{Tube}\, s}. \tag{40}$$

  Therefore, the total increase of volume ΔV_{Tot} is

$$\Delta V_{Tot} = \Delta V_{Tube} + \Delta V_{Oil} = V_0 \frac{\Delta p\, d}{E_{Tube}\, s} + V_0 \frac{\Delta p}{B_{Oil}} = V_0 \frac{\Delta p}{B'_{Tot}}. \tag{41}$$

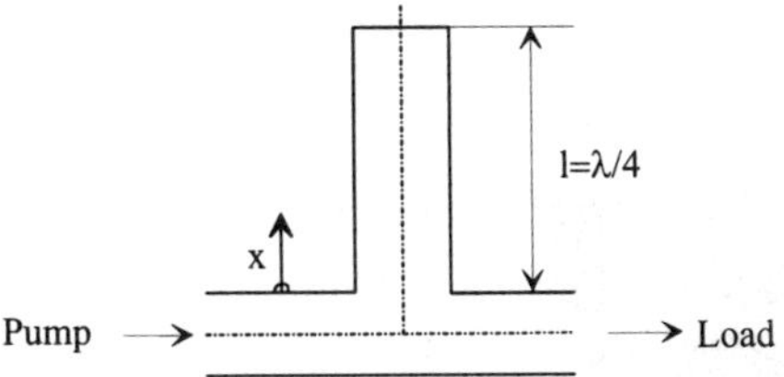

(a) Principal arrangement

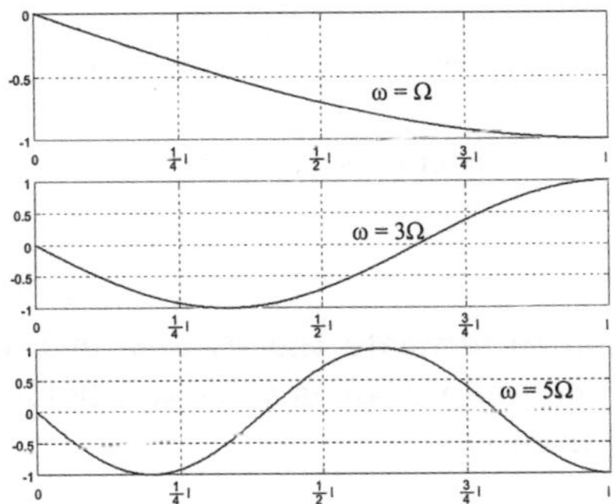

(b) Pressure mode shapes for $\omega = \Omega$, $\omega = 3\,\Omega$ and $\omega = 5\,\Omega$

Figure 14. Arrangement and pressure mode shapes for a $\lambda/4$ side branch resonator

Hence, the bulk modulus B'_{Tot}, considering both effects is given by

$$B'_{Tot} = \frac{B_{Oil}}{1 + \frac{B_{Oil}}{E_{Tube}}\frac{d}{s}}, \tag{42}$$

which results in a natural frequency f of the resonator of

$$f = \frac{1}{4\,l}\sqrt{\frac{B'_{Tot}}{\rho}} = \frac{1}{4\,l}\sqrt{\frac{1}{\rho}\,\frac{B_{Oil}}{1 + \frac{B_{Oil}}{E_{Tube}}\frac{d}{s}}}. \tag{43}$$

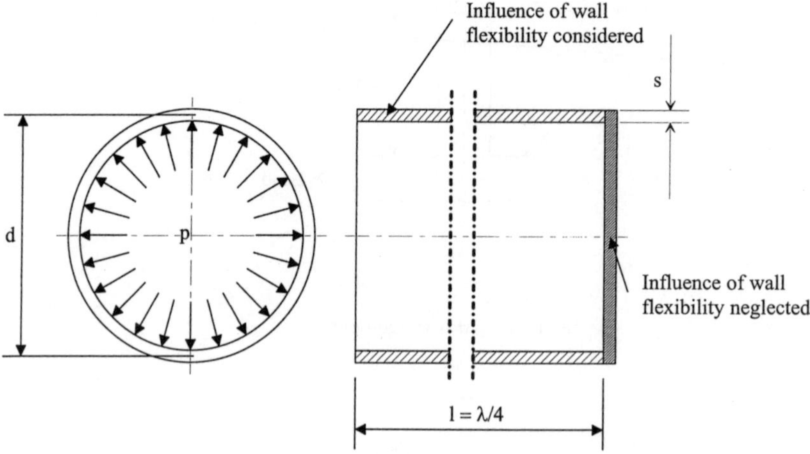

Figure 15. $\lambda/4$ Side branch resonator considering the influence of wall stiffness

A simple equation to estimate the stresses in a cylindrical vessel with diameter d and wall thickness s due to a homogeneous pressure distribution p is given by

$$\sigma = \frac{p\,d}{2\,s}.\tag{44}$$

- Flexible elements in the resonator chamber

  Another way to build more compact resonators is by inserting *flexible elements* into the resonator chamber. One possible arrangement of such a system is depicted in Fig. 16.

  As in the previous paragraph of this section, the insertion of a flexible element into the resonator chamber results in a reduced bulk modulus B'_{Tot} of the resonator and consequently in a shorter length of the resonator chamber for a given frequency.

  This time neglecting the influence of wall stiffness, the equations governing B'_{Tot} are as follows

$$\Delta V_{Tot} = \Delta V_{Oil} + \Delta V_{Elem} = \frac{V_{Oil}\,\Delta p}{B_{Oil}} + \frac{V_{Elem}\,\Delta p}{E_{Elem}}$$
$$= \frac{V_{Oil}\,\Delta p}{B'_{Tot}}\tag{45a}$$

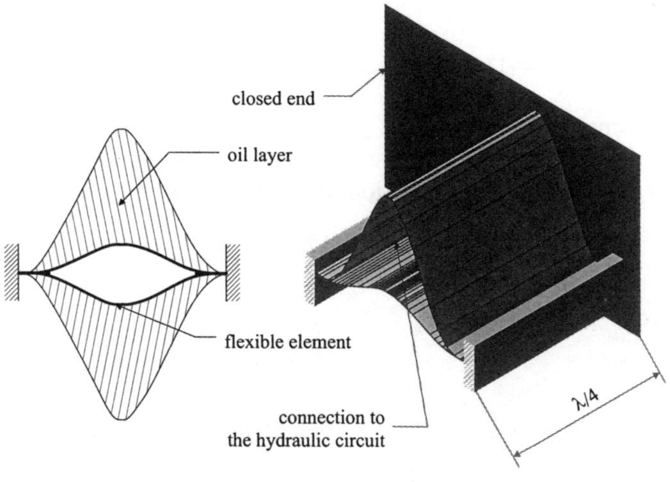

(a) Total system

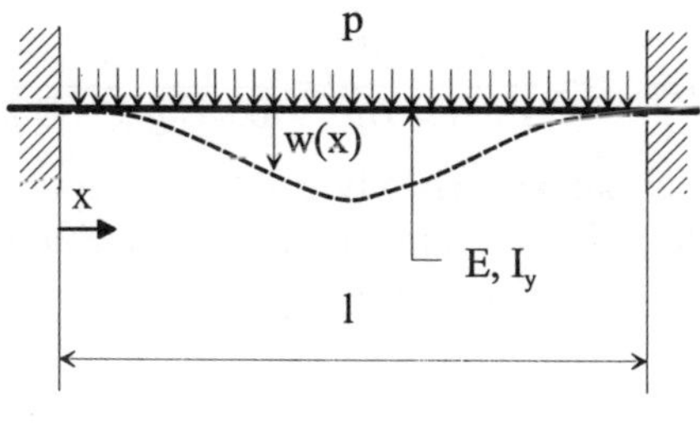

(b) Front view

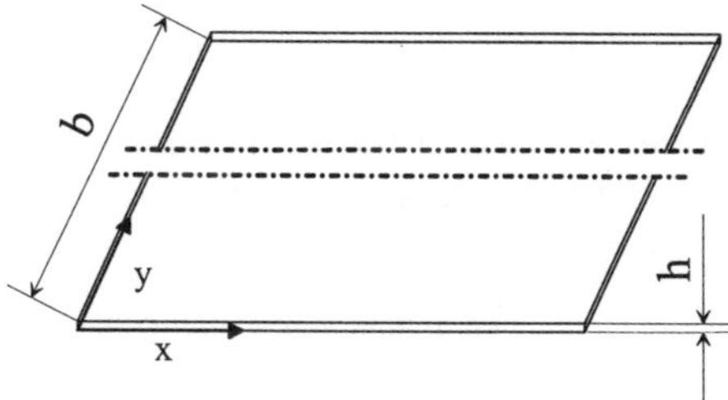

(c) Isometric view

Figure 16. $\lambda/4$ resonator with flexible elements in the resonator chamber

where

$$E_{Elem} = 1/\beta_{Elem} \quad \text{and} \quad \beta_{Elem} = \frac{1}{V_{Elem}} \frac{\partial p}{\partial V}. \tag{45b}$$

Neglecting the boundaries at $y = 0$ and $y = \lambda/4$, the deformation of the element due to p may be approximated using the simple *beam theory*. Doing so, the deflection of an element due to pressure p is

$$w(x) = \frac{1}{24} \frac{p \, b \, l^4 (\frac{x^2}{l^2} - 2 \frac{x^3}{l^3} + \frac{x^4}{l^4})}{E \, I_y}, \tag{46a}$$

where

$$I_y = \frac{b \, h^3}{12}. \tag{46b}$$

Thus the displaced volume V of one "half-element" due to p is

$$V = b \int_{x=0}^{l} w(x) \, dx = \frac{b \, p \, l^5}{60 \, E \, h^3} \tag{47}$$

and the coefficient of compression β_{Elem} for the element may be written as

$$\beta_{Elem} = \frac{1}{V_{0Elem}} \frac{\partial V}{\partial p} = \frac{b \, l^5}{60 \, V_{0Elem} \, E \, h^3}. \tag{48}$$

This approach is particularly vivid if we assume that the flexible elements are shaped according to the deformation due to a static pressure p_{Stat}. In that case, the un-deformed volume of one half-element is

$$V_{0Elem} = \frac{b \, p_{Stat} \, l^5}{60 \, E \, h^3} \tag{49}$$

and hence the value of $E_{Elem} = 1/\beta_{Elem}$ is simply given by

$$E_{Elem} = \frac{1}{\beta_{Elem}} = p_{Stat}. \tag{50}$$

Furthermore, if we assume that the oil volume in the resonator chamber V_{Oil} is related to the volume covered by the flexible element V_{0Elem} by

$$V_{Oil} = \delta \, V_{0Elem}, \tag{51}$$

the combined bulk modulus B'_{Tot} may be written as

$$B'_{Tot} = \frac{\delta}{\frac{\delta}{B_{Oil}} + \frac{1}{p_{Stat}}}. \tag{52}$$

Since the value of δ is crucial for B'_{Tot}, the qualitative behaviour of B'_{Tot} over δ is depicted in Fig. 17.

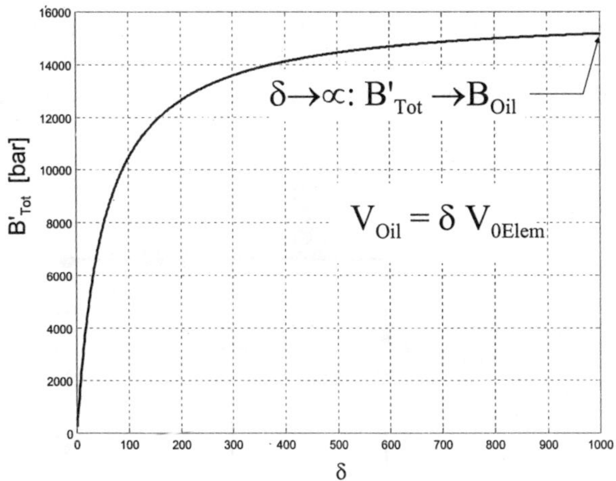

Figure 17. Combined bulk modulus B'_{Tot} over δ for $B_{Oil} = 16000$ bar and $p_{Stat} = 300$ bar

As in the previous paragraph, the required length of the resonator l for a given resonance frequency f may be calculated using Eq. (43).

Since the flexible elements, as well as the surrounding resonator tube are exposed to the hydraulic system pressure, the mechanical stresses in these elements are of interest. Again making use of the beam theory, the maximum normal stress in the flexible element due to the system pressure p_{Sys} is

$$\sigma_{max} = \frac{p_{Sys}\, l^2}{2\, h^2}.\tag{53}$$

However, since the shape of the tube surrounding the resonance chamber is non–circular, the stresses in this element due to p_{Sys} require a more detailed investigation.

4 Illustrative example and discussion

In order to highlight the merits and shortcomings of different solid body compensators, the following example will be discussed: Suppose two hydraulic systems, one low pressure system with $p_{Sys} = 40$ bar and one conventional hydraulic system with $p_{Sys} = 200$ bar, are driven by a 9 piston pump with a nominal flow of $Q_{Nom} = 50$ l/min running at a speed of $n_{Nom} = 1450$ rpm. The hydraulic circuit consists of a pump, one compensator, a pressure line (modelled by its hydraulic resistance, capacity and inductivity) and a hydraulic load, modelled by a constant pressure boundary condition $p_{Back} = 50$ bar.

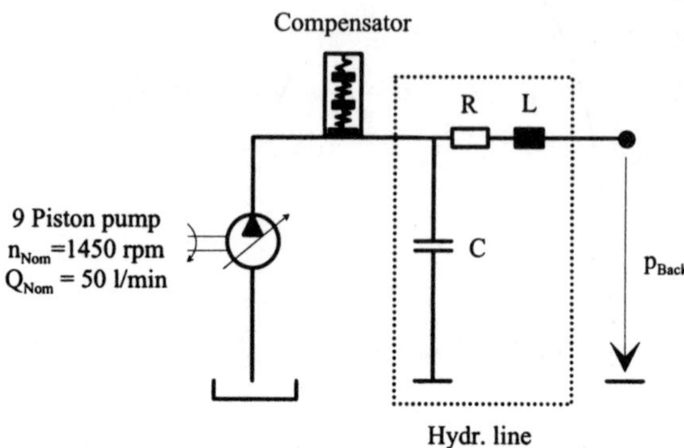

Figure 18. Structure of application example

In case of the multi DOF mass-spring compensator, a simulation study will be carried out in order to evaluate the performance of the device.

4.1   Multi degree-of-freedom mass-spring compensator

The first N harmonics of the incoming flow pulsations may be attenuated by a mass-spring oscillator of order N. Assuming a third order system ($N = 3$) and a mass $m = 0.6$ kg, the masses $m_1 = m$, $m_2 = m/2$ and $m_3 = m/3$ are given accordingly. Since the base harmonic $f = 435$ Hz (see Eq. (3) and Tab. 2) and hence $\Omega = 2\pi f = 2733$ rad/s, the spring stiffness $c_N = c$ (see Eq. (22c)) is fixed to $c = \Omega^2 m = 4.482$ kN/mm with $c_3 = c$, $c_2 = 2c$ and $c_1 = 3c$.

Since no experimental work has yet been carried out by the author, a simulation study was performed in order to estimate noise attenuation characteristics of this device. For this purpose, a model was developed in MATLAB/SIMULINK (see Fig. 19 and [7]).

For the motion of the piston, the sealing gasket acts as a damper assuming Newtonian behaviour of the hydraulic fluid in the gap between m_N and the

Table 2. Parameters of application example

Parameter		Unit	Value	Description
Hydr. Pump	n_{Pist}	-	9	Number of pistons
	n	rpm	1450	Pump speed
	Q_N	l/min	50	Rated pump flow
Hydr. Fluid	B	bar	16000	Bulk modulus
	ρ_F	kg/m^3	850	Mass density
	ν_{Fluid}	mm^2/s	46	Kinematic viscosity
System pressure	p_{Sys}	bar	40 and 200	Maximum system pressure
Compensator material	E	N/mm^2	2.1E5	Young's modulus of steel
	ν	-	0.3	Poisson's ratio
	ρ	kg/m^3	7800	Mass density of steel
	σ_{max}	N/mm^2	500	Maximum permissible stress
Pressure line	l_L	m	1	Length of pressure line
	d_L	mm	50	Diameter of pressure line
	R_H	$\frac{bar}{l/min}$	4×10^{-3}	Hydraulic resistance

cylinder surface. The damping ratio d due to the sealing gap is

$$d_3 = \underbrace{\rho\,\nu_{Fluid}}_{\eta_{Fluid}} \frac{D\,\pi\,l}{dsp} \qquad \text{where} \qquad dsp = \frac{D_{Cyl} - D_{Piston}}{2}. \qquad (54)$$

Because the magnitude of the damping ratio d_3 is crucial for the performance of the compensator, the influence of different dimensionless damping ratios $\bar{\zeta}$ (see Eq. (25)) has been investigated.

As can be seen in Tab. 3, good noise attenuation levels L_Q (see [2])

$$L_Q = 10\log\left[\frac{Q_i}{Q_{i\,ref}}\right]^2 = 20\log\frac{Q_i}{Q_{i\,ref}} \qquad (55)$$

may be achieved even with fairly high damping ratios $\bar{\zeta}$, where Q_i is the attenuated pulsation at the ith harmonic (compensator in operation) and $Q_{i\,ref}$ is the pulsation without compensation.

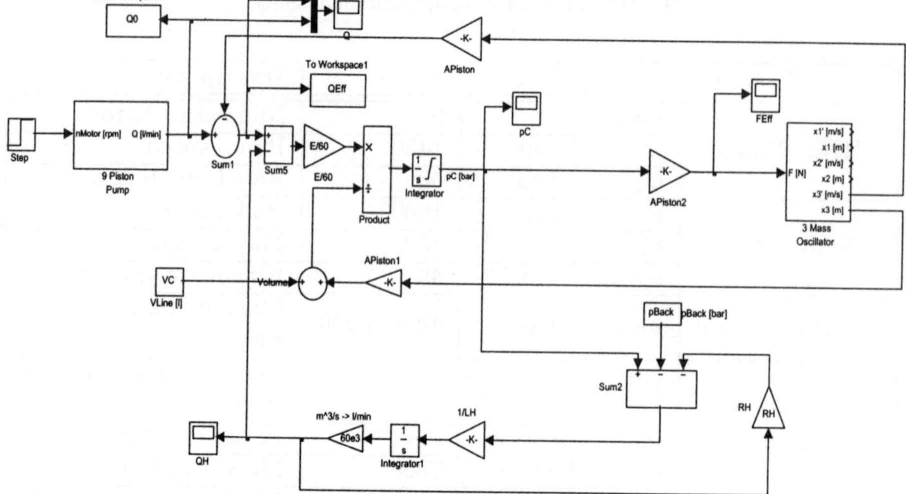

Figure 19. Simulation model in MATLAB

	L_Q [dB]	
	$\bar{\zeta} = 0.01$	$\bar{\zeta} = 0.1$
Ω_1	38.7	19.3
Ω_2	44.6	25.1
Ω_3	33.7	15.4

Table 3. Attenuation performance of a multi DOF mass spring resonator with $\bar{\zeta} \in \{0.01, 0.1\}$

Evaluation

$\oplus$ Adjustment of resonant frequency is simple

$\oplus$ Ability to compensate one base harmonic and $N - 1$ higher order harmonics by utilising one compensator

$\ominus$ Sealing element between m_N and the cylinder wall causes damping

$\ominus$ Expensive

4.2 Compensator based on plate/shell element

To facilitate a natural frequency of $f = 435$ Hz and to limit the equivalent stresses at the boundary to $\sigma_{max} = 500$ N/mm^2, the Eqs. (34) – (38) need to be solved for the plate thickness h and the plate radius a. Doing so results in the following geometries

p_{max} [bar]	a [mm]	h [mm]
40	13.4	1
200	150	24.5

Evaluation

$\oplus$ Simple mechanical design

$\ominus$ Compensator is tuned to one frequency only

$\ominus$ Plates become large and bulky for high system pressures and low frequencies of the pulsating pump flow

4.3 Compact $\lambda/4$ side-branch resonator

- Influence of wall flexibility

  Assuming a diameter of the resonator of $d = 40$ mm, the required wall thickness s for $\sigma_{max} = 500$ N/mm^2 is given by Eq. (44) and results in

p_{max} [bar]	s [mm]
40	0.16
200	0.8

  The frequency condition in Eq. (43) may be fulfilled by making the length of the resonator tube to

p_{max} [bar]	l [mm]
40	463
200	671

Evaluation

$\oplus$ Simple mechanical design

$\oplus$ Attenuation of one base harmonic and odd higher order harmonics

$\ominus$ Device becomes rather long for high system pressures

---

- Flexible elements in the resonator chamber

Again, assuming a diameter of the resonator of $d = 40$ mm and a pre-deformation of the element according to the deformation at $p_{Stat} = 300$ bar, the necessary thickness h of the flexible element due to p_{Sys} may be calculated using Eq. (53). In case of the low pressure system ($p_{Sys} = 40$ bar), the required thickness for $\sigma_{max} = 500$ N/mm^2 is $h = 2.5$ mm and for the conventional hydraulic system ($p_{Sys} = 200$ bar), the required thickness $h = 5.6$ mm. By making the maximum width of the oil layer surrounding the element to 5 mm (this can be achieved by making $\delta = 7$ for the low pressure system and $\delta = 75$ for the conventional hydraulic system – see Fig 20), the combined bulk modulus B'_{Tot} may be calculated according to Eq. (52).

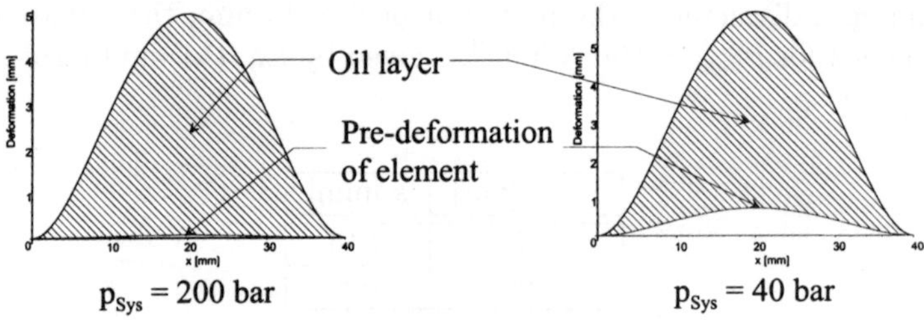

Figure 20. Element shape and oil layer for $p_{Sys} \in \{40, 200\}$ bar

In order to achieve a resonance frequency of $f = 435$ Hz, the length of the device may be calculated to

$$p_{Sys} = 200 \quad bar \qquad l = 603 \quad mm$$
$$p_{Sys} = 40 \quad bar \qquad l = 268 \quad mm.$$

Evaluation

$\oplus$ Attenuation of one base harmonic and odd higher order harmonics

$\ominus$ Device becomes rather long for high system pressures

5 Conclusions

In this contribution, several compensators, such as multi degree-of-freedom mass spring oscillators, compact $\lambda/4$ side branch resonators and compensators based on plate/shell elements, for the attenuation of fluid flow pulsations were investigated. Amongst these concepts, two designs appear particularly promising. First of all, multi DOF mass spring oscillators allow the attenuation of a base harmonic and higher order harmonics of it in a single device. Although a solution for the frequency tuning of multi DOF systems was presented by the author, the required spring stiffnesses may be problematic for standard spring designs. Secondly, $\lambda/4$ side branch resonators, especially with additional flexible elements in the resonator chamber, build fairly compact and allow the attenuation of a base harmonic and odd higher order harmonics in a single device.

In general, even at this present time, solid body compensators appear to be a vital and cost effective alternative to existing conventional devices, such as accumulators, especially at frequencies $f \geq 400$ Hz. It is believed that with the advent of switching techniques (switching frequencies of up to 1 kHz appear to be feasible in the near future) in hydraulics, solid body compensators will become a ubiquitous part of hydraulics.

Acknowledgements

The author wishes to thank Prof. R. Scheidl, without whom this work would not have been possible. Furthermore, he wants to thank M. Garstenauer, N. Krimbacher and B. Manhartsgruber for many fruitful discussions.

Nomenclature

a	Radius of plate	m
A	Area	m^2
b	Width	m
B	Bulk modulus	N/m^2
β	Compression coefficient	m^2/N
c	Spring stiffness	N/m
C	Acoustic capacity	m^5/N
c_S	Speed of sound	m/s
d	Damping ratio	$\frac{N}{m/s}$
δ	Ratio of V_{Oil}/V_{0Elem}	$-$
E	Young's modulus	N/m^2
η_{Fluid}	Dynamic viscosity	$N\ s/m^2$
f	Frequency	Hz
F	Force	N
h	Thickness of plate	m
l	Length	m
L	Acoustic inductivity	kg/m^4
λ	Wave length	m
L_Q	Logarithmic measure of noise attenuation	dB
m	Mass	kg
M	Bending moment per unit length	$\frac{N\ m}{m}$
n	Pump speed	rpm
N	Order of system	$-$
n_{Pist}	Number of pistons	$-$
ν	Poisson's ratio	$-$
ν_{Fluid}	Kinematic viscosity	mm^2/s
p	Pressure	Pa
Q	Volume flow	m^3/s
Q_{spec}	Specific pump flow	m^3/rpm
ρ	Specific mass	kg/m^3
s	Wall thickness	m
σ	Mechanical stress	N/m^2
V	Volume	m^3
ω	Angular frequency	rad/s
x	Displacement	m
ζ	Dimensionless damping ratio	$-$

References

1. Backé, W. and Murrenhof, H. (1994) *Grundlagen der Ölhydraulik*, Lecture notes: Institute for Fluidpower, RWTH Aachen, Germany.
2. Beranek, L. L. and Vér, I. L. (1992) *Noise and Vibration Control Engineering: principles and applications*, John Wiley, New York.
3. Dodson, J. M., Dowling, D. R. and Grosh, K. (1998) "Experimental investigation of quarter wavelength silencers in large-scale hydraulic systems", *Noise Control Engineering Journal*, **46**(1), 15.
4. Edge, K. and Darling, J. (1988) "A theoretical model of axial piston pump flow ripple", *Bath Workshop on Power Transmission and Motion Control*, University of Bath, UK.
5. Garstenauer, M., Grammer, S. and Scheidl, R. (1996) "The Resonance converter - A novel method for hydraulic fluid power control", *Proceedings of Mechatronics '96*, University of Skövde, Sweden.
6. Kojima, E. and Ichiyanagi, T. (1998) "Development research of new types of multiple volume resonators", *Bath Workshop on Power Transmission and Motion Control*, University of Bath, UK.
7. The MathWorks Inc. (1996) *Using Matlab*, Natick, Massachusetts.
8. Mikota, J. (1999) "Frequency tuning of chain structure multi-body oscillators to place the natural frequencies at Ω_1 and $N-1$ integer multiples $\Omega_2 \ldots \Omega_N$", *Proceedings of GAMM*, Germany.
9. Mikota, J. and Scheidl, R. (1999) "Solid body compensators for the filtering of fluid flow pulsations in hydraulic systems", *Mechatronics and Robotics'99*, TU Brno, Czech Republic.
10. Parkus, H. (1988) *Mechanik der festen Körper*, Springer, Germany.
11. Pilkey, W. D. (1994) *Formulas for stress, strain and structural matrices*, John Wiley and Sons, New York.
12. Scheidl, R. and Riha, G. (1999) "Energy efficient switching control by a hydraulic resonance converter", *Bath Workshop on Power Transmission and Motion Control*, University of Bath, UK.
13. Strang, G. (1988) *Linear Algebra and its applications*, Third Ed., Harcourt College Publishing, ISBN 0155510053.
14. Ziegler, F. (1991) *Mechanics of solids and fluids*, Springer.
15. Zollner, M. and Zwicker, E. (1998) *Elektroakustik*, Springer.
16. Zurmühl, R. and Falk, S. (1986) *Matrizen und ihre Anwendungen*, Springer, Germany.

SOME ASPECTS OF WASHING COMPLEX NON-LINEAR DYNAMICS

M. BOLTEŽAR

*University of Ljubljana, Faculty of Mechanical Engineering,
Aškerčeva 6, 1000 Ljubljana, Slovenia
E-mail: miha.boltezar@fs.uni-lj.si*

A nonlinear planar centrifugally excited oscillatory system was studied in its
steady-state domain. The integral of the correlation dimension and Lyapunov exponents were used to describe the motion of the model in the phase space. Power
spectral and bispectral analyses have further been used to analyse the behaviour
of the model in the frequency domain. The experimental work has been under-
taken on the washing-machine washing complex dynamics to verify the theoretical
approach.

1 Introduction

Non-linear modelling of vibrations is receiving more and more attention in
both theoretical and applied research. New findings in the theory of deterministic chaos over the last two decades are the result of research in several
scientific disciplines. In many cases the signal-processing tools that are being developed by the mathematical community are beginning to penetrate
technical areas.

In our approach we will concentrate simultaneously on two topics of nonlinear dynamics: phase-space analyses and higher-order spectral analyses.
Each topic has a significant area of application in various technical subjects.
A number of applications for the theory of deterministic chaos exist in mechanical engineering, these include cutting dynamics, [1,2,3], the modelling of
dry friction as a dissipation mechanism [4], which includes stick-slip vibrations
[5] and impact oscillations [6,7]. Ideas from the theory of non-linear dynamical systems are widely used in medicine, applications include analysing EEG
signals [8,9,10,11] and heart period dynamics [12,13].

In the last 20 to 30 years the power spectrum has come to be seen as a
powerful tool for analyzing both deterministic and stochastic technical processes. When estimating the power spectrum the assumption is made that the
frequency components are uncorrelated, thus compressing the phase relations
between them. This information is sufficient for a statistical description of a
signal with a Gaussian probability density function.

In many situations, however, we must deal with nonlinear systems, where
second-order spectral analysis is insufficient. In these cases higher-order spec-

tra, also known as polyspectra, must be used. Special cases of higher-order spectra are bispectra and trispectra, these spectra are based upon third- and fourth-order Fourier transformations and contain information on the skewness and kurtosis of the probability distribution.

Over the last ten years the use of bispectra has increased. Bispectra are used to provide an additional insight into nonlinear processes, in particular for the identification of non-Gaussian processes, filtering out Gaussian noise, the identification of non-minimal phase systems, the identification of certain types of non-linearities, estimating the number of harmonics, amplitudes and phases in a process, time-delay estimations between measurement sensors and estimating transfer functions.

Bispectra are now finding uses in the fields of speech recognition [14], sonars [15], radars [16], geophysics [17,18] and image processing [19]. Important and growing areas of application for higher-order spectral analysis are condition monitoring [20,21,22,23,24,25,26] and cutting dynamics [27,28]. Bicoherence analyses can also be found in medical applications [29,30].

Our approach tends to combine both phase-space analyses and bispectral analyses. Such a combined approach has so far been reported in medicine [31], fluid dynamics [32] and mechanical non-linear oscillations [33,34,35].

The aspects of non-linear dynamics of washing complex presented in this chapter are the result of many years of research dealing with the modelling of washing-machine dynamics. Our earlier research originally included a simpler model [36], which was followed by the study of spin-up through resonance [37] and the modelling of the washing machine's housing [38]. The first analyses in the phase space were further reported in 1997 [39] and 1998 [40]. Bispectral analyses of washing complex dynamics were reported in [41,42].

The chapter is organised into two major sections, with theoretical modelling described in the first. The theory covers the evolution of the equations of motion, their numerical integration, the calculation of correlation dimensions in real and reconstructed phase spaces as well as the estimation of the largest Lypunov exponents. In the second part an experimental verification of the numerically computed results is given. The same measures were estimated on an experimental time series. To finish, a comparison between the numerical and experimental results is presented.

2   Theoretical modelling

The modelling of nonlinear dynamics has been attracting increasing attention in recent years. Modern analytical tools - specifically the analyses in the phase space developed primarily in the basic scientific disciplines - are now entering into the applied technical sciences [43,44,45,4,46,47,48,49,50,40,33]. On the other hand, bispectral analysis is a part of the rapidly expanding field of Higher Order Statistics and can provide information on the quadratic type nonlinearities in the signal. The two major sections of this chapter endeavour to combine both approaches, using them firstly on a theoretical model and secondly on real technical application.

Dry friction frequently occurs in real technical systems. Its impact on the dynamics of the single DOF system is shown in [43,45,4,51,52] and on systems with several DOF in [44,46,49,50,40,53]. The possibility of using the integral of correlation dimension (ICD) as a mean of quantitative phase space analysis to interpret aperiodic time series of the system with dry friction is shown in [44]. The comparison of the correlation dimension between the real and reconstructed phase space of the mechanical model with several DOF is given in [46,49,40]. Lyapunov exponents represent another measure for estimating the divergence of trajectories in phase space. They are used to detect the chaotic behaviour in single [43] and in several [44,40] DOF models.

The statistical stability of the bispectrum estimate is achieved by dividing the time series into the segments for averaging. Because the variance of the bispectrum's estimate is dependant upon the second order properties [15,54,55], the bispectrum is normalised into the skewness function [15] or bicoherence [14]. When normalising, attention should be paid to the problem of division by small numbers. One way of dealing with this problem is to add a low level white Gaussian noise to the signal [14]. In this chapter both the numerator and the denominator of the bicoherence estimate are compared to the numerical threshold for zero value, thus allowing for a check on the possibility of dividing by small number.

In the experimentation, a washing machine was used as the application from the real engineering world. The washing machine's suspension optimisation was conducted [53] on the washing complex modelled as the rigid body with 6 DOF. Our intention is to compare the results of the theoretical modelling of the washing machine's washing complex dynamics, as described in this section, with the measured responses in the time, frequency and the phase space domains as described in Experiment section.

2.1 *Description of the model*

The model (see Figure 1) consists of two rigid bodies. The first rigid body, with its centre of gravity at point **T**, is called the basic body. Its links to the surroundings consist of a linear spring, linear viscous damper and element with implemented dry friction. All of the elements of vibroisolation are collinear with the axis of the coordinate system $X - Y$. The second rigid body, with its centre of gravity at point $\mathbf{T}_e$, is called the rotor. It is driven by torque $M(t)$ and attached to the basic body by bearings at point $\mathbf{S}$.

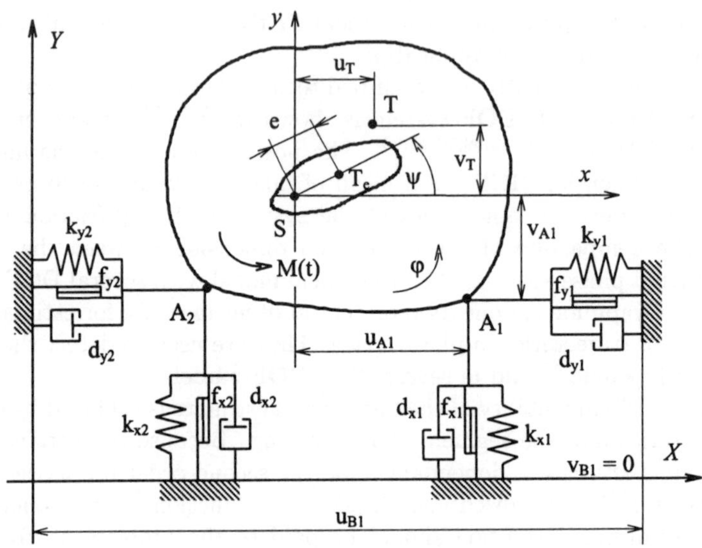

Figure 1. The model.

The model has four degrees of freedom. Coordinates x, y and φ are needed to determine the motion of the basic body. The x and y describe the horizontal and vertical translation of the basic body, respectively, and the φ describes its rotation. To characterise the rotation of the rotor the additional coordinate ψ is used. The φ and the y start at X axis. In order to obtain equations of motion, the Lagrangian equations of second order were used, Eq. (1). The motion takes place in configuration space, hence $q_1 = x$, $q_2 = y$, $q_3 = \varphi$, $q_4 = \psi$ and so

$$\frac{\mathrm{d}}{\mathrm{d}t}\frac{\partial T}{\partial \dot{q}_j} - \frac{\partial T}{\partial q_j} = Q_j - \frac{\partial V}{\partial q_j} \quad , \qquad j = 1,\ldots,4 \tag{1}$$

where T denotes kinetic energy, V denotes potential energy and Q_j are generalised nonconservative forces. The definitions of the used functions are

$$\begin{aligned} h_{xi} &= u_{A_i}\cos\varphi - v_{A_i}\sin\varphi & h_{yi} &= v_{A_i}\cos\varphi + u_{A_i}\sin\varphi \\ t_x &= u_T\cos\varphi - v_T\sin\varphi & t_y &= v_T\cos\varphi + u_T\sin\varphi \end{aligned} \tag{2}$$

where u's and v's are described in figure 1. The kinetic energy of the system can be written as

$$T = \tfrac{1}{2}J_T\dot\varphi^2 + \tfrac{1}{2}J_{T_e}\dot\psi^2 +$$
$$\tfrac{m}{2}\left[\dot x^2 + \dot y^2 + (u_T^2 + v_T^2)\dot\varphi^2 + 2\dot\varphi(t_x\dot y - t_y\dot x)\right] + \tag{3}$$
$$\tfrac{m_e}{2}\left[\dot x^2 + \dot y^2 + e^2\dot\psi^2 + 2e\dot\psi^2(\dot y\cos\psi - \dot x\sin\psi)\right]$$

and the potential energy as

$$V = mg(y + t_y) + m_e g(y + e\sin\psi) +$$
$$\tfrac{1}{2}\sum_{i=1}^{2} k_{xi}(L_{xoi} - L_{xi})^2 + \tag{4}$$
$$\tfrac{1}{2}\sum_{i=1}^{2} k_{yi}(L_{yoi} - L_{yi})^2$$

where L_{xoi} and L_{yoi} are the undeformed lengths of the springs in the X and Y directions as denoted by their indices, and L_{xi} and L_{yi} are deformed lengths of the same springs at time t:

$$L_{xi} = \begin{cases} L_{xoi}, & L_{xi} \geq L_{xoi} \\ |\, u_{B_i} - x - h_{xi} \,|, & L_{xi} < L_{xoi} \end{cases}$$
$$L_{yi} = \begin{cases} L_{yoi}, & L_{yi} \geq L_{yoi} \\ |\, v_{B_i} - y - h_{yi} \,|, & L_{yi} < L_{yoi} \end{cases} \tag{5}$$

The generalised non-conservative forces consist of the driving torque, friction forces and viscous damping forces. The torque is given by

$$M(t) = a\psi^b, \quad a > 0, \quad b < 0 \tag{6}$$

and represents the driving force of the model, reflecting the real characteristics of the electromotor. The friction forces are given according to the discontinuous Coulomb model of dry friction as

$$F = N\mu_k \, \text{sgn}(\nu) \tag{7}$$

where N denotes normal force on the surface, F is the friction force, ν is relative velocity between surfaces in contact, and μ_k the kinetic coefficient of friction. The excitation is modelled as non-ideal and not explicitly time dependent; hence the model is autonomous.

The parameter identification was done mostly by experimentation and is described precisely in the reference [49]. Hence the parameters of the vibroisolation were determined experimentally in both horizontal and vertical directions. Consequently, the chosen configuration of the suspension simply follows the experimental approach. The details of the model's parameters can be found in table 1.

Table 1. The parameters of the model

Rotor			
Rotor	m_e	6.2	kg
	J_{T_e}	0.1939	kgm^2
	e	0.01719	m
Basic body	m_t	42.805	kg
	J_T	2.6811	kgm^2
	u_t	-0.00854	m
	v_t	0.04387	m
	u_{A_1}	0.239	m
	v_{A_1}	-0.166	m
	u_{A_2}	-0.239	m
	v_{A_2}	-0.166	m
Torque	a	119623.5	/
	b	-2.45023	/
Vibroisolation	k_{x_1}	7484.591	N/m
	k_{x_2}	14969.182	N/m
	k_{y_1}	9701.845	N/m
	k_{y_2}	10990.797	N/m
	d_{x_1}	79.3	Ns/m
	d_{x_2}	79.3	Ns/m
	d_{y_1}	107.5	Ns/m
	d_{y_2}	107.5	Ns/m
	N_{x_1}	12.222	N
	N_{x_2}	15.959	N
	N_{y_1}	116.281	N
	N_{y_2}	151.837	N
	μ_k	1	/

2.2   The results of numerical simulation

The set of four equations of motion Eq.(1) has been transformed into a system of eight ordinary differential equations of first order. The numerical integration of the latter was carried out with Runge-Kutta method with error estimation of $\mathcal{O}(h^5)$, where h denotes the constant integration step. The numerical integration always starts at the same initial conditions, which are defined by stable static equilibrium and zero velocities. For a certain set of geometrical and material parameters of the model, the time histories of accelerations in horizontal (x) and vertical (y) directions are presented in figure 2.

The time histories of integrated accelerations show that motion of the model has one dominant frequency; hence the frequency of spin dry. The signals of accelerations in horizontal and vertical direction are shifted by $\pi/2$; hence the model's response follows the centrifugal excitation.

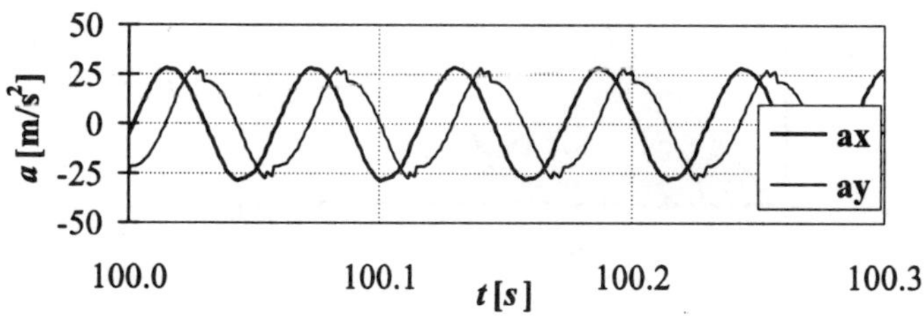

Figure 2. Numerically integrated time histories of accelerations; a_x is horizontal and a_y is vertical acceleration.

Spectral analysis

Upon assuming that $x(n)$, $n = 0, \pm1, \pm2, \ldots$, is a real, stationary and random process, the discrete third order cumulant spectrum or bispectrum $B(2\pi f_1, 2\pi f_2)$ of $x(n)$ is defined as [19]

$$B(2\pi f_1, 2\pi f_2) = \sum_{\tau_1=-\infty}^{\tau_1=+\infty} \sum_{\tau_2=-\infty}^{\tau_2=+\infty} c_3(\tau_1, \tau_2) e^{-j(2\pi f_1 \tau_1 + 2\pi f_2 \tau_2)} \tag{8}$$

where $c_3(\tau_1, \tau_2)$ is the third order cumulant of $x(n)$. The alternative approach is to construct a consistent bispectrum estimate $\hat{B}$ using the DFT $X(k)$ of a signal $x(n)$, $n = 0, 1, \ldots, N-1$ [15]. When using this technique, the signal of length N points is subdivided into K segments having M points ($N = K \cdot M$). The DFT of each segment is calculated and averaged over K segments:

$$\hat{B}(k_1, k_2) = \frac{1}{K} \sum_{i=0}^{i=K-1} X_i(k_1) X_i(k_2) X_i^*(k_1 + k_2) \tag{9}$$

where k_1 and k_2 denote indices of frequency, i is the index of the segment and X^* is the complex conjugate of X. If the highest frequency component f_{max} of the process $x(n)$ complies with Nyquist's sampling condition ($f_{max} \leq f_N = f_s/2 = 1/2\,\Delta t$), the principal domain of $B(2\pi f_1, 2\pi f_2)$ is composed of inner (IT) and outer (OT) triangles [56] (see Figure 3).

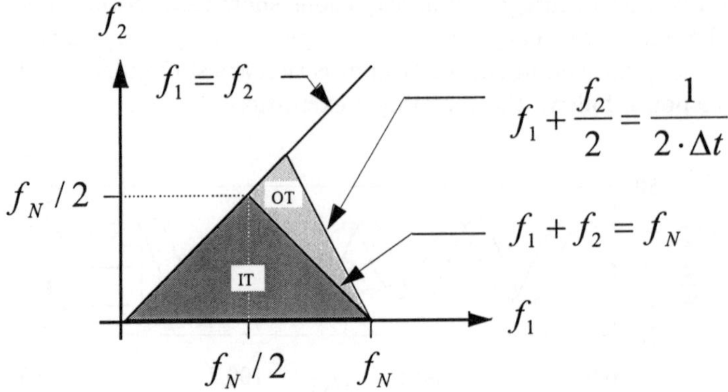

Figure 3. The principal domain of the discrete bispectrum.

The variance of the bispectrum estimate for stochastic processes can be estimated using the following expression [15]:

$$\mathrm{var}\left\{\hat{B}(k_1, k_2)\right\} \propto M P(k_1) P(k_2) P(k_1 + k_2) \tag{10}$$

where P stands for the power spectrum. The bispectrum estimate is therefore sensitive to both second and third order properties. To eliminate the sensitivity of the bispectrum estimate to second order properties a new measure, called the skewness function was introduced [15]. However, for the bispectra of signals, conforming to the cosine model [54,14]:

$$x(n\,\Delta t) = \sum_i A_i \cos(2\pi f_i n\,\Delta t + \phi_i) + noise_{Gauss}(n) \tag{11}$$

an alternative normalisation, called bicoherence is often used [55,18]:

$$\hat{b}^2(k_1, k_2) = \frac{|\hat{B}(k_1, k_2)|^2}{\frac{1}{K}\sum_{i=0}^{i=K-1} |X_i(k_1) X_i(k_2)|^2 \, \hat{P}(k_1 + k_2)} \tag{12}$$

In equation (11), A_i and ϕ_i denote random amplitude and random phase, generated at the beginning of each segment, and f_i denotes the frequency of the ith mode.

One useful property of the bicoherence is that it is bounded between zero and unity, a property which the skewness function does not share. The bicoherence $\hat{b}(k_1, k_2)$ can be interpreted as a portion of power due to the quadratic phase coupling (QPC) of frequency components k_1 and k_2 [55]; thus when QPC takes place bicoherence should be close to unity. To distinguish between low, but nonzero, bicoherence values and truly zero bicoherence values, the minimal significant value of the bicoherence must be calculated

$$\hat{b}^2_{sig} = -\frac{\ln(1 - T_{\alpha\%})}{K} \tag{13}$$

where $T_{\alpha\%}$ is the significance in %. The $\hat{b}^2_{sig}$ is based on the assumption that the process involved is Gaussian, in which case the bicoherence should be approximately χ^2 distributed with two degrees of freedom [17].

When dealing with signals, conforming to the sinusoidal model, the assumption of phase randomisation between various sinusoidal components is assumed. If the phases of the various sinusoidal components are constant for all segments, bicoherence can peak even if no QPC takes place [14]. To overcome this problem Fackrell proposed testing the phase of the bispectrum. It was shown [14] that in the case of QPC, the phase of the bispectrum is zero. Assuming that the phase of the bispectrum is approximately normally distributed around the true phase [18], the maximal phase for a given significance level $T_{\theta\%}$ can be determined. The $T_{\theta\%}$ is determined from the standard normal tables $N(\mu, \sigma)$, where μ denotes mean value and σ a standard deviation:

$$\theta_{sig}(k_1, k_2) = \frac{T_{\theta\%})}{\sqrt{2K}} \sqrt{\frac{1}{\hat{b}^2(k_1, k_2)} - 1} \tag{14}$$

Thus only the bicoherence values greater than equation Eq.(13) and lower than equation Eq.(14) are considered for QPC detection.

The power spectra were calculated using 4096 FFT points per segment. Overlapping of 5% was used to obtain 128 segments and the Hamming window was applied in the time domain. The power spectra of horizontal and vertical acceleration are presented in figures 4 and 5. In the horizontal direction the spin dry frequency of 17.57 Hz is clearly visible. All other frequencies contain significantly lower power. In the vertical direction (see Figure 5) the spin dry frequency of 17.57 Hz and its higher harmonics at 52.73 Hz ($3\times$) and 87.4 Hz ($5\times$) can be extracted. All other frequencies have a power lower than -70 dB. It is interesting to note that only odd harmonics appear. This is due to the fact that the signal meets the requirement of symmetry of the third kind in Fourier analysis.

The bicoherences of horizontal and vertical accelerations were calculated using 1024 FFT points per segment, giving a frequency resolution of 1.95 Hz. Overlapping of 5% was used to obtain 513 segments for averaging. Because of its ability to resolve QPC peaks [14,57] Hamming window was applied in the time domain. The number of segments and the selected significance level

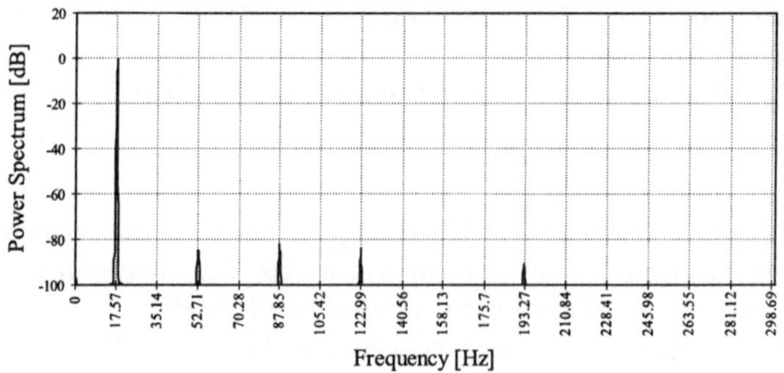

Figure 4. Power spectrum of horizontal model's acceleration.

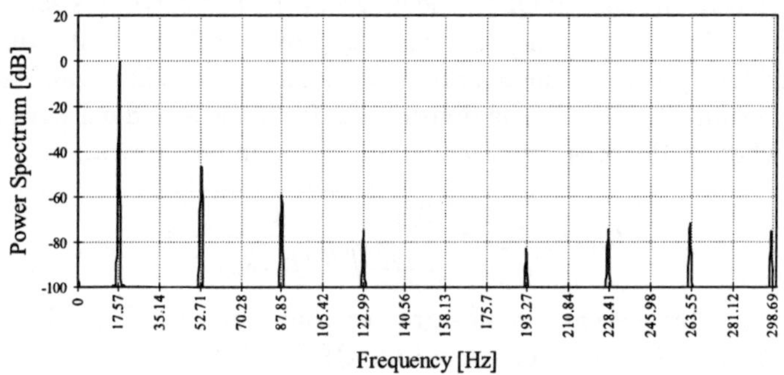

Figure 5. Power spectrum of vertical model's acceleration.

of $T_{\alpha\%}$ defined the minimal significant bicoherence value of 0.0058. The significance level of $T_{\theta\%}$ was used to determine the maximal allowed phase of the bispectrum. The bicoherences of horizontal and vertical acceleration are presented in figures 6 and 7.

In both figures a "wall" of bicoherence values ≈ 0.35, ≈ 0.23 is clearly visible. This is the result of the normalising process. When normalising, the bispectrum is divided by the power spectrum and by another factor, equation (12). When the denominator is small, the problem of the division by small number occurs and can result in high bicoherence values. Because of the division, the bicoherence can peak even if the numerator is very small or close to zero. We must bear in mind that the bicoherence gives the proportion of the

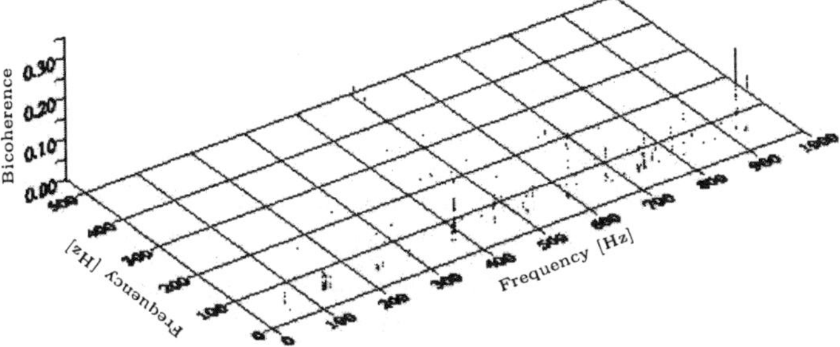

Figure 6. Bicoherence surface plot of horizontal model's acceleration.

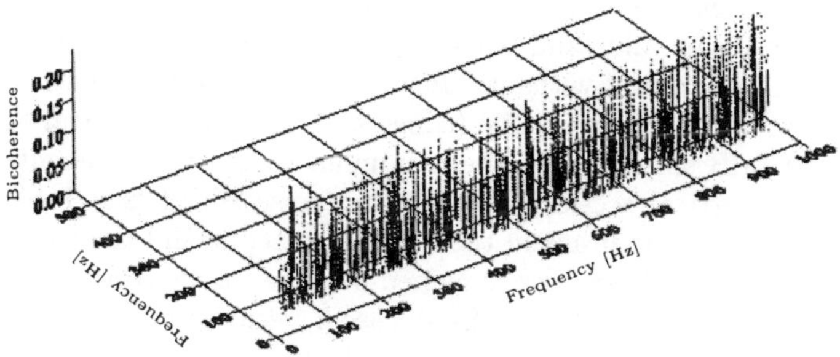

Figure 7. Bicoherence surface plot of vertical model's acceleration.

power due to the QPC of components k_1 and k_2, irrespective of the power of
these two components. If there is a large amount of QPC between frequencies
with small power, the QPC between frequencies with high power would be
among the large number of high bicoherence values. In other words, no matter
how small may be the power of components k_1 and k_2, if the phases of these
two components quadratically couple, the bicoherence exhibits a peak. The
question is: should the bicoherence peak if the power of the components k_1

and k_2 is significantly lower than the power of the largest components? If the power spectrum is composed of several distinct high values and has virtually no power elsewhere, the problem of the division by small number can occur even more often. One way of tackling this problem is to add low level white Gaussian noise to the signal [14]. This ensures that there is always some power in the spectral component at each frequency, and so even if the bispectrum is zero, the denominator is always greater than zero. The drawback of this approach is that it reduces the bicoherence values.

In this chapter an alternative approach to determining the numerical threshold for zero value has been implemented. Whenever the numerator is smaller than this threshold the bicoherence is set to zero. If the numerator is greater, and the denominator smaller than this threshold, a small, constant and positive value is added to the denominator. By increasing the numerical threshold the number of divisions by small number is decreased. However, it is difficult to determine when to stop increasing the numerical threshold.

In order to compare both approaches, the bicoherence of the vertical accelerations has been estimated twice; first on a signal with added Gaussian noise with zero mean value and 0.5 variance, ... SNR = 29 dB (called case a), and second on a signal without added Gaussian noise but with 1E-10 as a numerical threshold for zero value (called case b). The numerical threshold of 1E-10 is 182 dB smaller than the maximum value of the magnitude bispectrum of the vertical accelerations with no noise added (see Table 2). In case a the number of the peaks in the wall decreased as did the values of the bicoherence. The "wall" however remained (see Figure 8). Applying only numerical threshold for zero value of 1E-10, case b, resulted in a great reduction in the number of bicoherence peaks: the "wall" could hardly be distinguished, while the highest bicoherence value only slightly decreased (≈ 0.22); compare Figures 7 and 9. Moreover, some of the highest peaks in the bicoherence (see Table 3) were at the same frequency as the peak of the magnitude bispectrum (see Table 2). The effectiveness of our approach is clearly evident.

Table 2. Magnitude bispectrum values for horizontal and vertical acceletations of the model

Horizontal	0.21 at (17.57, 17.57) (Hz)
Vertical	0.13 at (17.57, 17.57) (Hz)

The magnitude bispectra of both horizontal and vertical accelerations have also been calculated. For both accelerations only QPC at (17.57, 17.57) Hz were found (see Table 2). Thus the QPC of the basic harmonic partly generates second harmonics, the power of which is here negligible

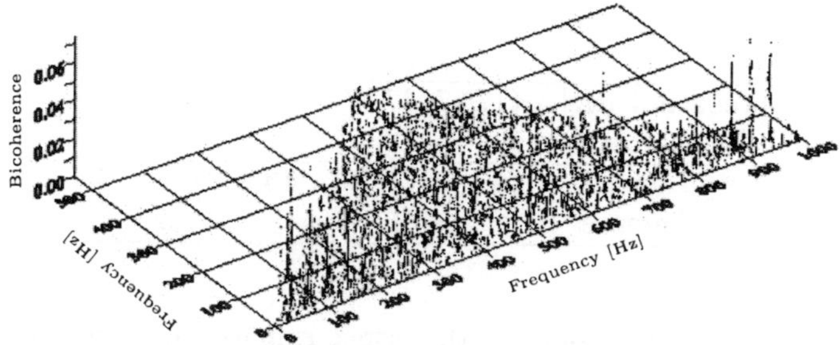

Figure 8. Bicoherence surface plot of vertical model's acceleration. Gaussian noise with zero mean value and variance 0.5 was added to signal.

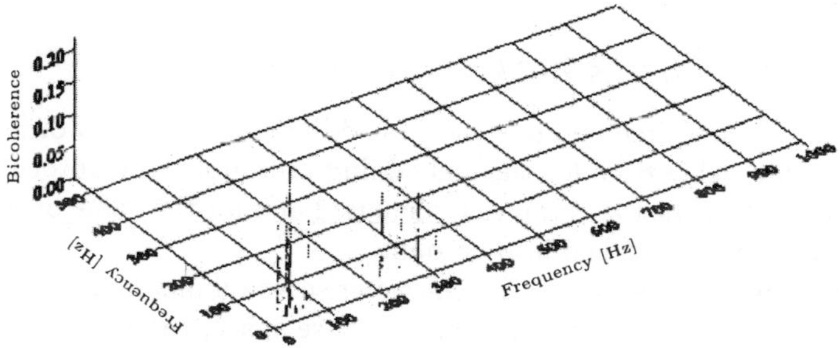

Figure 9. Bicoherence surface plot of vertical model's acceleration. No added Gaussian noise. Numerical treshold for zero value was 1E-10.

(see Figures 4 and 5).

The magnitude bispectra were calculated because it has been found useful to check magnitude bispectrum for QPC [57] when the number of bicoherence peaks is large. This is due to the fact that no division is needed to calculate the bispectrum, thus eliminating the problem of division by a small number.

Table 3. Bicoherence values of vertical accelerations. Numerical threshold for zero value 1E-10

Frequency [Hz]	(17.57, 17.57)	(52.73, 35.17)
Bicoherence	0.221	0.221
Frequency [Hz]	(87.89, 35.17)	(332.03, 35.17)
Bicoherence	0.221	0.222

However, the statistical stability of the bispectrum estimate requires that the bispectrum's number of degrees of freedom be larger than 120 [14,57].

Phase-space analysis

Although the integration of the full set of equations of motion allows one to create the real phase space, our interest lies also in the reconstruction process [58,59] on the basis of a single numerically integrated time history of the system's dynamics. From a single measured time history $q(t)$ one can create delay vectors

$$\mathbf{X}(t) = \{q(t), q(t-\tau), \ldots, q(t-(d-1)\tau)\} \tag{15}$$

where τ is the delay time and d is the embedding dimension. In order to find the appropriate delay time for the embedding procedure, the autocorrelation function of the time history $q(t)$ was calculated and for the value of τ the first decorrelation time was taken into account.

Correlation dimension. To characterise the dimensionality of an attractor in real as well as in reconstructed phase space, the Integral of Correlation Dimension (ICD) was estimated. It is defined as [60,61]

$$ICD(N,\ell) = \frac{2}{N(N-1)} \sum_{i=1}^{N-1} \sum_{j=i+1}^{N} \mathcal{H}\left(\ell - \|\mathbf{X}_i - \mathbf{X}_j\|\right) \tag{16}$$

where N denotes the number of points of the phase space, $\mathcal{H}$ is the Heaviside function, ℓ characteristical length and $\|\mathbf{X}_i - \mathbf{X}_j\|$ denotes the distance between two points $\mathbf{X}_i$ and $\mathbf{X}_j$ of the phase space. This distance can be defined as

$$\|\mathbf{X}_i - \mathbf{X}_j\| = \max \mid (\mathbf{X}_i)_k - (\mathbf{X}_j)_k \mid , \qquad k = 1, \ldots, d \tag{17}$$

It was shown [60,61] that for most chaotic attractors and for small ℓ-s ICD scales like

$$ICD(N,\ell) \propto \ell^{\nu} \tag{18}$$

where ν denotes the correlation dimension of the attractor in d-dimensional phase space.

The extraction of the correlation dimension in the real phase space of the model resulted in an estimate for the correlation dimension where $\nu = 1.00 \pm 0.00$ (see Figure 10). It was then possible to compute the integrals of correlation dimension after forming delay vectors from a single numerically integrated time history. For vertical acceleration, the Integrals of correlation dimension for the three embedding dimensions are shown in figure 11. The values of the correlation dimension estimates for the different time histories are shown in table 4. They differ from 3 to 15% compared with those obtained in real phase space. The most significant difference between real and reconstructed phase space in the correlation dimension estimation is found in horizontal displacement due to decaying transient vibrations visible only in this signal [49]. This results from the inherent property of the driving torque model that was used to model the electromotor characteristics. The driving torque reaches zero value at infinity, and consequently the rotational speed reaches maximum value at infinity. Hence, the model never really reaches the steady state in the very strict theoretical sense, but after some time the growth of velocity becomes negligible from the technical point of view.

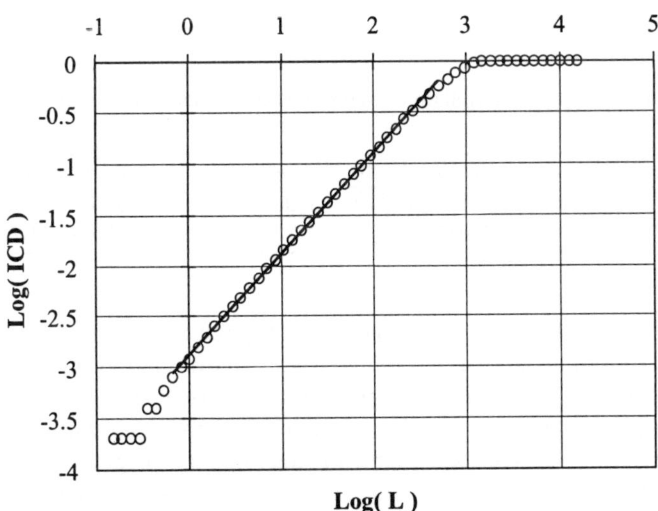

Figure 10. Integral of correlation dimension, real phase space of the model.

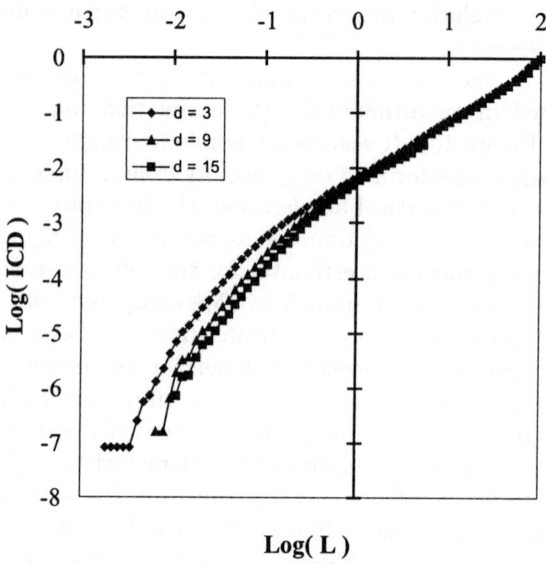

Figure 11. Integral of correlation dimension, reconstructed phase space on vertical acceleration of the model.

Table 4. Estimation of the correlation dimension of the model based on different numerically integrated time histories

	displacement	velocity	acceleration
horizontal	1.152±0.010	1.031±0.003	1.043±0.002
vertical	1.087±0.005	1.030±0.003	1.055±0.007

In the case of the real phase space, 10000 points were used (see Figure 10). All of the results in table 4 and plots in figure 11 are based on a computation using 5000 points [62] when reconstructing the phase space.

Lyapunov exponents. The Lyapunov exponents measure the exponential divergence (positive exponents) or convergence (negative exponents) of two initially neighbouring trajectories in the phase space; hence the chaotic or regular behaviour of the system under consideration could be detected [58,63,64,65]. We monitor the long-term evolution of infinitesimal n-dimensional sphere of initial conditions in n-dimensional phase space. The sphere will become an n dimensional ellipsoid. The ith Lyapunov exponent in n-dimensional phase space is defined [65] as

$$\lambda_i = \lim_{t \to \infty} \frac{1}{t} \frac{p_i(t)}{p_i(0)}, \qquad i = 1, \ldots, n \tag{19}$$

where $p_i(t)$ is the length of the ellipsoidal principal axis and $p_i(0)$ is the length of the infinitesimal sphere's axis. If at least one of the Lyapunov exponents is greater than zero, then the system exhibits chaotic dynamics.

The technique [65] for computing the complete Lyapunov spectrum of exponents directly from explicitly known equations of motion has been used. The method calculates the Lyapunov spectrum by numerical integration of n non-linear equations of motion for some post transient initial conditions and the n linearised equations of motion for n different initial conditions that define the arbitrary co-ordinate system defined by n orthonormal base vectors. Because of the exponential divergence of trajectories of the chaotic system each base vector will diverge in magnitude and tend to fall into a local direction of most rapid growth, the Gram-Schmidt reorthonormalisation [65] procedure should be used repeatedly.

The largest Lyapunov exponent was estimated as $\lambda_{max} = -0.68$ (see Figure 12); hence the model's motion is regular and the model's attractor is a limit cycle.

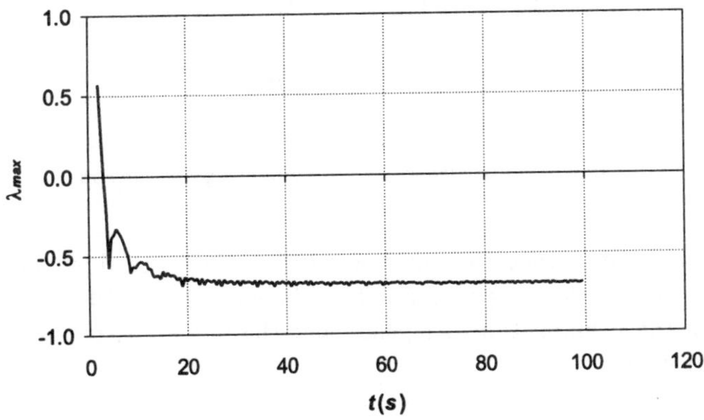

Figure 12. Convergence of the model's largest Lyapunov exponent.

Attractor visualization. The attractor visualisation enables one to visualise the attractor's shape in such a way that a trajectory in the phase space in presented in the Euclidean space of the same dimension. The major drawback of the method is that we are only able to present three-dimensional objects. In the case of multi-dimensional phase space one could present the attractor's shape in the form of two- or three-dimensional slices of multi-dimensional Euclidean space.

The cross section of real phase space of the time history of co-ordinates x and y presents an in-plane motion of the model's point **S** during spin dry (see Figure 13).

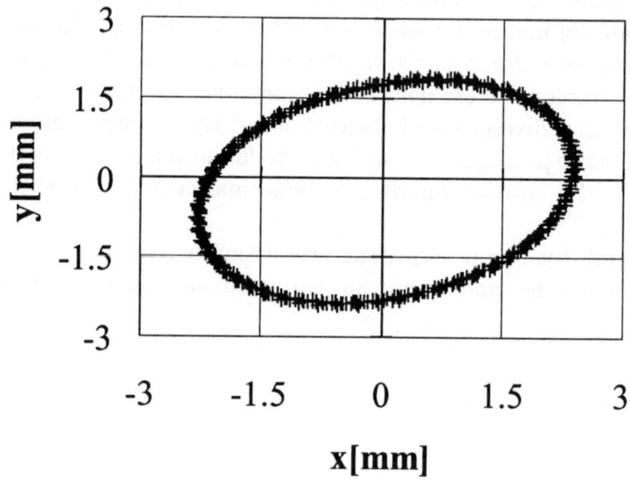

Figure 13. Visualization of the model's plane motion.

The visualised attractor reconstructed from the vertical velocity time history in three-dimensional space, $d = 3$, is shown in figure 14.

Each of the visualisations suggests that the model's attractor is likely to be the limit cycle.

2.3   Conclusions to the theoretical modelling

In this section the steady-state responses, extracted from the non-linear centrifugally excited planar oscillatory system were studied. Our work is aimed at finding a better understanding and modelling of non-linear behaviour in machine dynamics.

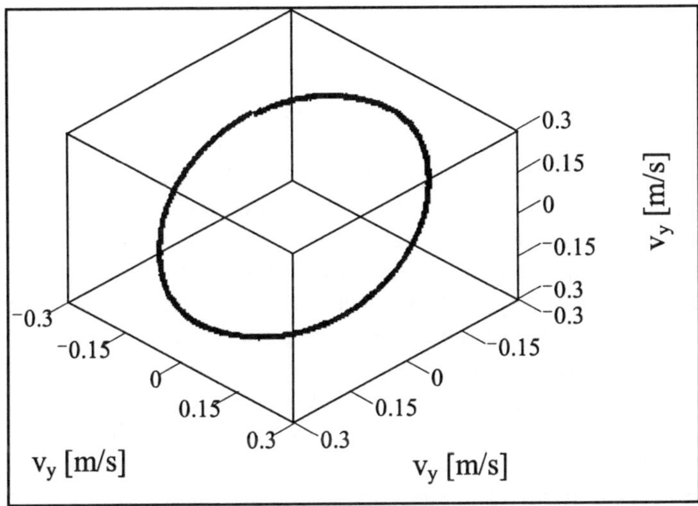

Figure 14. Visualization of the model's attractor reconstructed from the vertical velocity time history.

We have concentrated on the planar model of the washing-machine washing complex due to its negligible motions of the washing complex in the direction perpendicular to the plane of motion.

The computation of the correlation dimension of the real phase space and reconstructed phase space, based on different simulated time histories, has been carried out. The correlation dimension of the model's real phase space and of the reconstructed ones shows good agreement even though the model never really reaches the steady state in the very strict theoretical sense, but after some time the growth of velocity becomes negligible from the technical point of view. This phenomenon is the consequence of the torque model used, and is reflected in the slightly higher estimation of the correlation dimension from the signal of the horizontal displacement due to decaying transient vibration. The estimated values of correlation dimension of the model's attractor point towards the limit cycle as being the attractor's shape.

The estimated negative value of the largest Lyapunov exponent, computed directly from explicitly known equations of motion, is reflected in the model's regular nature of motion. It is impossible to visualise the complete model's attractor on account of its eight dimensional phase space. The visualisation of the section of the real phase space and the sections of the reconstructed one reveals the attractor as a limit cycle, thus showing good agreement with visualisations of the model's real phase space and of the reconstructed ones. Each of the conducted phase space analyses confirms that the model's dynamics is regular and that its attractor is a limit cycle.

The power spectral and bispectral analysis was used next to analyze the dynamical model of the washing complex. The dominance of the first mode has been found, while other modes have significantly lower power values. When calculating bicoherences, attention should be paid to the sensitivity to division by small numbers. This sensitivity can greatly influence the calculated results. In our study this effect was interpreted as the main reason behind the large number of the bicoherence peaks ("wall"). To overcome this problem other authors proposed adding Gaussian noise to the signal. However, this resulted in lower bicoherence values.

We propose an alternative approach. A numerical threshold for zero value was introduced, and both numerator and denominator were checked for near zero values. If the numerator was lower than this threshold, bicoherence was set to zero. If the numerator was greater than the threshold, the denominator was checked for the possibility of division by small number. If it was smaller than the threshold, a small positive constant was added. This approach significantly reduced the number of divisions by small number.

3　Experiment

In this section, experimental results are given to verify the deduced theoretical model from Theoretical modelling section.

The problem in experimental investigations of frictional systems lies mainly in the very complex frictional mechanism [44,52,49]. The stick-slip phenomenon was discussed in [52] in the case of a single DOF system with one and two stops per half-cycle. Another experimental example is described in [4] of the chaotic dynamics of a harmonically forced spring-mass system with dry friction designed to vary linearly with displacement.

For oscillating systems with several degrees of freedom, the influence of the degree of non-linearity to the integral of correlation dimension (ICD) was shown experimentally in the case of a series of coupled rigid bodies [66] in planar motion. On the other hand, the values of the correlation dimension were shown to be dependent on the cutting force in cutting processes where dry friction plays an important role [50]. It was also shown that fault diagnosis of the rolling element bearings may be made by using the correlation dimension [48].

To establish the chaotic nature of the processes, including the friction forces, the Lyapunov exponents were computed [44,40] from a single measured time history. It was shown that Lyapunov exponents may not be enough to distinguish between the strange chaotic and non-chaotic attractors [44].

Bicoherence measures the phase coherency among three harmonics [19,55], thus detecting quadratic phase coupling (QPC) in a signal. Because machine faults are often associated with some non-linear mode of operation which transfers energy between components of the harmonics, the bicoherence can be used for condition monitoring [67,68,24]. Bicoherence can be further used to discriminate between phase coupled and randomly excited harmonics and for estimation of the fraction of power due to the QPC [55].

In this section bicoherence has been used to provide additional insight into the washing machine dynamics. Special attention has been given to the statistical stability of the bicoherence estimator and the problem of the division by small number.

3.1 *Experimental set-up*

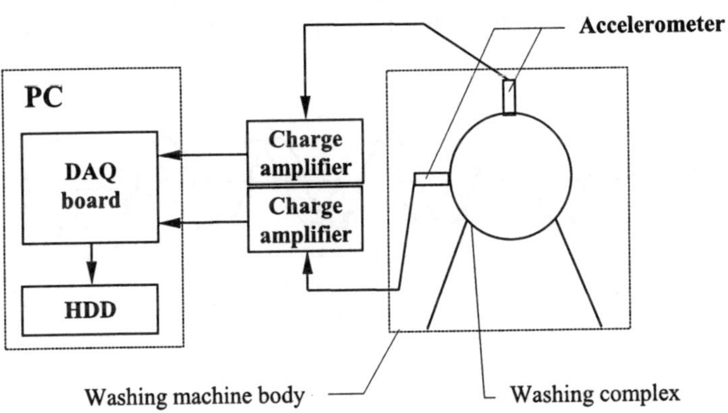

Figure 15. Experimental set-up.

The experiment was performed on a washing machine (see Figure 15). As a part of the washing machine, a washing complex is made up of the tub in which the drum is rotating, additional weights and the electromotor attached to the tub and the suspension arms holding the tub as vibroisolation. In the drum of the washing complex at maximum radius an excentrical mass was fixed. In that way the unbalance of the rotating parts was provided and the worst possible laundry distribution was simulated. The system accelerations in vertical and horizontal directions were measured simultaneously using two B&K type 4367 accelerometers with B&K type 2635 and 2626 charge amplifiers. The signals were acquired by NI AT-MIO-16E-1 data acquisition board and stored on HDD of PC for further analysis. The charge amplifier's filter was used for signal conditioning. The low-pass filtering was set at 3 kHz on both charge amplifiers (B&K 2635 and B&K 2626). The sampling frequency was 3 kHz due to aliasing. The signals were resampled to 1 kHz for phase space analysis.

3.2   *Experimental results analysis*

The time series of the horizontal (x) and vertical (y) accelerations are presented in figure 16. The time histories of two simultaneously measured accelerations show that the motion of the washing complex has one dominant frequency, hence the frequency of spin dry. The signals of accelerations in horizontal and vertical direction are approximately shifted by $\pi/2$, hence the washing complex response follows the centrifugal excitation. It was also found that the speed control was able to keep the spin dry frequency within the range of $\pm 1\%$, compared to the set spin dry frequency (see Figure 17).

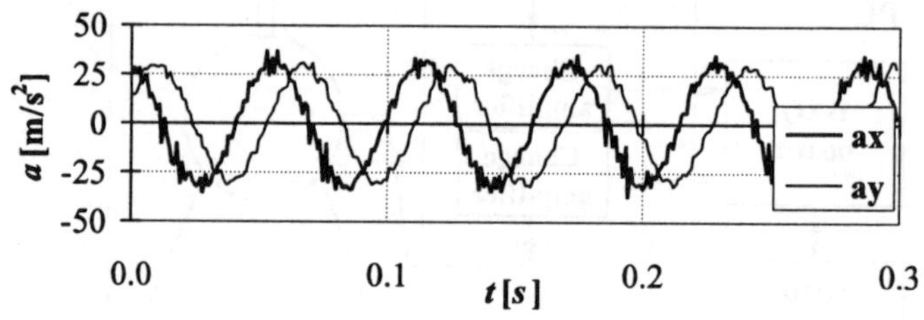

Figure 16. Measured steady state accelerations; a_x is horizontal and a_y is vertical acceleration.

Spectral analysis

Power spectra were calculated using 4096 FFT points. Overlapping of 5% was used to obtain 128 segments, and the Hamming window was applied in the time domain. A fourth order lowpass elliptic digital filter was used with a cut-off frequency of 300 Hz with approximately linear phase characteristics. The power spectra of horizontal and vertical acceleration are presented in figures 18 and 19. In both the horizontal and vertical direction the spin dry

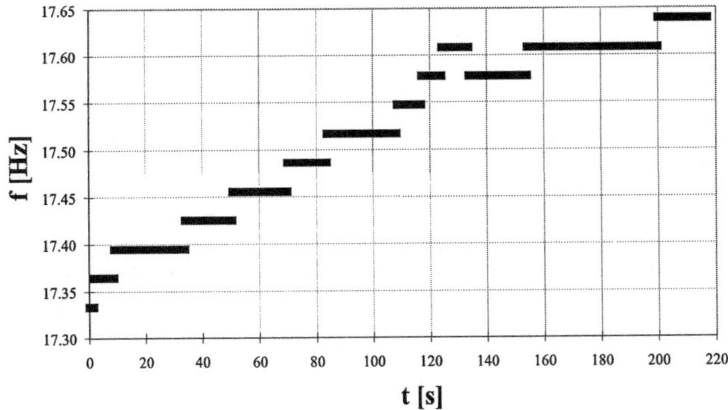

Figure 17. Spin dry frequency deviations.

frequency of 17.58 Hz is clearly visible. The third harmonic, generated by the three-arm support of the drum, has already 70 dB lower power. In figure 19, peaks at 145.02 and 208.01 Hz can be observed. The cause of these two peaks is unknown. At other frequencies the spectral power is significantly lower.

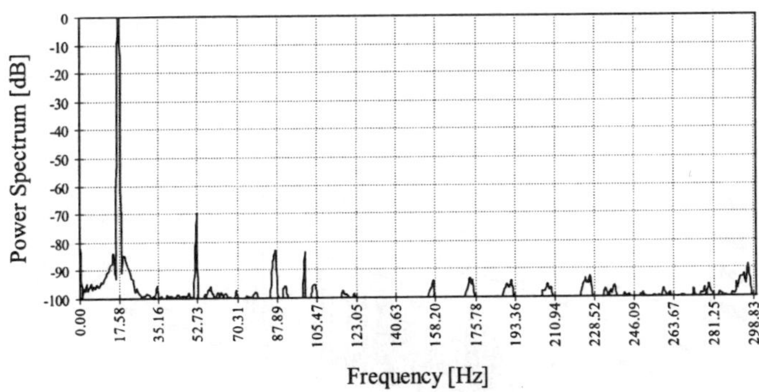

Figure 18. Power spectrum of the measured horizontal acceleration.

The bicoherences of horizontal and vertical measured accelerations were calculated using the same parameters as described previous section, and are presented in figures 20 and 22. The number of FFT points per segment was 1024; the 5% overlapping was used to obtain 513 segments. Because of its

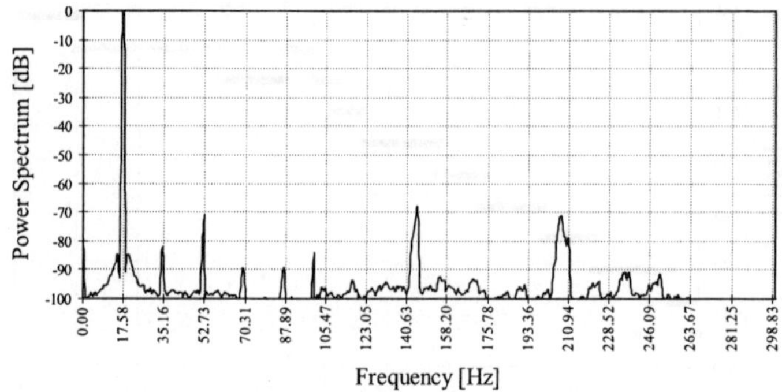

Figure 19. Power spectrum of the measured vertical acceleration.

ability to resolve QPC peaks [14,57], the Hamming window was applied in the time domain. The bicoherence significance of $T_{\alpha\%}$ was used to determine the minimal significant bicoherence level. Biphase significance was not used because it was found that the probability distribution of the biphase did not comply with the normal distribution. The numerical zero threshold was set to 1E-10.

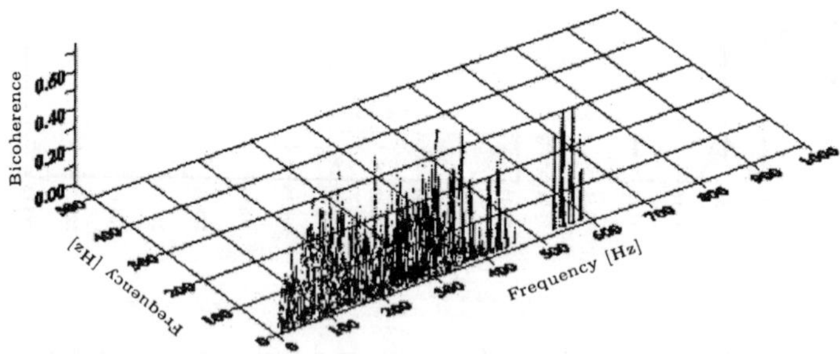

Figure 20. Bicoherence surface plot of the measured horizontal acceleration.

One can see that the bicoherence of the horizontal acceleration has multiple peaks, scattered over the most part of the IT. The numerous peaks are

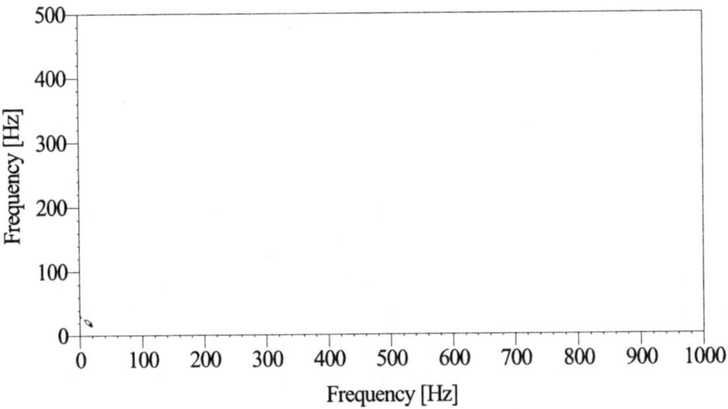

Figure 21. Magnitude spectrum contur plot of the measured vertical acceleration. The maximum value is 1.45, minimum nad maximum contur values are 0.2 and 1.4

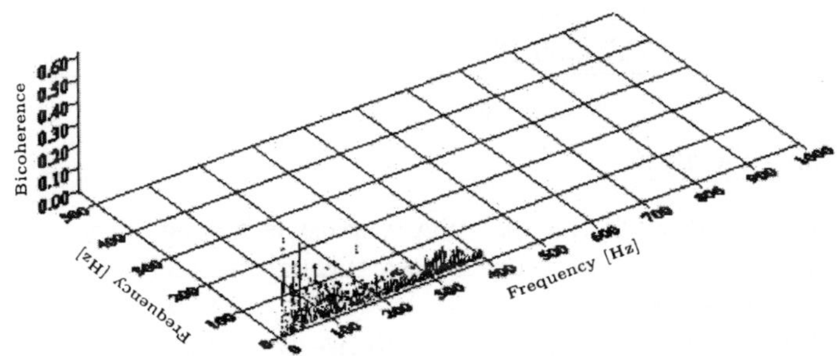

Figure 22. Bicoherence surface plot of the measured vertical acceleration.

probably caused by the division with small number, especially because the magnitude bispectrum exhibits only one peak of 1.45 at (17.58, 17.58) Hz (see Figure 21. Although the variance of the bispectrum is dependent upon the signal's second order properties, the large number of segments should minimise this influence. The bicoherence of the vertical accelerations gives a much clearer picture of QPC. Four distinct bicoherence values stand out (see Table 5). The second harmonic has very little power and can be observed only on the power spectrum of the vertical accelerations in figure 19. The little power it has, is 63% due to the QPC of the spin dry frequency. The third harmonic (≈ -70 dB) is 49% generated by the three-arm support of the drum and 51% generated by QPC of the spin dry frequency and second harmonic. Similarly, 41% of the fourth and 31% of the fifth harmonic's power is due to the QPC. These values confirm that the process involved is not linear and that quadratic non-linearities are present. In figure 20, a peak in the higher frequency domain (up from 300 Hz) can be seen. This is due to the magnitude characteristic of the digital filter used and division by small number when normalising. The filter's magnitude response starts to fall at 300 Hz, and only reaches values close to zero at 500 Hz. In this case both the numerator and denominator of the bicoherence are larger than the zero threshold, so that the algorithm can not filter out these values. This translates into the peak, seen in the figure 20.

Table 5. The highest bicoherence values of vertical accelerations

Bicoherence	Frequency [Hz]	Frequency sum [Hz]	Harmonic
0.63	(17.58, 17.58)	35.16	2.
0.51	(35.16, 17.58)	52.73	3.
0.41	(52.73, 17.58)	70.31	4.
0.31	(52.73, 35.16)	87.89	5.

Phase-space analysis

In experimental work the real phase space is almost always unknown, and the phase space could be reconstructed through the embedding procedure described in previous section.

Correlation dimension. To characterise the dimensionality of an attractor in reconstructed phase space the integral of correlation dimension was used. The formulas are presented in Theoretical modelling section.

The estimates of the correlation dimension were computed by employing a reconstruction procedure on the experimental time histories of the horizontal and vertical washing complex accelerations. The correlation dimensions were extracted as $\nu = 1.51\pm0.02$ and $\nu = 1.24\pm0.01$ respectively. The differences in the correlation dimension estimates are due to the fact that each signal carries the signature of the entire dynamics of the system, although the dynamics of the measured direction has a greater importance. It was shown [49] that the dynamics in the horizontal direction is richer than in the vertical direction. In both cases estimates of the correlation dimension show that the attractor is topologically a fractal object.

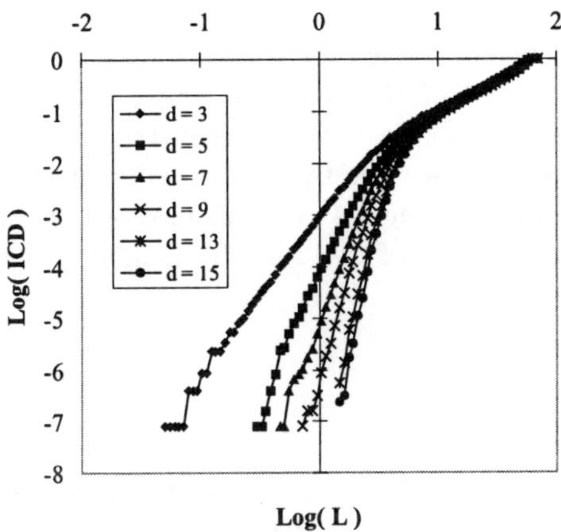

Figure 23. ICD on the measured horizontal acceleration.

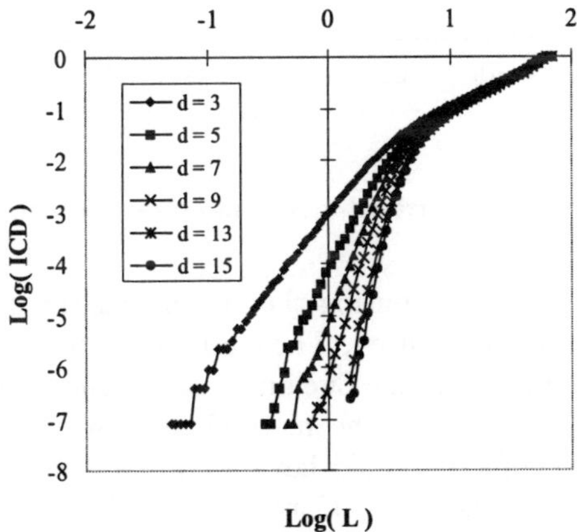

Figure 24. ICD on the measured vertical acceleration.

The first log-log diagram of the integral of the correlation dimension based on the horizontal acceleration is shown in figure 23, and the second one based on the vertical acceleration is shown in figure 24. The interesting part of

the diagrams lies above the point where the lines of the integral of correlation dimension computed with different embedding dimensions converge. The
lower part of the diagrams, with regard to the characteristic distance ℓ, shows
the typical dependence of correlation dimension on white noise hidden in the
measured signal [58].   All of the plots in figures 23 and 24 are based on a
computation using 5000 points [62] when reconstructing the phase space.

Lyapunov exponents. The Lyapunov exponents measure the exponential divergence (positive exponents - chaotic motion) or convergence (negative
exponents - regular motion) of two initially neighbouring trajectories in the
phase space.

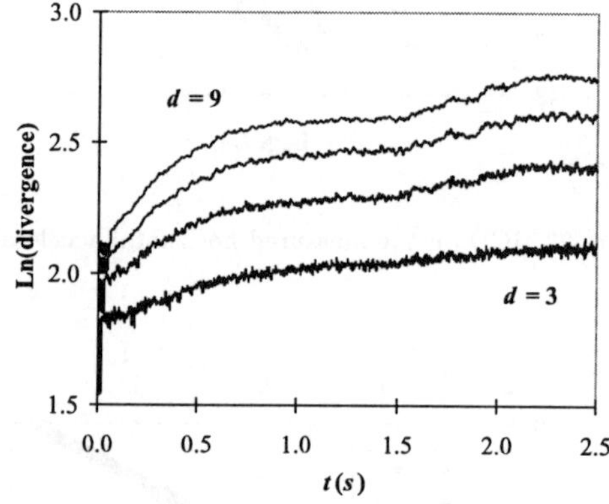

Figure 25. Divergence of trajectories on measured horizontal acceleration.

The technique [64] for computing the largest positive Lyapunov exponent
from small data sets has been used. The method locates the nearest neighbour
to the reference point using the Euclidean norm and so defining the initial
distance between the points.   The method imposes additional constraints,
hence the nearest neighbour should have a temporal separation greater than
the mean period of the time series. Consequently, the two neighbouring points
could be treated as nearby initial conditions for two different trajectories. The
largest Lyapunov exponent is then estimated as the mean rate of separation
of the neighbours.   The algorithm is unable to compute negative Lyapunov
exponents.

The estimates of the largest Lyapunov exponent were computed by employing a reconstruction procedure on the experimental time histories of the
horizontal and vertical washing complex's accelerations.   The largest Lya-

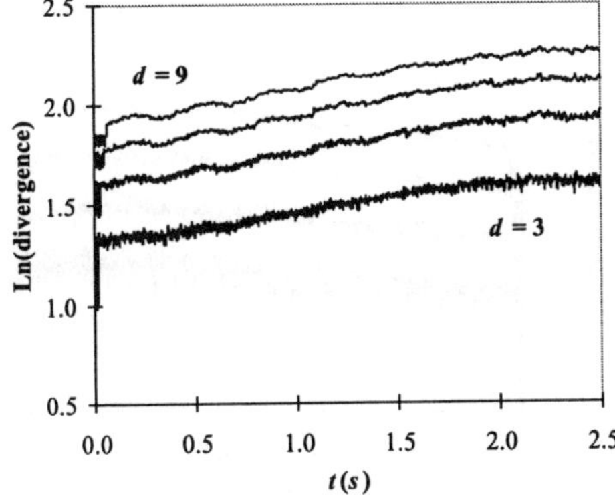

Figure 26. Divergence of trajectories on measured vertical acceleration.

punov exponents were estimated as $\lambda_{max,x} = 0.20 \pm 0.00$ and $\lambda_{max,y} = 0.16 \pm 0.00$ respectively, as the inclination of the slope of the linear regression line imposed on plots in figures 25 and 26, respectively, for $d = 9$. The differences in Lyapunov exponent estimates are due to the same phenomena as described in case of the correlation dimension estimation. It is shown again that the processes in the horizontal direction are more 'turbulent' than those in vertical direction.

To check for spurious Lyapunov exponents the time flow of processes was reversed [63]. It would be normal, if the computed Lyapunov exponents were not spurious, to expect negative estimates of the largest Lyapunov exponents of both processes when time is reversed. Since the method is unable to compute negative Lyapunov exponents, the almost zero value of both the largest Lyapunov exponents of reverse processes was estimated (see Figures 27 and 28). Hence the estimation of the largest Lyapunov exponents of accelerations' time histories shows that they are not spurious and the dynamics of the washing complex appears to be chaotic.

All of the plots in figures 25, 26, 27 and 28 are based on a computation using 5000 points of reconstructed phase space [62].

Attractor visualization. In the case of experimental data the visualisation could be done through the embedding procedure as described in the Theoretical modelling section.

The visualised attractor from the horizontal acceleration's time history in three-dimensional space, $d = 3$, is shown in figure 29, and the visualised

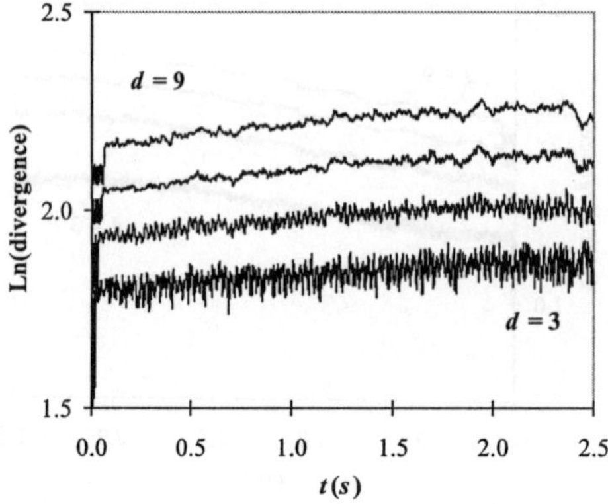

Figure 27. Divergence of trajectories on reversed signal of measured horizontal acceleration.

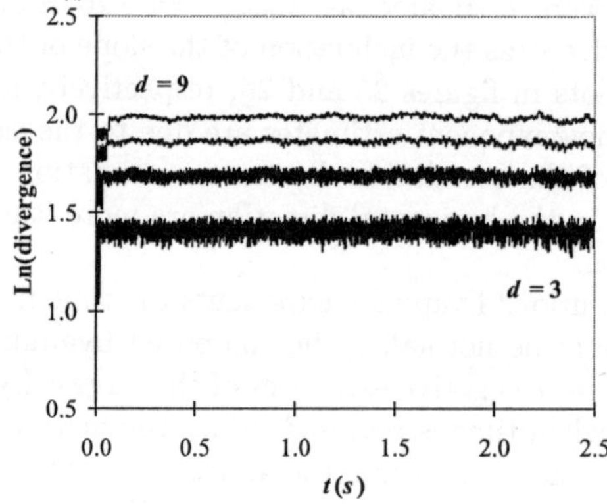

Figure 28. Divergence of trajectories on reversed signal of measured vertical acceleration.

attractor from vertical acceleration's time history in three-dimensional space, $d = 3$, is shown in figure 30.

The washing complex attractor is torus-shaped; this is due to the one predominant harmonic which is driven by spin dry. It is also visible that the dynamics in horizontal direction are widely dispersed around the predominant harmonic's limit cycle than in the vertical direction. This is in agreement

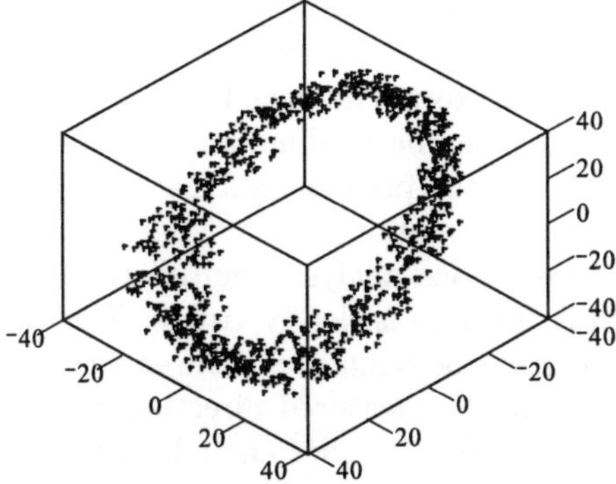

Figure 29. Visualization of the washing complex attractor from the measured horizontal acceleration.

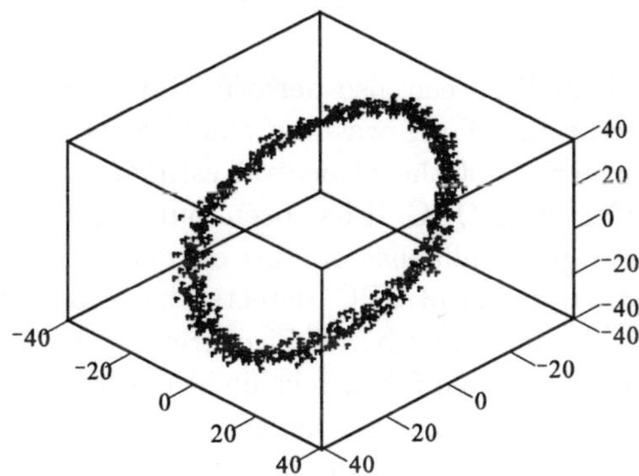

Figure 30. Visualization of the washing complex attractor from the measured vertical acceleration.

with results of analysis of the attractor's dimensionality and of the Lyapunov exponents.

3.3   Conclusions to the experimental work

The steady-state responses of the washing-machine washing complex oscillations were analysed. The results of the simulations and experiment are compared.

The estimates of the correlation dimension are revealed on the fractal structured attractor of the washing complex dynamics. This is consistent with the estimates of the largest Lyapunov exponent. Its positive value reveals the chaotic nature of the attractor. The visualisation of the attractor shows its torus-shape, which is due to one predominant harmonic driven by spin dry, and the dispersion around the predominant harmonic's limit cycle, which is due to the chaotic nature of motion.

In all cases of phase space analysis the difference between dynamics in vertical and horizontal direction is clearly visible. The difference is due to the fact that each signal carries the signature of the entire dynamics of the system, although the dynamics of the measured direction has a greater importance. Nevertheless each of the signals carries enough of the system's dynamics to enable all phase space analysis, regardless of the direction of the processed time history, to show the same nature of the washing complex motion.

In spite of the chaotic nature of washing complex motion, the power spectra do not appear to be continuous due to the predominant harmonic. The power spectra of such processes could be easily mistaken for a quasi-periodic one.

Bispectral analysis has been also performed on the measured horizontal and vertical accelerations of the washing complex oscillations. Due to the poor statistical properties of the bispectral estimate, bicoherence has been used as a mean of detecting QPC. It has been found that the division by small numbers can have a great influence on this estimate. The division by small number increases the number of QPC, detected by bicoherence estimate, and is demonstrated in the large number of bicoherence peaks. To overcome this problem, a numerical threshold for zero value has been introduced. Whenever the numerator of the bicoherence is smaller than this threshold, bicoherence is set to zero. If on the other hand the numerator is greater and the denominator is smaller than the threshold, a small constant value is added to the denominator. It has also been found useful to check the magnitude bispectrum.

All the bicoherences presented are to the some degree influenced by the division by small numbers. When comparing the results of the model and experiment, the latter suggest that the theoretical model is not complex enough and exhibits less QPC. The experimental results show that the third, fourth and fifth harmonic are not independent, but are to the some extent the result of the QPC.

The discrepancy between the model and experiment could be attributed to our not being able to model in every single detail such a complex device as the washing machine is. For example, the real electromotor is capable of holding a constant spin dry speed in the range of $\pm 1\%$ of the declared speed.

The supports contain some elastomeric parts that were modelled only by their frictional properties. In our case we found that even the nature of motion of the model and the washing complex is not the same, but that the model well reflects the steady-state motion of the washing complex in the frequency and time domains.

4   Conclusions

A non-linear planar centrifugally excited oscillatory system has been studied in its steady-state domain. The dynamic behaviour in the phase space and in the frequency domain has been analysed by a model based on the numerical integration of non-linear equations of motion and by experimentally measured washing complex dynamics. The integral of the correlation dimension and Lyapunov exponents were used as a quantitative measure to describe the motion of the model. The estimates of the correlation dimension for the values in real phase space and for those obtained by embedding numerical time histories of the model show good agreement. The largest positive Lyapunov exponent and the non-integer value of the correlation dimension of the attractor confirm the chaotic nature of the washing complex dynamics. Power spectral and bispectral analyses have also been used to analyse the behaviour of the model and the real system in the frequency domain. The dominance of the first mode was found in both cases, while other modes have significantly lower power. The quadratic phase coupling between the second and fifth harmonic is present in the washing complex dynamic responses.

References

1. Gradišek, J. and Govekar, E. and Grabec, I. (1996) "A chaotic cutting process and determining optimal cutting parameter values using neural networks" *International Journal of Machine Tools & Manufacture* **36**, 1161-1172.
2. Gans, R.F. (1995) "When is cutting chaotic?" *Journal of Sound and Vibration* **188**, 75-83.
3. Tansel, I.N. (1992) "The chaotic characteristics of 3-dimensional cutting" *International Journal of Machine Tools & Manufacture* **32**, 811-827.
4. Feeny, B. (1994) "Chaos in a forced dry-friction oscillator - experiments and numerical modeling" *Journal of Sound and Vibration* **170**, 303-323.
5. Galvanetto, U. (1999) "Dynamics of a simple damped oscillator undergoing stick-slip vibrations" *Meccanica* **34**, 337-347.

6. Begley, C.J. (1998) "Impact response and the influence of friction" *Journal of Sound and Vibration* **211**, 801-818.

7. Blazejczyk-Okolewska, B. and Kapitaniak, T. (1996) "Dynamics of impact oscillator with dry friction" *Chaos Solitons & Fractals* **7**, 1455-1459.

8. Kim, D.J. and Jeong, J. and Chae, J.H. and Park, S. and Kim, S.Y. and Go, H.J. and Paik, I.H. and Kim, K.S. and Choi, B. (2000) "An estimation of the first positive Lyapunov exponent of the EEG in patients with schizophrenia" *Psychiatry Research-Neuroimaging* **98**, 177-189.

9. Lehnertz, K. (1999) "Non-linear time series analysis of intracranial EEG recordings in patients with epilepsy - an overview" *International Journal of Psychophysiology* **34**, 45-52.

10. Jeong, J. and Kim, S.Y. and Han, S.H. (1998) "Non-linear dynamical analysis of the EEG in Alzheimer's disease with optimal embedding dimension" *Electroencephalography and Clinical Neurophysiology* **106**, 220-228.

11. Micheloyannis, S. and Flitzanis, N. and Papanikolaou, E. and Bourkas, M. and Terzakis, D. and Arvanitis, S. and Stam, C.J. (1998) "Usefulness of non-linear EEG analysis" *Acta Neurologica Scandinavica* **97**, 13-19.

12. Yum, M.K. and Kim, N.S. and Oh, J.W. and Kim, C.R. and Lee, J.W. and Kim, S.K. and Noh, C.I. and Choi, J.Y. and Yun, Y.S. (1999) "Non-linear cardiac dynamics and morning dip: an unsound circadian rhythm" *Clinical Physiology* **19**, 56-67.

13. Bettermann, H. and Van Leeuwen, P. (1998) "Evidence of phase transitions in heart period dynamics" *Biological Cybernetics* **78**, 63-70.

14. Fackrell, J. (1996) "Bispectral analysis of speech signals", *University of Edinburg*.

15. Hinich, M.J. (1982) "Testing for gaussianity and linearity of a stationary time series" *Journal of Time Series Analysis* **3**, 169-176.

16. Jouny, I. and Garber, F.D. and Moses, R.L. (1995) "Radar target identification using bispectrum: A comparative study" *IEEE Transactions on Aerospace and Electronic Systems* **31**, 69-77.

17. Haubrich, R.A. (1965) "Earth Noise, 5 to 500 Milicycles per second" *Journal of Geophysical Research* **10**, 1415-1427.

18. Elgar, S. and Sebert, G. (1989) "Statistic of bicoherence and biphase" *Journal of Geophisical Research* **94**, 10993-10998.

19. Nikias, C.H. and Petropolu, A.P. (1993) *Higher-order spectra analysis, a nonlinear processing framework*, Prentice Hall.

20. Parker, B.E. and Ware, H.A. and Wipf, D.P. and Tompkins, W.R. and Clark, B.R. and Larson, E.C. and Poor, H.V. (2000) "Fault diagnostics using statistical change detection in the bispectral domain" *Mechanical*

Systems and Signal Processing **14**, 561-570.

21. McCormick, A.C. and Nandi, A.K. (1999) "Bispectral and trispectral features for machine condition diagnosis" *IEE Proceedings-Vision Image and Signal Processing* **146**, 229-234.

22. Howard, I.M. (1997) "Higher-order spectral techniques for machine vibration condition monitoring" *Proceedings of the Institution of Mechanical Engineers Part G- Journal of Aerospace Engineering* **211**, 211-219.

23. Shan, J.M. and Stockton, R. (1997) "Vibration analysis of rotating machinery using a bispectrum" *Journal of Sound and Vibration* **200**, 533-539.

24. Chow, T.W.S. and Fei, G. (1995) "Three phase induction machines asymmetrical faults identification using bispectrum" *IEEE Transactions on Energy Conversion* **10**, 688-693.

25. Jeffries, W.Q. and Chambers, J.A. and Infield, D.G. (1998) "Experience with bicoherence of electrical power for condition monitoring of wind turbine blades" *IEE Proceedings-Vision Image and Signal Processing* **145**, 141-148.

26. Li, C.J. and Ma, J. and Hwang, B. (1996) "Bearing condition monitoring by pattern recognition based on bicoherence analysis of vibrations" *Proceedings of the Institution of Mechanical Engineers Part C-Journal of Mechanical Engineering Science* **210**, 277-285.

27. Davies, M.A. and Balachandran, B. (2000) "Impact dynamics in milling of thin-walled structures" *Nonlinear Dynamics* **22**, 375-392.

28. Boltežar, M. and Gradišek, J. and Simonovski, I. and Govekar, E. and Grabec, I. and Kuhelj, A. (1999c), "Bispectral analysis of the cutting process", *Proceedings of the IEEE Signal Processing Workshop on Higher-Order Statistics.*

29. Bullock, T.H. and Achimowicz, J.Z. and Duckrow, R.B. and Spencer, S.S. and Iragui-Madoz, V.J. (1997) "Bicoherence of intracranial EEG in sleep, wakefulness and seizures" *Electroencephalography and Clinical Neurophysiology* **103**, 661-678.

30. Shils, J.L. and Litt, M. and Skolnick, B.E. and Stecker, M.M. (1996) "Bispectral analysis of visual interactions in humans" *Electroencephalography and Clinical Neurophysiology* **98**, 113-125.

31. Schmidt, K. and Schwab, M. and Abrams, R.M. and Witte, H. (1999) "Nonlinear analysis of the fetal ECoG: predictability and bispectral measures" *Theory in Biosciences* **118**, 219-230.

32. Narayanan, S. and Hussain, F. (1996) "Measurements of spatiotemporal dynamics in a forced plane mixing layer" *Journal of Fluid Mechanics* **320**, 71-115.

33. Boltežar, M. and Hammond, J.K. (1999a) "Experimental study of the vibrational behaviour of a coupled non-linear mechanical system" *Mechanical Systems and Signal Processing* **13**, 375-394.

34. Jakšić, N. and Boltežar, M. and Simonovski, I. and Kuhelj, A. (1999) "Dynamical behaviour of the planar non-linear mechanical system - Part I: Theoretical modelling" *Journal of Sound and Vibration* **226**, 923-940.

35. Boltežar, M. and Jakšić, N. and Simonovski, I. and Kuhelj, A. (1999b) "Dynamical behaviour of the planar non-linear mechanical system - Part II: Experiment" *Journal of Sound and Vibration* **226**, 947-953.

36. Kuhelj, A. and Boltežar, M. and Godler, I., (1988) "Dynamic simulation of the washing machine behavior (Dinamična simulacija pralnika)", *Kuhljevi dnevi '88*.

37. Jakšić, N. and Boltežar, M., (1995) "Some properties of centrifugally excited non-linear planar system (Nekatere lastnosti nelinearnega modela centrifugalno vzbujanega ravninskega sistema)", *Kuhljevi dnevi '95*.

38. Simonovski, I. and Jakšić, N. and Boltežar, M., (1996) "Dynamics of an extended system of rigid bodies at centrifugal excitation. (Dinamika razširjenega modela centrifugalno vzbujanega sistema togih teles.)", *Kuhljevi dnevi '96*.

39. Boltežar, M. and Jakšić, N. and Kuhelj, A., (1997) "Quantitative phase space analysis of the non-linear planar oscillatory system", *GAMM, Wissenschaftliche Jahrestagung, Regensburg*.

40. Boltežar, M. and Jakšić, N. (1998) "Quantitative phase space analysis of the nonlinear planar oscillatory system" *Zeitschrift für Angewandte Mathematik und Mechanik* **78**, 287-288.

41. Boltežar, M. and Simonovski, I. and Jakšić, N. and Kuhelj, A., (1998) "Bispectral Analysis of the Planar Non-linear Mechanical System", *1st Congress of Slovenian Acoustical Society*.

42. Simonovski, I. and Boltežar, M. and Kuhelj, A. (1999) "Theoretical Background of the Bispectral Analysis" *Journal of Mechanical Engineering* **45**, 12-24.

43. Narayanan, S. and Jayaraman, K. (1991) "Chaotic Vibration in a Nonlinear Oscillator with Coulomb Damping" *Journal of Sound and Vibration* **146**, 17-31.

44. Barron, R. and Wojewoda, J. and Brindley J. and Kapitaniak T. (1993) "Interpretation of Aperiodic Time Series: a New View of Dry Friction" *Journal of Sound and Vibration* **162**, 369-375.

45. Feeny, B. (1992) "A nonsmooth Coulomb friction oscillator" *Physica D* **59**, 25-38.

46. Boltežar, M. and Jakšić, N., (1996) "Phase space analysis of the non-

linear centrifugally excited planar oscillatory system", *EUROMECH - 2nd European Non-linear Oscillation Conference.*

47. Logan, D. and Mathew, J. (1996a) "Using the correlation dimension for vibration fault diagnosis of rolling element bearings - I. Basic concepts" *Mechanical Systems and Signal Processing* **10**, 241-250.

48. Logan, D. and Mathew, J. (1996b) "Using the correlation dimension for vibration fault diagnosis of rolling element bearings - II. Selection of experimental parameters" *Mechanical Systems and Signal Processing* **10**, 251-264.

49. Jakšić, N. (1997) "Vibration analysis of the non-linear centrifugally excited planar system (in Slovene)", *Faculty for Mechanical Engineering, Ljubljana.*

50. Grabec, I. (1988) "Chaotic dynamics of the cutting process" *International Journal of Machine Tools and Manufacturing* **28**, 19-32.

51. Shaw, S.W. (1986) "Dynamic Response of a System with Dry Friction" *Journal of Sound and Vibration* **108**, 305-325.

52. Den Hartog, J.P. (1931) "Forced Vibrations With Combined Coulomb and Viscous Friction" *Transactions of The ASME* **53**, 107-115.

53. Türkay, O.S. and Kiray, B. and Tuğcu, A.K. and Sümer, I.T. (1995) "Formulation and implementation of parametric optimisation of a washing machine suspension system" *Mechanical Systems and Signal Processing* **9**, 359-377.

54. Chandran, V. and Elgar, S.L. (1992) "Mean and Variance Estimate of the Bispectrum of a Harmonic Random Process-An Analysis Including Leakage Effects" *IEEE Transactions on Signal Processing* **39**, 2640-2651.

55. Kim, Y.C. and Powers, E. (1979) "Digital Bispectral Analysis and Its Application to Nonlinear Wave Interactions" *IEEE Transactions on Plasma Science* **PS-7**, 120-131.

56. Hinich, M.J. and Messer, H. (1995) "On the Principal Domain of the Discrete Bispectrum of a Stationary Signal" *IEEE Transactions on Signal Processing* **43**, 2139-2134.

57. Simonovski, I. (1998) "Third Order Spectra Analysis of Nonlinear Dynamical Systems (in Slovene)", *Faculty for Mechanical Engineering, Ljubljana.*

58. Eckmann, J.P. and Ruelle, D. (1985) "Ergodic theory of chaos and strange attractors" *Reviews of Modern Physics* **57**, 617-655.

59. Takens, F. (1980) "Detecting strange attractors in turbulence", *Lecture notes in mathematics: Proceedings of Dynamical Systems and Turbulence* **898**.

60. Grassberger, P. and Procaccia, I. (1983a) "Characterisation of strange

attractors" *Physical Review Letters* **50**, 346-349.

61. Grassberger, P. and Procaccia, I. (1983b) "Measuring the strangeness of strange attractors" *Physica 9D* , 189-208.

62. Eckmann, J.P. and Ruelle, D. (1992) "Fundamental limitations for estimating dimensions and Lyapunov exponents in dynamical systems" *Physica D* **56**, 185-187.

63. Parlitz, U. (1991) "Indentification of true and spurious Lyapunov exponents from time series" *International Journal of Bifurcation Chaos* **2**, 155-165.

64. Rosenstein, M.T. and Collins, J.J. and De Luca, C.J. (1993) "A practical method for calculating largest Lyapunov exponents from small data sets" *Physica D* **65**, 117-134.

65. Wolf, A. and Swift, J.B. and Swinney, H.L. and Vastano, J. (1985) "Determining Lyapunov exponents from a time series" *Physica 16D* , 285-317.

66. Boltežar, M. and Hammond, J.K., (1994) "Experimental analysis of a multi-degree-of-freedom nonlinear mechanical system", *Fifth International Conference on Recent Advances in Structural Dynamics.*

67. Fackrell, J.W. and White, P.R. and Hammond, J.K. and Pinnington, R.J. and Parsons, A.T. (1995) "The Interpretation of the Bispectra of Vibration Signals-II. Experimental Results and Application" *Mechanical Systems and Signal Processing* **9**, 267-274.

68. Barker, R.W. and Hinich, M.J, (1993) "Statistical Monitoring of Rotating Machinery by Cumulant Spectral Analysis", *IEEE Signal Processing Workshop on Higher-Order Statistic.*

ANALYSIS AND NONLINEAR CONTROL OF HYDRAULIC SYSTEMS IN ROLLING MILLS

R. M. NOVAK

Christian Doppler Laboratory for Automatic Control of Mechatronic Systems in Steel Industries, c/o Rainer Novak, Altenbergerstreet 69, A-4040 Linz / Austria, E-mail: novak@mechatronik.uni-linz.ac.at

Abstract

Hydraulic actuators are used in a wide range of industrial applications but hydraulic systems are nonlinear in their nature. The steadily rising demands for tighter thickness tolerances in rolling mills gave the motivation for the development of a nonlinear controller based on the exact Input–Output–Linearization. In this contribution it will be shown that the dynamical behavior of a hydraulic adjustment system can be improved by such a nonlinear control concept. One can also find a detailed root locus analysis of the linearized system of such a plant as well as an FFT–identification of a rolling mill.

1 Introduction

Modern rolling mills are typical Multi-Input-Multi-Output (MIMO) systems that contain several hydraulic piston actuators. The actual development in this industry is towards tighter thickness tolerances and this demands can only be satisfied, if the nonlinearities in the plant are taken into account.

This work was done in cooperation with the VOEST Alpine Industrieanlagenbau GmbH, a well known manufacturer of machinery for the steel industry. The motivation for this work was the feedback of their setup engineers, that the hydraulic actuators behave very sensitive when they work near the edges and the objective was, to find a controller that can cope with this problem. Traditional linear controllers handle this phenomenon by gain scheduling, this means for the practical implementation that the maintenance staff must specify several ranges and several gains. The effect is that the controller software becomes overloaded and the behavior of the whole plant becomes less understandable.

In the first step we tried to solve this problem by a well established method in nonlinear control theory, namely the Input–Output–Linearization. Practically this concept suffers from the disadvantage that one must know the whole state, although in most cases only a part of the state can be measured. This means that in real world applications observers are used to overcome this difficulty, but the dynamics of the observer influence the static feedback of the Input–Output–Linearization and the separation theorem does not hold in the nonlinear case.

Anyone who has ever simulated a hydraulic system knows that this is a subtle problem, although it seems to be innocuous at the first glance. A nonlinear controller for a hydraulic actuator, that ensures uniform dynamical behavior in the whole operating range was originally developed by Kugi[1]. Later this approach led to a systematic methodology for a general type of systems, which was developed by Schlacher[2]. The objective of this contribution is to get a general idea of the latest developments. The reader should be familiar in this context with the methods developed by Isidori[3] or Nijmeijer[4]. The idea of the method due to Schlacher[2] is to find an output so that the nonlinear feedback depends only on measurable quantities and to linearize the system on this output.

It should be emphasized that the most important point in the whole analysis is the understanding of the plant. If a rolling mill is considered with the main actuator plus balancing- and bending actuators including their valves it makes no problem to derive the state equations which give a high order model. To gain a deeper insight it is necessary to find a reduced order model, which includes the relevant dynamics. We will see in subsection 3.1 and section 5 that it is possible, to explain the system behavior with a simple 3rd order model.

All equations are derived for a single acting piston, because it provides the basis to understand more complicated hydraulic systems. The contribution is organized as follows. Section 2 introduces different definitions in connection with the isothermal bulk modulus, in section 3 models for a hydraulic single acting piston are presented. Additionally a comprehensive analysis based on traditional control methods to understand some hydraulic phenomenons can also be found in this sections. The analysis prerequisite is easily met by Chen[5] or Kailath[6]. In section 4 a model of a servovalve is presented as well as an algorithm, which can be used to determine the relevant parameters for such a device from the data sheets. Section 5 shows the identified model of a rolling mill and section 6 demonstrates the benefits of the developed nonlinear controller.

2   The Isothermal Bulk Modulus E

Consider, for example, a closed insulated cylinder as shown in figure 1. The pressure P in the fluid is a function of the mass density ρ, if the piston is slowly compressed an infinitesimal distance causing a decrease in the volume

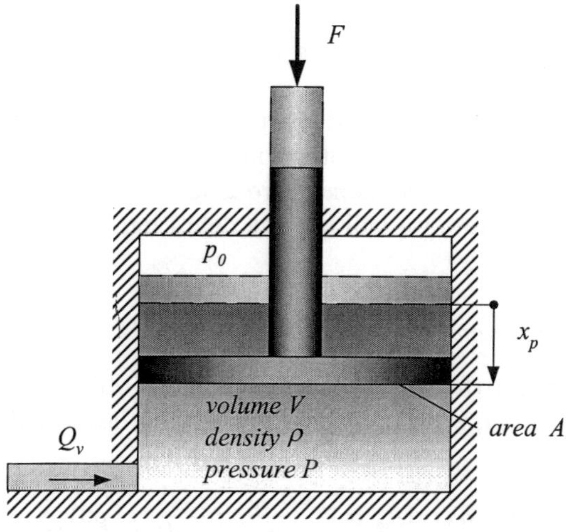

Figure 1. Control volume with compressible fluid.

dV and changes dP. The pressure can also be changed by a mass flow Q_v from or to the cylinder.

To find a mathematical model for this behavior we introduce an essential parameter in hydraulics, namely the isothermal bulk modulus[a] E, and one can find the following definitions for this parameter in the literature

Literature	Definition	Solution
Truckenbrodt[7] p. 6	$\dfrac{1}{E} = \dfrac{1}{\rho}\left(\dfrac{\partial \rho}{\partial P}\right)_T$	$\rho = \rho_0 \exp\left(\dfrac{P - P_0}{E}\right)$
Johnson[8] p. 8.2, or Lide[9] p. 6.137	$\dfrac{1}{E} = -\dfrac{1}{V}\left(\dfrac{\partial V}{\partial P}\right)_T$	$V = V_0 \exp\left(-\dfrac{P - P_0}{E}\right)$
Murrrenhoff[10] p. 118	$\dfrac{1}{E} = -\dfrac{1}{V_0}\left(\dfrac{\partial V}{\partial P}\right)_T$	$V = V_0 - \dfrac{V_0(P - P_0)}{E}$

with the density of the fluid ρ, the actual pressure in the fluid P, the atmospheric pressure P_0, the density of the fluid at atmospheric pressure ρ_0, the control volume V, the initial volume V_0 and the temperature T. The right

---

[a]One can find different definitions for this parameter in the literature, some books define the reciprocal value $\kappa = 1/E$ which is called the isothermal compressibility

column in above table is the solution of the corresponding differential equation, i.e., it gives the pressure P as a function of the mass density ρ or the volume V.

Mostly E is defined at a fixed temperature and for a fixed mass m in the control volume, because the isothermal bulk modulus is measured in this way. In the considered applications there is always a flow from the valve to the cylinder and therefore we will use the definition of Truckenbrodt[7] in the following sections.

2.1   The State Equation

The state equation for the system of figure 1 can be obtained from the conservation of mass theorem, which may be expressed mathematically as 'continuity equation':

$$\int_{A_S} \rho\, v_n\, \mathrm{d}A_s = -\frac{\mathrm{d}m_{cv}}{\mathrm{d}t} = -\frac{\mathrm{d}}{\mathrm{d}t}\int_V \rho \mathrm{d}V. \tag{1}$$

with the density at a point ρ, the velocity normal to the surface (+ directed outward) V_n, the area element of control surface $\mathrm{d}A_s$, the element of control volume $\mathrm{d}V$, the mass in the control volume m_{cv}. Assuming

1. a constant pressure in the control volume and on its surface

2. a constant temperature of the fluid

3. a homogenous fluid

one obtains with the indicated sign convention

$$\rho Q_v = -\frac{\mathrm{d}}{\mathrm{d}t}\left(\rho V\right). \tag{2}$$

and by using the definition in Truckenbrodt[7]

$$\frac{d}{dt}\left(\rho V\right) = \frac{\partial \rho}{\partial P}\frac{\partial P}{\partial t}V + \rho\frac{\partial V}{\partial t} = \rho\frac{V}{E}\frac{\partial P}{\partial t} + \rho\dot{x}_p A. \tag{3}$$

The first assumption means that the flow Q_v to the chamber has the same density then the density of the control volume itself. The second and the third assumption mean, that the pressure field depends only on the density, i.e., $\rho = \rho\left(P\right)$. Because the density can be eliminated, the state equation is

$$\frac{dP}{dt} = \frac{E}{V}\left(-\dot{x}_p A + Q_v\right). \tag{4}$$

In the excellent book of Merritt[11] p. 15–18 the influence of entrapped air in the fluid is shown as well as the influence of a non–rigid cylinder. Both effects can be modeled by a reduction of the bulk modulus.

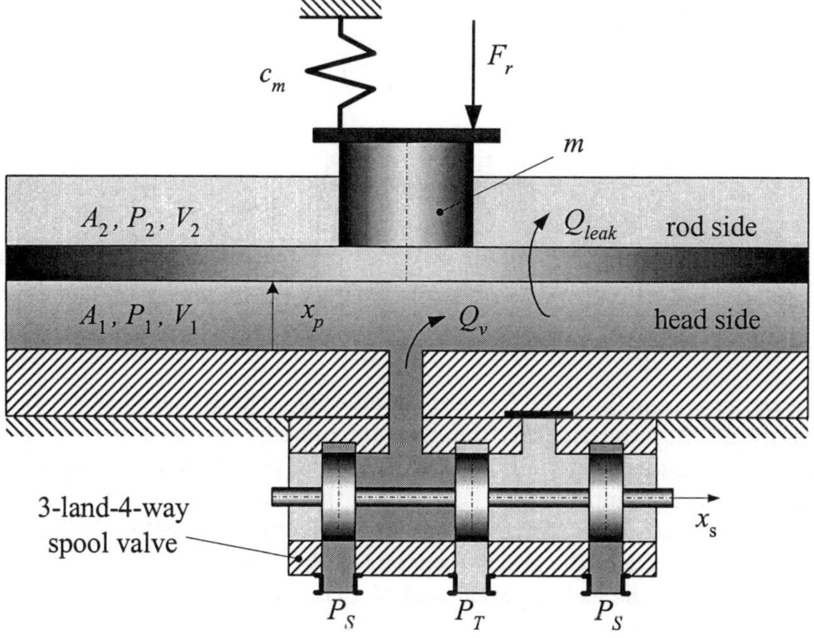

Figure 2. Single Acting Cylinder with an external load.

3   Model of a Single Acting Cylinder

The model to be studied in this section is depicted in figure 2. It's a valve controlled single acting piston working against a spring-mass-damper system. All the following considerations are based on this model. In subsection 3.8 it will be shown, how the equations can be further reduced.

The goal of this section is to develope a model that can be used to control the piston position x_p or the hydraulic force $F_h = P_1 A_1$ through the valve flow Q_v. The force F_r is considered as an unknown external disturbance onto the system. In this section a simple nonlinear model for the actuator will be developed, friction effects are neglected. The parameter c_m describes an external linear spring and d is the damping coefficient due to a viscous friction force between the piston and the cylinder. One obtains the state equations for the system under consideration with the continuity equation and the definition of the isothermal bulk modulus, see previous section,

$$E = \rho \left(\frac{\partial P}{\partial \rho} \right)_T \tag{5}$$

as

$$\frac{\mathrm{d}P_1}{\mathrm{d}t} = \frac{E\left(-v_p A_1 + Q_v - C_l P_1\right)}{V_p + x_p A_1} \tag{6a}$$

$$\frac{\mathrm{d}x_p}{\mathrm{d}t} = v_p \tag{6b}$$

$$\frac{\mathrm{d}v_p}{\mathrm{d}t} = \frac{P_1 A_1 - P_2 A_2 - c_m x_p - F_r - d v_p - mg}{m}, \tag{6c}$$

with the flow from/into the forward chamber Q_v, the piston position x_p, the pressure on the head side P_1, the constant pressure on the rod side P_2, the piston area on the head (rod) side A_1 (A_2), the pipe volume V_p, the volume of the forward chamber $V_1 = V_p + x_p A_1$, the mass of all moving parts m, the external disturbance force F_r, the supply (return) pressure P_S (P_T), the piston velocity v_p and the gravity constant g. A laminar leakage $Q_{leak} = C_l P_1$ is assumed with the leakage coefficient C_l. The influence of this parameter is considered in subsection 3.5, in all other sections $C_l = 0$.

The measurable quantities are the piston position x_p and the piston pressure P_1. For the purpose of this analysis it is assumed, that the dynamics of the valve (see section 4) are sufficiently fast, so that the spool displacement from null x_s, can be used as the reference input for the plant. The valve flow Q_v can be calculated, see, e.g., Merritt[11] p. 81, Murrenhoff[10] p. 69, for regular operating conditions

$$\begin{array}{ll} x_s \geqslant 0 & x_s < 0 \\ Q_v = K_d x_s \sqrt{P_S - P_1} & Q_v = K_d x_s \sqrt{P_1 - P_T} \end{array} \tag{7}$$

with the valve coefficient K_d and the spool position from null x_s.

For control tasks we neglect the leakage flows and the friction. Under these assumptions the model due to eqs. (6) contains two essential nonlinearities, namely the dependency of the pressure on the volume eq. (6a) and the square root function of the input due to eqs. (7).

3.1   Analysis of the Linearized System

The purpose of this section is to get a revealing insight for some physical phenomena of hydraulic systems by classical control technique methods. Anyone who has experience in the simulation of hydraulic models knows that these are stiff systems and the control of eqs. (6) is a very challenging task although it seems to be an innocuous problem at the first glance. The problems that occur during the simulations give the motivation for the analysis of the linearized system.

There is one thing in eq. (4) that seems to be mysterious if one looks at this equation at the first time, because it is not obvious why the right hand side contains the velocity $\dot{x}_p = v_p$? Figure 1 contains no mass and for $Q_v = 0$ it is expected from the engineering point of view, that the system will behave like a spring if one exerts the force F on the system. But why has the mathematical formulation eq. (4) the form of a 'differentiated spring'

$$\dot{F} = \dot{P}A = -\frac{EA^2}{V}\dot{x}_p,\ \text{see also subsection 3.8.}$$

The model due to eqs. (6) becomes linear by assuming a constant volume $V_0 = V_p + \bar{x}_p A_1$ in an operating point, with the mean displacement $\bar{x}_p$. The reference input to the system is Q_v and the disturbance input is F_r. To simplify the following expressions we introduce another relevant parameter, namely the hydraulic stiffness $c_h = \dfrac{A_1^2 E}{V_0}$ (compare eq. (6a) and eq. (6c), see Merritt[11] p. 151, or Murrenhoff[10] p. 146). Please keep in mind for the following investigations, that this parameter depends on the piston position.

3.2   Disturbance and Reference Behavior

Both disturbance transfer functions are 2nd order

$$G_{d,x}(s) = \frac{x_p(s)}{F_r(s)} = \frac{-1}{m\,s^2 + d\,s + c_h + c_m} \tag{8a}$$

$$G_{d,F}(s) = \frac{F_h(s)}{F_r(s)} = \frac{c_h}{m\,s^2 + d\,s + c_h + c_m}, \tag{8b}$$

and one can see they have the same denominator and differ only in their gain. This systems can be replaced by a pure mechanical spring–mass–damper system, with the parallel circuit of the two springs c_h and c_m, see figure 3. The left column of this figure depicts the manifested physical scheme with inputs and outputs, the middle column illustrates the equivalent linearized systems. Because the dynamics of the hydraulic subsystems are included in c_h, which is a constant, the linearized systems are driven with incompressible oil.

To analyze the system behavior concerning the input Q_v we split eqs. (6) into the mechanical part eq. (6b) and eq. (6c), with F_h as system input and x_p as system output and into the hydraulic part eq. (6a), with Q_v as system

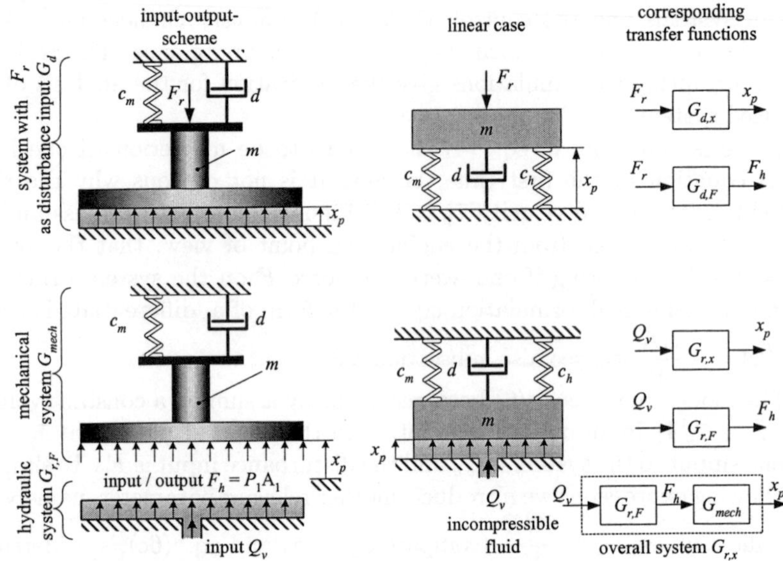

Figure 3. Disturbance and reference behavior of a single acting piston, linear case on the right side.

input and F_h as system output, see figure 3.

$$G_{mech} = \frac{x_p(s)}{F_h(s)} = \frac{1}{(ms^2 + ds + c_m)} \tag{9a}$$

$$G_{r,F}(s) = \frac{F_h(s)}{Q_v(s)} = \frac{c_h}{A_1} \frac{(ms^2 + ds + c_m)}{s(ms^2 + ds + c_h + c_m)} \tag{9b}$$

$$G_{r,x}(s) = \frac{x_p(s)}{Q_v(s)} = \frac{c_h}{A_1} \frac{1}{s(ms^2 + ds + c_h + c_m)} \tag{9c}$$

It is clear from the order of the eqs. (6), that the denominator of G_{mech} must be contained in the numerator of $G_{r,F}$. There is one interesting question: starting with a pure mechanical system eq. (6b) and eq. (6c) with F_h as system input and x_p as system output: how are the poles of G_{mech} changing, if the pressure satisfies a hydraulic equation, namely eq. (6a), i.e., what happens if one starts with eqs. (6b) and (6c) and adds eq. (6a)?

The answer can be given by the root locus of the transfer function $G_{r,x}$, eq. (9c) with c_h as parameter. We see that the overall system contains a pole

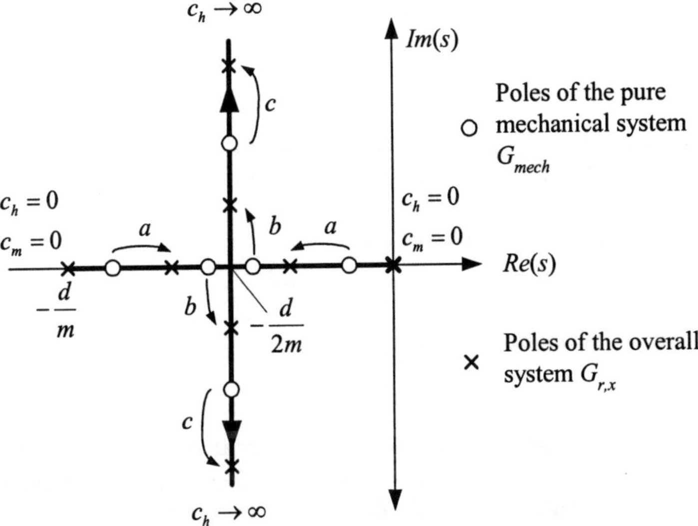

Figure 4. Poles of the mechanical system and the overall system.

at zero and the relation for roots of $\left(ms^2 + ds + c_h + c_m\right)$ and the roots of $\left(ms^2 + ds + c_m\right)$ can be seen in figure 4. For example, if

$$d^2 < 4 \cdot m \cdot c_m$$

then the mechanical system has complex poles (see case c) below and they are shifted along the imaginary axis with increasing c_h. This effect is the reason, why hydraulic systems are inherent stiff systems, as one gets a pole at 0 and the other poles are shifted towards $\pm j \infty$ because the hydraulic spring c_h has in general, a very high numerical value. So if one needs a discrete controller for the linearized system, see, e.g., Aström[12], there may arise problems, whenever the system works near the edges. Near the edges $V_0 \approx 0$ and so we get for $\lim c_h \to \infty$

$$G_{r,F,\text{lim}} = \frac{ms^2 + ds + c_m}{A_1 s}$$

$$G_{r,F,\text{lim}} = \frac{1}{A_1 s}$$

a force transfer function which is not proper, and a simple integrator for the position transfer function.

For purpose of illustration we consider the following 3 cases for figure 4 and adopt the parameters

case a)

$$m = 40 \cdot 10^3 \, [kg] \qquad E = 1.6 \cdot 10^9 \, [Pa] \qquad V_0 = 3.5 \cdot 10^{-2} \, [m^3]$$
$$c_m = 7 \cdot 10^9 \, [N/m] \qquad d = 7 \cdot 10^7 \, [kg/s] \qquad A = 0.66 \, [m^2]$$

case b) Same parameters of case a), but

$$d = 5 \cdot 10^7.$$

case c) Same parameters of case a), but

$$d = 4 \cdot 10^6.$$

Eq. 9a can be used to identify the mechanical system G_{mech}, if the pressure and the position can be measured. The effect of the hydraulic equation is to add an integrator and a parallel spring to the mechanical system and the next question is: Is there a way back to the original system G_{mech}? The answer is yes, one can also find a nice relation between the poles of the pure mechanical system G_{mech} and the poles of the overall system $G_{r,x}$ by studying the influence of an external laminar leakage $Q_{leak} = C_l P_1$ (see eq. (6a) and subsection 3.5).

3.3   The High Gain Effect

But what happens from the control point of view with G_{mech}, because figure 4 shows only the open loop behavior? If eq. (6a) is added, the input F_h to the mechanical system depends also on the system output x_p. This means that a physical feedback is induced, and what exactly happens can be seen in figure 5. Because of the high numerical value of the hydraulic spring, the influence of the hydraulic equation (6a) on the mechanical system can be interpreted as a physical implemented high gain feedback for the mechanical system. The dotted lines with the block C_l are only active, if a laminar leakage is considered. For the further analysis it will be very helpful to keep in mind the underlying feedback configuration of the actuator. And now it is clear, that eq. (9b) and eq. (9c) must have the same denominator. Compare also figure 5 with figure 3.

In the next subsections the mechanical system will be a spring–mass–damper system in combination with a single acting cylinder, because it provides the basis to understand more complex systems. In practical applications the mechanical system will be more complicated, e.g., a hydraulic actuator

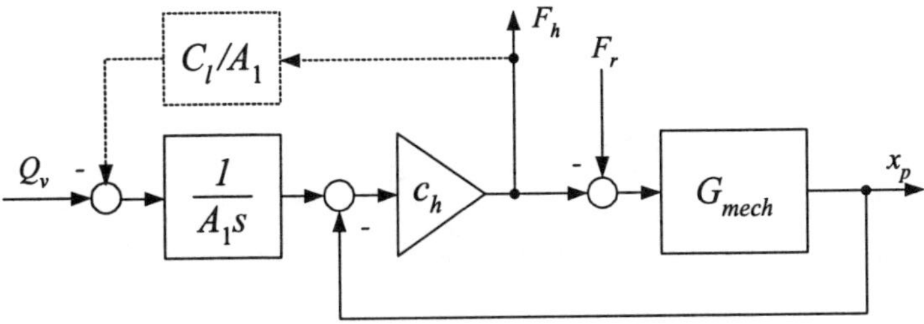

Figure 5. The influence of a hydraulic equation on a mechanical system.

in combination with balancing and bending cylinders. These systems are for itself hydraulic systems with independent control loops. In such a case the mechanical system is given by

$$G_{mech} = \frac{n_{mech}}{d_{mech}} \tag{10}$$

and the influence of the hydraulic equation can be seen very easily in figure 5. For the more general case, the disturbance and reference transfer functions become

$$G_{d,x}(s) = \frac{x_p(s)}{F_r(s)} = \frac{-n_{mech}}{d_{mech} + c_h n_{mech}} \tag{11a}$$

$$G_{d,F}(s) = \frac{F_h(s)}{F_r(s)} = \frac{c_h n_{mech}}{d_{mech} + c_h n_{mech}} \tag{11b}$$

and

$$G_{r,x}(s) = \frac{x_p(s)}{Q_v(s)} = \frac{1}{sA_1}\frac{c_h n_{mech}}{d_{mech} + c_h n_{mech}} \tag{12a}$$

$$G_{r,F}(s) = \frac{F_h(s)}{Q_v(s)} = \frac{1}{sA_1}\frac{c_h d_{mech}}{d_{mech} + c_h n_{mech}}. \tag{12b}$$

The reader familiar with high gain concepts may suspect what problems must arise in the control of hydraulic systems. It can be said in advance, that the mechanical system in rolling mills due to eq. (10) contains several zeros, which are determined by the closed loop behavior of the other control loops. Such a closed loop can be the dynamical behavior of the rod pressure, because P_2 is held at a constant value by means of a pressure control valve or the dynamical behavior of the bending cylinder.

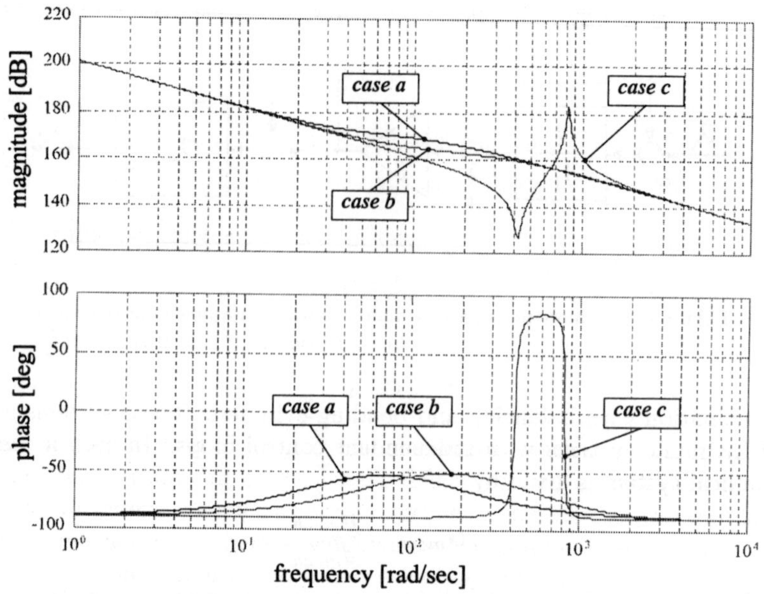

Figure 6. Bodeplots for the force transfer function $G_{r,F}$.

It is a well known fact from the root locus theory, that for $\lim c_h \to \infty$ the poles of the closed loop in eq. (12b) coincide with the zeros of the open loop, see, e.g., Skodestad[13]. Because of this structural feedback one can expect that the pressure transfer function is a system which contains an integrator plus several zeros and poles close to each other similar to figure 6. The position transfer function eq. (12a) can be approximated fairly well with an integrator. This is indeed what one will find in the identification, see also section 5.

3.4   *Pressure Transfer Function*

A very important issue for hydraulic systems that needs to be addressed is the force transfer function $G_{r,F}$ due to eq. (9b), see, e.g., Alleyne[14]. Therefore we will analyze it more detailed in this subsection. It can be assumed that in general the mechanical system G_{mech} is damped very low. This is considered in case c) in the previous subsection. The bodeplots for all 3 cases can be seen in figure 6. An engineering interpretation for case c) is: The system behavior can be approximated by an integrator, but there exists a frequency band where the system has differentiating behavior and this range becomes infinite for $\lim c_h \to \infty$. Concerning rolling mills the case where no material is in the roll gap should also be taken into account, i.e. $c_m = 0$. But in this case, a force control of the actuator makes no sense.

3.5 Influence of Leakages

The influence of leakage flows on the dynamics of the linearized system due to eq. (6) is investigated in this subsection. We distinguish between external laminar leakages (see figure 2) and internal leakages caused by the valve. A linear leakage model (see section 3, figure 2) of the type $Q_{leak} = C_l P_1$ is used in the first step, the physical reason for this leakage may be a radial clearance between the piston and the cylinder or non-perfect seals. With the abbreviation $\widetilde{C}_l = (E\,C_l)/V_0$ it can be shown, that the disturbance transfer functions due to eqs. (9c) and (9b) are modified in the way

$$G_{d,x}^{leak}(s) = -\frac{s + \widetilde{C}_l}{s\left(ms^2 + ds + c_h + c_m\right) + \left(ms^2 + ds + c_m\right)\widetilde{C}_l} \tag{13a}$$

$$G_{d,F}^{leak}(s) = \frac{s\,c_h}{s\left(ms^2 + ds + c_h + c_m\right) + \left(ms^2 + ds + c_m\right)\widetilde{C}_l}, \tag{13b}$$

the corresponding reference transfer functions become

$$G_{r,x}^{leak}(s) = \frac{c_h}{A_1}\frac{1}{s\left(ms^2 + ds + c_h + c_m\right) + \left(ms^2 + ds + c_m\right)\widetilde{C}_l} \tag{14a}$$

$$G_{r,F}^{leak}(s) = \frac{c_h}{A_1}\frac{\left(ms^2 + ds + c_m\right)}{s\left(ms^2 + ds + c_h + c_m\right) + \left(ms^2 + ds + c_m\right)\widetilde{C}_l}. \tag{14b}$$

The zero in eq. (13a) is very interesting as we will see in section 3.7. In section 3.1 the mechanical system G_{mech} eq. (9a) and the overall system $G_{r,x}$ eq. (9c) has been defined. Using the knowledge of this section one can see the relation between both systems in eqs. (14) by the root locus. For the purpose of a physical interpretation it is interesting to note that the same root locus is obtained, if the pressure transfer function $G_{r,F}$ due to eq. (9b) is controlled by a P–controller, but this is closed loop behavior. The reader may check, that the orders of the disturbance transfer functions have changed, compare eqs. (13) and (8) and that the denominators of eqs. (14a) and (14b) become different if $c_m = 0$.

Without any leakages $C_l = 0$ the poles of $G_{r,x}^{leak}$ coincide with the poles of $G_{r,x}$, for $C_l \to \infty$ the poles of $G_{r,x}^{leak}$ coincide with the poles of the pure mechanical system eq. (9a). Figure 7 shows the different root loci for the

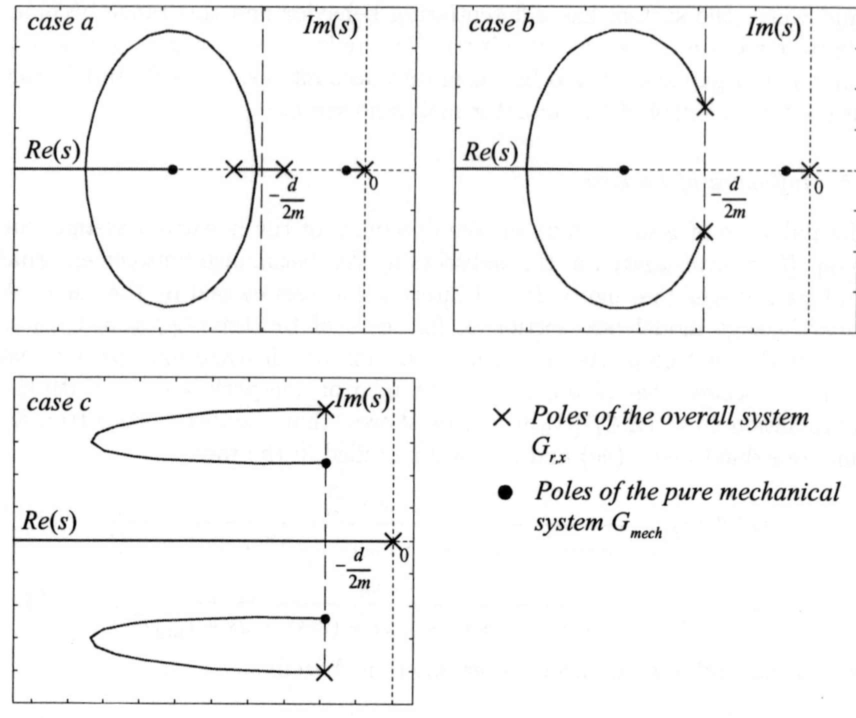

Figure 7. Poles of $G_{r,x}^{leak}$ as root locus with $C_l > 0$.

three cases given in section 3.1, see also figure 4. Considering the different cases in figure 7 it is obvious, that the statement 'more leakage means more damping' is not right, which is proposed in some papers.

At the first glance there is the surprising effect that for $c_m \neq 0$ system eq. (14a) looses it's integrating behavior. Nevertheless for realistic values of C_l one obtains a pole very close to the origin because the third pole is shifted along the real axis into the negative direction (compare figure 7). In this case the system can be approximated fairly well by a first order transfer function with a large time constant. Because the root loci depend continuously on the parameters we conclude that for realistic values of C_l also the other poles of eq. (14a) stay close to ones of eq. (9c). This means that this kind of leakage does not effect the dynamics very much, but what is the influence on the steady state behavior? If the systems eqs. (14a) and (14b) are controlled by

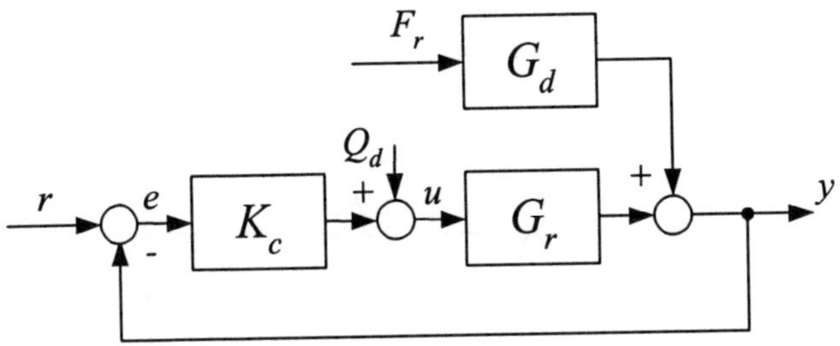

Figure 8. One degree of freedom control configuration using eqs. (14) and (13).

a P–controller with gain K_c in a one degree of freedom scheme due to figure (8), with disturbances Q_d and F_r the steady state outputs for a reference step signal are

$$x_p = \frac{A_1}{C_l\, c_m + K_{c,x}\, A_1} \left(K_{c,x}\, x_{p,ref} - \frac{C_l}{A_1} F_r + Q_d \right) \tag{15a}$$

$$F_h = \frac{A_1}{C_l + K_{c,F}\, A_1} \left(K_{c,F}\, F_{h,ref} + Q_d \right), \tag{15b}$$

and they are independent of E and V_0, because these expressions appear in the same way in c_h and in $\widetilde{C}_l$. From constructional point of view eqs. (15) show also why hydraulic systems are so popular in industrial applications, the steady state behavior depends mainly on A_1 and K_c.

The next interesting question is, what happens if the laminar leakage changes it's sign? The reason could be, that because of radial clearances in the valve there is a flow $Q_{leak} = C_l\,(P_S - P_1)$ to the head chamber. In this case we obtain again eqs. (14a, 14b) with the difference, that C_l becomes negative. The root locus in this case can be seen in figure 9 and obviously this is the more critical case. Considering case c) we obtain a highly instable system. This behavior is the reason, why hydraulic systems in rolling mills are never driven open loop.

In most books the system input is not the flow Q_v but the spool position x_s. For such an analysis the eqs. (7) are linearized in the form

$$Q_v = K_p\, x_s + K_x\, P_1$$

with the flow–pressure–gain K_x and the pressure sensitivity K_p. In this case the parameter C_l should be replaced by

$$C_l := C_l - K_p$$

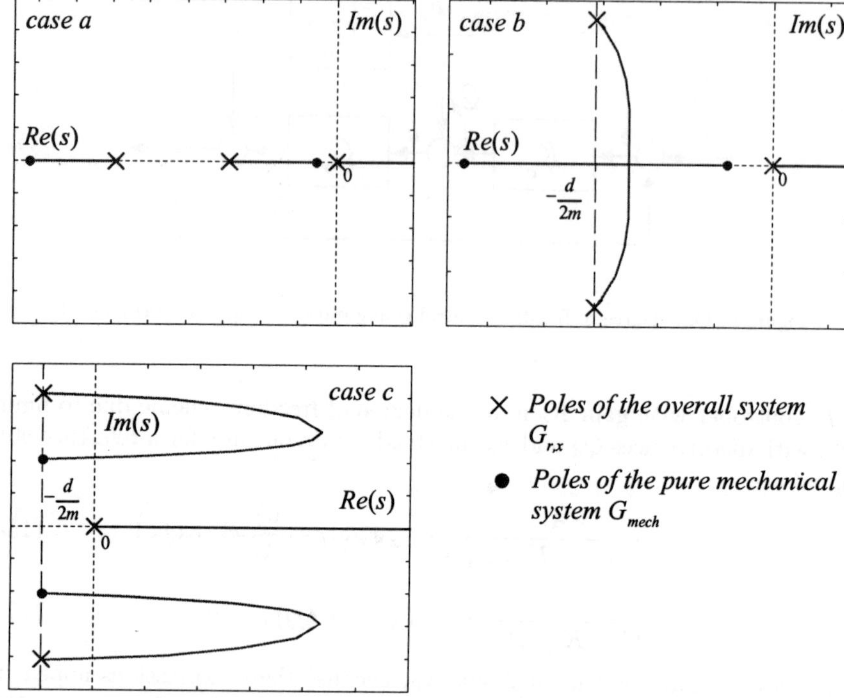

Figure 9. Poles of $G_{r,x}^{leak}$ with $C_l < 0$, i.e., laminar leakage flow from the valve to the piston.

in eqs. (13 and 14), where K_p has always a negative value and the numerator must be multiplied with K_x.

Next we consider nonlinear valve leakages. Eq. (7) gives the flow of a spool-servovalve where x_s is the spool position. For small valve openings, i.e., small values of x_s, this equation should be modified to take into account some nonlinear phenomena like leakages due to annular gaps in a multi stage servovalve or worn valve edges (see figure 10) caused by abrasive oil ingredients. Both effects give an additional turbulent flow Q_d from/to the piston, which depends also on the piston pressure P_1. These effects are time dependent and the simplest way to fit simulation results with measured data is to use an underlaped spool valve in the simulations.

Industrial valves also have some other problems, like offset errors in the spool position (see section 4), some measurements can be found in Lee[15].

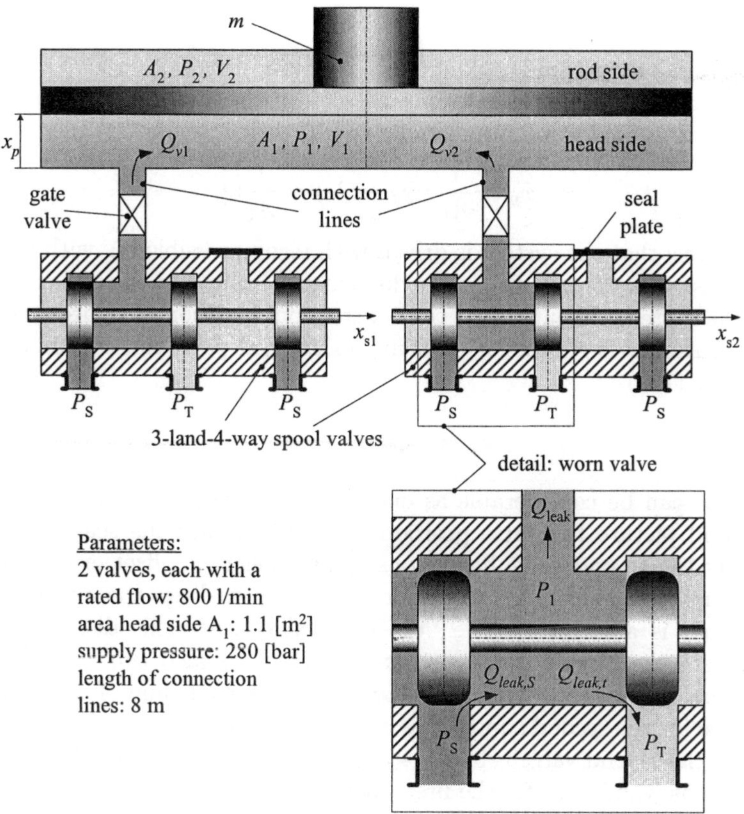

Figure 10. Single acting piston with cut–in–circuit and detail of a worn valve.

Another source for this kind of leakage can be seen in figure 10. The considered system contains a single–acting ram that is connected through the connection lines with two spool valves which are driven parallel in a so called cut–in–circuit (see figure 10). This configuration is often used in rolling mills because of two reasons:

- the used piston actuators have enormous dimensions and so 2 valves are necessary to obtain a sufficient flow

- a standstill of a rolling mill causes high costs (also for the following production lines) and if one valve breaks down, the system can continue with only one valve.

Hydraulic actuators in rolling mills are never driven open loop and so the influence of valve leakages can only be seen in the closed loop. To find the steady state behavior we approximate eqs. 6 with

$$\dot{x}_p = \frac{1}{A_1}\left(Q_d + Q_v\right) = \frac{1}{A_1}Q, \tag{16}$$

which means that the system is driven with incompressible oil, with the valve flow Q_v as system input, Q_d as the disturbance input and the piston position x_p as the system output. If, in the control configuration of figure 8 the gain of the controller is K_c and it contains no integral part, the output becomes due to eq. (15a)

$$x_p = x_{p,ref} + \frac{1}{K_c}Q_d. \tag{17}$$

This error can be considerable as one can see in figure 11. In the first and the second picture the valve position is printed versus the scaled time and obviously the second valve has a considerable negative offset error. Because of this error, the reference value for the valves to keep the system in it's equilibrium is about 2.5 %, i.e., the first valve needs this displacement to compensate the flow of the second valve. After a scaled time of 0.2 we see a reference step in the output to $50\,\mu m$, after a scaled time of about 0.5 the gate valve for the second supply line is closed (see figure 10) and the reference value for the second valve is 0. We see that after this time the reference value which is now valid only for the first valve changes to 1 %, the piston position moves about $0.12\,mm$. A steady state error in the piston position of about $0.12\,mm$ is much in this kind of application. There are two possibilities to avoid this

1. an integral part is incorporated in the controller, in this case the open loop due to eq. (16) becomes a double integrator

2. the leakage is taken into account by the spool offset value (see, e.g., Kugi[1]).

After a scaled time of 0.7 we can see the next $50\,\mu m$ reference step.

3.6   Influence of the Mill Stretch

Objective of this subsection is, to investigate what happens, when the model due to eqs. (6) contains an additional spring. For example the millstretch

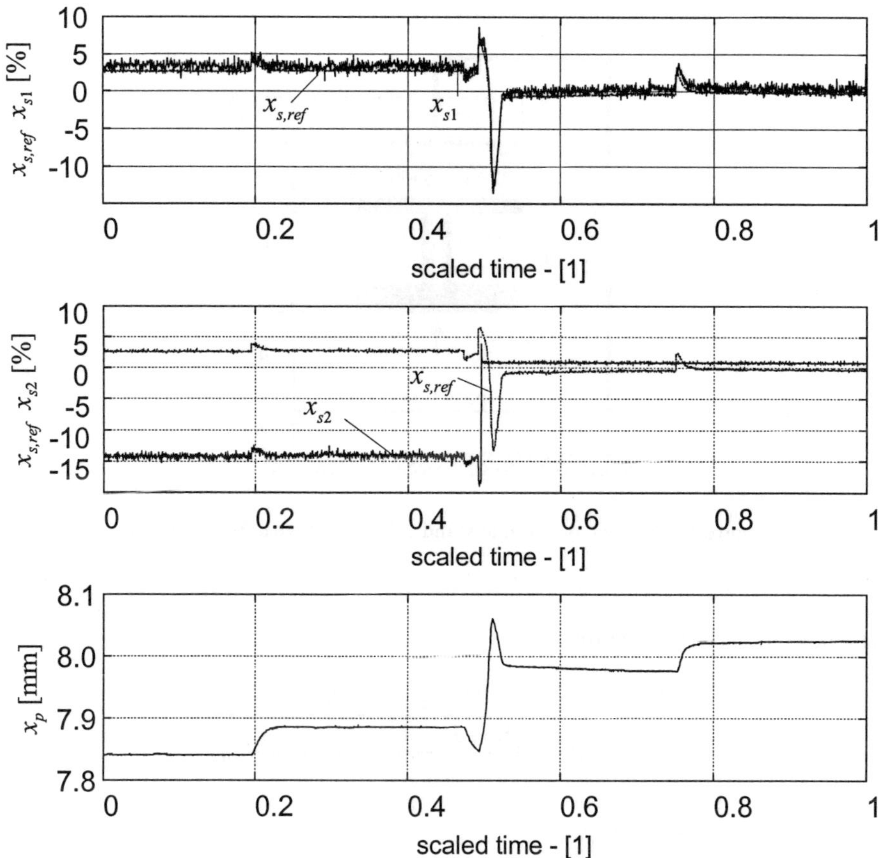

Figure 11. Leakage behavior in a cut–in–circuit.

coefficient c_s is an important parameter for rolling mills, see figure 12. The question is: how does this parameter influence the dynamics of eq. (6) ? With

$$h_{ex} = \frac{F_h}{c_s} - x_p + h_0 \tag{18}$$

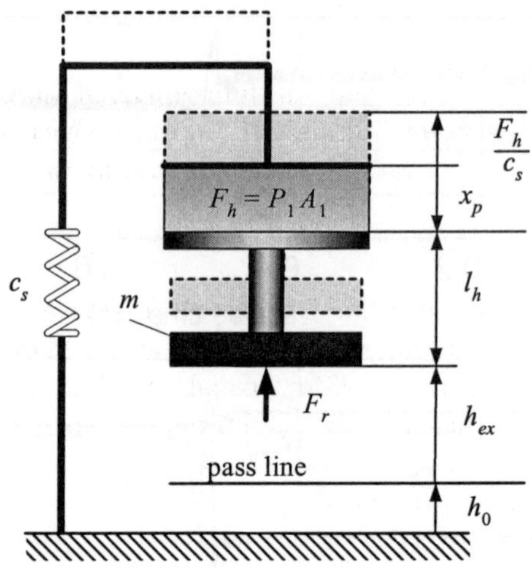

Figure 12. Scheme of a single stand rolling mill with flexible stand.

the state equations become

$$\frac{dP_1}{dt} = \frac{E\left(v_{ex}A_1 + Q_v\right)}{V_p + \dfrac{(E + P_1)\, A_1^2}{c_s} - A_1\left(h_{ex} - l_h\right)} \tag{19a}$$

$$\frac{dh_{ex}}{dt} = v_{ex} \tag{19b}$$

$$\frac{dv_{ex}}{dt} = \frac{-P_1 A_1 + P_2 A_2 - d v_{ex} + F_r - mg}{m}. \tag{19c}$$

Here the constant h_0 in eq. (18) is necessary to incorporate the influence of the arbitrary initial values $F_h|_{t=0} = P_{10}A_1$, $h_{ex}|_{t=0} = h_{ex0}$, $x_p|_{t=0} = x_{p0}$

$$h_0 = h_{ex0} + x_{p0} - \frac{P_{10}A_1}{c_s} \tag{20}$$

with the initial pressure in the forward chamber P_{10}. This value is determined physically by a slow electromechanic adjustment system. The influence of the pressure P_1 in the denominator may be neglected because $E = 1.6 \cdot 10^9$ $[Pa]$ and P_1 is in a range of $0..3 \cdot 10^7$ $[Pa]$. In this case the system due to (19) has the same structure then that of eq. (6), with the difference that the mill stretch coefficient c_s creates an additional volume

$$V_{c_s} = \frac{EA_1^2}{c_s}.\tag{21}$$

Because of this, the observations for the linearized system (see section 3.1) can be extended easily to this case by substituting

$$V_0 := V_0 + V_{c_s}\tag{22}$$

i.e., replacing c_h by the modified stiffness

$$c_h := \frac{c_h c_s}{c_h + c_s}.\tag{23}$$

Another difference is, that the damping d is here proportional to the exit velocity and no longer proportional to the relative velocity between piston and chamber. But for the following sections in the models of single stand rolling mills we replace with $V_0 = V_p + A_1 \left(h_{ex0} + x_{p0}\right)$, such that eq. (19a) looks like

$$\frac{dP_1}{dt} = \frac{E\left(v_{ex}A_1 + Q_v\right)}{V_0 + \dfrac{EA_1^2}{c_s} - A_1 h_{ex}}.$$

As mentioned in the previous section, traditional control concepts for rolling mills are based on eq. (44) and eq. (23).

3.7 The Hydraulic Spring

We have seen in the previous sections, that the input-output behavior of a single acting cylinder is determined by 3rd order transfer functions, see eqs. (14a), (14b), but 3rd order equations are very unhandy. In this section the equations will be rewritten and we will gain additional insights into the system dynamics. Here we will take advantage from the fact, that not only input and output can be measured, but also one state is explicitly known. If the mechanical system is more complicated than a spring-mass-damper system, remember eqs. (12) and (11), this knowledge can be used for the development of reduced order equations.

Considering the position transfer functions eqs. (9c) and (14a), the influence of the hydraulic system eq. (6a) on the mechanical system,

$$x_p\left(s\right) = G_{mech}\left(s\right) F_h\left(s\right)\tag{24}$$

where the input F_h is coupled with the output x_p through a dynamical system, can be summarized as

1) without any leakages eq. (6a) adds an integrator

2) without any leakages eq. (6a) adds in the denominator of G_{mech} a the spring c_h parallel to the system

3) eq. (14a) describes a relation between the overall system and the mechanical system

One should keep in mind that in the case of a laminar leakage all transfer functions eqs. (14a), (14b), (13a), (13b) have the same denominator, differences occur only if $c_m = 0$. In this case the denominators are determined by a 2nd order equation or an integrator plus a 2nd order equation.

The output x_p of the overall system due to the input Q_v, can be visualized physically as

$$\underbrace{x_p\,(s)}_{\substack{\text{actual} \\ \text{piston} \\ \text{position}}} \;=\; \underbrace{\frac{1}{sA_1}Q_v\,(s)}_{\substack{\text{movement} \\ \text{due to incom-} \\ \text{pressible fluid}}} \;-\; \underbrace{\frac{V_0}{EA_1}P_1\,(s)}_{\substack{\text{movement} \\ \text{due to fluid} \\ \text{compressibility}}} \;-\; \underbrace{\frac{C_l}{sA_1}P_1\,(s)}_{\substack{\text{movement} \\ \text{due to} \\ \text{leakages}}}.$$

$$\tag{25}$$

Eq. (25) is analyzed in the following way:

1) If an incompressible fluid is used, i.e., $\lim E \to \infty$ and the system has no leakages then the piston position is determined exclusively by the integrated fluid $\dfrac{Q_v}{sA_1}$. In this case the output is completely independent of the piston pressure P_1. Because of the high gain effect, the loop transfer function in figure 5 becomes unity. If a leakage is taken into account the analysis can be done using singular perturbation techniques.

2) To see the influence of c_h on the output we set $Q_v = 0$ and $C_l = 0$. If a force F_r is applied on the piston, it will move because of the fluid compressibility. Now the output x_p and the input F_h of the original mechanical system eq. (24) are coupled by eq. (8a), which is 2nd order.

3) If $Q_v = 0$ and $\lim E \to \infty$ the system behavior is similar to 1).

All these properties give the idea that the linearized version of eqs. (6)

can be approximated by

$$\frac{dP_1}{dt} = \frac{E\left(Q_v - C_l P_1\right)}{V_0} \tag{26a}$$

$$\frac{dx_p}{dt} = v_p \tag{26b}$$

$$\frac{dv_p}{dt} = \frac{P_1 A_1 - P_2 A_2 - \left(c_m + c_h\right) x_p - F_r - dv_p - mg}{m}. \tag{26c}$$

Here the hydraulic spring appears explicitly in the mechanical system, i.e.,

$$G_{mech,ch} = \frac{A_1}{\left(ms^2 + ds + c_h + c_m\right)}. \tag{27}$$

Computing the position transfer function with $C_l = 0$ one obtains exactly eq. (9c), but if a leakage is considered it becomes

$$G_{r,x}^{leak,s}\left(s\right) = \frac{x_p\left(s\right)}{Q_v\left(s\right)} = \frac{c_h}{A_1} \frac{1}{\left(ms^2 + ds + c_h + c_m\right)\left(s + \dfrac{EC_l}{V_0}\right)}, \tag{28}$$

which is an approximation for eq. (14a). The 3rd pole appears also in the numerator of eq. (13a). If the force transfer function is computed it becomes

$$G_{r,F}^{leak,s} = \frac{x_p\left(s\right)}{Q_v\left(s\right)} = \frac{G_{r,x}}{G_{mech}} = \frac{E}{V_0} \frac{1}{s + \dfrac{EC_l}{V_0}}. \tag{29}$$

Comparing it with eq. (14b) and looking at the bode plots fig. 6, one can see what happens - the mechanical system is completely cancelled, which is obvious, as it must appear in the numerator of $G_{r,F}$. So eq. (6a) adds more than a simple spring plus an integrator. Also the feedback induced by this equation is important because it gives a structural property. The meaning of the pole at

$$s_{0,fixed} = -\frac{EC_l}{V_0} \tag{30}$$

becomes clear if the following experiment is executed. Assume the system due to figure 1 is in an equilibrium and the piston is fixed mechanically. The pressure in the chamber is P_{10} and at time $t = 0$ a port is opened, so that a laminar leakage flow $Q = C_l P$ can flow out of the system. The state equation in the case is given by

$$\frac{dP}{dt} = -\frac{EC_l}{V_0} P. \tag{31}$$

It is important in this experiment, that the piston cannot move. If the same experiment is performed for the system due to figure 2, the piston will move, because of the finite value of the external spring c_m. The dynamics in this case is determined by the homogenous part of eq. (6), with the initial condition $x_0^T = [P_{10}, 0, 0]$. The Laplace transform of this signal is

$$P_1(s) = \frac{V_0}{E} G_{r,F} P_{10} \tag{32}$$

which can be approximated by its quasi–static behavior with $m = 0$, $d = 0$ by

$$P_1(s) = \frac{c_m}{s(c_m + c_h) + \dfrac{EC_l}{V_0} c_m} P_{10} \tag{33}$$

with a pole at

$$s_0 = -\frac{EC_l}{V_0} \frac{c_m}{c_m + c_h}. \tag{34}$$

With realistic parameters eq. (34) is indeed a very good approximation for the 3rd pole of eq. (14b) if $c_m \neq 0$. The simplified version eq. (30) can be obtained from eq. (34) by $\lim_{c_m \to \infty}(s_0)$. So for the analysis of more complicated hydraulic systems the denominator

$$s\left(ms^2 + ds + c_h + c_m\right) + \left(ms^2 + ds + c_m\right) \frac{E}{V_0} C_l \tag{35}$$

will be approximated with

$$\left(ms^2 + ds + c_m + c_h\right)(s - s_0). \tag{36}$$

This means that a leakage influences only the integrating behavior, e.g., the real pole of the system, but not the poles of the mechanical system. For a double acting cylinder one obtains with this approximation instead of a 4th order equation the product of two 2nd order equations. Figure 13 shows the disturbance behavior for a low damped mechanical system when the system is approximated by the dominating pole, see eq. (34). Due to eqs. (26) the disturbance transfer functions

$$G_{d,x,s}(s) = \frac{x_p(s)}{F_r(s)} = \frac{-1}{m s^2 + d s + c_h + c_m} \tag{37a}$$

$$G_{d,F,s}(s) = \frac{F_h(s)}{F_r(s)} = 0 \tag{37b}$$

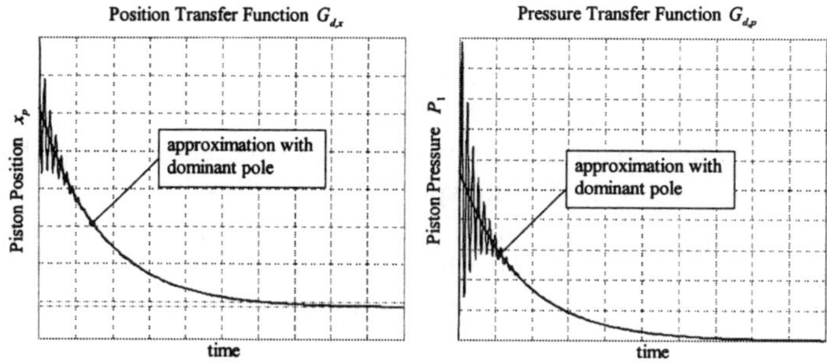

Figure 13. Step responses of disturbance transfer functions $G_{d,x}$ and $G_{d,F}$.

become independent of the leakage. Considering the position transfer functions eqs. (9b) and (14b) one can find similar to eq. (25) a relation for the pressure

$$\underbrace{P_1(s)}_{\substack{\text{actual} \\ \text{piston} \\ \text{pressure}}} = \underbrace{\frac{s}{s + \dfrac{EC_l}{V_0}} \frac{E}{V_0}}_{\substack{\text{leakage} \\ \text{dynamics}}} \left(P_{1,fluid} \underbrace{\frac{Q_v(s)}{s}}_{\substack{\text{pressure} \\ \text{due to} \\ \text{fluid flow}}} - P_{1,piston} \underbrace{A_1 x_p(s)}_{\substack{\text{pressure due} \\ \text{to piston} \\ \text{movement}}} \right).$$

$$\text{actual piston pressure} = \text{leakage dynamics} \quad (\text{pressure due to fluid flow} - \text{pressure due to piston movement})$$

$$(38)$$

It should be noted, that eqs. (25) and (38) do not contain any approximations, eqs. (14a) and (14b) can be obtained by using eq. (24). Eq. (38) can be interpreted as follows.

1) For a rigidly fixed piston i.e., $\lim c_m \to \infty$ the pressure is determined by $P_{1,fluid}$. Because of the leakage the dynamics are determined by eq. (30), see also eq. (31).

2) If the flow $Q_v = 0$ and the piston is movable, one obtains again a pole at eq. (30).

To avoid any confusions it should be repeated that if the input is Q_v or F_r the dominant pole of the system can be approximated by eq. (34), see also eq. (36). If the piston position x_p is taken as input, the dominant pole is at eq. (30), without any approximation.

3.8   Different Loads

In this section we will consider model reductions based on physical considerations. In Merritt[11], p. 295 - 309, one can find different load situations and
this section should give a brief overview. The reader should keep the issues of
the previous subsection in mind, especially the effect if a mechanical system
is driven by a hydraulic subsystem. Starting from eqs. 6 and neglecting the
influence of the laminar leakages, the following special cases can be derived:

Mass dominant load: The state equations are

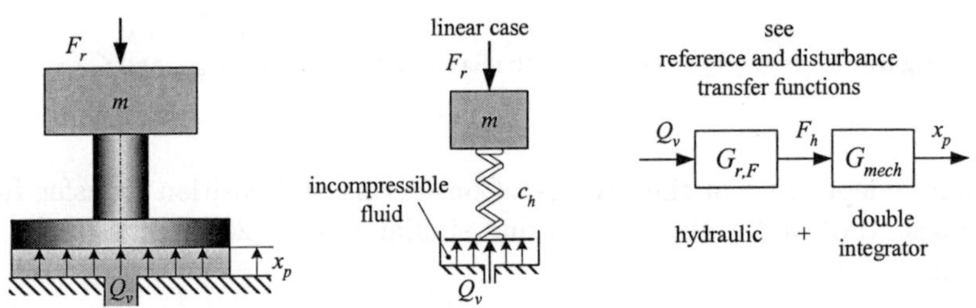

Figure 14. Hydraulic system with mass dominant load.

$$\frac{dP_1}{dt} = \frac{E\left(-v_p A_1 + Q_v\right)}{V_p + x_p A_1} \tag{39a}$$

$$\frac{dx_p}{dt} = v_p \tag{39b}$$

$$\frac{dv_p}{dt} = \frac{P_1 A_1 - P_2 A_2 - F_r - mg}{m}. \tag{39c}$$

The corresponding transfer functions are

$$G_{mech} = \frac{1}{ms^2}$$

$$G_{r,F}\left(s\right) = \frac{c_h}{A_1}\frac{ms}{ms^2 + c_h} \qquad G_{d,F}\left(s\right) = \frac{c_h}{ms^2 + c_h} \tag{40}$$

$$G_{r,x}\left(s\right) = \frac{c_h}{A_1}\frac{1}{s\left(ms^2 + c_h\right)} \qquad G_{d,x}\left(s\right) = \frac{-1}{ms^2 + c_h}.$$

As one can see, a double integrator driven by a hydraulic becomes an un-
damped mechanical system with poles on the imaginary axis plus an integrator.

Viscous force dominant load: The state equations are

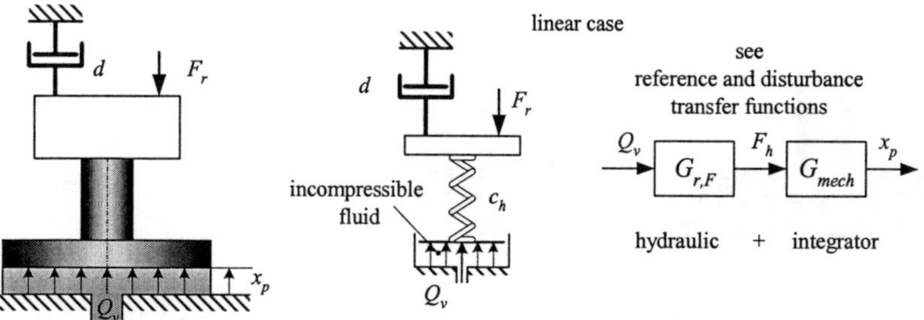

Figure 15. Hydraulic system with viscous force dominant load.

$$\frac{dP_1}{dt} = -\frac{EA_1\left(P_1 A_1 - P_2 A_2 - F_r\right)}{d\left(V_p + x_p A_1\right)} + \frac{EQ_v}{\left(V_p + x_p A_1\right)} \tag{41a}$$

$$\frac{dx_p}{dt} = \frac{1}{d}\left(P_1 A_1 - P_2 A_2 - F_r\right). \tag{41b}$$

The corresponding transfer functions are

$$G_{mech} = \frac{1}{ds}$$

$$G_{r,F}\left(s\right) = \frac{c_h}{A_1}\frac{d}{ds + c_h} \qquad G_{d,F}\left(s\right) = \frac{c_h}{ds + c_h} \tag{42}$$

$$G_{r,x}\left(s\right) = \frac{c_h}{A_1}\frac{1}{s\left(ds + c_h\right)} \qquad G_{d,x}\left(s\right) = \frac{-1}{ds + c_h}.$$

As one can see, the integrator of the mechanical system is transformed into a first order transfer function and the hydraulic equation adds another integrator.

Spring force dominant load: The state equation is

$$\frac{dx_p}{dt} = \frac{EA_1}{EA_1^2 + c_m\left(V_p + x_p A_1\right)}Q_v \tag{43a}$$

$$\text{or} \tag{43b}$$

$$\frac{dP_1}{dt} = \frac{Ec_m}{EA_1^2 + c_m\left(V_p + x_p A_1\right)}Q_v.$$

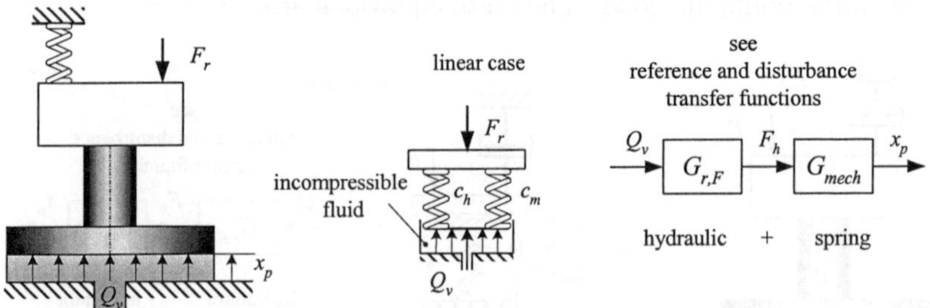

Figure 16. Hydraulic system with spring force dominant load.

The corresponding transfer functions are

$$G_{mech} = \frac{1}{c_m}$$

$$G_{r,p}(s) = \frac{c_h}{A_1} \frac{c_m}{s\,(c_h + c_m)} \qquad G_{d,F}(s) = \frac{c_h}{c_h + c_m} \qquad (44)$$

$$G_{r,x}(s) = \frac{c_h}{A_1} \frac{1}{s\,(c_h + c_m)} \qquad G_{d,x}(s) = \frac{-1}{c_h + c_m}.$$

Traditional approaches for the thickness or force control of rolling mills are based on the linearized versions of eqs. (43) namely eqs. (44). It is an interesting exercise to see what happens if a leakage is included in the transfer function due to section 3.5.

4   Model of a Servovalve

Objective of this section is to derive a simple analytical model of a servovalve that is sufficient for a simulation. There exist many sophisticated models in the literature for this devices (see, e.g., Merritt [11], Murrrenhoff[16] or for concrete measurements Lee [15]), because a servovalve is a highly complex non-linear system that contains many dirty effects like friction, hysteresis, offset errors, saturation and so on. Unfortunately most constructive parameters of such a valve are not available from the manufacturers and it is hard to identify them, so we restrict to the case that only measured step responses are known. In this chapter we give a short algorithm to approximate the dynamics of a multi stage servovalve with a simple nonlinear model.

An example of a two-stage critical center servovalve can be seen in figure

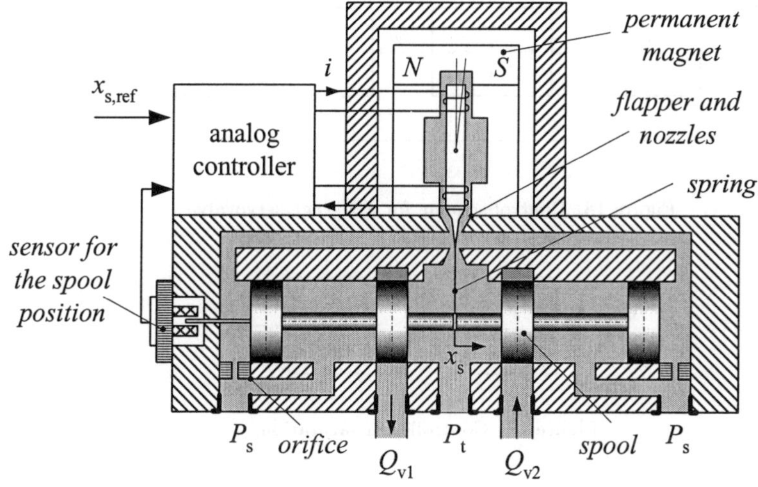

Figure 17. Scheme of a two-stage servovalve.

17. Most of this valves use an integrated analog controller which is designed
to work in saturation for high input amplitudes.

The input to this plant is the reference spool position $x_{s,ref}$, the output
is the actual spool position x_s. We will not perform a detailed analysis of such
a system here, this can be found in Merritt[11] or Blackburn[17]. The dynamics
can be approximated with an integrator to model the spool behavior plus
the dynamics for the solenoid-flapper-configuration. Here, the last part is
assumed to be a first order transfer function, the complete control scheme can
be seen in figure 18, the analog controller is of P-type.

Referring to the step responses of typical data sheets one can see that
during a certain time interval of the step response the system is in saturation,
see figure 21. In this case the dynamics are determined by the integrating
behavior of the main spool, i.e., the system is driven open loop.

In this simple model there are three parameters that must be determined
by the step response, P, V and a and different criterions may be used to choose
them. The following way is suggested: assuming that the control loop is in
saturation for steps higher than 20% of the maximum spool stroke, we can
use a simplified open loop structure to determine V and a (see fig. 19) The

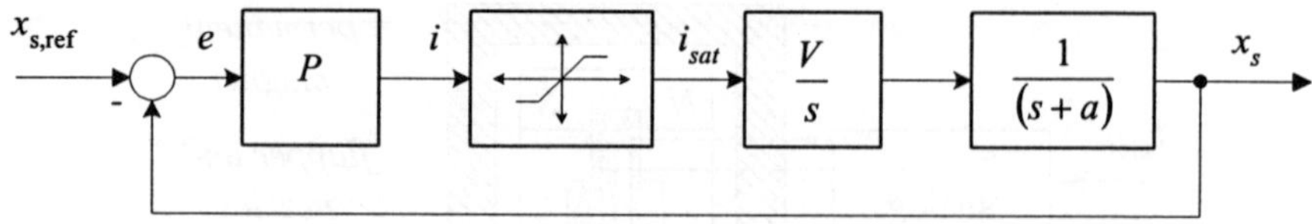

Figure 18. Control scheme for a 2-stage-servovalve.

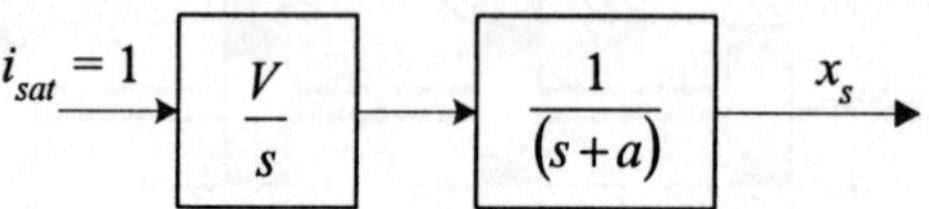

Figure 19. Controller in saturation.

step response is

$$x_s\left(t\right) = \frac{V}{a^2}\left(at - \left(1 - \exp\left(-at\right)\right)\right),\tag{45}$$

the plot fig. 20, shows eq. (45) with evaluated parameters

$$V = 3$$
$$a = 10$$

Comparing fig. 20 with existing data sheets the parameter $\frac{1}{a}$ can be determined very simple. To determine the parameter V one must find a range in the step response where x_s can be assumed to be

$$x_s\left(t\right) = \frac{V}{a}t$$

then the approximation of the time derivative by $\dot{x}_s\left(t\right) \approx \dfrac{\Delta x_s}{\Delta t}$ gives the way to compute V with eq. (46)

$$V = a\frac{\Delta x_s}{\Delta t}.\tag{46}$$

The 3rd parameter P is determined by the saturation condition eq. (47)

$$\text{valve in saturation at: } Pe \geqslant 1\tag{47}$$

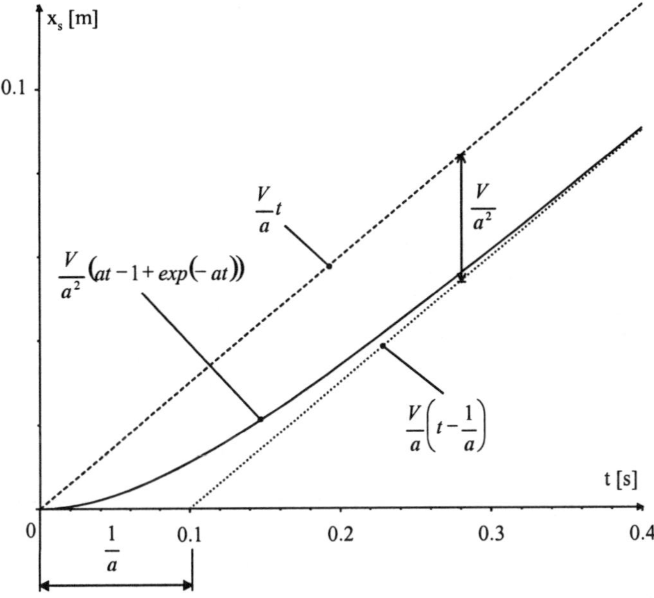

Figure 20. Step response of open loop.

If we assume that the valve works like a linear system for errors $e \leq 0,2\, x_{s,\max}$ the parameter P is

$$P = \frac{1}{0.2\, x_{s,\max}} = \frac{5}{x_{s,\max}}. \tag{48}$$

A valve is always used in combination with an actuator, see, e.g., figure 2. The input to such an actuator is the flow which is determined by the spool position x_s and the pressure due to eq. 7. But for small valve displacements this equation must be modified, because in this case the flow is dominated by leakage flows. The reason for this leakages is that a practical valve has radial clearances and worn spool edges because of abrasive oil ingredients, see Lee[15]. This should be taken into account in the simulation because of the influence on the steady state behavior of the piston.

5   Identification

In the preceding sections different hydraulic systems where considered and analyzed with the objective to get a deeper understanding for the relevant

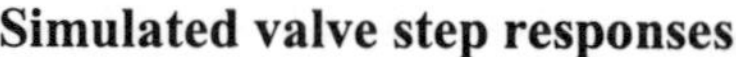

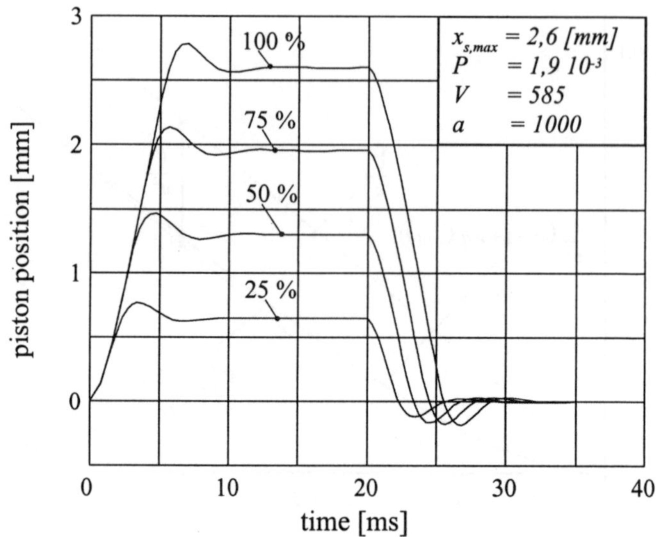

Figure 21. Simulation: step response of closed loop.

parameters. In this section we will compare the analytical linearized model of a rolling mill with sampled data.

The model under consideration can be seen in figure 22. It is a single acting cylinder with a pressure control valve for the rod side. The equations for this system are not derived in this contribution and the reader is referred to [18]. There a spring force dominant case is assumed. A mill stand consists of two adjustment systems which are often called operator side (OS) and drive side (DS), or sometimes north and south side. Each side can be controlled individually with a servovalve. Mechanically they are coupled through the backup and work rolls and through the rod side pressure, because both rod sides are connected with the same pressure reduction valve. In the following bode plots, which are based on the FFT transformation, this coupling will be neglected and one will find, that both sides behave nearly equal.

Because of the influences in section 3.5 it is impossible to drive the system open loop and so, the plots are based on the sampled data of the closed loop system. The used position and force controllers are nonlinear and they are presented in section 6.

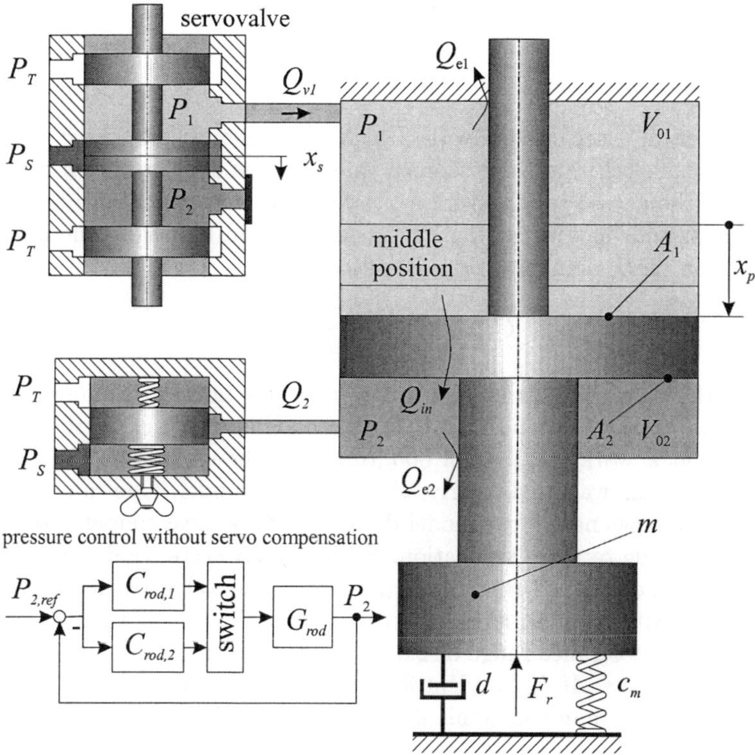

Figure 22. Double acting cylinder with pressure control valve on the rod side.

The detailed analysis of the measurements shows further, that the system contains an additional important nonlinearity, namely, the pressure control valve for the rod side. Because this valve has only a mechanical feedback, the square root function which determines the flow cannot be compensated trough a nonlinear controller. Another problem, which is natural for hydraulic systems, takes effect in this case and can be seen in figure 22 - the non-smooth right hand side due to eq. (7). To include this in the simulation models one should use different gains for both directions. In the analyzed configuration the rod side pressure is for itself an independent closed loop dynamical system, as it is sketched in figure 22.

Because of all these nonlinearities, the system response is dependent from the operating point. As operating points we will use no load, i.e., $c_m = 0$, a load of 800 [*tons*] and 1600 [*tons*].

5.1 *Servovalve*

Due to section 4 the servovalve is a highly complex dynamical system. The input to this system is the reference value for the spool position $x_{s,ref}$, the output is the actual spool position x_s. Nothing is known about the internal controllers, and, looking at the FFT bode plots in figure 23, the dynamical behavior in the z–domain is approximated by

$$G_{valve}\left(z\right) = \frac{x_s}{x_{s,ref}} = \frac{1}{z^3}.$$

Further the data shows, that the square root function for the flow eq. (7) is a sufficient approximation. Figure 24 shows a step response of the hydraulic force F_{h1} at a work roll load of 800 [*tons*]. In the simulation the measured spool position x_s was used to drive the analytical model. One can see that for operation regimes near 0 the model does not fit the measurements because of the leakages due to valve truncation. So for steady state conditions the valve leakages should also be taken into account. The valve parameter K_d was determined with least squares algorithms. The identified value is for both sides within a tolerance range of 2.5 [%] of it's nominal value, which is given in the data sheets. If K_d is identified in operating points near 0 it becomes two times higher than the nominal value.

5.2 *Millstretch*

A parameter which is often measured in rolling mills is the millstretch, see also subsection 3.6. How is it measured? In the system due to figure 12 the upper work–rolls are pressed against the lower ones. In a quasi–static approach the hydraulic force F_h of OS and DS is measured as a function of the piston displacement x_p. A measure millstretch calibration curve can be seen in figure 25 and obviously the relation is a linear one.

5.3 *Coulomb Friction Load*

In this section a rolling mill is identified with no load in the roll gap i.e., $c_m = 0$ and it was impossible to obtain a convincing plot for the force transfer function. As a matter of fact the spring force dominant load is not a good approximation in this case because of $G_{r,F} = 0$. With no load the system is always driven in position control mode. The following analysis is based on figure 27, see also figure 5. Due to Merritt[11] the input-output characteristic

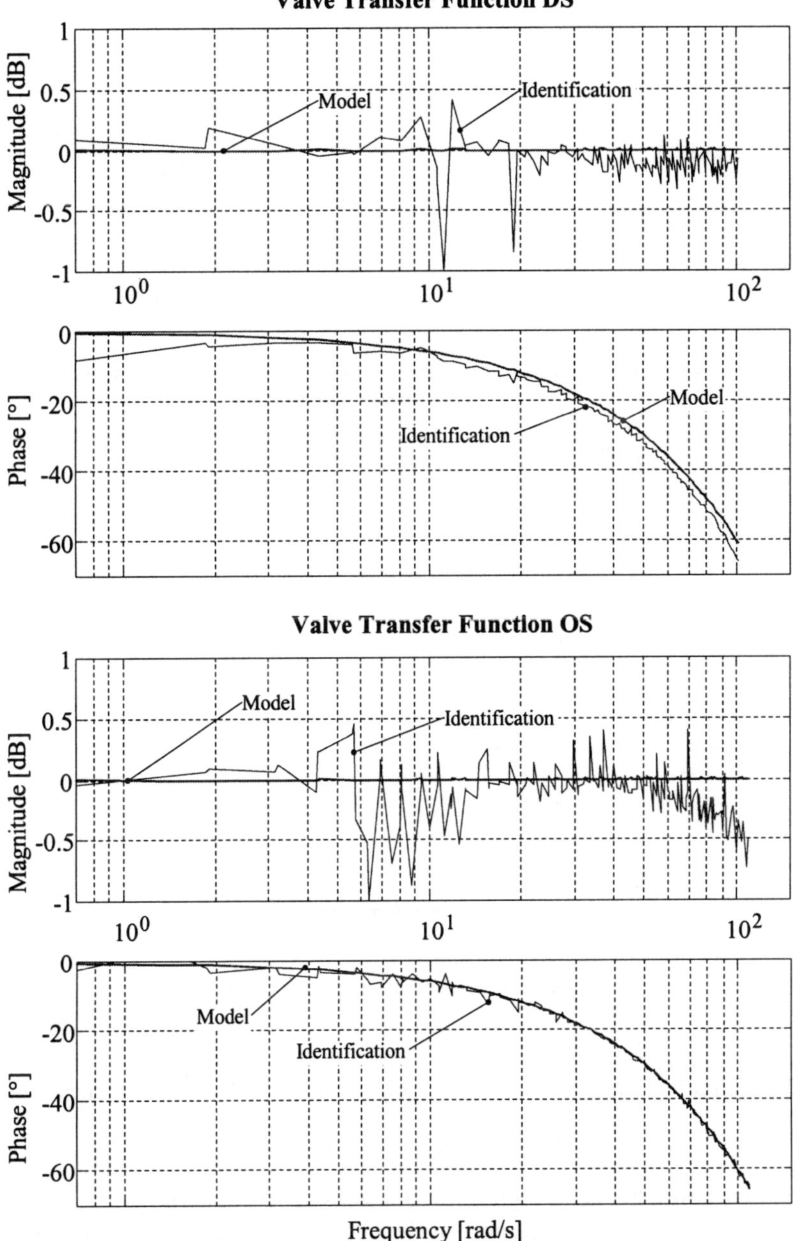

Figure 23. Identification of the servovalve.

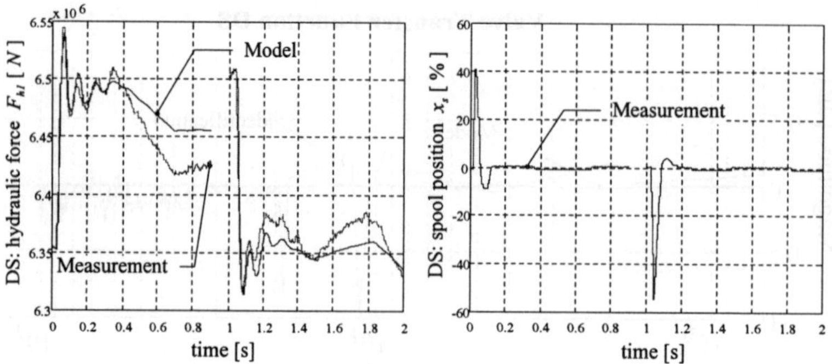

Figure 24. Valve behaviour near 0, hydraulic force F_{h1}, measurement and simulation.

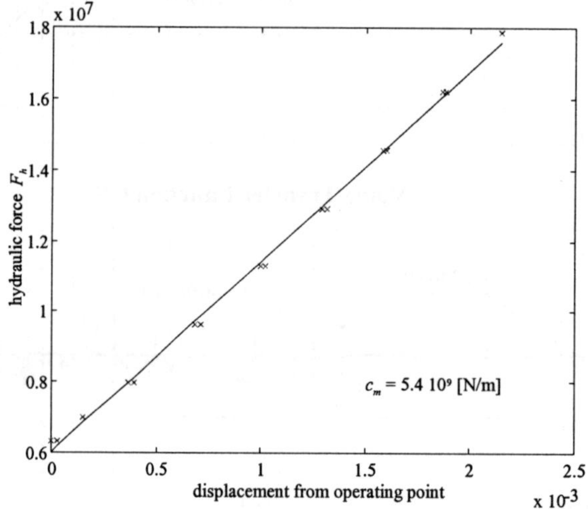

Figure 25. Measured millstretch curve.

of the mechanical system G_{fric} with spring dominant load and a friction force

$$F_{fric} = F_c \text{sign}(v_p) + (F_s - F_c) \exp(-c|v_p|) \text{sign}(v_p)$$

is given by figure 27. Although $c_m = 0$ it is assumed that the influence of the rod side pressure plus the influence of the bending system can be approximated by a spring $\tilde{c}_m$, and so we assume again a spring force dominant case. It should

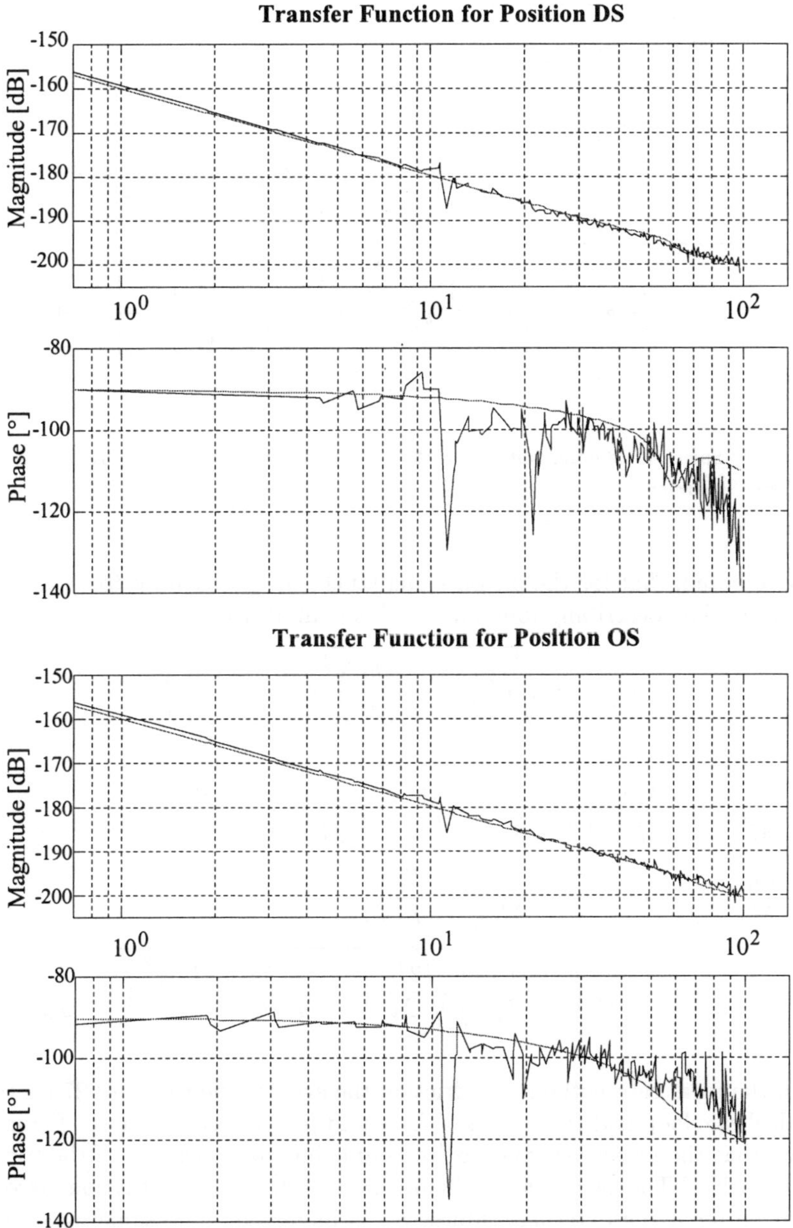

Figure 26. Identification with no load: position transfer functions for DS and OS.

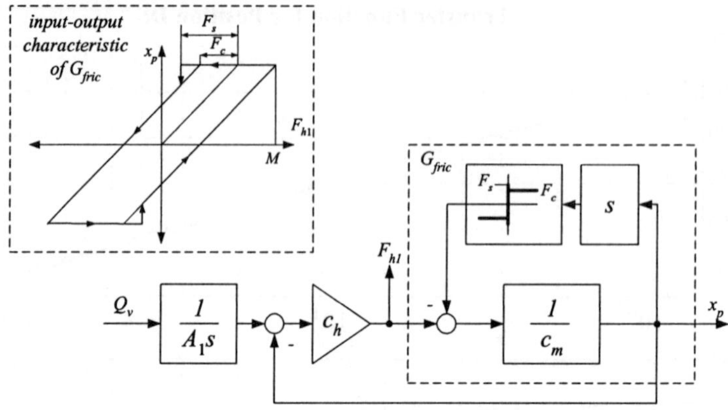

Figure 27. Spring force dominant case with Coulomb friction load and input–output characteristic of G_{fric}.

also be noted, that for the piston being stuck, the position feedback becomes also zero. The describing function for this nonlinearity is

$$G_{fric} = \sqrt{\left(\frac{a_1}{M}\right)^2 + \left(\frac{b_1}{M}\right)^2} \angle \tan^{-1}\left(\frac{a_1}{b_1}\right) \tag{49}$$

where

$$\frac{\pi a_1}{M} = 4\left(\frac{F_c}{M}\right)^2\left(1+\frac{F_s}{F_c}\right) - 4\frac{F_c}{M} - \left(\frac{F_c}{M}\right)^2\left(1+\frac{F_s}{F_c}\right)^2$$

$$\frac{\pi b_1}{M} = \frac{\pi}{2} + \sin^{-1}\left[1 - \frac{F_c}{M}\left(1+\frac{F_s}{F_c}\right)\right]$$

$$+ \left[1 - \frac{F_c}{M}\left(3 - \frac{F_s}{F_c}\right)\right]\sqrt{\frac{2F_c}{M}\left(1+\frac{F_s}{F_c}\right) - \left(\frac{F_c}{M}\right)^2\left(1+\frac{F_s}{F_c}\right)^2} \tag{50}$$

So in the ideal case without any friction the system due to figure 27 contains an algebraic loop and the signals of force and position are in phase. Figure 28 shows that the measurements are far from being ideal, because of the friction nonlinearity. The phase lag in figure 28 is $-50\,[°]$, the situation becomes much better if the external stiffness c_m is increased. The same experiment was performed with a work roll load of 800 [*tons*]. There the phase lag was only $-16\,[°]$, and because friction is less important in such a situation one

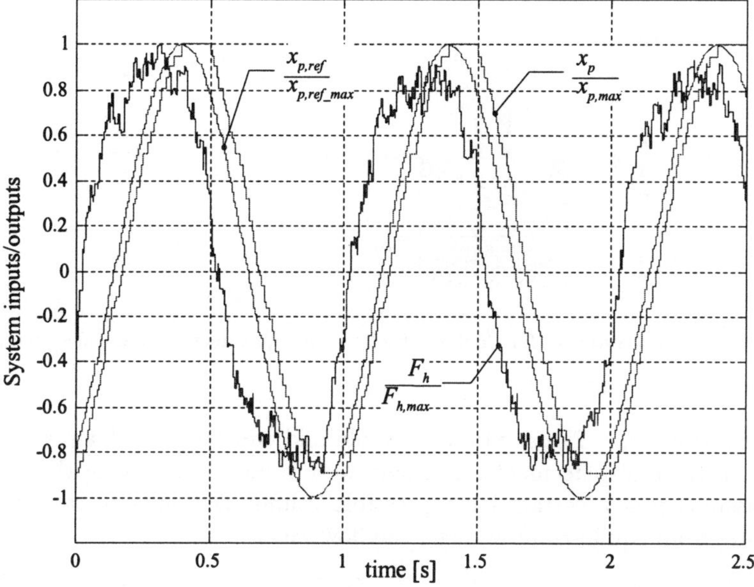

Figure 28. Hydraulic force F_h and piston position x_p with sinus–exitation.

will find more bode plots in the next subsections. The identification of the friction parameters from eq. (49) is costly and here we propose a simpler way.

For the following we assume that an integrator is a sufficient good approximation for the piston position and the rest of the system should be described by some characteristics. Typical identification schemes for friction are based on a scheme as depicted in figure 29. Here as nonlinearity appears the simplified version of figure 28. So in this case a spring dominant load is assumed and it is necessary to measure the spring position x_2 as a function of x_1. Inspired from the figure 3 it is assumed that the flow from the valve to the chamber is incompressible. Now the piston position x_p has the same meaning than x_2 and can be plotted versus the integrated flow, which is an approximation of x_1, see figure 29 for more explanation. We refer to system eqs. (6) with $m = d = c_m = 0$ and a piston is being stuck, because of Coulomb friction. The pressure becomes

$$\frac{\mathrm{d}P_1}{\mathrm{d}t} = \frac{E}{V_p + x_p A_1} Q_v \tag{51}$$

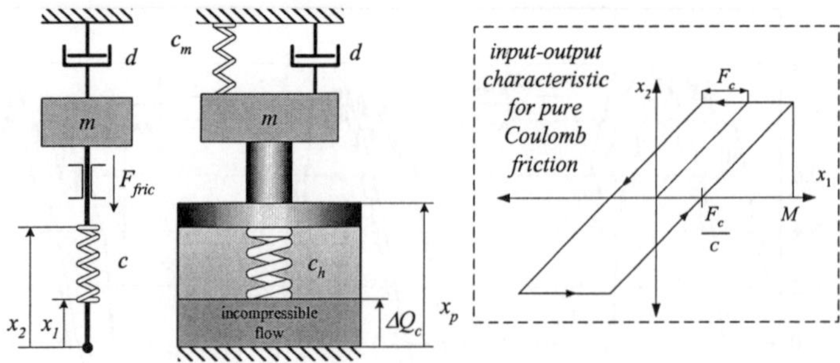

Figure 29. Hysteresis for Coulomb friction and equivalent mechanical system.

with constant x_p. The solution is simple and can be given for positive and negative valve displacements. Actually we are not interested in the pressure and assume that a certain incompressible volume flow goes to the chamber. Starting from equilibrium we need the flow ΔQ_v

$$\frac{E}{V_p + x_p A_1} \int Q_v \mathrm{d}t = \frac{E K_d}{V_p + x_p A_1} \int \sqrt{P_S - P_1} x_s \mathrm{d}t = \frac{E \, \Delta Q_v}{V_p + x_p A_1} = F_c \quad (52)$$

to overcome the sticking condition. After this the pressure is

$$P_1 = \frac{F_c}{A_1} \quad (53)$$

and the piston position must be

$$x_p = \frac{1}{A_1} \int Q_v \mathrm{d}t = \frac{K_d}{A_1} \sqrt{P_S - \frac{F_c}{A_1}} \int x_s \mathrm{d}t. \quad (54)$$

So we get the same input-output characteristic of figure 29. For negative valve displacements and/or $c_m \neq$ a similar analysis can be done and we conclude, that eqs. (52) and (54) can be used to measure the system friction. With the definition of the integrated flow per area

$$x_{p,C} = \frac{1}{A_1} \int Q_v \mathrm{d}t \quad (55)$$

and the intersections of the hysteresis curve with the x–axis from eq. (52) is

$$x_{p,C0} = \frac{\Delta Q_v}{A_1}. \quad (56)$$

Physically this gives a position, namely the amount of volume per area which is necessary to overcome the sticking condition for the piston. The diagrams

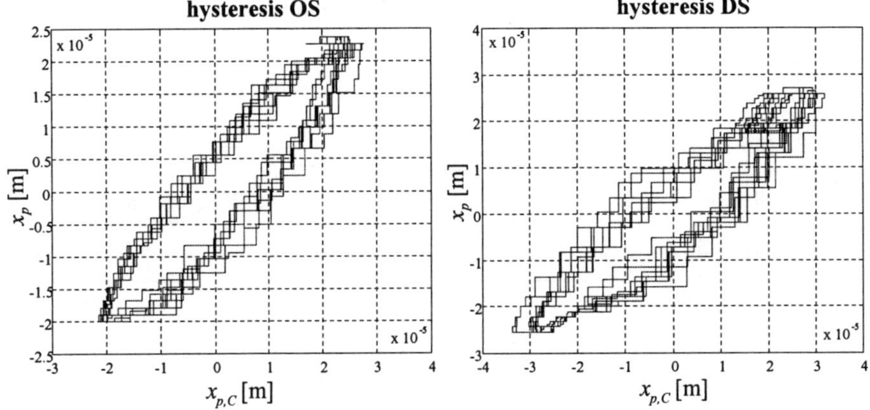

Figure 30. Input-output behavior for OS and DS with Coulomb friction.

for OS and DS can be seen in figure 30. In both cases a flow of about $7.5\,\mu m \cdot A_1$ is needed to overcome sticking.

5.4 Work Roll Load — 800 tons

In this section the identification is based on closed loop system responses where the work rolls are pressed against each other to give a spring force dominant load. A mill stand consists of two adjustment systems which are often called operator side (OS) and drive side (DS), or sometimes north side and south side. Both sides are identified, because the gain of the controller was different for both sides. For the drive side the gain of the P–controller in the outer loop is about twice as high as for the operator side.

5.5 Work Roll Load — 1500 tons

In this section operator and drive side are identified, similar to subsection 5.4 with the difference that both sides have the same controller gain. Unfortunately the excitation was smaller than in the previous case but one can see, that the identified model fits the equations very well. During the excitation the work rolls were driven at a constant speed of $\omega_r = 1.3\ [rad/s]$ to avoid damages. The influence of this disturbance is also visible in the figures. The FFT–identification under rolling conditions is also possible, but as the excitation is very small, one gets a strong reduced frequency range in this case.

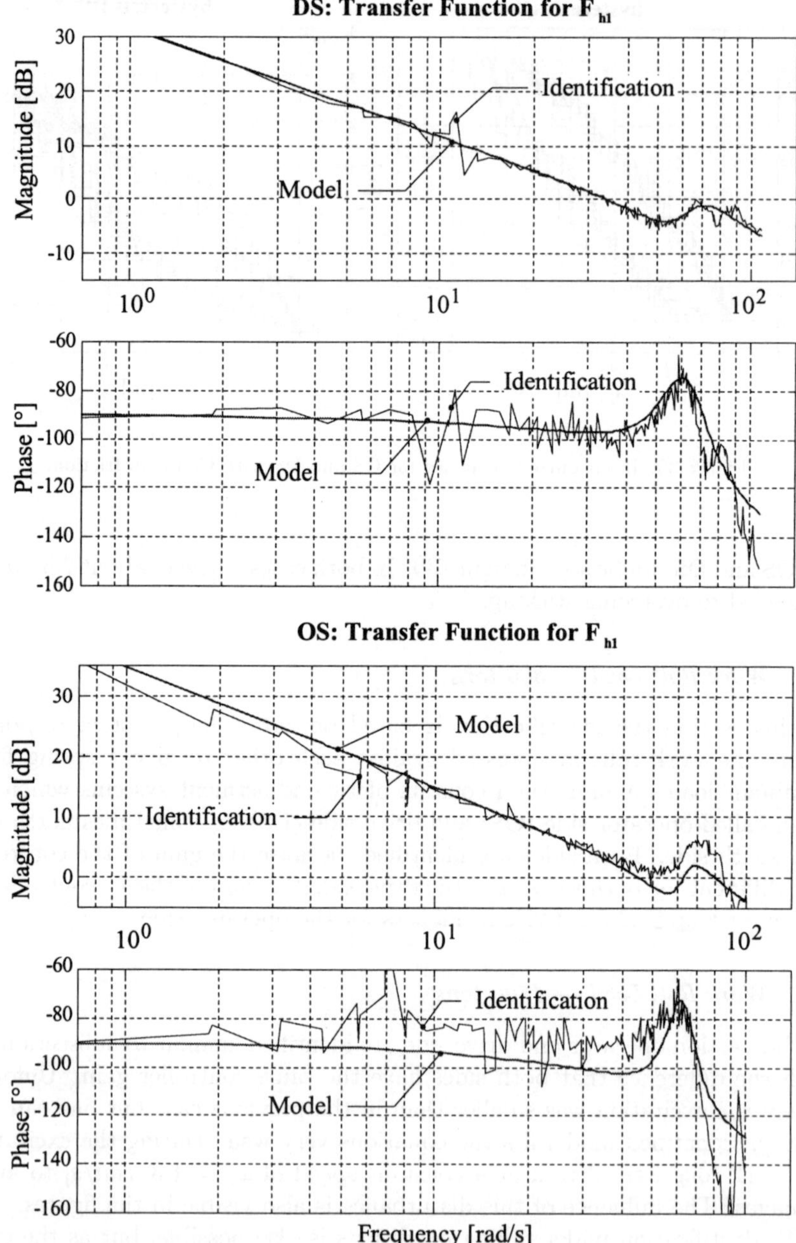

Figure 31. Identification of force transfer function with work roll load of 800 tons.

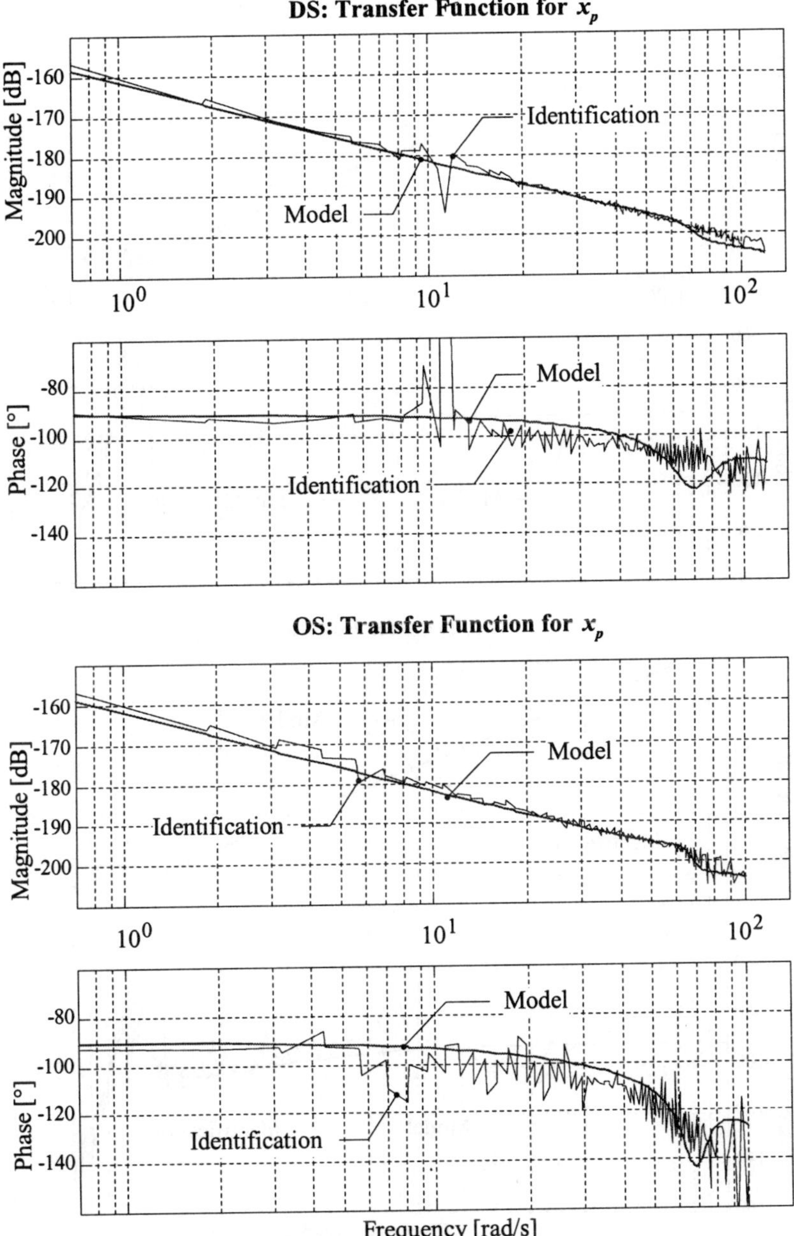

Figure 32. Identification of position transfer function with work roll load of 800 tons.

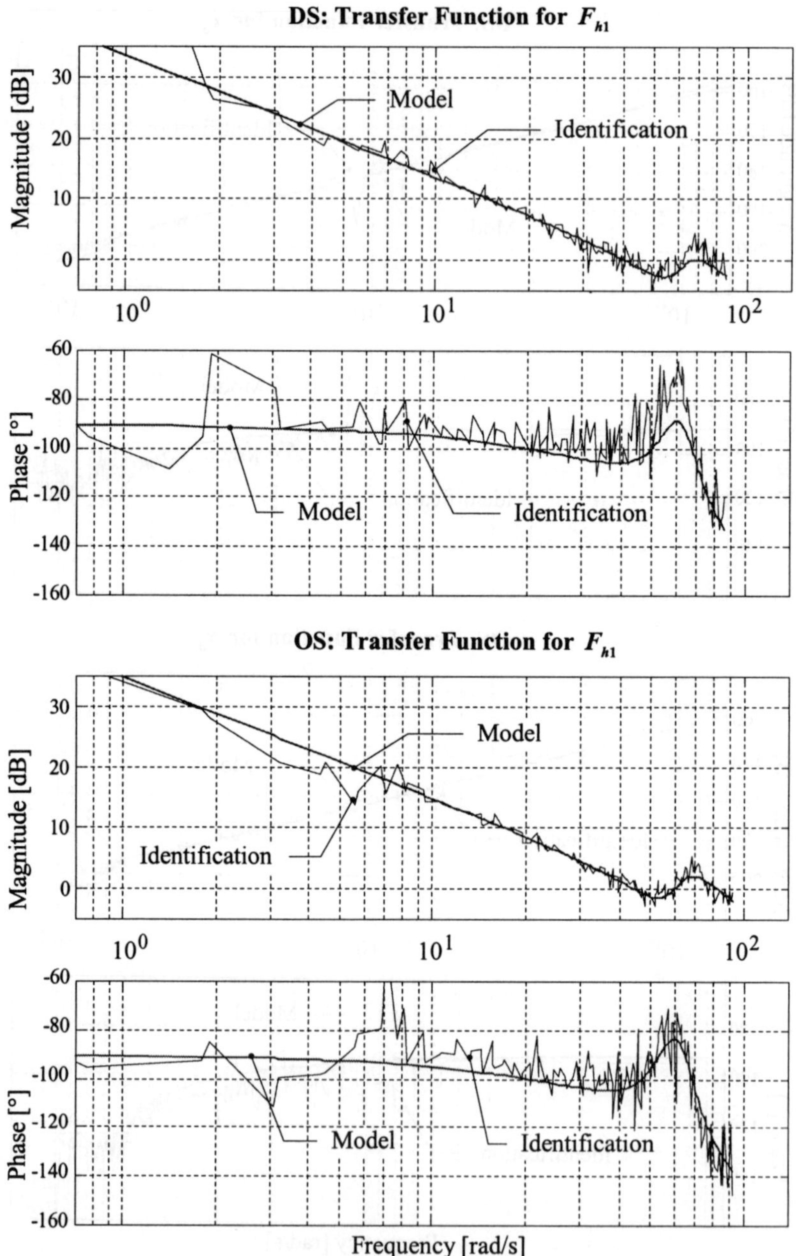

Figure 33. Identification of force transfer function with work roll load of 1500 tons.

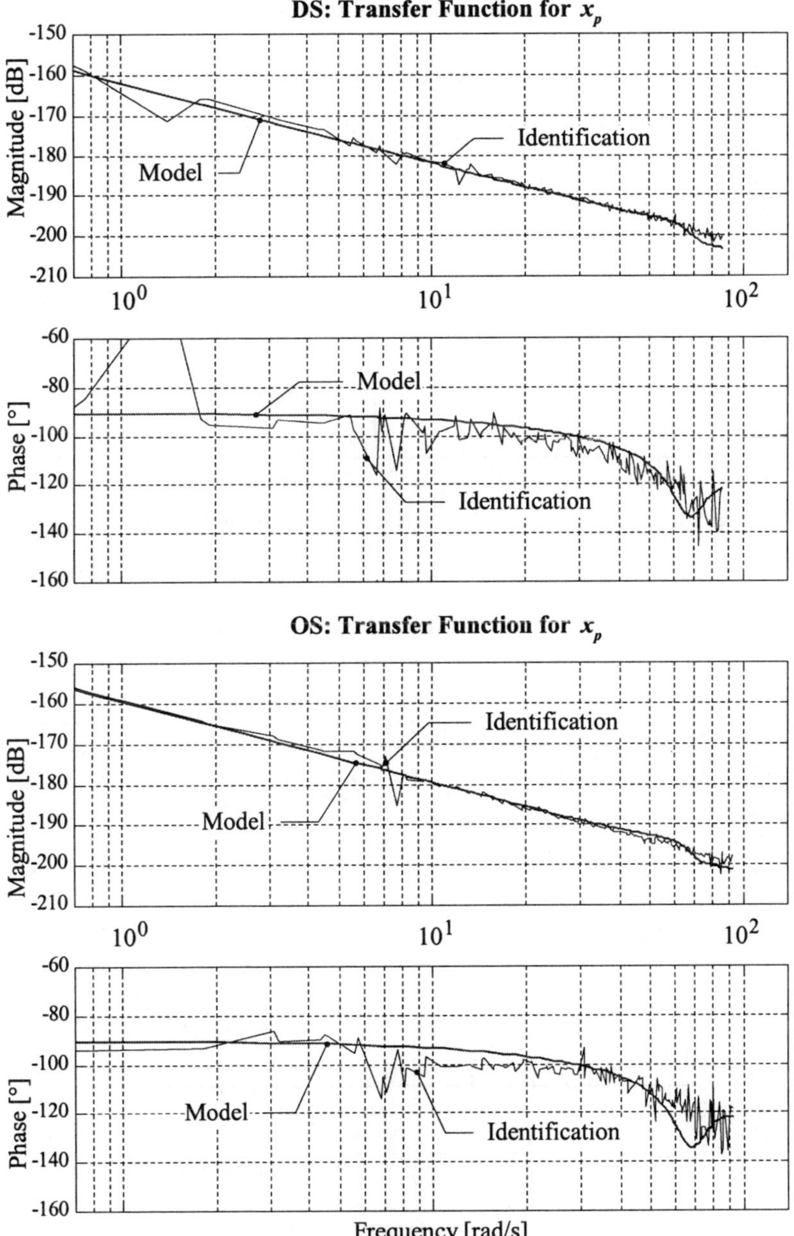

Figure 34. Identification of position transfer function with work roll load of 1500 tons.

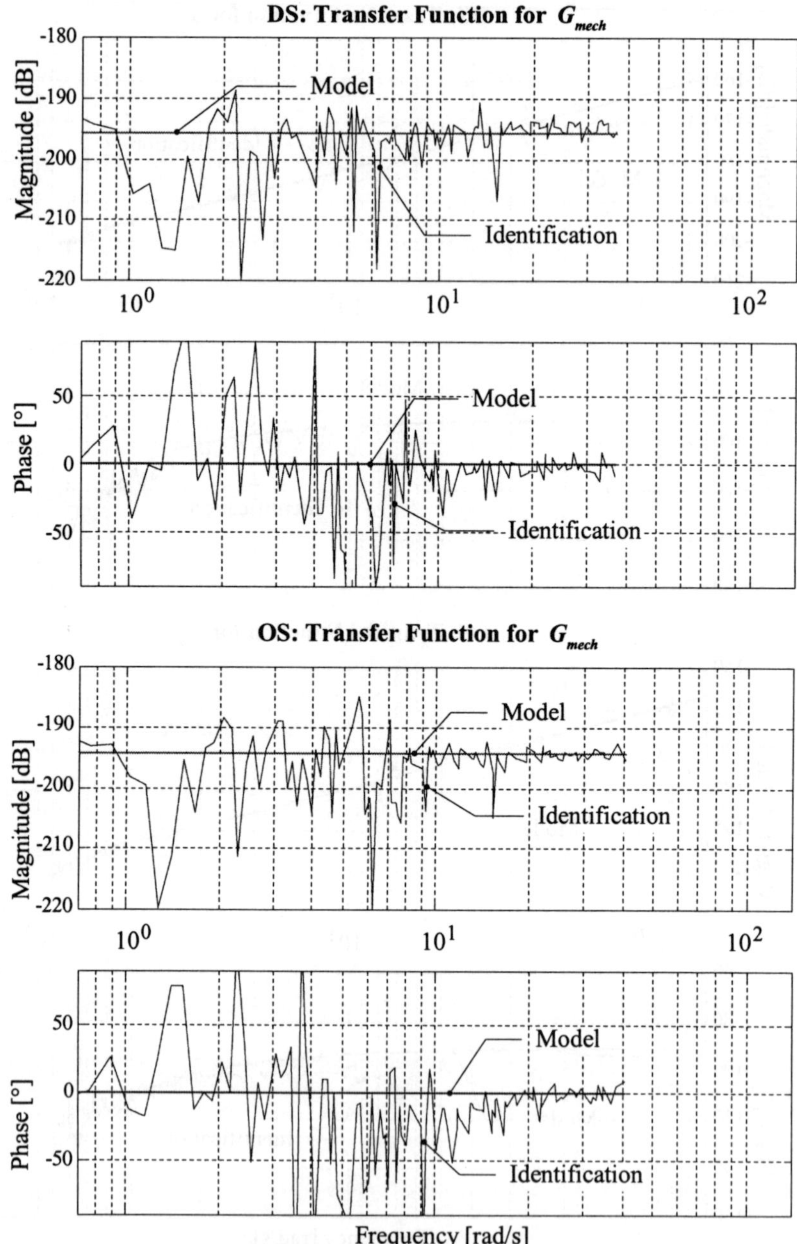

Figure 35. Identification of eq. (9a) with work roll load of 1500 tons.

The analysis shows further, that the resonance in figure 33 should be taken into account by a notch filter.

6 Nonlinear Control

An intuitive solution to get a linear behavior in eq. (6a) is to set

$$Q_v = \frac{(V_0 + x_p A_1)}{E} v + v_p A_1 \tag{57}$$

with v as new system input. Unfortunately one needs the velocity signal to cancel the influence of the mechanical system v_p in the first equation. One well established method in nonlinear control theory is the method of Input–State–Linearization or Input–Output–Linearization, see, e.g., Isidori[3], Nijmeijer[4]. It is easy to show that the system eq. (6) with the output x_p is flat, i.e., an Input–State–Linearization is possible in this case. Using the coordinates

$$z_1 = x_p \tag{58}$$
$$z_2 = v_p$$
$$z_3 = \frac{1}{m} \left(P_1 A_1 - P_2 A_2 + mg - dv_p - c_m x_p \right)$$

the equations become

$$\frac{dz_1}{dt} = v_p = z_2 \tag{59}$$
$$\frac{dz_1}{dt} = \frac{1}{m} \left(P_1 A_1 - P_2 A_2 + mg - dv_p - c_m x_p \right) = z_3$$
$$\frac{dz_1}{dt} = -\frac{\left(c_m z_2 + dz_3 \right) \left(V_p + z_1 A_1 \right) + A_1^2 E z_2 - A_1 E Q}{m \left(V_p + z_1 A_1 \right)}$$

where we have neglected the laminar leakage $Q_{leak} = C_l P_1$. Denoting the Lie derivative of a scalarfield h along the integral curve of the vectorfield f with $L_f h(x)$ and its r–times repeated derivative with $L_f^r h(x)$ the right hand side of eq. (58) can be generally expressed as

$$\frac{dz_i}{dt} = L_f^i h(x), \qquad i = 1, .., n-1 \tag{60}$$
$$\frac{dz_n}{dt} = L_f^n h(x) + L_g L_f^{n-1} h(x)$$

where n is the order of the system. Via the input transformation

$$Q = -\left((v - a_0 z_1 - a_1 z_2 - a_2 z_3) \frac{m\,(V_p + z_1 A_1)}{A_1 E} - \right.$$

$$\left. (c_m z_2 + d z_3)\,(V_p + z_1 A_1) - A_1^2 E z_2 \right), \tag{61}$$

generally written in the form

$$u = \frac{v - \sum_{k=0}^{n-1} a_k L_f^k h\,(x)}{L_g L_f^{n-1} h\,(x)} \tag{62}$$

one obtains the linear system

$$\frac{dz_1}{dt} = z_2 \tag{63}$$

$$\frac{dz_2}{dt} = z_3$$

$$\frac{dz_3}{dt} = -a_0 z_1 - a_1 z_2 - a_2 z_3 + v.$$

The problem in this case is that the new coordinates z_2 and z_3 contain the velocity signal v_p, but usually this signal is not measured. Motivated by eq. (9b) one can choose for instance

$$a_0 = 0, \quad a_1 = \frac{c_m}{m}, \quad a_2 = \frac{d}{m}$$

and eq. (61) gives directly eq. (57). So there appears always the problems, that the influence of the mechanical system on the pressure, eq. (6a), is given through he velocity. Because the input transformation is expressed in x-coordinates there are 2 possibilities to solve the problem. Either an additional velocity sensor is used, which is expensive, or an observer for the coordinate v_p is designed. It is a known problem of the Input–Output Linearization that the whole state is needed. Controllers that are used in rolling mills suffer also from the problem that the measured signals contain considerable transducer and quantization noise and therefore we do not want to use an observer and suggest a different way namely, the Input–Output–Linearization with constrained measurement, see Schlacher[2]. With this method it is possible to determine an output z, such that the nonlinear feedback due to eq. (62) does not contain the velocity signal. For the single acting cylinder due to eqs. (6) the output z can be written in the form

$$z = h\,(P_1, x_p) = f\left(\frac{(V_p + x_p A_1)}{A_1} \exp\left(\frac{P_1}{E} \right) \right) \tag{64}$$

where f is a continuous function. For the purpose of a physical interpretation of this output we use as function $f = E \ln (.)$ and obtain

$$z = P_1 - E \ln \left(\frac{V_0}{V_0 + x_p A} \right).$$ (65)

Using V_0 in the numerator of the logarithm is necessary to include the initial value of the pressure. Performing the Input–Output–Linearization for this output z we get the nonlinear state feedback for positive valve displacements $x_s \geqslant 0$

$$x_s = \frac{(\alpha z + v)(V_0 + x_p A)}{(E K_d \sqrt{P_S - P_1})}$$ (66)

and for negative valve displacements $x_s < 0$

$$x_s = \frac{(\alpha z + v)(V_0 + x_p A)}{(E K_d \sqrt{P_1 - P_T})}$$ (67)

with the new plant input v and $\alpha > 0$. This gives the closed loop system which is linear from the new input v to the output z

$$\dot{z} = -\alpha z + v.$$ (68)

For the artificial output z, eqs. (64) and (65), one can find a nice physical interpretation. Using the definition of the isothermal bulk modulus due to eq. (5) we can find the solution in the form

$$P - P_0 = E \ln \left(\frac{\rho}{\rho_0} \right).$$ (69)

With the assumption:

$$\rho = \frac{m_{cv}}{V} = \frac{\rho_0 \left(V_0 + \int Q_v \right)}{V_0 + x_p A_1}$$ (70)

and setting

$$\dot{w} = Q_v$$ (71)

we obtain

$$P_1 = P_0 + E \ln \left(\frac{V_0 + w}{V_0 + x_p A_1} \right)$$ (72a)

$$\dot{w} = Q_v$$ (72b)

and find

$$z = E \ln \left(1 + \frac{w}{V_0} \right) = P_{valve}. \tag{73}$$

This means that an output that is independent of the piston velocity is for example the pressure produced by the valve, or a related quantity, i.e., the density in the control volume produced by the valve, as the pressure is a function of the density due to eq. (5). In Kugi[19] or Novak [20] one can find measurements of a successful implementation of this nonlinear hydraulic controller as well as the equations for the general, a double acting piston.

6.1 Nonlinear Control — Step Responses

The controller developed in eq. (64) has become the standard controller of the VAI for hydraulic actuators in rolling mills. It was implemented in several rolling mills and figures 36 and 37 show measured step responses of a single stand hot strip mill with a linear and the proposed nonlinear controller. The nonlinear controller uses also a P–controller in the outer loop. Figure 37 contains also position steps from a cold rolling mill, here the work rolls are pressed against each other to give an external stiffness c_m. One can see from both plots that the system behaves linear from the reference position $x_{p,ref}$ to the measured actual position x_p, all nonlinearities are cancelled by the spool position x_s. Figure 37 shows also the hydraulic force of the forward chamber F_{h1}, but this system contains also a dynamical system for the rod side pressure. The identification of the linearized system of this rolling mill for Operator and Drive Side can be found in section 5. Here only step responses for the piston position x_p are shown, but the controller can also be used for force control. Unfortunately the measurements with force control do not contain any information about the spool position, so they are not plotted here. One may wonder, why the hydraulic force in figure 37 is different for positive and negative steps. The reason is, that the work rolls are driven at a constant speed ω_r to avoid damages during the position steps. The first stand contains a massive eccentricity in the backup rolls and this disturbance is the reason for this behavior.

7   Conclusions

A model of a hydraulic actuator was presented, analyzed and verified by a FFT–identification. The measurements in the last section show that it is possible to improve the dynamics of existing hydraulic systems by means of

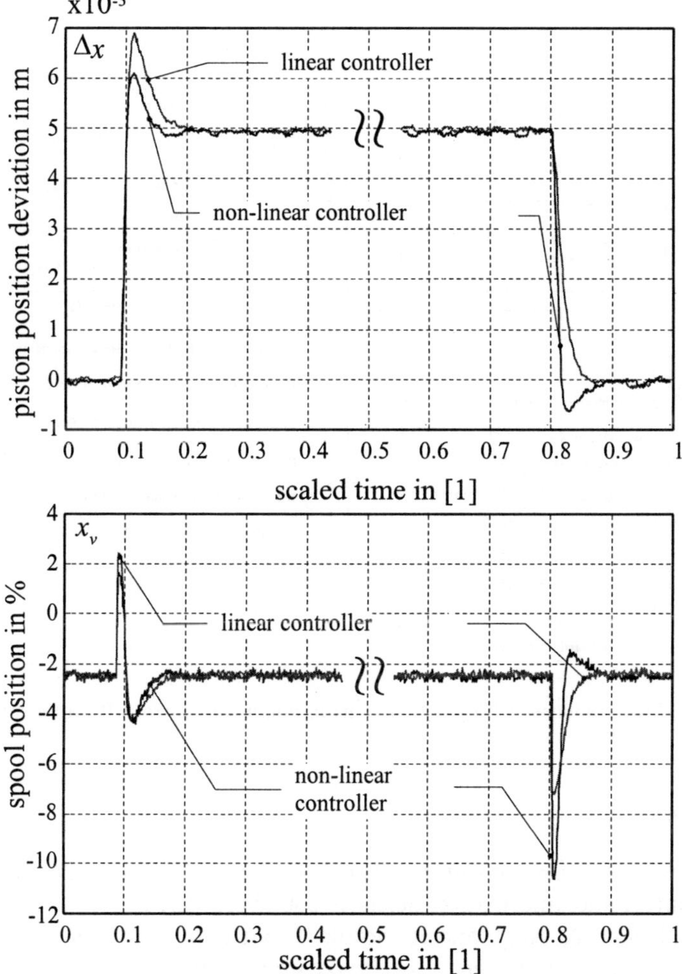

Figure 36. Step responses with position control, linear and non–linear controller.

nonlinear control concepts. Even in high automated plants like rolling mills modern control theory can be used to improve the product quality.

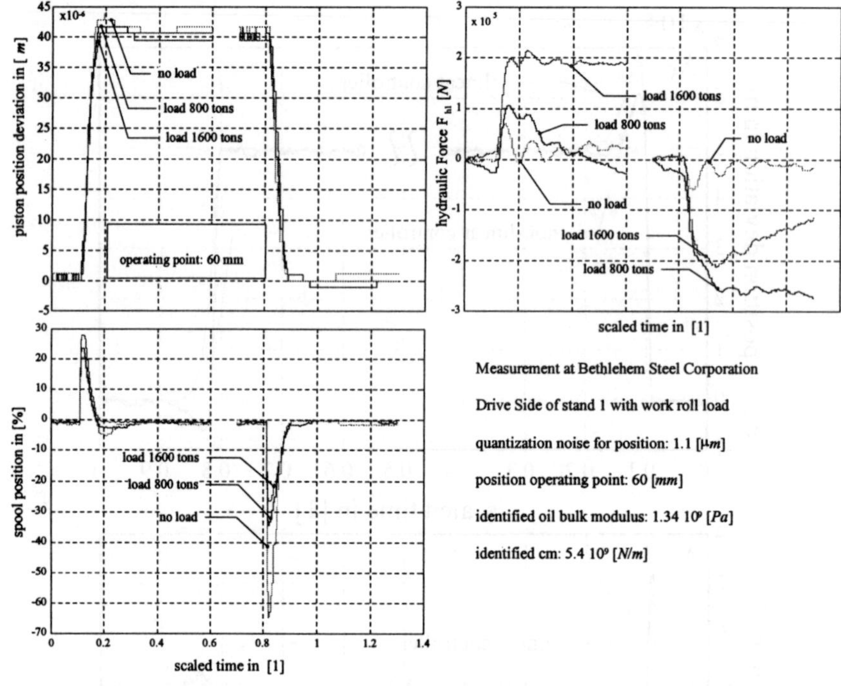

Figure 37. Step responses with position control, nonlinear control with different loads.

References

1. A. Kugi. *Non-Linear Control Based on Physical Models.* Lecture Notes in Control and Information Sciences 260. Springer, 2001.
2. K. Schlacher, A. Kugi, and R. Novak. Input to output linearization with constrained measurement. In *Proceedings of the 5th IFAC Symposium Nonlinear Control Systems: NOLCOS 01*, Saint-Petersburg, Russia, 4-6 July 2001.
3. A. Isidori. *Nonlinear Control Systems.* Springer, 3rd edition, 1996.
4. H. Nijmeijer and A. J. van der Schaft. *Nonlinear Dynamical Control Systems.* 1991.
5. C. T. Chen. *Linear System Theory and Design.* Oxford, New York, 1984.
6. T. Kailath. *Linear Systems.* Prentice Hall, 1980.
7. E. Truckenbrodt. *Fluidmechanik Grundlagen und Elementare Strö-*

mungsvorgänge Dichtebeständiger Fluide, volume 1.   Springer Verlag, 3rd edition, 1989.

8. R. W. Johnson, editor. *The Handbook of Fluid Dynamics*. CRC Press LLC, 1998.

9. D. R. Lide, editor. *Handbook of Chemistry and Physics*. CRC Press, Inc., 76 edition, 1995.

10. H. Murrenhoff. *Grundlagen der Fluidtechnik*, volume 1 of *Fluidtechnik*. Aachen, 1998.

11. H. E. Merritt. *Hydraulic Control Systems*. 1967.

12. K. J. Aström and B. Wittenmark. *Computer Controlled Systems: Theory and Design*. Prentice-Hall Inc., Englewood Cliffs, 1990.

13. S. Skogestad and I. Postlethwaite. *Multivariable Feedback Control*. J. Wiley & Sons Ltd, England, 1996.

14. A. Alleyne and R. Liu.   On the limitations of force tracking control for hydraulic servosystems. *Journal of Dynamic Systems, Measurement, and Control*, 121:184 – 190, 1999.

15. Kyo-Il Lee. *Dynamisches Verhalten der Steuerkette Servoventil-Motor-Last*. PhD thesis, Technische Hochschule Aachen, December 1977.

16. H. Murrenhoff. *Servohydraulik*.   Verlag Mainz, Aachen, fluidtechnik edition, 1998.

17. J. F. Blackburn, G. Reethof, and J. L. Shearer. *Fluid Power Control*. The Technology Press of M.I.T. and John Wiley & Sons Inc., 1960.

18. R. M. Novak. *Analysis and Non-linear Control of Hydraulic Systems in Rolling Mills*. PhD thesis, Johannes Kepler University Linz/Austria, December 2001.

19. A. Kugi, R.M. Novak, and K. Schlacher.   Non-linear control in rolling mills: A new perspective. In *Conference Record of the 2000 IEEE Industry Applications Conference: IAS 2000*, 2000.

20. R.M. Novak, K. Schlacher, A. Kugi, and H. Frank.   Nonlinear hydraulic gap control: A practical approach. In *Proceedings of the IFAC Control System Design: CSD 2000*, pages 605–609, Bratislava, Slovak Republik, 18-20 June 2000.

MATHEMATICAL MODELLING AND NONLINEAR CONTROL OF A TEMPER ROLLING MILL

STEFAN FUCHSHUMER AND KURT SCHLACHER

Christian Doppler Laboratory for Automatic Control of Mechatronic Systems in Steel Industries, University of Linz, Altenbergerstrasse 69, A-4040 Linz, Austria
E-mail: [fuchshumer,schlacher]@mechatronik.uni-linz.ac.at

ANDREAS KUGI

Department of Automatic Control and Control Systems Technology,
University of Linz, address like above
E-mail: kugi@mechatronik.uni-linz.ac.at

This contribution is devoted to the mathematical description of a temper rolling mill and a nonlinear controller design based on physical considerations. The plant consists of a four-high mill stand, unwinder, rewinder, and bridle rolls in the entry and exit section of the mill. Special emphasis is laid on the mathematical modelling of the rollgap, since it establishes the interconnection between the strip tensions, the elongation coefficient, the roll force and the slip conditions. The nonlinear control concept is characterized by the fact that the torques of the bridle rolls are intended as control inputs for the elongation and master speed control, whereas the hydraulic adjustment system and the main mill drive are used to take effect on the strip tensions.

1 Introduction

Temper rolling, also referred to as skin pass rolling, is the last process step in the overall rolling production route. The aim of this rolling concept, where just a slight strip thickness reduction under the action of the hydraulic force and appropriate strip tensions is performed, is to establish certain material characteristics and a desired roughness of the strip surface. Instead of measuring the thickness reduction of the strip by thickness transducers as in cold rolling, the information is usually obtained indirectly by sensing the elongation of the strip using the strip velocity signals near the rollgap.

To cope with the high quality demands on the rolled strip, particular attention has to be paid to the control system, which is intended to adjust the desired elongation coefficient and the strip tensions prescribed by the metallurgists. In order to design the control system, it is important to investigate the underlying physics of the mill and to derive an adequate mathematical description. The development of an analytical model, which incorporates the main nonlinear dynamics and physical interconnections of the system, constitutes the first part of this contribution. The second part is dedicated to the controller design using differential geometric methods.

2   Mathematical Modelling

The skin pass mill configuration considered within this contribution is schematically depicted in Figure 1. The elongation of the strip is performed in the rollgap of the four-high mill stand under the action of the force exerted by the hydraulic actuator and the strip entry (backward) tension σ_{en} and the exit (forward) tension σ_{ex}. The unwinder (pay-off reel) and the rewinder (exit tension reel) are connected to the entry and exit bridle rolls via elastic strip elements.

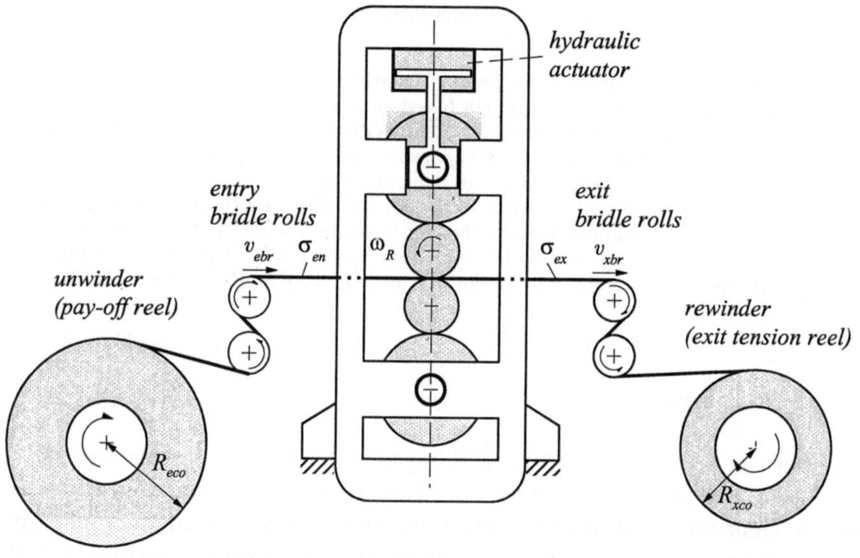

Figure 1. Configuration of the skin pass mill under consideration.

First, particular attention will be paid to a detailed treatment of the rollgap model, since it covers the interconnections between the reduction (or, equivalently, the elongation coefficient) of the steel strip, the backward and forward tension, the roll force, the slip conditions and the material parameters of the strip and the elastic work rolls. Thus, by means of the rollgap model, the interactions between the bridle roll dynamics and the mill stand dynamics can be described using an elastic spring model for the connecting strip elements.

Then, the focus will be directed towards the characterization of the dynamical behavior of the mill stand by means of a simple spring-mass-damper system. The effect of the elastic mill stretch can be measured during a calibration process and will be taken into account in the model by an appropriate spring. The nonlinear differential equations capturing the dynamics of the hydraulic actuator will be discussed on the basis of a single–acting piston.

The motion of the bridle rolls and the winders will also be described by differential equations, and the strip elements connecting the winders to the bridle rolls, and the bridle rolls to the rollgap, respectively, will be modelled as massless linear elastic springs. Then, the dynamical models of the different components together with the implicit algebraic equations of the rollgap model will be put together to form the entire multi–input nonlinear mathematical model of the skin pass mill. This model serves as the basis for the controller design based on nonlinear design techniques, in particular on the exact input/output–linearization theory.

2.1 Non–Circular Arc Rollgap Model

An important property of the classical rollgap models for the cold rolling case (see, e.g., Bland [1]) is the assumption that the work rolls remain circular under the action of the rolling load, though with a larger radius, referred to as Hitchcock's equivalent radius (see Hitchcock [10]). This approximation is appropriate for the typical cold rolling regime. But in the temper rolling scenario it turns out that the work roll deformations are so significant that the assumption of a circular roll is no longer valid. Thus, the elastic roll deformations have to be investigated in detail, which is the topic of the following discussions.

Very interesting research results for the temper rolling case have been presented in the metal forming literature (see, e.g., Fleck [4,5], Domanti [2], Jortner [13]) commonly referred to as non–circular arc rollgap models, which also constitute the basis for the rollgap model used within this contribution.

a) Elastic Roll Deformations

Different approaches for handling the elastic work roll deformations occurring in the temper rolling case can be found in the literature. In the paper of Fleck [5], the roll is assumed to behave like an elastic half–space in the vicinity of the rollgap (see also the book of Johnson [12]). This is an assumption well met in the rolling scenario. On the other hand, Jortner [13] investigates the radial deformations of the surface points of an elastic cylinder caused by a diametrically equivalent rolling pressure distribution.

These approaches are characterized by a common starting point, namely Flamant's problem (see, e.g., Johnson [12]), which is illustrated in Figure 2. The problem is to calculate the stresses, strains and deformations of the three–dimensional elastic half–space caused by a normal force. This force of intensity P per unit length is distributed along the z–axis and is acting in the direction of the y–axis. Under the assumption of plane strain, which is well met due to the infinite extension of the elastic half–space in the z–direction, the problem can be treated as a two–dimensional one.

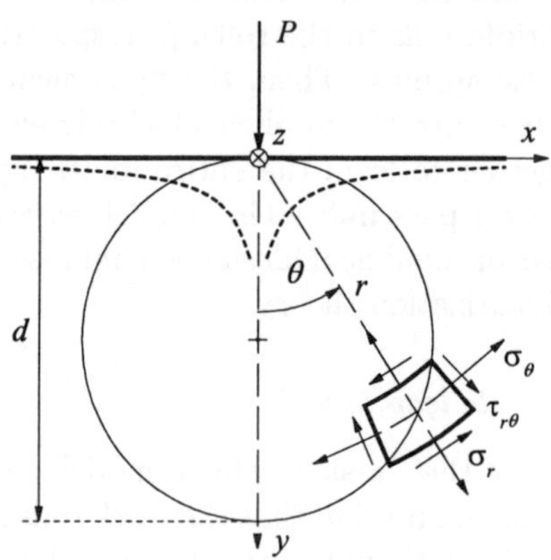

Figure 2. Flamant's problem: Elastic half–space loaded by a force of intensity P per unit length distributed along the z–axis.

Since it is convenient to introduce polar coordinates for handling this problem as well as for the following study of the loaded work roll, the basic equations of the linearized theory of elasticity for plane strain will be briefly reviewed in polar coordinates (see, e.g., Timoshenko [26], Ziegler [28]). In absence of internal stress sources, the stress components $\sigma_r(r, \theta)$, $\sigma_\theta(r, \theta)$, $\tau_{r\theta}(r, \theta)$ (radial, tangential and circumferential stress, see also Figure 2) have to satisfy the equilibrium equations

$$\frac{\partial \sigma_r}{\partial r} + \frac{1}{r}\frac{\partial \tau_{r\theta}}{\partial \theta} + \frac{\sigma_r - \sigma_\theta}{r} = 0 \,,$$

$$\frac{1}{r}\frac{\partial \sigma_\theta}{\partial \theta} + \frac{\partial \tau_{r\theta}}{\partial r} + \frac{2\tau_{r\theta}}{r} = 0 \,.$$

$$(1)$$

The strain components $\varepsilon_r(r, \theta)$, $\varepsilon_\theta(r, \theta)$ and $\gamma_{r\theta}(r, \theta)$ are connected with the displacements $u(r, \theta)$ and $v(r, \theta)$ in radial and tangential direction via the (linearized) relations

$$\varepsilon_r = \frac{\partial u}{\partial r}\,, \qquad \varepsilon_\theta = \frac{u}{r} + \frac{1}{r}\frac{\partial v}{\partial \theta}\,, \qquad \gamma_{r\theta} = \frac{1}{r}\frac{\partial u}{\partial \theta} + \frac{\partial v}{\partial r} - \frac{v}{r}\,. \tag{2}$$

Assuming linear elastic behavior and a plane–strain scenario with E and ν denoting Young's modulus and Poisson's number, we get

$$\gamma_{rz} = \gamma_{\theta z} = 0 \qquad\qquad \rightarrow \qquad \tau_{rz} = \tau_{\theta z} = 0\,,$$

$$\varepsilon_z = \frac{1}{E}\left[\sigma_z - \nu\left(\sigma_r + \sigma_\theta\right)\right] = 0 \quad \rightarrow \quad \sigma_z = \nu\left(\sigma_r + \sigma_\theta\right)$$

by applying the three–dimensional constitutive law of Hooke. Thus, the equations of the plane–strain equivalent of Hooke's law read as

$$\varepsilon_r = \frac{1}{E}\left[\left(1 - \nu^2\right)\sigma_r - \nu\left(1 + \nu\right)\sigma_\theta\right] = \frac{1}{E^*}\left[\sigma_r - \nu^*\sigma_\theta\right]\,,$$

$$\varepsilon_\theta = \frac{1}{E}\left[\left(1 - \nu^2\right)\sigma_\theta - \nu\left(1 + \nu\right)\sigma_r\right] = \frac{1}{E^*}\left[\sigma_\theta - \nu^*\sigma_r\right]\,, \tag{3}$$

$$\gamma_{r\theta} = \frac{1}{G}\tau_{r\theta}\,,$$

where $E^* = E/\left(1 - \nu^2\right)$ denotes the plane–strain Young modulus and $\nu^* = \nu/\left(1 - \nu\right)$ the plane–strain Poisson ratio. In addition to the equilibrium equations (1), the stress components also have to satisfy the compatibility condition

$$\left(\frac{\partial^2}{\partial r^2} + \frac{1}{r}\frac{\partial}{\partial r} + \frac{1}{r^2}\frac{\partial^2}{\partial \theta^2}\right)\left(\sigma_r + \sigma_\theta\right) = 0\,. \tag{4}$$

A convenient and commonly used way to deal with such linear elasticity problems is to introduce Airy's stress function $\phi\left(r,\theta\right)$ (G.B. Airy, 1862),

$$\sigma_r = \frac{1}{r}\frac{\partial \phi}{\partial r} + \frac{1}{r^2}\frac{\partial^2 \phi}{\partial \theta^2}, \quad \sigma_\theta = \frac{\partial^2 \phi}{\partial r^2}, \quad \tau_{r\theta} = -\frac{\partial}{\partial r}\left(\frac{1}{r}\frac{\partial \phi}{\partial \theta}\right) \tag{5}$$

in accordance with the equilibrium equations (1). Expressing the compatibility condition (4) in terms of the Airy stress function, we obtain the partial differential equation

$$\left(\frac{\partial^2}{\partial r^2} + \frac{1}{r}\frac{\partial}{\partial r} + \frac{1}{r^2}\frac{\partial^2}{\partial \theta^2}\right)\left(\frac{\partial^2 \phi}{\partial r^2} + \frac{1}{r}\frac{\partial \phi}{\partial r} + \frac{1}{r^2}\frac{\partial^2 \phi}{\partial \theta^2}\right) = 0\,. \tag{6}$$

Thus, the problem of calculating the stress distribution of a linear–elastic problem is equivalent to solve the partial differential equation (6), where the stress function $\phi\left(r,\theta\right)$ also has to satisfy the given boundary conditions.

It is easy to see, that the stress function candidate

$$\phi\left(r,\theta\right) = -\frac{P}{\pi} r\theta \sin\theta \qquad (7)$$

for the problem due to Figure 2 satisfies the partial differential equation (6). This solution was given by Flamant in the year 1892 and is usually referred to as *simple radial stress distribution*. Thus, using Eq. (5), we get the stress components in the form

$$\sigma_r\left(r,\theta\right) = -\frac{2P}{\pi}\frac{\cos\theta}{r}\ , \qquad \sigma_\theta\left(r,\theta\right) = 0\ , \qquad \tau_{r\theta}\left(r,\theta\right) = 0 \qquad (8)$$

in accordance with the given boundary condition such that the surface of the elastic half–space is unloaded except at the point of application of the load P. The points of equal radial stress intensity are located on a circle with diameter d, which passes through the point where the force P acts on the half–space (see Figure 2). The radial stress at those points is

$$\sigma_r = -\frac{2P}{\pi d}\ .$$

Once the stress distribution is obtained, the strains $\varepsilon_r\left(r,\theta\right)$, $\varepsilon_\theta\left(r,\theta\right)$, $\gamma_{r\theta}\left(r,\theta\right)$, and the deformations $u\left(r,\theta\right)$, $v\left(r,\theta\right)$ can be calculated using Eqns. (2) and (3). The derivation of these formulas is omitted here, since they are not needed for the further explanations. The displacements of surface points in the y–direction, i.e., $v\left(r,\theta = \pm\frac{\pi}{2}\right)$ are indicated in Figure 2 by the dashed line.

The next step towards the calculation of the elastic work roll deformations is to consider the problem due to Figure 3 according to the investigations of Jortner [13]. Due to the fact that the width of the work rolls is very large and that the rolling load is assumed not to vary in the axial direction, the assumption of a plane–strain scenario is appropriate.

Based on Flamant's solution (8), the stress distributions caused by the diametrically applied forces are obtained separately in terms of the coordinate systems (r_1,θ_1) and (r_2,θ_2), which are located in the points A and B,

$$\sigma_{r_2}^A = -\frac{2P}{\pi}\frac{\cos\theta_2}{r_2}\ , \quad \sigma_{\theta_2}^A = 0\ , \quad \tau_{r_2\theta_2}^A = 0\ ,$$

$$\sigma_{r_1}^B = -\frac{2P}{\pi}\frac{\cos\theta_1}{r_1}\ , \quad \sigma_{\theta_1}^B = 0\ , \quad \tau_{r_1\theta_1}^B = 0\ . \qquad (9)$$

Then, both stress profiles are transformed to a common polar coordinate system (r,θ), which is located in the center of the cylinder. Since the linearized

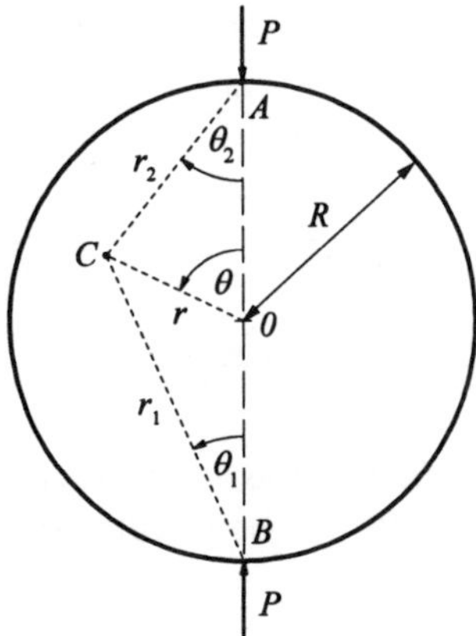

Figure 3. Elastic cylinder loaded by diametrically equivalent forces.

theory of elasticity is considered, both solutions can be superimposed to give the resulting stress distribution. Carrying out the indicated computations, we get the following result in the coordinate system (r, θ):

$$\sigma_r\left(r, \theta\right) = -\frac{2P}{\pi}\left[\frac{(R-r\cos\theta)(r-R\cos\theta)^2}{(R^2+r^2-2Rr\cos\theta)^2} + \frac{(R+r\cos\theta)(r+R\cos\theta)^2}{(R^2+r^2+2Rr\cos\theta)^2}\right] ,$$

$$\sigma_\theta\left(r, \theta\right) = -\frac{2P}{\pi}\left[\frac{(R-r\cos\theta)R^2\sin^2\theta}{(R^2+r^2-2Rr\cos\theta)^2} + \frac{(R+r\cos\theta)R^2\sin^2\theta}{(R^2+r^2+2Rr\cos\theta)^2}\right] ,$$

$$\tau_{r\theta}\left(r, \theta\right) = -\frac{2P}{\pi}\left[\frac{(R-r\cos\theta)(r-R\cos\theta)R\sin\theta}{(R^2+r^2-2Rr\cos\theta)^2} - \frac{(R+r\cos\theta)(r+R\cos\theta)R\sin\theta}{(R^2+r^2+2Rr\cos\theta)^2}\right] .$$

$$(10)$$

The evaluation of these equations on the surface of the cylinder shows an inconsistency with the given boundary condition, namely that the surface is unloaded except for the points A and B,

$$\sigma_r\left(R, \theta\right) = -\frac{P}{\pi R} , \qquad \sigma_\theta\left(R, \theta\right) = -\frac{P}{\pi R} , \qquad \tau_{r\theta}\left(R, \theta\right) = 0 . \qquad (11)$$

Thus, to fulfill the boundary conditions, the term $P/\left(\pi R\right)$ is added to the radial and circumferential stress of Eq. (10) in accordance with Eqns. (1)

and (4). The solution of the problem due to Figure 3 reads as:

$$\sigma_r\,(r,\theta) = -\frac{2P}{\pi}\left[\frac{(R-r\cos\theta)(r-R\cos\theta)^2}{(R^2+r^2-2Rr\cos\theta)^2} + \frac{(R+r\cos\theta)(r+R\cos\theta)^2}{(R^2+r^2+2Rr\cos\theta)^2} - \frac{1}{2R}\right],$$

$$\sigma_\theta\,(r,\theta) = -\frac{2P}{\pi}\left[\frac{(R-r\cos\theta)R^2\sin^2\theta}{(R^2+r^2-2Rr\cos\theta)^2} + \frac{(R+r\cos\theta)R^2\sin^2\theta}{(R^2+r^2+2Rr\cos\theta)^2} - \frac{1}{2R}\right],$$

$$\tau_{r\theta}\,(r,\theta) = -\frac{2P}{\pi}\left[\frac{(R-r\cos\theta)(r-R\cos\theta)R\sin\theta}{(R^2+r^2-2Rr\cos\theta)^2} - \frac{(R+r\cos\theta)(r+R\cos\theta)R\sin\theta}{(R^2+r^2+2Rr\cos\theta)^2}\right].$$

$$(12)$$

The radial displacement $u\,(R,\theta)$ of the points on the cylinder surface according to the stress profile of Eq. (12) can be calculated by means of Eqns. (2) and (3),

$$u\,(R,\theta) = \int_0^R \varepsilon_r\,(r,\theta)\ \mathrm{d}r = \frac{1}{E^*}\int_0^R (\sigma_r\,(r,\theta) - \nu^*\sigma_\theta\,(r,\theta))\ \mathrm{d}r\ ,$$

$$u\,(R,\theta) = \frac{P}{\pi E^*}\left\{2 + \cos\theta\ln\left(\frac{1-\cos\theta}{1+\cos\theta}\right) -\right.$$

$$\left. - (1-\nu^*)\sin\theta\left[\arctan\left(\frac{1+\cos\theta}{\sin\theta}\right) + \arctan\left(\frac{1-\cos\theta}{\sin\theta}\right)\right]\right\},$$

$$(13)$$

and is schematically depicted in Figure 4a. The singularity of the stress and the displacement at the point $\theta = 0$ and $\theta = \pi$ is due to the action of the concentrated loads. Eq. (13) can be further simplified by considering the identity

$$\arctan\left(\frac{1+\cos\theta}{\sin\theta}\right) + \arctan\left(\frac{1-\cos\theta}{\sin\theta}\right) = \frac{\pi}{2}\operatorname{sign}(\sin\theta)$$

to get the solution

$$u\,(R,\theta) = \frac{P}{\pi E^*}\Lambda\,(\theta) \tag{14}$$

with the abbreviation

$$\Lambda\,(\theta) = 2 + \cos\theta\ln\left(\tan^2\left(\frac{\theta}{2}\right)\right) - (1-\nu^*)\frac{\pi}{2}\sin\theta\ \operatorname{sign}(\sin\theta)\ . \tag{15}$$

Now, Eqns. (14) and (15) can be used to solve the original problem, namely, to calculate the radial deformations of the elastic work roll loaded with an arbitrary, but diametrically equivalent pressure profile. As already encountered in the previous considerations, the assumption of a diametrically equivalent

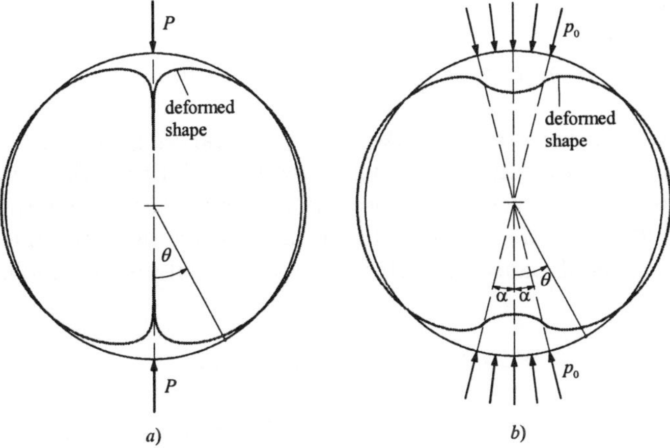

Figure 4. Elastic cylinder loaded with diametrically applied forces P (a) and a constant pressure profile p_0 (b).

load simplifies the analysis significantly, and can be easily accepted due to St.Venant's principle, although the work roll/backup roll contact pressure is not exactly equal to the rollgap pressure distribution.

Given an arbitrary, but diametrically equivalent pressure distribution $p(\beta)$, $\theta_{en} \leq \beta \leq \theta_{ex}$, with θ_{en} and θ_{ex} denoting the rollgap entry and the exit angle, the radial displacements of the points on the cylinder surface are obtained by carrying out the following integration over the roll/strip contact region,

$$u(R, \theta) = \frac{R}{\pi E^*} \int_{\theta_{en}}^{\theta_{ex}} \Lambda(\theta - \beta)\, p(\beta)\, \mathrm{d}\beta \, . \tag{16}$$

The analytical evaluation of this integral might be a hard task for arbitrary $p(\beta)$. Thus, with regard to a further numerical implementation, it is convenient to perform a spatial discretization of the pressure profile. The individual deformations caused by the pressure elements can be superimposed to give the entire deformation scenario of the work roll.

Figure 4*b* shows the elastic cylinder loaded with a constant pressure profile p_0, which is applied over a finite angle 2α. The integral expression of Eq. (16)

184   *S. Fuchshumer, K. Schlacher and A. Kugi*

simplifies to

$$u\left(R,\theta\right) = \frac{p_0 R}{\pi E^*} \int_{-\alpha}^{\alpha} \Lambda\left(\theta - \beta\right) \mathrm{d}\beta \ . \tag{17}$$

Nevertheless, it is necessary to take care when carrying out the indicated integration because the function $\Lambda\left(\theta\right)$ involves a singularity at the point $\theta = 0$. Especially for the case when the "observation point" θ lies inside the loaded region, i.e. $|\theta| < \alpha$, the calculations are a little bit delicate and have to be done by means of Cauchy principal values. For notational simplicity, the abbreviation

$$\Xi\left(\theta, \alpha\right) = \int_{-\alpha}^{\alpha} \Lambda\left(\theta - \beta\right) \mathrm{d}\beta \tag{18}$$

is arranged, where a superscript o or i is added to $\Xi\left(\theta, \alpha\right)$ in order to indicate the scenario being considered. The superscript o is used for the case $|\theta| > \alpha$ (the "observation point" θ is outside the loaded region), and i (the angle θ is inside the loaded region) to cover the case $|\theta| \leq \alpha$. The computations for $|\theta| > \alpha$ are straightforward and give the result

$$\Xi^o\left(\theta, \alpha\right) = \sin\left(\varphi\right) \ln\left(\tan^2\left(\frac{\varphi}{2}\right)\right)\Big|_{\varphi=\theta-\alpha}^{\varphi=\theta+\alpha} +$$
$$+ \left(1 - \nu^*\right) \frac{\pi}{2} \left[\cos\left(\varphi\right) \operatorname{sign}\left(\sin\left(\varphi\right)\right)\right]\Big|_{\varphi=\theta-\alpha}^{\varphi=\theta+\alpha} \ . \tag{19}$$

To cope with the singularity of $\Lambda\left(\theta\right)$ at the point $\theta = 0$ when evaluating the function $\Xi^i\left(\theta, \alpha\right)$, the integral of Eq. (18) is split into two parts, followed by a limit operation

$$\Xi^i\left(\theta, \alpha\right) = \lim_{\varepsilon \to 0} \left[\int_{-\alpha}^{\theta-\varepsilon} \Lambda\left(\theta - \beta\right) \mathrm{d}\beta + \int_{\theta+\varepsilon}^{\alpha} \Lambda\left(\theta - \beta\right) \mathrm{d}\beta\right] \ .$$

As the result for the case $|\theta| \leq \alpha$, we get

$$\Xi^i\left(\theta, \alpha\right) = \Xi^o\left(\theta, \alpha\right) - \left(1 - \nu^*\right) \pi \ . \tag{20}$$

In the metal forming literature, the Eqns. (19) and (20) in combination with (17) and (18) are commonly referred to as Jortner's influence functions.

At this point, we are ready to calculate the elastic work roll deformations caused by a diametrically applied, piecewise constant pressure distribution p_j, $j = 1, \ldots, N$, which is obtained by a spatial discretization of the rolling load $p(\beta)$, $\beta \in [\theta_{en}, \theta_{ex}]$. Here, N denotes the number of discretization nodes, and the discretization angle $\Delta\theta$ is

$$\Delta\theta = \frac{\theta_{ex} - \theta_{en}}{N - 1}. \tag{21}$$

Throughout the further discussions, the subscripts R and S are added to the variables and parameters in order to distinguish between work roll parameters (subscript R) and strip parameters (subscript S). The undeformed roll radius will be denoted with R_0 in order to prevent confusions.

Using Jortner's influence function, we can calculate the distance from the roll center to a roll surface point at an arbitrary angle θ, i.e., the deformed work roll profile, in the form

$$R(\theta) = R_0 \left[1 + \frac{1}{\pi E_R^*} \sum_{j=1}^{N} \Xi\left(\theta - \theta_j, \frac{\Delta\theta}{2}\right) p_j \right]. \tag{22}$$

Evaluating Eq. (22) at a discretization node $k \in \{1, \ldots, N\}$, we get

$$R_k \doteq R(\theta_k) = R_0 \left[1 + \frac{1}{\pi E_R^*} \sum_{j=1}^{N} \Xi_{k-j}\, p_j \right], \tag{23}$$

with the abbreviations

$$\Xi_{k-j} = \Xi^i\left(\theta_k - \theta_j, \tfrac{\Delta\theta}{2}\right) \quad \text{for} \quad k = j,$$

$$\Xi_{k-j} = \Xi^o\left(\theta_k - \theta_j, \tfrac{\Delta\theta}{2}\right) \quad \text{for} \quad k \neq j. \tag{24}$$

b) Characterization of the Strip Behavior in the Rollgap

Now, we will turn our attention towards the mathematical description of the strip behavior in the rollgap. In contrast to the classical cold rolling scenario, we also have to count for the elastic deformation of the strip in the temper rolling case. Thus, we will apply the constitutive law of Prandtl–Reuß in order to cover the behavior of the elastic–plastic body (see, e.g., Pawelski [22], Ziegler [28]).

Following the metal forming literature, the rollgap is divided into four distinct zones (see Figure 6). The name "backward slip" is used to indicate regions where the strip is moving slower than the work roll surface, and in the "forward slip" region, i.e. the exit region of the rollgap, the strip is moving faster than the work roll surface. After passing the elastic compression zone at the entry section, the strip is deformed elastic–plastically, where the regions of backward and forward slip are separated by the so-called neutral section. At this neutral section, the strip velocity equals the circumference speed $v_R = \omega_R R_0$ of the work rolls. The elastic recovery zone, which should not be neglected in the temper rolling case, is located in the exit region of the rollgap. Rolling regimes characterized by a central flat zone without any slip between the work roll surface and the strip, i.e. rolling of very thin strip (see, e.g., Fleck [4,5], Domanti [2]) are not considered within this contribution.

As a prerequisite for the following discussions we will briefly discuss the yield criterion of v.Mises (outlined in Carthesian coordinates x, y, z), which is included in the Prandtl–Reuß equations. The constitutive law of v.Mises is based on the assumption of an isochoric plastic deformation of the isotropic material.

With σ_{ij} $(i,j = x, y, z)$ denoting the stress components and δ_{ij} the Kronecker symbols ($\delta_{ij} = 1$ for $i = j$, $\delta_{ij} = 0$ otherwise), we get the deviatoric stress components s_{ij} as

$$s_{ij} = \sigma_{ij} - m\delta_{ij} \qquad \text{with} \quad m = \frac{1}{3}\left(\sigma_{xx} + \sigma_{yy} + \sigma_{zz}\right), \tag{25}$$

where m denotes the mean value of the normal stresses. For the sake of notational simplicity we will drop the arguments (x, y, z) of the stress and strain variables. By means of the second invariant of the deviatoric stress tensor,

$$J_2 = \sum_{i,j} s_{ij}^2 , \tag{26}$$

the relation

$$f_{Mises}\left(J_2, k_f\right) = \frac{1}{2}J_2 - \frac{1}{3}k_f^2 \leq 0 , \tag{27}$$

is proposed by v.Mises. The material parameter k_f, which might also be dependent on the strain ε_{ij}, the strain rate $\dot{\varepsilon}_{ij}$ and the temperature, is the yield stress of the one–dimensional tensile test. In the case $f_{Mises}\left(J_2, k_f\right) < 0$ the material is in the non–plastic state, whereas plastic flow occurs in the case $f_{Mises}\left(J_2, k_f\right) = 0$. Thus, the yield criterion of v.Mises is given by

$$J_2 = \frac{2}{3}k_f^2 , \tag{28}$$

or, equivalently, outlined in terms of the stress components σ_{ij},

$$\frac{1}{2}\left[\left(\sigma_{xx} - \sigma_{yy}\right)^2 + \left(\sigma_{xx} - \sigma_{zz}\right)^2 + \left(\sigma_{yy} - \sigma_{zz}\right)^2\right] + 3\left(\sigma_{xy}^2 + \sigma_{xz}^2 + \sigma_{yz}^2\right) = k_f^2 . \tag{29}$$

The deviatoric stresses s_{ij} are linked to the strain rates $\dot{\varepsilon}_{ij}$ by the equation

$$\dot{\varepsilon}_{ij} = \dot{\lambda} s_{ij} \tag{30}$$

(see, e.g., Pawelski [22]) with the value $\dot{\lambda} > 0$ to be determined.

In order to compare the three–dimensional plastic deformations of the body with the one–dimensional tensile test, usually the equivalent deformation φ and the associated deformation rate $\dot{\varphi}$ is introduced,

$$\sum_{i,j} \sigma_{ij} \dot{\varepsilon}_{ij} = k_f \dot{\varphi} \ . \tag{31}$$

Now, the yield stress k_f of the material, which is obtained by the standard one–dimensional tensile test, can be expressed in terms of φ, $\dot{\varphi}$ and the temperature T,

$$k_f = k_f \left(\varphi, \dot{\varphi}, T \right) \ . \tag{32}$$

By substituting the Eqns. (26), (28) and (30) into Eq. (31) and by using the identity $s_{xx} + s_{yy} + s_{zz} = 0$, we get

$$k_f \dot{\varphi} = \sum_{i,j} \sigma_{ij} \dot{\varepsilon}_{ij} = \dot{\lambda} \sum_{i,j} \sigma_{ij} s_{ij} = \dot{\lambda} \sum_{i,j} s_{ij}^2 = \dot{\lambda} J_2 = \dot{\lambda} \frac{2}{3} k_f^2 \ ,$$

and, thus,

$$\dot{\lambda} = \frac{3}{2} \frac{\dot{\varphi}}{k_f} \ . \tag{33}$$

Equivalently, by means of the Eqns. (30), (31), (33) and the relation $\dot{\varepsilon}_{xx} + \dot{\varepsilon}_{yy} + \dot{\varepsilon}_{zz} = 0$ due to the isochoric plastic deformation, we get

$$k_f \dot{\varphi} = \sum_{i,j} \sigma_{ij} \dot{\varepsilon}_{ij} = \sum_{i,j} s_{ij} \dot{\varepsilon}_{ij} = \frac{1}{\dot{\lambda}} \sum_{i,j} \dot{\varepsilon}_{ij}^2 = \frac{2}{3} \frac{k_f}{\dot{\varphi}} \sum_{i,j} \dot{\varepsilon}_{ij}^2 \ ,$$

and, thus,

$$\dot{\varphi} = \sqrt{\frac{2}{3} \sum_{i,j} \dot{\varepsilon}_{ij}^2} \quad \text{and} \quad \dot{\lambda} = \frac{1}{k_f} \sqrt{\frac{3}{2} \sum_{i,j} \dot{\varepsilon}_{ij}^2} \ . \tag{34}$$

Within the concept of the constitutive law of Prandtl–Reuß, the total increment $\Delta\varepsilon_{ij}$ of the strain is composed of an elastic strain component $\Delta\varepsilon_{ij}^{elast}$ due to Hooke's law and a component $\Delta\varepsilon_{ij}^{plast}$ due to the plastic flow,

$$\Delta\varepsilon_{ij} = \Delta\varepsilon_{ij}^{elast} + \Delta\varepsilon_{ij}^{plast} \ .$$

Thus, using the incremental form of Hooke's law and applying the plastic flow rule (30) of v.Mises, we obtain the incremental form of the Prandtl–Reuß law,

$$\Delta\varepsilon_{ij} = \frac{1+\nu}{E} \Delta\sigma_{ij} - \frac{\nu}{E} \left(\sum_{k} \Delta\sigma_{kk} \right) \delta_{ij} + \Delta\lambda \, s_{ij} \ . \tag{35}$$

Again, plastic flow, i.e., $\Delta\lambda > 0$, occurs in the case $f_{Mises}\left(J_2, k_f\right) = 0$ (see the yield criterion of Eq. (28) or (29)). In the case $f_{Mises}\left(J_2, k_f\right) < 0$, the Prandtl–Reuß equations reduce to Hooke's law.

The objective of the following considerations is to calculate the rolling pressure distribution $p\left(x\right)$ according to a given roll/strip contact profile with $h\left(x\right)$ denoting the strip thickness at the point x, see Figure 5. For the sake of convenience the strip equations are outlined in the Carthesian coordinate system (x, y, z).

The combination of the strip equations with the formulas covering the elastic work roll deformations, i.e., Jortner's equations, is done by the simple transformation $\Delta x = R_0\Delta\theta$, which qualifies as an appropriate approximation for the rolling scenario.

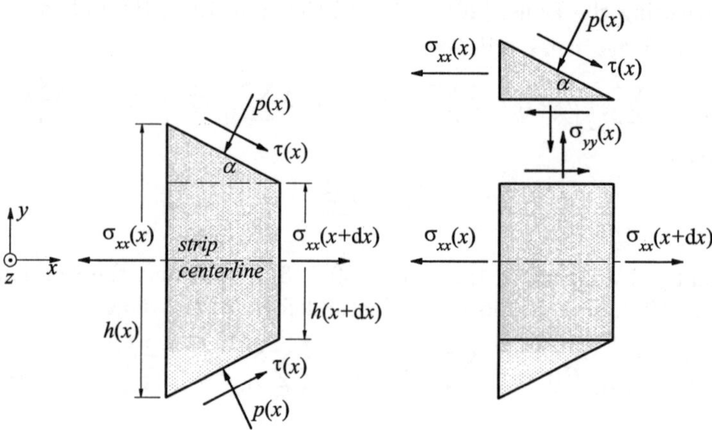

Figure 5. Geometry and stresses of a strip element deformed in the rollgap.

Following the classical elementary slab theory of rolling (see, e.g., Pawelski [22]), we will assume that vertical plane sections of the strip remain plane throughout the rollgap (see Figure 5) and that the deformation of the strip is homogeneous. Additionally, the stresses σ_{xx}, σ_{yy} and σ_{zz} are treated as principal stresses and the assumption of plane strain is arranged. As a consequence, all stresses and strains are functions of x only, i.e., $\sigma_{xx}\left(x\right)$, $\sigma_{yy}\left(x\right)$, etc. Since both work rolls are assumed to have the same radius R_0 and equal angular velocities ω_R, the geometry of the problem is symmetric with respect to the strip center–line.

Investigating the balance of forces for the outlined strip element in the x–direction (see Figure 5), we get the differential equation

$$\frac{\mathrm{d}}{\mathrm{d}x}\left(\sigma_{xx}\left(x\right)h\left(x\right)\right) - 2p\left(x\right)\tan\left(\alpha\left(x\right)\right) + 2\tau\left(x\right) = 0 \tag{36}$$

using the abbreviation

$$\tan\left(\alpha\left(x\right)\right) = -\frac{1}{2}\frac{\partial h\left(x\right)}{\partial x} \ .$$

The balance of forces in the y–direction gives

$$\sigma_{yy}\left(x\right) + p\left(x\right) + \tau\left(x\right)\tan\left(\alpha\left(x\right)\right) = 0 \ . \tag{37}$$

A realistic description of the roll/strip interface is a difficult problem. A simple approach usually encountered in the analysis of rolling is the assumption of a constant Coulomb friction coefficient μ in the roll/strip contact region. We will also make use of this simple approximation, although the incorporation of more sophisticated friction laws into the proposed rollgap model is possible. Thus, the frictional traction is given by $\tau\left(x\right) = \gamma\mu p\left(x\right)$, with $\gamma = +1$ in the regions with backward slip, and $\gamma = -1$ in the zones characterized by forward slip. For convenience of the numerical implementation of the rollgap model, this function is approximated by the arctan–function.

Due to the short length of contact between the work rolls and the strip compared to the work roll radius, usually the approximation $p\left(x\right) = -\sigma_{yy}\left(x\right)$ is arranged. Then, the differential equation (36) reads as

$$\frac{\mathrm{d}}{\mathrm{d}x}\sigma_{xx}\left(x\right) = \frac{2}{h\left(x\right)}\left[\left(\sigma_{xx}\left(x\right) - \sigma_{yy}\left(x\right)\right)\tan\left(\alpha\left(x\right)\right) - \tau\left(x\right)\right] \ . \tag{38}$$

By means of the spatial discretization of the rollgap introduced via Eq. (21), we get

$$\sigma_{xx}\left(x + \Delta x\right) = \sigma_{xx}\left(x\right) +$$
$$+ \frac{1}{h\left(x\right)}\left[\left(\sigma_{yy}\left(x\right) - \sigma_{xx}\left(x\right)\right)\left(h\left(x + \Delta x\right) - h\left(x\right)\right) - 2\tau\left(x\right)\Delta x\right] \tag{39}$$

using the simple Euler integration method and the approximate relation $\Delta x = R_0\Delta\theta$.

Accordingly, the constitutive equations (35) of Prandtl–Reuß are outlined in terms of increments. Following the assumptions of the elementary slab theory introduced above (with a plane strain deformation, and σ_{xx}, σ_{yy} and

σ_{zz} regarded as principal stresses), the Eqns. (35) and (29) reduce to

$$\Delta\varepsilon_{xx} = \frac{1}{E_S}\Delta\sigma_{xx} - \frac{\nu_S}{E_S}\Delta\sigma_{yy} - \frac{\nu_S}{E_S}\Delta\sigma_{zz} + (2\sigma_{xx} - \sigma_{yy} - \sigma_{zz})\frac{\Delta\varphi}{2k_f(\varphi)} ,$$

$$\Delta\varepsilon_{yy} = \frac{1}{E_S}\Delta\sigma_{yy} - \frac{\nu_S}{E_S}\Delta\sigma_{xx} - \frac{\nu_S}{E_S}\Delta\sigma_{zz} + (2\sigma_{yy} - \sigma_{xx} - \sigma_{zz})\frac{\Delta\varphi}{2k_f(\varphi)} ,$$

$$0 = \frac{1}{E_S}\Delta\sigma_{zz} - \frac{\nu_S}{E_S}\Delta\sigma_{xx} - \frac{\nu_S}{E_S}\Delta\sigma_{yy} + (2\sigma_{zz} - \sigma_{xx} - \sigma_{yy})\frac{\Delta\varphi}{2k_f(\varphi)} ,$$

$$\tag{40}$$

and

$$(\sigma_{xx} - \sigma_{yy})^2 + (\sigma_{xx} - \sigma_{zz})^2 + (\sigma_{yy} - \sigma_{zz})^2 = 2k_f^2 , \tag{41}$$

where the argument x has been omitted for notational convenience. By means of the linearized theory of elasticity, the strain increment $\Delta\varepsilon_{yy}$ is given by

$$\Delta\varepsilon_{yy} = \frac{h(x + \Delta x) - h(x)}{h(x)} . \tag{42}$$

c) Outline of the Rollgap Model Structure

The work roll deformation formulas (23), (24), the algebraic strip equations (39), (40), (41) and (42) together with the additional constraints at the rollgap entry and exit point,

$$R(\theta_{en})\cos(\theta_{en}) - A + \frac{h_{en}}{2} = 0 , \qquad R(\theta_{ex})\cos(\theta_{ex}) - A + \frac{h_{ex}}{2} = 0 , \tag{43}$$

constitute the mathematical rollgap model. Here, A denotes the distance between the work roll center and the strip center–line. Finally, the rollgap model yields the form

$$0 = g_1(p_1, \ldots, p_N, \sigma_{en}, \sigma_{ex}, h_{en}, h_{ex}, k_f, E_R^*, \nu_R^*, E_S, \nu_S, \mu, R_0) ,$$
$$\vdots\ \vdots \qquad\qquad\qquad\qquad \vdots \qquad\qquad\qquad\qquad \tag{44}$$
$$0 = g_N(p_1, \ldots, p_N, \sigma_{en}, \sigma_{ex}, h_{en}, h_{ex}, k_f, E_R^*, \nu_R^*, E_S, \nu_S, \mu, R_0) .$$

The stresses σ_{en} and σ_{ex} denote the entry (backward) and exit (forward) strip tensions, and h_{en} and h_{ex} cover the strip entry and exit thickness, respectively.

Thus, a set of implicit algebraic equations is obtained by the discretization of the roll/strip contact arc, which leads to a differential–algebraic system (DAE) when combined with the dynamics of the mill stand and the other plant components.

A main difficulty of this rollgap model is the fact that the entry and exit angle of the rollgap as well as the parameter A are not known a priori.

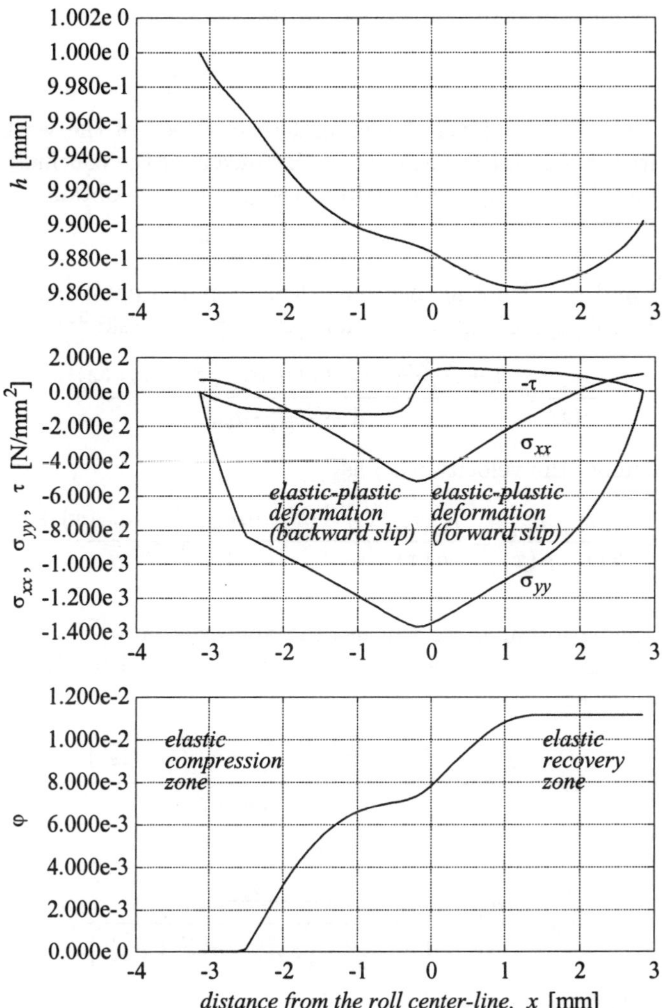

Figure 6. Typical roll–strip contact profile h, stresses σ_{xx}, σ_{yy}, τ, and equivalent deformation φ encountered in the temper rolling case.

d) Slip Conditions in the Rollgap

For calculating the slip conditions in the rollgap, we have to take the elastic volume compression due to the mean stress m (see Eq. (25)) into account (see, e.g., Pawelski [22]). As already mentioned, the plasticity concept of v.Mises considers the plastic deformation to be isochoric. The backward (entry) slip ς_{en} and the forward (exit) slip ς_{ex} are defined as

$$\varsigma_{en} = \frac{v_{en}}{v_R} \; , \qquad \varsigma_{ex} = \frac{v_{ex}}{v_R} \; . \tag{45}$$

Here, v_R, v_{en} and v_{ex} cover the circumference speed of the work rolls, the rollgap entry speed and the exit speed. By means of the continuity equation with $v(x)$ as the velocity of the strip at the point x, we obtain

$$\rho(x) h(x) v(x) = \rho_{en} v_{en} h_{en} \; , \tag{46}$$

with $\rho(x)$ and ρ_{en} denoting the mass density at the point x and the entry point of the rollgap. Using the relation (see, e.g., Pawelski [22])

$$\frac{\rho_{en}}{\rho(x)} = \frac{1 + \frac{1-2\nu_S}{E_S}\left(\sigma_{xx}(x) + \sigma_{yy}(x) + \sigma_{zz}(x)\right)}{1 + \frac{1-2\nu_S}{E_S}\sigma_{en}} \; , \tag{47}$$

we can calculate the velocity $v(x)$ as

$$v(x) = \frac{h_{en} v_{en}}{h(x)} \frac{\rho_{en}}{\rho(x)} = \frac{h_{en} v_{en}}{h(x)} \frac{1 + \frac{1-2\nu_S}{E_S}\left(\sigma_{xx}(x) + \sigma_{yy}(x) + \sigma_{zz}(x)\right)}{1 + \frac{1-2\nu_S}{E_S}\sigma_{en}} \; . \tag{48}$$

In the following formulas we will use the abbreviation $\sigma_{ii}^{(n)} = \sigma_{ii}(x_n)$. The evaluation of Eq. (48) at the neutral point $x = x_n$ with $v(x_n) = v_R$ and $h(x_n) = h_n$ gives

$$v_R = \frac{h_{en} v_{en}}{h_n} \frac{1 + \frac{1-2\nu_S}{E_S}\left(\sigma_{xx}^{(n)} + \sigma_{yy}^{(n)} + \sigma_{zz}^{(n)}\right)}{1 + \frac{1-2\nu_S}{E_S}\sigma_{en}} \; ,$$

and, thus,

$$v_{en} = \frac{h_n}{h_{en}} v_R \frac{1 + \frac{1-2\nu_S}{E_S}\sigma_{en}}{1 + \frac{1-2\nu_S}{E_S}\left(\sigma_{xx}^{(n)} + \sigma_{yy}^{(n)} + \sigma_{zz}^{(n)}\right)} \; . \tag{49}$$

Arranging the abbreviation

$$\bar{h}_n = \frac{h_n}{1 + \frac{1-2\nu_S}{E_S}\left(\sigma_{xx}^{(n)} + \sigma_{yy}^{(n)} + \sigma_{zz}^{(n)}\right)} \; , \tag{50}$$

we can calculate the backward slip ς_{en} as

$$\varsigma_{en} = \frac{\bar{h}_n}{h_{en}}\left(1 + \frac{1-2\nu_S}{E_S}\sigma_{en}\right) . \tag{51}$$

Equivalently, by means of the continuity relation

$$\rho\left(x_n\right)h_n v_R = \rho_{ex}v_{ex}h_{ex} ,$$

see also Eq. (46), the forward slip ς_{ex} is obtained as

$$\varsigma_{ex} = \frac{\bar{h}_n}{h_{ex}}\left(1 + \frac{1-2\nu_S}{E_S}\sigma_{ex}\right) . \tag{52}$$

The elongation coefficient ε, which is one of the key variables of the temper rolling process, is defined as

$$\varepsilon = \frac{v_{ex}}{v_{en}} - 1 = \frac{\varsigma_{ex}}{\varsigma_{en}} - 1 = \frac{h_{en}}{h_{ex}}\frac{\left(1 + \frac{1-2\nu_S}{E_S}\sigma_{ex}\right)}{\left(1 + \frac{1-2\nu_S}{E_S}\sigma_{en}\right)} - 1 . \tag{53}$$

e) Numerical Example of the Rollgap Scenario in the Temper Rolling Case

To illustrate a typical rollgap scenario for the temper rolling regime, a numerical example using typical parameters from the literature (see, e.g., Krause [15], Beitz [3], Weber [27], Domanti [2], Fleck [4,5]) is investigated. The parameters are given as follows: $R_0 = 220\text{mm}$, $k_f = 750\text{N/mm}^2$, $\mu = 0.12$, $E_R = E_S = 2.1 \times 10^5\text{N/mm}^2$, $\nu_R = \nu_S = 0.3$, $h_{en} = 1\text{mm}$, $\varepsilon = 0.01$, $\sigma_{en} = 70\text{N/mm}^2$ and $\sigma_{ex} = 100\text{N/mm}^2$. Figure 6 shows the roll/strip contact arc, the stress distribution in the strip and the equivalent deformation φ. This example should emphasize the necessity of taking the elastic work roll deformations and the elastic–plastic behavior of the strip into account.

2.2 Mill Stand Dynamics and Hydraulic Actuator

Since the mechanical construction of a four–high mill stand is rather complicated, the calculation of the stresses and elastic deformations due to the roll force F_r and the force of the hydraulic adjustment system F_h is usually done via finite element techniques. For the purpose of a controller design this approach is unsuitable. Thus, the dynamical behavior is approximately described by a simple equivalent mechanical model consisting of concentrated springs, masses and dampers.

The elastic stretching of the mill stand due to the hydraulic force F_h is described by the measured mill stretch calibration curve $f_{str}\left(F_h\right)$. The approximation of this mill stretch curve by means of a linear spring points out to be appropriate. This spring coefficient c_g is also known as "mill stretch coefficient" and can be obtained in a calibration process during the mill setup.

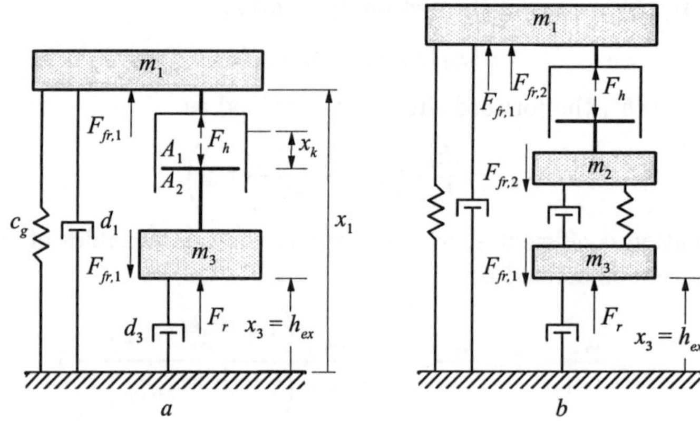

Figure 7. Simple lumped–parameter mill stand models.

Several approaches to model the mill stand dynamics for control purposes have been presented in the metal forming literature. We will briefly discuss two lumped–parameter spring–mass–damper models, which have been proposed by Kugi [17,18], see Figure 7. The center–line of the strip is assumed to be kept constant, thus, it coincides with the inertial frame. In both cases of Figure 7 the mass of the mill housing is taken into account by a single mass m_1. The two concepts just differ in the representation of the masses of the moving parts. In Figure 7a the total mass of all moving parts (i.e., hydraulic piston, upper work roll, upper backup roll, chocks) is represented by m_3, while in Figure 7b the moving masses are subdivided into two parts, m_2 and m_3. Here, m_3 covers the mass of the work roll and the associated chocks, while m_2 counts for the mass of the backup roll, the chocks and the hydraulic piston.

The 3–mass model of Figure 7b was designed for simulating eccentricity effects caused by the work and/or backup rolls. Since the problem of eccentricity compensation, which has been addressed by Kugi [16,18,19], is beyond the scope of this contribution, we will restrict the following investigations to the simple mill stand model depicted in Figure 7a. The friction forces between the roll chocks and the mill housing $F_{fr,1}$ and $F_{fr,2}$ are included in the simulation model but they are not considered in the model for the controller design.

The differential equations of motion of the mill stand model due to Figure 7a read as

$$\dot{x}_1 = v_1 \, ,$$

$$\dot{v}_1 = \frac{1}{m_1} \left(-c_g \left(x_1 - l_1 \right) - d_1 v_1 + F_h - m_1 g \right) \, ,$$

$$\dot{x}_3 = v_3 \, ,$$

$$\dot{v}_3 = \frac{1}{m_3} \left(F_r - F_h - d_3 v_3 - m_3 g \right) \, ,$$

(54)

where g denotes the constant of gravity, F_r the roll force, d_1 and d_3 the damping coefficients, and l_1 the length of the unloaded mill spring c_g. The abbreviation $\dot{x}$ indicates the derivative of x with respect to the time t. The symbols h_{ex} and x_3 are used synonymously for the strip exit thickness.

Given a constant roll force F_r^s, the stationary hydraulic force F_h^s and the corresponding equilibrium position x_1^s read as

$$F_h^s = F_r^s - m_3 g \, , \qquad x_1^s = l_1 + \frac{F_h^s - m_1 g}{c_g} \, .$$

(55)

The piston position x_k (see Figures 7a and 8) is given by

$$x_k = \left(x_1 - x_1^s \right) - \left(x_3 - x_3^s \right)$$

(56)

with x_3^s denoting an arbitrary stationary strip exit thickness.

The force F_h is provided by a hydraulic actuator in a single–acting piston configuration as shown in Figure 8. The return chamber is loaded with a constant pressure P_2 acting on the effective piston area A_2. The following considerations concerning the modelling as well as the control concept can be extended to a double–acting double–ended piston configuration, as discussed in detail by Kugi [16].

The pressure in the forward chamber with the effective piston area A_1 is denoted by P_1, and V_0 is the volume of the forward chamber for $x_k = 0$. The variable Q describes the flow from the servo valve to the forward chamber, and the leakage flows are taken into account by Q_{leak}.

Under the assumptions that the supply pressure P_S is constant; the servo valve is rigidly connected to the supply pressure pump; the temperature T of the oil is constant, and the oil is isotropic with $\rho_{oil}\left(P_1 \right)$ as the mass density, we get for the continuity equation

$$\frac{\mathrm{d}}{\mathrm{d}t} \left(\rho_{oil} \left(P_1 \right) \left(V_0 + A_1 x_k \right) \right) = \rho_{oil} \left(P_1 \right) \left(Q - Q_{leak} \right) \, .$$

(57)

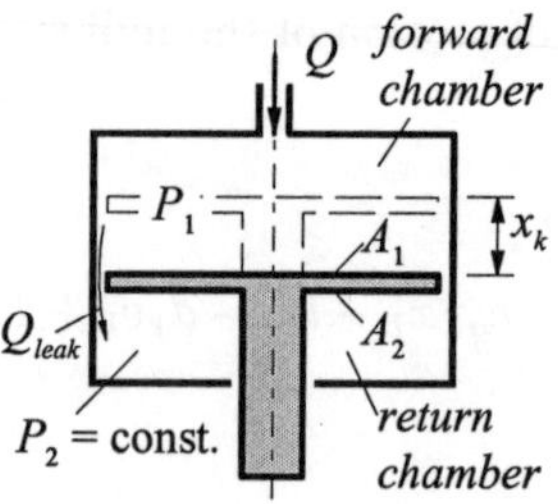

Figure 8. Scheme of the hydraulic actuator in a single–acting piston configuration.

Using the definition of the isothermal bulk modulus E_{oil} of the oil,

$$\frac{1}{E_{oil}} = \frac{1}{\rho_{oil}} \left(\frac{\partial \rho_{oil}}{\partial P_1} \right)_{T=\text{const}.} \tag{58}$$

and the notation $v_k = \dot{x}_k$, we can rewrite Eq. (57) in the form

$$\dot{P}_1 = \frac{E_{oil}}{V_0 + A_1 x_k} \left(-A_1 v_k + Q - Q_{leak} \right) . \tag{59}$$

Analogously, the differential equation for the hydraulic force $F_h = P_1 A_1 - P_2 A_2$ reads as

$$\dot{F}_h = \frac{E_{oil} A_1}{V_0 + A_1 x_k} \left(-A_1 v_k + Q - Q_{leak} \right) . \tag{60}$$

The dynamics of the servo valve will be neglected, because they are very fast compared to the dynamics of the other parts of the hydraulic actuator. Thus, we will use the displacement x_v of the servo valve as the control input.

In order to describe the (static) valve behavior, two cases have to be distinguished, namely, the case $x_v \geq 0$, where the supply pressure P_S is connected to the forward chamber via the valve orifice, and the case $x_v < 0$, where the connection of the forward chamber to the tank (with the tank pressure P_T) is established. With K_v denoting the valve coefficient, the equations

$$Q = K_v x_v \sqrt{P_S - P_1} \qquad \text{for} \qquad x_v \geq 0$$

$$\tag{61}$$

$$Q = K_v x_v \sqrt{P_1 - P_T} \qquad \text{for} \qquad x_v < 0$$

are obtained.

Although the abbreviation F_r has been used to count for the roll force in Eq. (54), we have to keep in mind that there is an implicit interconnection

between F_r, the state variable $x_3 = h_{ex}$, the strip entry thickness h_{en}, the strip
tensions σ_{en} and σ_{ex}, the geometry of the rolls and the material parameters,
as described by the rollgap model of Eq. (44). Thus, the set of N implicit
algebraic equations (44) has to be added to the ordinary differential equations
(54) obeying the relation

$$F_r = B_S R_0 \Delta\theta \sum\nolimits_{j=1}^{N} p_j \, , \tag{62}$$

with B_S denoting the strip width. Therefore, the dynamics of the overall system are captured by a set of differential–algebraic equations (DAE), reflecting
the fact that the dynamics of the rollgap phenomena are neglected. In other
words, the rollgap behavior is treated in a quasi-static manner. This quasi-
static consideration of the rollgap as outlined in Subsection 2.1 is indeed a
valid approximation, since the dynamics of the deformation phenomena are
much faster than those of the mill stand.

The control design problem for nonlinear differential algebraic systems
(see, e.g., Schlacher [24]) in the rolling industry as well as the aspects of the
numerical simulation, outlined for the case of multi-stand cold rolling mills
using the rollgap model of Bland [1], are addressed by Grabmair [7].

2.3 Characterization of the Elastic Strip Elements

The mass of the elastic strip elements connecting the winders to the bridle
rolls, and the bridle rolls to the rollgap will be neglected. Thus, for modelling
the strip elements, we use a simple linear–elastic spring with a constant spring
coefficient. Here, E_S denotes the Young modulus of the strip.

Since the strip entering the skin pass mill was already preprocessed in a
cold rolling mill, it meets very tight thickness tolerances (see, e.g., Ginzburg [6]).
Thus, for the calculation of the effective strip tensions σ_{eco} (between the unwinder and the entry bridle rolls, with *eco* as an abbreviation for "entry coil")
and σ_{en} (between the entry bridle rolls and the rollgap), the strip elements
are assumed to exhibit a constant (nominal) entry thickness h_{en}^{nom}.

The plant operator usually tends to establish a given elongation coefficient over the entire strip. Therefore, the assumption of a constant strip exit
thickness h_{ex}^{nom} is arranged, too. The equations being obtained are exemplary
outlined for the effective strip tensions of the entry section of the mill (see

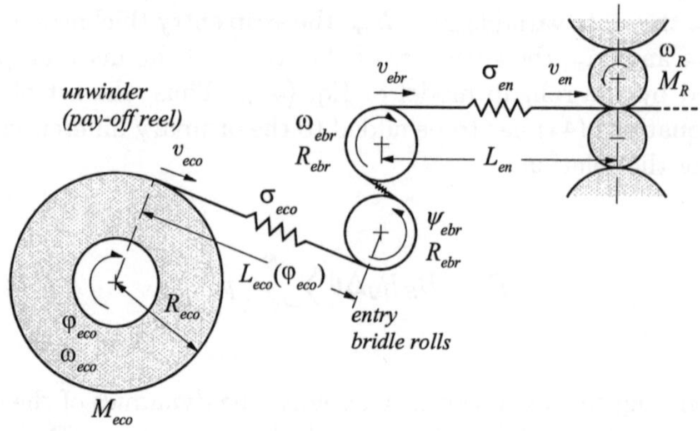

Figure 9. Elastic strip elements connecting the unwinder, the entry bridle rolls and the rollgap.

Figure 9 and Eq (49)),

$$\dot{\sigma}_{eco} = \frac{E_S}{\bar{L}_{eco}} \left(\psi_{ebr} R_{ebr} - \omega_{eco} R_{eco} \left(\varphi_{eco} \right) \right) ,$$

$$\dot{\sigma}_{en} = \frac{E_S}{L_{en}} \left(\frac{\bar{h}_n \left(\cdot \right)}{h_{en}} \left(1 + \frac{1 - 2\nu_S}{E_S} \sigma_{en} \right) \omega_R R_0 - \omega_{ebr} R_{ebr} \right) ,$$

$$(63)$$

with $\bar{h}_n \left(\cdot \right) \doteq \bar{h}_n \left(h_{en}, h_{ex}, \sigma_{en}, \sigma_{ex} \right)$ due to Eq. (50), whereas the dependence of $\bar{h}_n$ on the material, geometric and lubrication parameters has been supressed for notational convenience. The angle φ_{eco} denotes the angle of the unwinder, the variables $\omega_{eco} = \dot{\varphi}_{eco}$, ω_R, ψ_{ebr} and ω_{ebr} count for the angluar velocities of the unwinder (with radius $R_{eco} \left(\varphi_{eco} \right)$), the work rolls (with radius R_0), and the entry bridle rolls (with radius R_{ebr}).

Since the radius of the unwinder varies as a function of the angle φ_{eco}, the length L_{eco} is also a function of φ_{eco} (see Figure 9). In contrast to this, the distance L_{en} between the bridle roll center and the work roll center–line is constant. The change of the strip length $L_{eco} \left(\varphi_{eco} \right)$ has minor influence on the system dynamics. Thus, an average constant length $\bar{L}_{eco}$ is used in Eq. (63).

The equations characterizing the strip tensions between the rollgap and the exit bridle rolls (i.e., σ_{ex}), and between the exit bridle rolls and the rewinder (i.e., σ_{xco}, with xco as an abbreviation for "exit coil") are obtained in a similar way.

2.4 Bridle Roll Dynamics

Since the winders cannot generate the required backward and forward tensions, σ_{en} and σ_{ex}, the bridle rolls are included in the temper rolling mill configuration (see Figure 1).

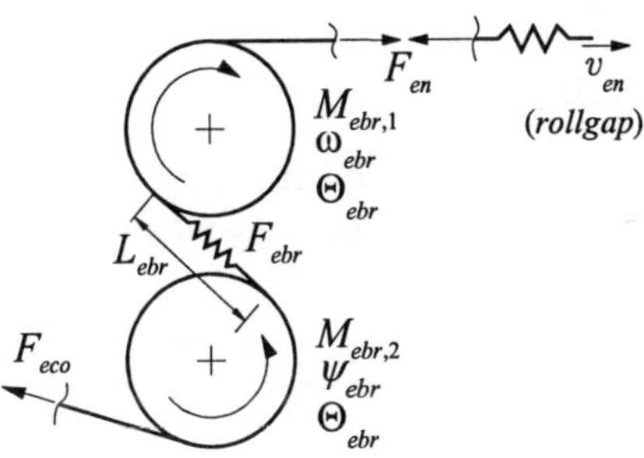

Figure 10. Entry Bridle Rolls.

The maximum force difference $F_{en} - F_{eco}$ between the bridle entry and exit strip element (see Figure 10) is limited by the wrap angle and can be calculated by Eytelwein's equation (see, e.g, Beitz [3]). But it is worth mentioning that this maximum value is only available in the case, when the strip is slipping on the rolls, which is of course not acceptable in our application. Thus, this value can only be thought of being the absolute limit and the control law has to keep the force difference below this critical value.

According to Figure 10, we are now going to develop the simple differential equations covering the dynamical behavior of a two–roll bridle configuration, exemplary outlined for the entry bridle rolls (index *ebr*). The motor torques $M_{ebr,1}$ and $M_{ebr,2}$ driving the bridle rolls will serve as input variables. The only quantity being measureable is the angular velocity ω_{ebr}. For the sake of simplicity we will assume that the moments of inertia Θ_{ebr} as well as the friction coefficients d_{ebr} of the two drives, including the rolls, the transmission shafts and the motors, are identical. The friction torques are given by $d_{ebr}\omega_{ebr}$ and $d_{ebr}\psi_{ebr}$.

The external forces acting on the bridle rolls are

$$F_{en} = \sigma_{en}B_S h_{en}^{nom} , \qquad F_{eco} = \sigma_{eco}B_S h_{en}^{nom} , \tag{64}$$

with B_S denoting the strip width. Assuming that there is no slip between the bridle rolls and the strip, we get the state space representation of the system in the form

$$\dot{\omega}_{ebr} = \frac{1}{\Theta_{ebr}} \left(M_{ebr,1} + (F_{en} - F_{ebr}) R_{ebr} - d_{ebr}\omega_{ebr} \right) ,$$

$$\dot{\psi}_{ebr} = \frac{1}{\Theta_{ebr}} \left(M_{ebr,2} + (F_{ebr} - F_{eco}) R_{ebr} - d_{ebr}\psi_{ebr} \right) , \qquad (65)$$

$$\dot{F}_{ebr} = c_{ebr} R_{ebr} \left(\omega_{ebr} - \psi_{ebr} \right) .$$

The radius of both rolls is denoted by R_{ebr}, the variable F_{ebr} covers the force due to the elastic stretching of the strip element establishing the coupling of the rolls. The parameter c_{ebr},

$$c_{ebr} = \frac{E_S B_S h_{en}^{nom}}{L_{ebr}} , \qquad (66)$$

is the associated spring constant and L_{ebr} denotes the length of the strip element.

The corresponding set of differential equations for the exit bridle rolls can be obtained in the same way. The corresponding quantities will be indicated by the subscript xbr.

2.5   Winder Dynamics

According to the previous discussion of a nominal strip entry and exit thickness (h_{en}^{nom}, h_{ex}^{nom}), the winder radii R_{eco} and R_{xco} can be calculated in terms of the winder angles φ_{eco} and φ_{xco}. See Figure 9 for the case of the unwinder. For the sake of convenience, the following equations are outlined with the index co capturing both the unwinder (index eco – "entry coil", $\kappa = +1$) and the rewinder (index xco – "exit coil", $\kappa = -1$) behavior. The winder radius $R_{co}(\varphi_{co})$ reads as

$$R_{co}(\varphi_{co}) = R_M + \left(N_{co}^0 - \kappa \frac{\varphi_{co}}{2\pi} \right) h_{co}^{nom} , \qquad (67)$$

with N_{co}^0 denoting the initial windings and R_M the radius of the empty mandrel. Then, the winder dynamics might be approximately described by the nonlinear differential equations

$$\dot{\varphi}_{co} = \omega_{co} ,$$

$$\dot{\omega}_{co} = \frac{M_{co} + \kappa F_{co} R_{co}(\varphi_{co}) - M_{fr,co}}{\Theta_{co}(\varphi_{co}) + \Theta_M} , \qquad (68)$$

with

$$\Theta_{co}\left(\varphi_{co}\right) = \frac{1}{2}\pi\rho_S B_S \left(R_{co}^4\left(\varphi_{co}\right) - R_M^4\right) . \tag{69}$$

Here, ρ_S denotes the mass density of the strip, and $\Theta_{co}\left(\varphi_{co}\right)$ and Θ_M are the moments of inertia of the coil and the transmission shaft. M_{co} is the winder drive torque, the force exerted on the winder by the strip is denoted with F_{co} ($F_{eco} = \sigma_{eco}B_S h_{en}^{nom}$, $F_{xco} = \sigma_{xco}B_S h_{ex}^{nom}$, see Eq. (64)), and a friction torque is taken into account by $M_{fr,co} = d_{co}\omega_{co}$.

2.6 *The Entire Mathematical Model of the Skin Pass Mill*

As a result of the preceding discussions we will summarize the equations covering the dynamical behavior of the skin pass mill configuration due to Figure 1. The dynamics of the lumped–parameter mill stand model of Figure 7a and the hydraulic actuator of Figure 8 are given by the Eqns. (54), (56), (60) and (61), whereas the rollgap scenario is covered by the set of algebraic equations (44) and (62).

The effective tensions in the strip elements connecting the rollgap to the entry and exit bridle rolls are described by the differential equations

$$\dot{\sigma}_{en} = \frac{E_S}{L_{en}}\left(\frac{\bar{h}_n\left(\cdot\right)}{h_{en}}\left(1 + \frac{1 - 2\nu_S}{E_S}\sigma_{en}\right)\omega_R R_0 - \omega_{ebr}R_{ebr}\right) ,$$

$$\dot{\sigma}_{ex} = \frac{E_S}{L_{ex}}\left(\omega_{xbr}R_{xbr} - \frac{\bar{h}_n\left(\cdot\right)}{x_3}\left(1 + \frac{1 - 2\nu_S}{E_S}\sigma_{ex}\right)\omega_R R_0\right) . \tag{70}$$

See Eq. (63), which also includes the differential equation for the tension σ_{eco}. The characterization of the entry and exit bridle roll dynamics is done via equations of the type of Eq. (65). The motion of the unwinder and rewinder is covered by Eq. (68). With Θ_R denoting the moment of inertia of the work and the backup rolls, the transmission shaft and the motor, M_R the main mill drive torque, $d_R\omega_R$ the friction torque, and $M_r\left(h_{en}, h_{ex}, \sigma_{en}, \sigma_{ex}\right)$ the roll torque, the main mill drive dynamics are given by the differential equation

$$\dot{\omega}_R = \frac{1}{\Theta_R}\left(M_R - M_r\left(h_{en}, h_{ex}, \sigma_{en}, \sigma_{ex}\right) - d_R\omega_R\right) . \tag{71}$$

Thus, by neglecting the dynamical behavior of the servo valve of the hydraulic adjustment system and the motor dynamics, we get the control inputs u of the skin pass mill,

$$u = \left[x_v, M_R, M_{ebr,1}, M_{ebr,2}, M_{xbr,1}, M_{xbr,2}, M_{eco}, M_{xco}\right]^T .$$

For the purpose of simulation as well as for testing the proposed control concept the entire mathematical model of the system is implemented in MATLAB/SIMULINK [20]. An efficient way to implement the mathematical model in MATLAB/SIMULINK is given by CMEX s–function programming techniques [20]. The simulation model also includes appropriate models of the drive characteristics, the servo valve dynamics, and of the position, velocity and pressure sensors.

3   Control of the Skin Pass Mill

The goal of the control concept is on the one hand to establish a certain elongation of the strip and on the other hand to control the strip tensions σ_{en} and σ_{ex}.

It is usual in the industrial practice that this task is done using the hydraulic adjustment system as an actuator for the elongation control loop, whereas the bridle torques are set in order to achieve, at least stationary, the desired strip tensions. Usually, the main mill drive serves as the master speed device, and in acceleration/deceleration situations, additional preset values are added to the bridle and winder torques in the sense of a feedforward strategy.

In contrast to this classical concept the proposed control approach is characterized by using the bridle roll torques as control inputs for the elongation and master speed control, whereas the hydraulic actuator and the main mill drive are used to control the strip tensions σ_{en} and σ_{ex}. At first we will discuss the properties and restrictions of the skin pass mill, which have to be taken into account in the controller design.

3.1   *Properties and Restrictions of the Plant*

Since the skin pass mill under consideration is not equipped with velocity sensors in the vicinity of the rollgap due to constructional reasons, the elongation coefficient of the rolled strip covered by Eq. (53),

$$\varepsilon = \frac{v_{ex}}{v_{en}} - 1 \, ,$$

is not directly measureable. As usual in temper rolling automation concepts, only the angular velocities of the bridle rolls (ω_{ebr}, ω_{xbr}), the main mill drive (ω_R) and the winders (ω_{eco}, ω_{xco}) can be obtained via incremental encoders. Due to this reason, the signal $\varkappa$,

$$\varkappa \doteq \frac{v_{xbr}}{v_{ebr}} - 1 = \frac{R_{xbr}}{R_{ebr}} \frac{\omega_{xbr}}{\omega_{ebr}} - 1 \, , \tag{72}$$

is defined to be the control variable, which corresponds exactly to the elongation coefficient ε in the case of constant strip tensions, $\dot{\sigma}_{en} = \dot{\sigma}_{ex} = 0$.

But it is obvious that ε and $\varkappa$ are different during transient changes of the strip tensions. In the following discussion the term *"elongation control"* will always refer to the control of the variable $\varkappa$, which will be also called *"elongation coefficient"*.

As already mentioned in the introduction of the temper rolling process, the elongation of the strip, and, therefore, the difference between the velocity signals v_{ebr} and v_{xbr} is very small. Thus, high–precision angle encoders have to be installed on the bridle roll shafts or the motor shafts, respectively.

The positions x_1 and $x_3 = h_{ex}$ as well as the corresponding velocities v_1 and v_3 of the mill stand model due to Figure 7a are not measureable. Only the piston position x_k and the pressure P_1 in the forward chamber can be obtained by sensors.

The absence of a measurement signal of the piston velocity v_k points out to be a significant restriction for the control design task, as will be discussed in the analysis of the nonlinear hydraulic gap controller.

We will assume that the plant is equipped with tension transducers for the strip tensions σ_{eco}, σ_{en}, σ_{ex} and σ_{xco}. For the validation of the control concept and to ensure that the controller is practically feasible, a significant corruption of the signals by transducer noise is taken into account in the simulation model.

Last but not least we have to bear in mind that the control law has to be implemented on a digital signal processor with a predefined minimal sampling time.

3.2   *Outline of the Proposed Control Concept*

The scheme of the proposed control concept for the temper rolling mill is depicted in Figure 11. The concept is organized as a cascaded system with the hydraulic gap control, the speed control of the bridle rolls and the main mill drive as inner control loops. The winder control to be presented in Subsection 3.7 is not included in Figure 11.

The following Subsections 3.3, 3.4 and 3.5 are concerned with the discussion of the inner control loops. In Subsection 3.6 the nonlinear controller for the strip tensions σ_{en} and σ_{ex} assuming an idealized behavior of the hydraulic adjustment system as well as of the main mill drive speed control loop is presented.

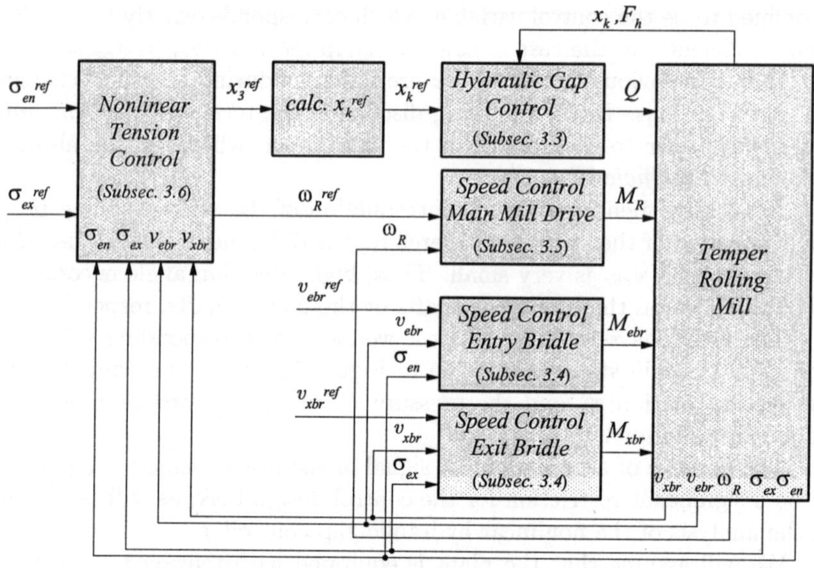

Figure 11. Scheme of the tension- and elongation control concept for the temper rolling mill. The reference values v_{ebr}^{ref} and v_{xbr}^{ref} for the bridle roll velocities are obtained using the reference elongation coefficient $\varkappa^{ref}$ and the reference master speed of the mill.

3.3   Nonlinear Hydraulic Gap Control

A key element of the entire control concept for the skin pass mill is the hydraulic gap control (HGC), i.e., the control of the piston position x_k. A detailed and concise treatment of the following discussion can be found in the contributions of Kugi [16,17,18]. For the convenience of the reader the key ideas of this concept are briefly summarized below.

Throughout the following considerations, we assume a roll force of the form $F_r = F_r^0 - c_3 x_3$ (with the unknown, but constant force F_r^0, and the material deformation coefficient c_3) instead of the rollgap model of Eq. (44). The system (54), (56), (60) and (61) exhibits the structure of an affine-input system (AI-system) with x_v as the control input. For this type of nonlinear systems the powerful control synthesis tools of exact input/state– and exact input/output–linearization are available (see, e.g., Isidori [11], Nijmeijer [21], Sastry [23], Khalil [14]). Generally, these techniques lead to control laws which require the information of the full state. A modified version of the exact input/output–linearization with constrained measurements can be found in the contribution of Schlacher [25]. This is also the theoretical background of the nonlinear controller being used later on.

It turns out that in our case the system is in fact exact input/state–linearizable. But a straightforward implementation of this nonlinear state feedback control law fails due to the fact that the velocity signal v_k is not directly measureable. Thus, it has to be obtained by approximate numerical differentiation of the position signal x_k. Due to the fact that the position signal is corrupted by considerable transducer and quantization noise this strategy is practically unsuitable.

Following the ideas of Kugi [16,17,18], we design a nonlinear controller based on the exact input/output–linearization concept using the artificial output z,

$$z = F_h - E_{oil} A_1 \ln \left(\frac{V_0}{V_0 + A_1 x_k} \right) . \tag{73}$$

In the contribution of Kugi [16] also the more general case of a double–acting double–ended piston configuration is discussed. Obviously, the relative degree (see again the literature cited above) for the output z of Eq. (73) is 1. In the following discussions regarding the controller design the leakage flow Q_{leak} is set to zero in Eq. (60). Thus, the control law can be obtained by calculating the derivative of z with respect to the time t,

$$\dot{z} = \dot{F}_h + \frac{E_{oil} A_1}{V_0 + A_1 x_k} A_1 v_k = \frac{E_{oil} A_1}{V_0 + A_1 x_k} Q = u , \tag{74}$$

with the new input u. To obtain a linear behavior from the new input u to the chosen output z, e.g., $\dot{z} = u$, the nonlinear control law is calculated in the form

$$Q = \frac{V_0 + A_1 x_k}{E_{oil} A_1} u . \tag{75}$$

Note, that as the central feature of this control concept, the velocity signal v_k is not needed for the calculation of the control input Q. For controlling the piston position x_k the control law

$$u = \alpha E_{oil} A_1 \ln \left(\frac{V_0 + A_1 x_{k,ref}}{V_0 + A_1 x_k} \right) , \tag{76}$$

and, thus,

$$Q = \alpha \left(V_0 + A_1 x_k \right) \ln \left(\frac{V_0 + A_1 x_{k,ref}}{V_0 + A_1 x_k} \right) \tag{77}$$

is proposed by Kugi [16], where α serves as a tuning parameter to adjust the desired closed loop dynamics. Including the static compensation of the servo valve function of Eq. (61), we get the nonlinear state feedback controller

$$x_v = \alpha \frac{V_0 + A_1 x_k}{K_v \sqrt{P_S - P_1}} \ln \left(\frac{V_0 + A_1 x_{k,ref}}{V_0 + A_1 x_k} \right) \tag{78}$$

for $x_v \geq 0$, and

$$x_v = \alpha \frac{V_0 + A_1 x_k}{K_v \sqrt{P_1 - P_T}} \ln \left(\frac{V_0 + A_1 x_{k,ref}}{V_0 + A_1 x_k} \right) \tag{79}$$

for $x_v < 0$. The proof of the global asymptotic stability for the parameter α satisfying a certain inequality condition is based on the Popov criterion and can be found in the book of Kugi [16].

This concept, which is successfully implemented in hot and cold rolling mills (Kugi [17]), will serve as a central part of the proposed control concept.

3.4   Speed Control of the Bridle Rolls / Elongation Control

As already discussed, the key observation leading to the control concept being presented is that the physical elongation coefficient ε (see Eq. (53)) characterizing the rollgap scenario is not directly measureable. Instead of ε the elongation coefficient $\varkappa$ (see Eq. (72)), which is obtained by measuring the angular velocities of the entry and exit bridle rolls is used as a control variable. Let us emphasize again that $\varkappa$ and ε are identical only in the case $\dot{\sigma}_{en} = \dot{\sigma}_{ex} = 0$. Thus, the bridle torques M_{ebr} and M_{xbr} serve as control inputs for the output $\varkappa$, whereas, in addition, either the entry or the exit bridle roll is used as a speed master of the plant.

Since the performance of the elongation control concept inherently depends on the dynamics of the bridle speed control, in particular on the disturbance rejection performance of this inner control loop, the control design task for the bridle rolls will be investigated in detail. Again we will point out the ideas on the basis of the entry bridle rolls (see Eqns. (64), (65), (66) and Figure 10). But of course this concept can be applied to the exit bridle rolls in a straightforward manner.

The key goal is to control the angular velocity ω_{ebr}, which is the only quantity measureable in the configuration being considered. In addition we wish to establish a certain strip force F_{ebr} in order to determine the load sharing between both rolls. Obviously, due to the absence of a force transducer for F_{ebr}, this quantity cannot be directly controlled. Nevertheless, we will show by using a simple input transformation that a desired transient behavior of the force F_{ebr} can be achieved at least indirectly for the nominal dynamics.

To cope with this problem, we will at first compensate for the torques exerted by the strip forces F_{en} and F_{eco},

$$M_{ebr,1} = u_1 - F_{en}R_{ebr} , \qquad M_{ebr,2} = u_2 + F_{eco}R_{ebr} , \qquad (80)$$

by introducing the new inputs u_1 and u_2. Then, the linear bridle roll dynamics of Eq. (65) read as

$$\dot{\omega}_{ebr} = \frac{1}{\Theta_{ebr}} \left(u_1 - F_{ebr}R_{ebr} - d_{ebr}\omega_{ebr} \right) ,$$

$$\dot{\psi}_{ebr} = \frac{1}{\Theta_{ebr}} \left(u_2 + F_{ebr}R_{ebr} - d_{ebr}\psi_{ebr} \right) , \qquad (81)$$

$$\dot{F}_{ebr} = c_{ebr}R_{ebr} \left(\omega_{ebr} - \psi_{ebr} \right) .$$

The following investigations will be done using the frequency domain equivalent of Eq. (81). Here, the notation $\hat{x}(s)$ is arranged to indicate the corresponding Laplace transform of $x(t)$. When there is no danger of confusion, we will drop the argument s for the sake of notational simplicity. Thus, we get

$$\hat{\omega}_{ebr} = \frac{\left(\Theta_{ebr}s^2 + d_{ebr}s + c_{ebr}R_{ebr}^2 \right) \hat{u}_1 + c_{ebr}R_{ebr}^2 \hat{u}_2}{\left(\Theta_{ebr}s + d_{ebr} \right) \left(\Theta_{ebr}s^2 + d_{ebr}s + 2c_{ebr}R_{ebr}^2 \right)} \qquad (82)$$

and

$$\hat{F}_{ebr} = \frac{c_{ebr}R_{ebr}}{\Theta_{ebr}s^2 + d_{ebr}s + 2c_{ebr}R_{ebr}^2} \left(\hat{u}_1 - \hat{u}_2 \right) . \qquad (83)$$

Motivated by the structure of Eq. (83), we will apply the regular input transformation

$$\begin{bmatrix} \hat{u}_1 \\ \hat{u}_2 \end{bmatrix} = \begin{bmatrix} \alpha & \beta \\ \gamma & \delta \end{bmatrix} \begin{bmatrix} \hat{u} \\ \hat{w} \end{bmatrix} \qquad (84)$$

with the new inputs $\hat{u}$, $\hat{w}$ and the parameters α, β, γ, δ to be determined such that $\hat{u}$ does not act on $\hat{F}_{ebr}$. This goal can be easily achieved by choosing $\gamma = \alpha$. By means of the new inputs $\hat{u}$ and $\hat{w}$, Eq. (82) yields the form

$$\hat{\omega}_{ebr} = \frac{\alpha}{\Theta_{ebr}s + d_{ebr}} \hat{u} + \frac{s \left(\Theta_{ebr}s + d_{ebr} \right) \beta + c_{ebr}R_{ebr}^2 \left(\beta + \delta \right)}{\left(\Theta_{ebr}s + d_{ebr} \right) \left(\Theta_{ebr}s^2 + d_{ebr}s + 2c_{ebr}R_{ebr}^2 \right)} \hat{w} .$$

By choosing, e.g., $\alpha = 1$ and $\beta = -\delta = R_{ebr}$, we obtain

$$\hat{F}_{ebr} = G_{\hat{F}_{ebr}/\hat{w}}(s) \, \hat{w} = \frac{2c_{ebr}R_{ebr}^2}{\Theta_{ebr}s^2 + d_{ebr}s + 2c_{ebr}R_{ebr}^2} \hat{w} \qquad (85)$$

and

$$\hat{\omega}_{ebr} = G_{\hat{\omega}_{ebr}/\hat{u}}\left(s\right)\hat{u} + G_{\hat{\omega}_{ebr}/\hat{w}}\left(s\right)\hat{w} =$$

$$= \frac{1}{\Theta_{ebr}s + d_{ebr}}\hat{u} + \frac{R_{ebr}s}{\Theta_{ebr}s^2 + d_{ebr}s + 2c_{ebr}R_{ebr}^2}\hat{w}\,. \tag{86}$$

The frequency domain representation of Eqns. (85), (86) clearly reveals the meaning of the new inputs $\hat{u}$ and $\hat{w}$. That is, u does not have an influence on the force F_{ebr}. In the case $w\left(t\right) = w_0 = \text{const}$, we have the relations

$$\lim_{t \to \infty} F_{ebr}\left(t\right) = w_0 \tag{87}$$

and

$$\hat{\omega}_{ebr} = G_{\hat{\omega}_{ebr}/\hat{u}}\left(s\right)\hat{u} = \frac{1}{\Theta_{ebr}s + d_{ebr}}\hat{u}\,. \tag{88}$$

The controller design based on Eqns. (85) and (86) can now be addressed using standard linear design techniques. Particularly, by means of a H_2–design (see, e.g., Haas [8]), which is based on the well–known Youla–parametrization, the decoupling of the tracking problem from the disturbance rejection problem is obtained.

3.5   *Speed Control of the Main Mill Drive*

Within the entire control concept of the skin pass mill, we will make use of a cascaded control concept with a speed controller for the main mill drive in the inner loop. As already mentioned in the previous subsection concerning the speed control of the bridle rolls, special emphasis should be laid on the disturbance rejection performance of the closed loop. Again, the H_2– approach is used to design an appropriate speed controller for the main mill drive dynamics of Eq. (71).

3.6   *Nonlinear Tension Control: An Exact Input/Output–Linearization Approach*

The control of the strip tensions σ_{en} and σ_{ex} is done using the hydraulic adjustment system and the main mill drive as actuators. The hydraulic actuator is driven in the position control mode, see Subsection 3.3, and for the main mill drive the speed is controlled as presented in the previous subsection.

Since the dynamics of the inner control loops are considered to be ideal, the position $x_3 = h_{ex}$ (see Figure 7a) and the angular velocity ω_R can be taken as control inputs (see Figure 12). However, this approach is justified because

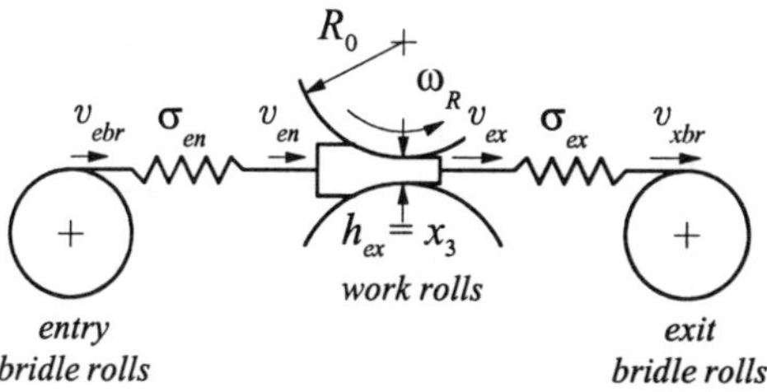

Figure 12. Nonlinear tension control: The simplified system according to Eq. (89).

the dynamics of the tension control loop are designed to be significantly slower than the actuator dynamics. Arranging the abbreviations

$$c_{en} = \frac{E_S}{L_{en}} , \qquad c_{ex} = \frac{E_S}{L_{ex}} , \qquad v_{ebr} = \omega_{ebr} R_{ebr} , \qquad v_{xbr} = \omega_{xbr} R_{xbr} ,$$

we can rewrite the differential equations (70) in the form

$$\dot{\sigma}_{en} = c_{en} \left(\frac{\bar{h}_n \left(h_{en}, x_3, \sigma_{en}, \sigma_{ex} \right)}{h_{en}} \left(1 + \frac{1 - 2\nu_S}{E_S} \sigma_{en} \right) \omega_R R_0 - v_{ebr} \right) ,$$

$$\dot{\sigma}_{ex} = c_{ex} \left(v_{xbr} - \frac{\bar{h}_n \left(h_{en}, x_3, \sigma_{en}, \sigma_{ex} \right)}{x_3} \left(1 + \frac{1 - 2\nu_S}{E_S} \sigma_{ex} \right) \omega_R R_0 \right) . \tag{89}$$

By means of an exact input/output–linearization approach we will design a static state feedback control law in order to obtain the decoupled and linear closed loop dynamics

$$\dot{\sigma}_{en} = -\alpha_1 \sigma_{en} + c_{en} u_1 ,$$

$$\dot{\sigma}_{ex} = -\alpha_2 \sigma_{ex} + c_{ex} u_2 . \tag{90}$$

Here, $\alpha_1 > 0$ and $\alpha_2 > 0$ serve as design parameters and u_1, u_2 constitute the new (artificial) input variables as schematically depicted in Figure 13. By solving the algebraic equations

$$\frac{\bar{h}_n \left(h_{en}, x_3, \sigma_{en}, \sigma_{ex} \right)}{h_{en}} \left(1 + \frac{1 - 2\nu_S}{E_S} \sigma_{en} \right) \omega_R R_0 - v_{ebr} = -\frac{\alpha_1}{c_{en}} \sigma_{en} + u_1 ,$$

$$v_{xbr} - \frac{\bar{h}_n \left(h_{en}, x_3, \sigma_{en}, \sigma_{ex} \right)}{x_3} \left(1 + \frac{1 - 2\nu_S}{E_S} \sigma_{ex} \right) \omega_R R_0 = -\frac{\alpha_2}{c_{ex}} \sigma_{ex} + u_2 ,$$

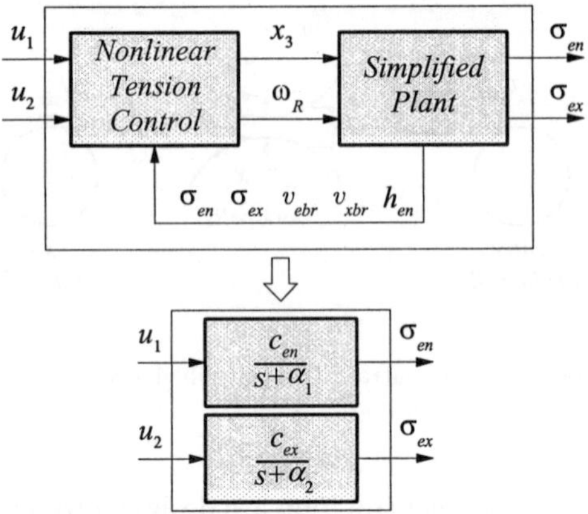

Figure 13. Nonlinear tension control (Eq. (91)) of the simplified system (Eq. (89)) and the corresponding linear and decoupled closed loop dynamics.

with respect to x_3 and ω_R, we get the nonlinear control law in the form

$$x_3 = h_{en} \frac{\left(1 + \frac{1-2\nu_S}{E_S}\sigma_{ex}\right) v_{ebr} + u_1 - \frac{\alpha_1}{c_{en}}\sigma_{en}}{\left(1 + \frac{1-2\nu_S}{E_S}\sigma_{en}\right) v_{xbr} - u_2 + \frac{\alpha_2}{c_{ex}}\sigma_{ex}},$$

$$\omega_R = h_{en} \frac{1}{\left(1 + \frac{1-2\nu_S}{E_S}\sigma_{en}\right)} \frac{v_{ebr} + u_1 - \frac{\alpha_1}{c_{en}}\sigma_{en}}{\bar{h}_n\left(h_{en}, x_3, \sigma_{en}, \sigma_{ex}\right) R_0}.$$

$$(91)$$

Additionally, due to the obvious physical constraint $0 < x_3 < h_{en}$ of the system, the inequality

$$0 < \frac{\left(1 + \frac{1-2\nu_S}{E_S}\sigma_{ex}\right) v_{ebr} + u_1 - \frac{\alpha_1}{c_{en}}\sigma_{en}}{\left(1 + \frac{1-2\nu_S}{E_S}\sigma_{en}\right) v_{xbr} - u_2 + \frac{\alpha_2}{c_{ex}}\sigma_{ex}} < 1 \qquad (92)$$

has to be satisfied. At this point it should be noted that the inner control loops of the Subsections 3.3, 3.4 and 3.5 also have to count for the constraints of the plant by means of anti–reset anti–windup strategies, see, e.g., Hippe [9].

Based on the linear and decoupled closed loop dynamics of Eq. (90) obtained so far, standard linear single–input single–output control design techniques are applied to design an outer control loop for the control inputs u_1 and u_2.

The previous investigations are characterized by the idealization of the main mill drive speed control loop and the hydraulic positioning control loop. In order to achieve certain reference strip tensions σ_{en}^{ref} and σ_{ex}^{ref}, the corresponding reference values x_3^{ref} and ω_R^{ref} have to be prescribed for the inner control loops. Here, the notation x_3^{ref}, ω_R^{ref} is used to indicate that the variables given by Eq. (91) are used as reference values for the inner control loops, see Figure 11.

Thus, we have to calculate a reference value x_k^{ref} for the nonlinear HGC of Subsection 3.3 due to x_3^{ref}. This is done by calculating the roll force $F_r\left(h_{en}, x_3^{ref}, \sigma_{en}^{ref}, \sigma_{ex}^{ref}\right)$ by means of the rollgap model, and the associated stationary hydraulic force $F_h^{ref} = F_r^{ref} - m_3 g$ (see Eq. (55)). The stationary elastic stretching $x_1^{ref} - x_1^s$ of the mill stand with respect to an equilibrium point indicated by the superscript s is given by

$$x_1^{ref} - x_1^s = \frac{F_h^{ref} - F_h^s}{c_g} = \frac{F_r\left(h_{en}, x_3^{ref}, \sigma_{en}^{ref}, \sigma_{ex}^{ref}\right) - F_r^s}{c_g} .$$

Thus, the reference value x_k^{ref} is

$$x_k^{ref} = \left(x_1^{ref} - x_1^s\right) - \left(x_3^{ref} - x_3^s\right) =$$

$$= \frac{F_r\left(h_{en}, x_3^{ref}, \sigma_{en}^{ref}, \sigma_{ex}^{ref}\right) - F_r^s}{c_g} - \left(x_3^{ref} - x_3^s\right) , \qquad (93)$$

see also Eqns. (55) and (56).

The mathematical proof of the stability of the closed loop tension control system when including the nonlinear HGC and the dynamics of the main mill drive speed control loop as depicted in Figure 11 is still an open problem and will be addressed in our future research.

In order to show the ability of the proposed concept to cope with "dirty" effects such as transducer and quantization noise and parameter inaccuracies, various numerical simulations have been performed (see Subsection 3.8).

3.7 Winder Control

In many practical applications the demand on reducing the costs of the skin pass mill is often accompanied by giving up the installation of the transducers for the winder tensions σ_{eco} and σ_{xco}. Therefore, in order to establish certain winder tensions a simple open loop strategy is applied. This also allows to compensate for the torques required during acceleration and deceleration situations in the sense of a feedforward strategy.

Nevertheless, in this contribution we assume that the mill is equipped with sensors for the winder tensions. Since the winder dynamics are covered by a nonlinear differential equation, see Subsection 2.5, we will apply the exact input/output–linearization technique to design a speed control loop for the winders.

a) Exact Input/Output–Linearization of the Winders as a Prerequisite for the Tension Control

Using the concepts of the exact input/output–linearization for the winder dynamics given by the Eqns. (67), (68) and (69), we find that the relative degree for the output function $v_{co} = R_{co}\left(\varphi_{co}\right)\omega_{co}$ is equal to 1,

$$\dot{v}_{co} = \frac{\partial R_{co}\left(\varphi_{co}\right)}{\partial \varphi_{co}}\omega_{co}^2 + R_{co}\left(\varphi_{co}\right)\dot{\omega}_{co} =$$

$$= -\kappa\frac{h_{co}^{nom}}{2\pi}\omega_{co}^2 + R_{co}\left(\varphi_{co}\right)\frac{M_{co} + \kappa F_{co}R_{co}\left(\varphi_{co}\right) - d_{co}\omega_{co}}{\Theta_{co}\left(\varphi_{co}\right) + \Theta_M} .$$

Obviously, v_{co} is the circumference speed of the coil, which is not measureable. Formulating the desired closed loop dynamics as

$$\dot{v}_{co} = -\alpha_{co}v_{co} + \alpha_{co}u_{co} , \tag{94}$$

with $\alpha_{co} > 0$ as a design parameter and u_{co} denoting the new input, we get the nonlinear static state feedback control law

$$M_{co} = -\kappa F_{co}R_{co}\left(\varphi_{co}\right) + d_{co}\omega_{co} +$$

$$+ \frac{\Theta_{co}\left(\varphi_{co}\right) + \Theta_M}{R_{co}\left(\varphi_{co}\right)}\left(-\alpha_{co}R_{co}\left(\varphi_{co}\right)\omega_{co} + \alpha_{co}u_{co} + \kappa\frac{h_{co}^{nom}}{2\pi}\omega_{co}^2\right) . \tag{95}$$

b) Control of the Winder Tensions σ_{eco} and σ_{xco}

Based on the linear winder dynamics of Eq. (94), we are now going to discuss a control concept for the winder tensions, exemplary outlined for the case of the unwinder. According to Eqns. (81), (85) and (86), the angluar velocity ψ_{ebr} is given by

$$\hat{\psi}_{ebr}\left(s\right) = \frac{1}{\Theta_{ebr}s + d_{ebr}}\hat{u}_{ebr}\left(s\right) - \frac{R_{ebr}s}{\Theta_{ebr}s^2 + d_{ebr}s + 2c_{ebr}R_{ebr}^2}\hat{w}_{ebr}\left(s\right) , \tag{96}$$

where the variables $\hat{u}$ and $\hat{w}$ have been augmented with the subscript *ebr* in order to prevent notational confusions. Here, we will discuss the case $w_{ebr}(t) = \text{const}$, thus, Eq. (96) simplifies to

$$\hat{\psi}_{ebr}(s) = \hat{\omega}_{ebr}(s) = \frac{1}{\Theta_{ebr}s + d_{ebr}}\hat{u}_{ebr}(s) \ , \tag{97}$$

see also Eq. (88). The unwinder dynamics (94) take the form

$$G_{\hat{v}_{eco}/\hat{u}_{eco}}(s) = \frac{\hat{v}_{eco}(s)}{\hat{u}_{eco}(s)} = \frac{1}{1 + s/\alpha_{eco}} \tag{98}$$

in the frequency domain. The frequency domain equivalent of Eq. (63) reads as

$$\hat{\sigma}_{eco}(s) = \frac{E_S}{\bar{L}_{eco}}\frac{\hat{\psi}_{ebr}(s)R_{ebr} - \hat{v}_{eco}(s)}{s} \ . \tag{99}$$

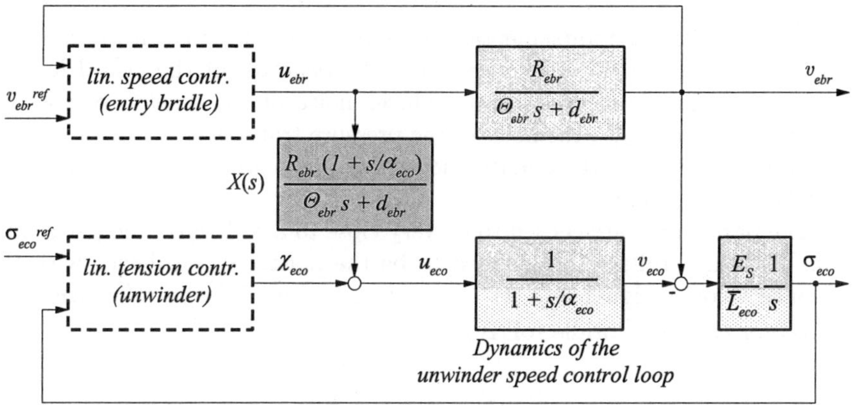

Figure 14. Scheme of the unwinder tension control and the entry bridle speed control (with the input $w_{ebr}(t) = \text{const}$).

In order to count for the effect of the bridle roll dynamics on the tension σ_{eco}, the compensation scheme

$$\hat{u}_{eco}(s) = \hat{\chi}_{eco}(s) + X(s)\hat{u}_{ebr}(s) \tag{100}$$

with the transfer function

$$X(s) = R_{ebr}\frac{1 + s/\alpha_{eco}}{\Theta_{ebr}s + d_{ebr}} \tag{101}$$

is proposed (see Figure 14). Thus, the problem of the winder tension control is reduced to design a controller for the simple linear system

$$\hat{\sigma}_{eco}(s) = -\frac{E_S/\bar{L}_{eco}}{s\left(1 + s/\alpha_{eco}\right)}\hat{\chi}_{eco}(s) \ . \tag{102}$$

By including the winder control strategy into the elongation/tension control scheme of the Subsections 3.2 − 3.6 in the proposed modular way, we gain a considerable flexibilty for changing the mill configuration. Thus, for example, a substitution of the winders with other strip processing lines requires just the modification of the control strategy for σ_{eco} and σ_{xco}, the other parts of the control concept are left unchanged.

3.8 Simulation Results

For illustrating the performance of the proposed control strategy, we will finish this contribution by demonstrating some selected simulation results (see Figures 15 and 16).

All controllers are implemented as discrete time systems with a sampling time $T_a = 5ms$. In order to investigate the behavior of the closed loop in presence of "dirty" effects, the transducer noise of the tension sensors for σ_{eco}, σ_{en}, σ_{ex} and σ_{xco}, the noise of the pressure transducer for the hydraulic pressure P_1, as well as the quantization noise of the piston position x_k (5μm) are taken into account.

The parameters are chosen to be very close to a real world temper rolling mill. The exit bridle roll is defined to be the master speed device with the reference velocity signal $v_{xbr}^{ref}(t)$. According to Eq. (72) the reference value $v_{ebr}^{ref}(t)$ for the entry bridle roll is given by

$$v_{ebr}^{ref}(t) = \frac{v_{xbr}^{ref}(t)}{1 + \varkappa^{ref}(t)} \ . \tag{103}$$

Here, $\varkappa^{ref}(t)$ denotes the reference elongation coefficient.

The numerical example of the rollgap scenario of Subsection 2.1 (see also Figure 6) is used to define the operating point. The other geometric and material parameters used in the simulation are as follows: Hydraulic actuator: $A_1 = A_2 = 0.6\text{m}^2$, $V_0 = 0.02\text{m}^3$, $P_2 = 3 \times 10^5\text{N/m}^2$, $E_{oil} = 1.6 \times 10^9\text{N/m}^2$; Mill stand parameters: $m_1 = m_3 = 50 \times 10^3\text{kg}$, $c_g = 4 \times 10^9\text{N/m}$, $d_1 = 5 \times 10^6\text{Ns/m}$, $d_3 = 5 \times 10^7\text{Ns/m}$, $\Theta_R = 4 \times 10^3\text{kgm}^2$, $d_R = 100\text{Nms/rad}$; Strip lengths: $L_{en} = L_{ex} = 3\text{m}$, $\bar{L}_{eco} = \bar{L}_{xco} = 3\text{m}$; Strip parameters: $B_S = 1\text{m}$, $\rho_S = 7.7 \times 10^3\text{kg/m}^3$; Bridle roll parameters: $R_{ebr} = R_{xbr} = 0.3\text{m}$, $\Theta_{ebr} = \Theta_{xbr} = 700\text{kgm}^2$, $d_{ebr} = d_{xbr} = 200\text{Nms/rad}$, $L_{ebr} = L_{xbr} = 1\text{m}$;

Winder parameters: $R_M = 0.3$m, $\Theta_M = 2 \times 10^3$kgm^2, $d_{co} = 300$Nms/rad, $N_{eco}^0 = 300$, $N_{xco}^0 = 0$.

In the example shown in the Figures 15 and 16 the reference signals for the master speed and the elongation coefficient are constant, $v_{xbr}^{ref}(t) = 0.5$m/s, $\varkappa^{ref}(t) = 0.01$. The reference signals for the strip tensions are chosen as

$$\sigma_{eco}^{ref}(t) = 20 \times 10^6 \text{N/m}^2 \ (1 + 0.2 \, \sigma(t - 0.5))$$
$$\sigma_{en}^{ref}(t) = 70 \times 10^6 \text{N/m}^2 \ (1 + 0.1 \, \sigma(t - 1.2))$$
$$\sigma_{ex}^{ref}(t) = 100 \times 10^6 \text{N/m}^2 \ (1 + 0.1 \, \sigma(t - 2))$$
$$\sigma_{xco}^{ref}(t) = 30 \times 10^6 \text{N/m}^2 \ (1 + 0.2 \, \sigma(t - 2.5))$$

with $\sigma(t)$ denoting the unit step, i.e., $\sigma(t) = 0$ for $t < 0$, and $\sigma(t) = 1$ for $t \geq 0$. In addition, a disturbance torque $M_r^{dist}(t)$,

$$M_r^{dist}(t) = 2 \times 10^3 \text{Nm} \ \sigma(t - 3) \ ,$$

is acting on the work rolls. The first and the second subplot of Figure 15 show the signals x_k^{ref} and ω_R^{ref} calculated by the nonlinear tension controller and the simulated plant outputs x_k and ω_R. The third subplot depicts the signals ε and $\varkappa$ (see also Eqns. (53), (72)), where the difference during transient changes of the strip tensions σ_{en} and σ_{ex} is getting apparent.

Figure 16 shows the reference signals for the strip tensions, the simulated plant outputs and the associated signals (thick lines) calculated by means of the nominal differential equations (90), (94) and (102).

4 Conclusions

Commencing with a detailed discussion of the nonlinear mathematical model of a temper rolling mill based on physical considerations, a nonlinear multi–input multi–output control concept has been proposed. The torques of the bridle rolls are used as control inputs for the elongation and master speed control, whereas the strip tensions are controlled by the hydraulic adjustment system and the main mill drive. The feasibility of the proposed nonlinear control concept in the presence of "dirty" effects such as transducer and quantization noise is illustrated by several simulation results.

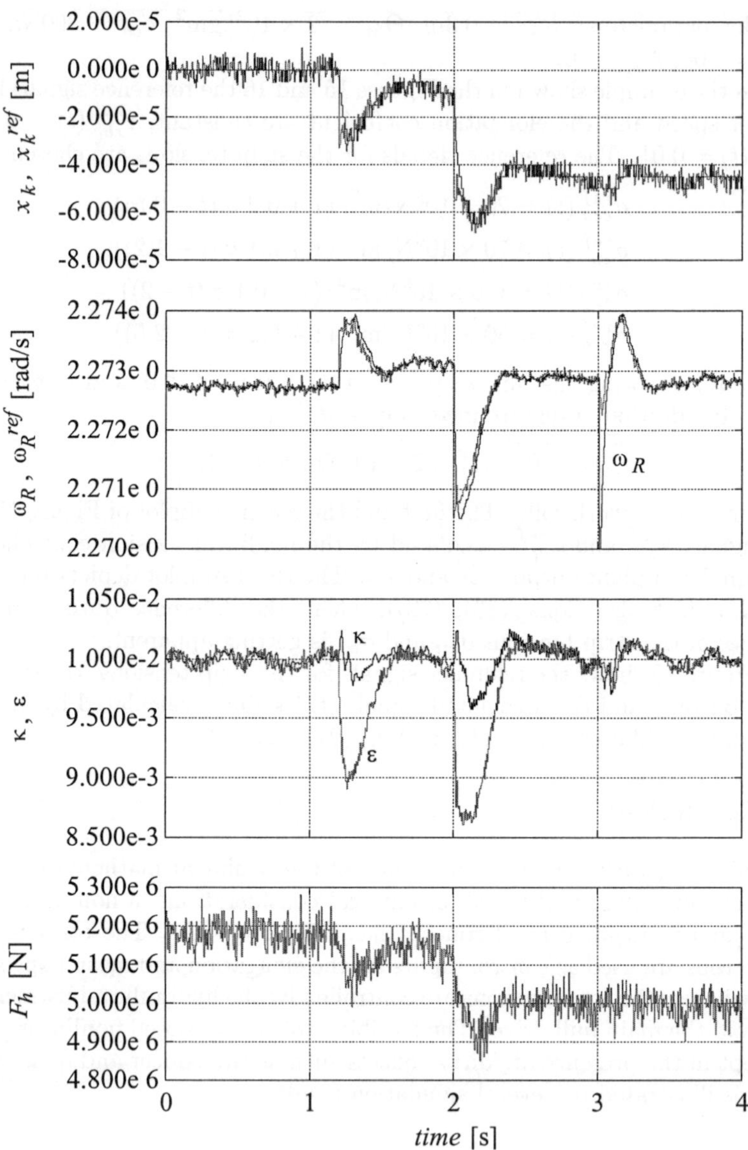

Figure 15. Simulation results (to be continued in Figure 16).

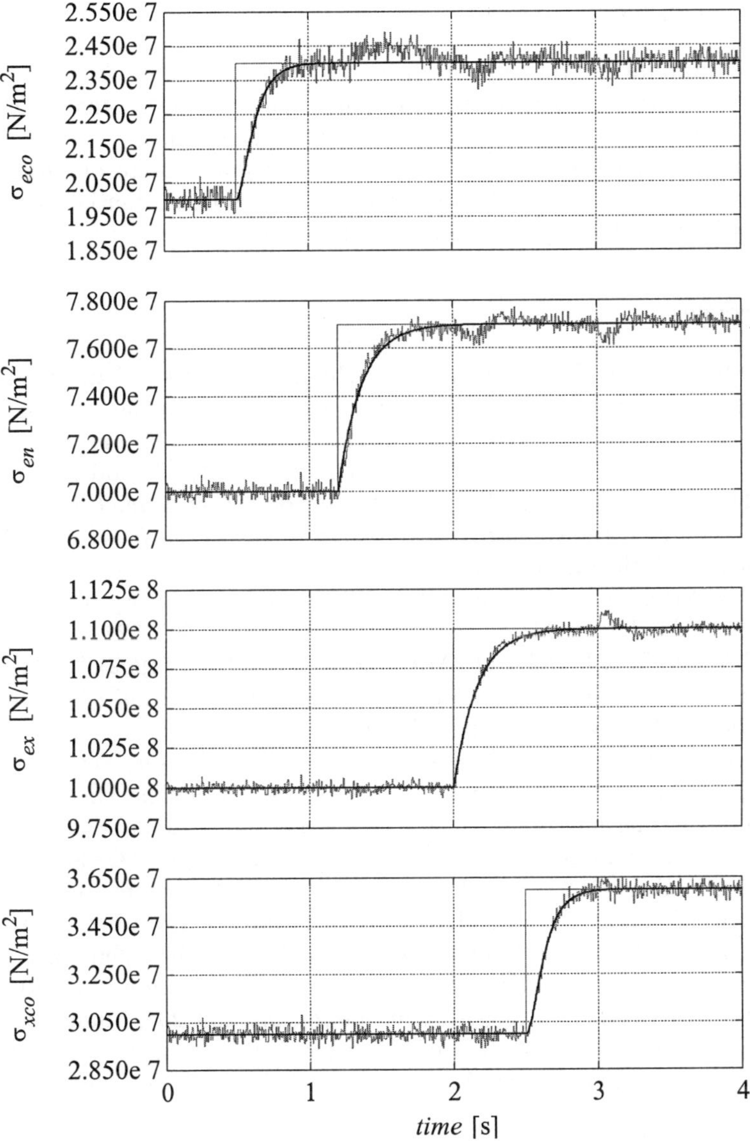

Figure 16. Simulation results (continued).

Acknowledgments

We are grateful for the support from our partner VOEST ALPINE INDUS-TRIEANLAGENBAU, Linz, Austria. In particular we want to thank Dipl.Ing. Georg Keintzel and Dipl.Ing. Michael Eder for their help.

Last but not least, we wish to express our gratitude to our friend and colleague Dipl.Ing. Gernot Grabmair for fruitful and substantial discussions and for his valuable suggestions for this contribution.

References

1. D.R. Bland, H. Ford (1952) "Cold rolling with strip tension, Part III: An approximate treatment of the elastic compression of the strip in rolling mills", *Journal of the Iron and Steel Institute*, pp. 245–249.
2. S.A. Domanti, W.J. Edwards, P.J. Thomas (1994) "A model for foil and thin strip rolling", *AISE Annual Convention, Cleveland, USA*.
3. W. Beitz, K.H. Küttner (1990) *Dubbel – Taschenbuch für den Maschinenbau*, 17th ed., Springer, Berlin.
4. N.A. Fleck, K.L. Johnson (1987) "Towards a new theory of cold rolling thin foil", *Int. J. Mech. Sci.*, Vol. 29, No. 7, pp. 507–524.
5. N.A. Fleck, K.L. Johnson, M.E. Mear, L.C. Zhang (1992) "Cold rolling of foil", In: *Proc. Instn. Mech. Engrs, Part B: Journal of Engineering Manufacture*, Vol. 206, pp. 119–194.
6. V.B. Ginzburg (1993) *High-quality steel rolling*, Dekker, New York.
7. G. Grabmair, K. Schlacher, A. Kugi (2000) "Coupling effects in multi stand rolling mills", In: *Proc. of the 8th International Conference on Metal Forming*, Krakow, Poland, Sept. 3-7, 2000, pp. 295–301.
8. W.Ch. Haas (1995) "H_2–Entwurf für Mehrgrößensysteme im Frequenzbereich", Ph.D. Thesis, Johannes Kepler University of Linz, Austria.
9. P. Hippe, C. Wurmthaler, A.H. Glattfelder, W. Schaufelberger (1995) "Regelung von Strecken mit Stellbegrenzung", In: *Entwurf nichtlinearer Regelungen* (Ed.: S. Engell), Oldenbourg, pp. 239–264, München.
10. J.H. Hitchcock (1935) "Roll neck bearings", Report of the *ASME Special Research Committee on Heavy-Duty Anti-Friction Bearings*, Appendix 1.2.
11. A. Isidori (1995) *Nonlinear Control Systems*, 3rd ed., Springer, London.
12. K.L. Johnson (1985) *Contact Mechanics*, Cambridge University Press, Cambridge.
13. D. Jortner, J.F. Osterle, C.F. Zorowski (1960) "An analysis of cold strip rolling", *Int. J. Mech. Sci.*, Vol. 2, pp. 179–194.

14. H.K. Khalil (1996) *Nonlinear systems*, 2nd ed., Prentice Hall, New Jersey.

15. U. Krause (1966) "Formänderungsfestigkeit der Werkstoffe beim Kaltumformen", In: *Grundlagen der bildsamen Formgebung*, Verlag Stahleisen GmbH, Düsseldorf.

16. A. Kugi (2000) *Non-linear control based on physical models*, Lecture Notes in Control and Information Sciences 260, Springer, London.

17. A. Kugi, K. Schlacher, R. Novak "Non-linear control in rolling mills: A new perspective", To appear in: *IEEE Transactions on Industry Applications*.

18. A. Kugi, K. Schlacher, G. Keintzel (1999) "Position control and active eccentricity compensation in rolling mills", *at–Automatisierungstechnik* 8/99, pp. 342–349.

19. A. Kugi, W. Haas, K. Schlacher, K. Aistleitner, H. Frank, G. Rigler (2000) "Active compensation of roll eccentricity in rolling mills", *IEEE Transactions on Industry Applications*, Vol. 36, No. 2, March/April 2000, pp. 625–632.

20. Matlab and Simulink, The Mathworks Inc.

21. H. Nijmeijer, A.J. Van der Schaft (1991) *Nonlinear dynamical control systems*, Springer, New York.

22. H. Pawelski, O. Pawelski (2000) *Technische Plastomechanik, Kompendium und Übungen*, Verlag Stahleisen GmbH, Düsseldorf.

23. S. Sastry (1999) *Nonlinear Systems, Analysis, Stability and Control*, Springer, New York.

24. K. Schlacher, A. Kugi "Control of nonlinear descriptor–systems, A computer algebra based approach", In: *Nonlinear Control in the Year 2000, Lecture Notes in Control and Information Sciences 259* (Eds.: A. Isidori, F. Lamnabhi–Lagarrigue, W. Respondek), Springer, pp. 379–395, London.

25. K. Schlacher, A. Kugi, R. Novak (2001) "Input to output linearization with constrained measurements", To be presented at the *5th IFAC Symposium on Nonlinear Control Systems (NOLCOS 2001)*, St.Petersburg, Russia, July 4–6, 2001.

26. S.P. Timoshenko, J.N. Goodier (1970) *Theory of Elasticity*, 3rd ed., McGraw-Hill, Singapore.

27. K.H. Weber (1959) *Metallformung – Berechnung von Walzkraft und Drehmoment beim Kaltwalzen mit und ohne Bandzug*, Freiberger Forschungshefte, Akademie–Verlag, Berlin.

28. F. Ziegler (1992) *Technische Mechanik der festen und flüssigen Körper*, 2nd ed., Springer, Wien, New York.

COMBINING CONTINUOUS AND DISCRETE ENERGY APPROACHES TO HIGH FREQUENCY DYNAMICS OF STRUCTURES

ALEXANDER K. BELYAEV

Department of Mechanics and Control Processes, State Technical University of St. Petersburg, 195251, St. Petersburg, Russia, E-mail: belyaev@ob4168.spb.edu

It is shown that the high frequency dynamics is a low frequency limit of thermodynamics and a high frequency limit of structural dynamics. It allows one to apply the conventional methods of mechanics and thermodynamics to description of complex engineering structures at high frequencies. The local principle in the vibrational conductivity approach to high frequency dynamics is suggested. The aim of the analysis is to demonstrate a way of combining Statistical Energy Analysis for modelling local dynamics of a structural member and the vibrational conductivity approach to high frequency dynamics for an integral description of complex engineering structures. The study has shown that all parameters of the present method can be determined in terms of the overall mechanical properties of the structure and local properties of the structural member attached to the complex structure.

1 Preface

1.1 High frequency dynamics

There exist many different types of dynamics. The types of dynamics related to this study are shown in Fig. 1 together with their frequency range of application. Historically, the dynamics of rigid bodies (I. Newton, 1687[1] and L. Euler, 1736[2]) was the first, and, from the point of view of vibration theory, this dynamics deals with zero eigenfrequencies. The dynamics of solids was developed a century later. Many names might be mentioned, however it was M. Duhamel, 1843[3], who began the systematic study of the vibration of solids. The frequencies, which this dynamics deals with, can be referred to as the natural frequencies of the solid under consideration. Thermodynamics should also be considered as a relevant dynamics in which the frequencies of thermal motions are essentially higher than the frequencies of mechanical vibrations. Classical thermodynamics was developed by J. Maxwell, 1867 and L. Boltzmann, 1872. They proposed a kinetic-statistical approach to thermal processes known as the "mechanical theory of heat" which linked thermodynamic processes to dynamics[4]. Quantum thermodynamics as developed by M. Planck, 1900, see e.g. [5], should also be considered as a relevant dynamics. The quantum effects are considerable only at very high frequencies, at least at the frequencies essentially higher than those of the thermal motion of the

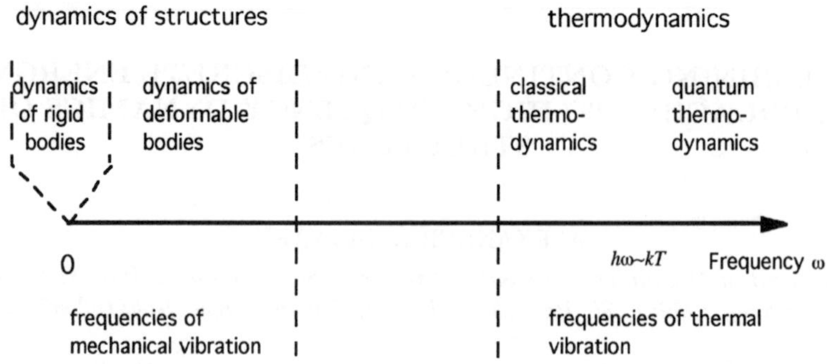

Figure 1. White patch on the map of dynamics

molecules.

The dynamics of rigid body can be derived from vibration theory since the rigid body motions correspond to zero eigenfrequencies within vibration theory. The equations governing the dynamics of solids can be obtained from rigid body dynamics provided that the solid is modelled by a regular spatial array of elemental mass-spring systems. In fact, one of the earliest works on the vibration of distributed, continuous systems performed by J. Lagrange in 1788 was based upon a discrete element model[6]. This means that the dynamics of rigid bodies and the dynamics of solids can be considered as two neighbouring dynamics. Classical thermodynamics and quantum thermodynamics may also be considered as two neighbouring dynamics. For example, the Planck formula for radiation links the Rayleigh-Jeans formula at low radiation frequencies and the Wien formula at high radiation frequencies when the quantum effects are considerable[5]. To the best of the author's knowledge, all attempts to derive the dynamics of solids directly from thermodynamics and visa versa have been rather awkward. One of a few successful applications of the vibration theory of solids to thermodynamics is the Debye theory of the low-temperature lattice heat capacity[5]. In this theory the discrete lattice was approximated as an elastic continuum in the long wavelength region and an appropriate number of long wavelength modes were taken to derive an equation for the heat capacity. Truesdell[7] uses the laws of thermodynamics to derive the constitutive laws and to find what kind of restrictions thermodynamics imposes on the mechanical properties of the continua.

The difficulties encountered when attempting to link the dynamics of solids to thermodynamics indicate that there exists a gap between these theories which "reserves" a place for a new dynamics, see Fig. 1. The latter is in

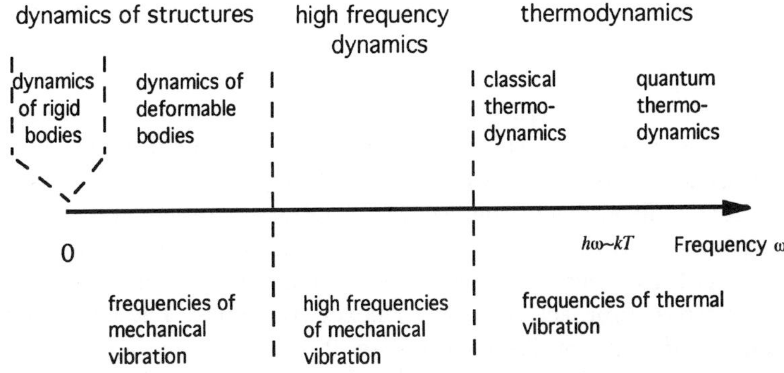

Figure 2. The different types of dynamics including high frequency dynamics

fact a low frequency limit of thermodynamics and a high frequency limit of the dynamics of solids. The new dynamics that fills this gap will be referred to here as high frequency dynamics, and the corresponding frequencies will be referred to as high frequencies of mechanical vibration or, for short, high frequencies, see Fig. 2. From a historical perspective, the development of dynamics is a kind of evolution of our knowledge about dynamic processes in natural science. Similar to the theory of human evolution, a missing link is seen to exist in the evolution of dynamics. At present this statement is no more than a mere allegation, however this allegation presents us with some conclusions. If high frequency dynamics is the high frequency limit of vibration theory and the low frequency limit of thermodynamics then the conventional methods of vibration theory and thermodynamics can be applied there. In particular, the governing equations of high frequency dynamics can be derived from these two different types of dynamics. Deriving the governing equations and properties of the new dynamics is a challenging problem since this implies the coupling of mechanical and thermal properties. As mentioned by Glimm[8] "Most theories become intractable when they require coupling between even two adjacent scales". An attempt will be made here to derive the boundary value problem for high frequency dynamics of complex engineering structures by methods of thermodynamics.

A detailed description of the vibration of an engineering structure is difficult, first, by the complexity of the structure's shape, then, by the assemblage of individual substructures and, finally, by the presence of various secondary systems attached to the primary structure. The vibration field of modern complex structures under broad-band excitation is a well-complicated function of both time and spatial coordinates, since so many modes are excited in the structure.

Outside of the region of these global resonances, however, the vibrations localize within each substructure or groups of substructures, thus different approaches are needed to deal with those vibrations. Historically, the Statistical Energy Analysis (SEA) was the first attempt to attack the problem of high frequency vibrations in engineering structures. In the framework of SEA any complex structure is viewed as a set of coupled substructures and one deals only with the spatially averaged vibration within each substructure, so SEA is a discrete approach to high frequency dynamics. The advantages and shortcomings of SEA are discussed in Sec. 2.

A number of continuous approaches have been offered[9] to describe the field of high frequency vibration of structures. The properties of a structure under consideration are reflected in these approaches in the form of certain averaged rigidity and averaged mass characteristics, and generalised spectra. For this reason, one refers to these approaches as integral methods. They enable one to avoid dipping into too many details of the structure and result in obtaining some generalised characteristics of the vibration field. This level of description can be considered sufficient for many cases. Nonetheless it is often necessary to know the vibration of a particular member of the structure. This cannot be managed with the integral methods alone, since the member itself is not represented in the dynamical model. Thus, to analyse the dynamics of a particular member one has to take into account both its individual mechanical characteristics and the nature of its interaction with other structural members. As has been already mentioned, trying to cover the entire structure makes this task completely hopeless due to the great computational difficulties. Thus a more realistic concept lies in the precise consideration of this member alone, e.g. by means of SEA, with the rest of the structure being described integrally.

In what follows, this idea is applied to modelling local dynamics of a structural member by combining discrete and continuous approaches.

1.2   *Inherent properties of engineering structures at high frequencies*

Complex mechanical structures like ships, buildings, spacecraft and aircraft are actually assemblies of substructures. Attempts to pedantically describe all details and peculiarities of real structures are doomed to failure for the following reasons. Firstly, at higher frequencies, the existence of many inherently uncontrolled factors plays a principle role. Review papers by Ibrahim[10] and Fahy[11] give a very deep insight into the problem of uncertainties in dynamics. In structural dynamics, uncertainties arise from stiffness, mass and damping fluctuations caused by variations in material properties as well as variations resulting from manufacturing and assembly. The latter factor causes vagueness in the boundary conditions for each structural member since the high-frequency dynamic properties of joints between structural members are especially uncertain. Secondly, the essential heterogeneity and presence of complicated interiors, i.e. secondary systems, have to be taken into account. Thirdly, even if it were possible to obtain an "exact" boundary-value problem in which all the complexity of the structure was taken into account and we could solve this problem, the very interpretation of this "exact" result would present great difficulty. The reason for this is that the field of vibration of a complex structure (for instance, under a broad-band excitation) is a very complicated function of time and spatial coordinates since a great many modes are excited in the structure. Summarising, one can say that the blind extension of conventional methods of modal analysis to higher frequencies at the expense of computational cost reaches a deadlock because the results become unreliable and "indigestible". Surprisingly, structural dynamicists "act" as if the methods and approaches of the traditional vibration theory are limitless, i.e. as if these methods may be extended to arbitrarily high frequencies. No such fundamental restriction as "axiom" of physics, which is "there is always a cut-off" seems to exist in vibration theory.

Another typical feature of complex structures is that all modern structures are weakly coupled. Any engineering structure is in fact an assembly of substructures attached to one another or to the framework only at several points. As a result, the complex structures possess such low global dynamic rigidity that only the first, the second and seldom the third global resonances are observed in such structures. Beyond the region of these few global resonances the vibrations localise within the structural members or groups of structural members. This phenomenon is known as strong vibration localization or normal mode localization and was first mentioned by Mandelstamm[12] in 1929. A detailed survey of the phenomena of mode localization in structures can be found in the review papers by Hodges and Woodhouse[13], Ibrahim[10], Li and Benaroya[14] and Benaroya[15]. Several representative studies are also worth mentioning, e.g. Pierre and Dowell[16], Cornwell and Bendiksen[17], Pierre and Cha[18,19], Ariaratnam and Xie[20] and Langley[21]. As shown in these papers, the phenomenon of mode localization occurs not only for disordered assemblies of weakly coupled subsystems, but also for nearly periodic structures like bladed-disk assemblies and truss structures. This effect has also been observed in solid-state physics, e.g. the metallic conductivity is reduced because of localization of the electron eigenstates, the so-called Anderson localization effect[22], 1958. On the other hand, the review of the literature in structural

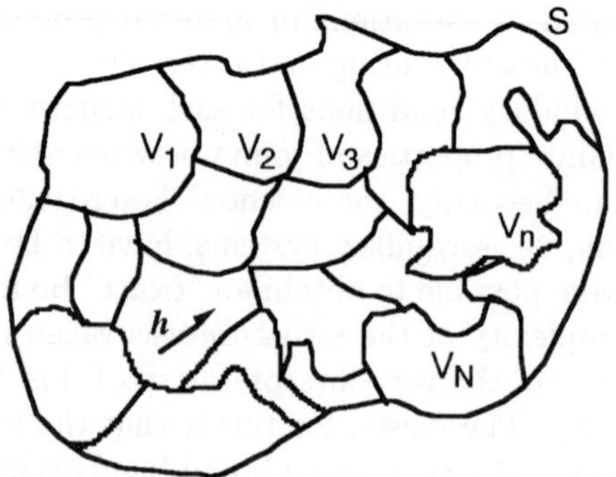

Figure 3. A schematic of a complex structure

dynamics suggests that present-day analytical methods are too idealised and do not adequately reflect the inherent complexity of structures. Complex engineering structures are often taken to consist of a few structural members of well-studied shapes (shells, beams etc.) otherwise lumped-mass models are applied. The consideration of secondary systems is commonly avoided in the literature. This avoidance is even more surprising taking into account the fact that the secondary systems actually comprise the major portion of the structural members of a unit. The main objective of the present study is to reveal the qualitative properties of structures at high frequencies and to make general and basic conclusions. An adequate modelling which allows for the heterogeneity, complexity and uncertainty of weakly-coupled complex mechanical structures is proposed in what follows.

2   Statistical Energy Analysis

2.1   *Preliminaries*

Let us consider an actual mechanical structure V which is assembled from a great number of structural members, each having volume V_n and boundary surface S_n $(n = 1, 2, ...N)$, Fig. 3. The structural members are attached to one another or to the framework. The sizes of the structural members V_n are much smaller than those of the entire structure V. Nevertheless, any structural member may be a complex dynamic system itself. Consider a representative structural member V_n, for short, a substructure V_n. Its dynamics is governed by the following equation

$$\mathbf{r} \in V_n, \quad \nabla \cdot \boldsymbol{\tau} + \mathbf{h} - \rho \frac{\partial^2 \mathbf{u}}{\partial t^2} = 0, \tag{1}$$

where $\mathbf{r}$ is the position vector, $\boldsymbol{\tau}$ is stress tensor, ρ is the mass density, $\mathbf{u}$ is the absolute displacement, $\mathbf{h}$ is a body force, ∇ is the Hamiltonian operator, $\mathbf{r}$ is the reference position and t is time. Multiplying Eq. (1) by velocity $\dot{\mathbf{u}}$ and integrating over the volume of substructure V_n one obtains

$$\int\limits_{V_n} \left(\nabla \cdot \boldsymbol{\tau} + \mathbf{h} - \rho \frac{\partial^2 \mathbf{u}}{\partial t^2} \right) \cdot \dot{\mathbf{u}} dV = 0. \tag{2}$$

Using the chain rule $(\nabla \cdot \boldsymbol{\tau}) \cdot \dot{\mathbf{u}} = \nabla \cdot (\boldsymbol{\tau} \cdot \dot{\mathbf{u}}) - \boldsymbol{\tau} \cdot \cdot (\nabla \dot{\mathbf{u}})$ and applying the Ostrogradsky-Gauss theorem and the rule of differentiation of an integral over the material volume with respect to time[23] yields

$$\int\limits_{V_n} \mathbf{h} \cdot \dot{\mathbf{u}} dV + \int\limits_{S_n} \mathbf{f} \cdot \dot{\mathbf{u}} dS - \int\limits_{V_n} \boldsymbol{\tau} \cdot \cdot \dot{\boldsymbol{\varepsilon}} dV - \dot{T} = 0, \tag{3}$$

where $\boldsymbol{\varepsilon} = (\nabla \mathbf{u})^s$ is the linear strain tensor, $\mathbf{f} = \mathbf{N} \cdot \boldsymbol{\tau}$ is the traction vector on the substructural surface S_n with exterior unit normal $\mathbf{N}$, $T = \frac{1}{2} \int\limits_{V_n} \rho \dot{\mathbf{u}} \cdot \dot{\mathbf{u}} dV$ denotes the kinetic energy of the subsystem and $\cdot\cdot$ stands for the double scalar product. The latter equation is in fact an integral form of the first law of thermodynamics[23], ensuring the power balance at any time instant. The power balance, averaged over the period P of a periodic process is given by

$$\frac{1}{P} \int\limits_t^{t+P} \left[\int\limits_{V_n} \mathbf{h} \cdot \dot{\mathbf{u}} dV + \int\limits_{S_n} \mathbf{f} \cdot \dot{\mathbf{u}} dS - \int\limits_{V_n} \boldsymbol{\tau} \cdot \cdot \dot{\boldsymbol{\varepsilon}} dV - \dot{T} \right] dt = 0. \tag{4}$$

It is clear that

$$\left\langle \dot{T} \right\rangle = \frac{1}{P} \int\limits_t^{t+P} \dot{T} dt = \frac{1}{P} [T(t+P) - T(t)] = 0, \tag{5}$$

where $\langle \rangle$ denotes the above time averaging.

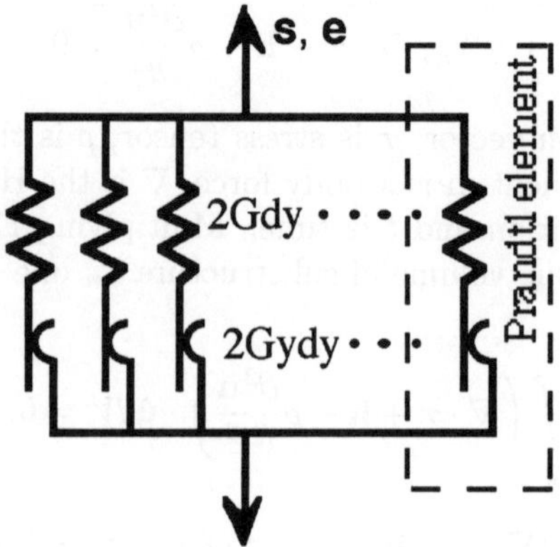

Figure 4. The Ishlinsky rheological model

2.2   *Energy dissipation modelling*

In order to account for the energy dissipation in the structure, let us make use of the theory of microplasticity[23]. This implies the following splitting the stress and strain tensors into spherical and deviatoric parts

$$\tau = \sigma\mathbf{E} + \mathbf{s}, \varepsilon = \frac{\vartheta}{3}\mathbf{E} + \mathbf{e}, \tag{6}$$

where $\mathbf{E}$ is the unity tensor. Here $\mathbf{s}$ and $\mathbf{e}$ are stress deviator and strain deviator, respectively, and σ and ϑ denote respectively the mean normal stress and the dilatation. The spherical parts of the stress and strain tensors exhibit elastic behaviour, that is, the constitutive equation for the mean normal stress and the dilatation is given by

$$\sigma = k\vartheta,$$

where k is the bulk modulus.

The deviatoric parts are known to be governed by constitutive equations of elastoplasticity. Palmov[23] has proven that the Ishlinsky rheological model is the most appropriate one to this aim. The model consists of an infinite number of the Prandtl elastoplastic elements in parallel, Fig. 4 and is governed by the following equation

$$s = 2G \left[\mathbf{e} - \int_0^\infty \mathbf{e}_y p(y) dy \right], \tag{7}$$

where G is the shear modulus, $p(y)$ is the distribution function of a continuous spectrum of nondimensional yield stress y. The plastic strain $\mathbf{e}_y$ in the Prandtl element with an yield stress y is as follows

$$\begin{cases} \dot{\mathbf{e}}_y = 0, & \sqrt{\frac{1}{2}(\mathbf{e} - \mathbf{e}_y) \cdot \cdot (\mathbf{e} - \mathbf{e}_y)} \leq y \\ \dot{\mathbf{e}}_y \neq 0, & y\dot{\mathbf{e}}_y/v_y = \mathbf{e} - \mathbf{e}_y \end{cases}, \tag{8}$$

where $v_y = \sqrt{\frac{1}{2}\dot{\mathbf{e}}_y \cdot \cdot \dot{\mathbf{e}}_y}$ is the intensity of shear strain rate, see [23] for detail.

The time averaging yields

$$\left\langle \int_{V_n} \tau \cdot \cdot \dot{\varepsilon} dV \right\rangle = \int_{V_n} \left(\left\langle \frac{d}{dt} \left[\frac{1}{2} k\vartheta^2 + 2G\mathbf{e} \cdot \cdot \mathbf{e} \right] \right\rangle - \left\langle 2G\dot{\mathbf{e}} \cdot \cdot \int_0^\infty \mathbf{e}_y p(y) dy \right\rangle \right) dV$$

$$= - \left\langle \int_{V_n} 2G\dot{\mathbf{e}} \cdot \cdot \int_0^\infty \mathbf{e}_y p(y) dy dV \right\rangle, \tag{9}$$

where the term with time rate of the conservative term vanishes by analogy with Eq. (5). Eq. (4) takes the form

$$\left\langle \int_{V_n} \mathbf{h} \cdot \dot{\mathbf{u}} dV \right\rangle + \left\langle \int_{V_n} 2G\dot{\mathbf{e}} \cdot \cdot \int_0^\infty \mathbf{e}_y p(y) dy dV \right\rangle + \left\langle \int_{S_n} \mathbf{f} \cdot \dot{\mathbf{u}} dS \right\rangle = 0. \tag{10}$$

The physical sense of the equation obtained is obvious, it expresses the average power balance. The first term is the input power, the second one gives the dissipated power, whereas the third describes the power, transmitted to the other subsystems.

Let us make a preliminary remark. The very fact that Eq. (10) is fulfilled for any periodic process implies that it is met in average for any harmonic which in turn means that the conventional methods of the theory of stationary random processes can be applied to further analysis. The correspondence principle[23] suggests a way for obtaining a boundary value problem for harmonic vibration of elastoplastic bodies. To incorporate the effects of internal friction one replaces the elastic moduli in the dynamic theory of elasticity by their complex values. Let us demonstrate how this principle works in the particular case of the dynamical elastoplasticity.

The effects of internal friction are described by Eqs. (7) and (8). To linearise a nonlinear tensorial constitutive equation (8) we assume that the strain deviators obey a harmonic law, i.e.

$$\mathbf{e} = \hat{\mathbf{e}}e^{i\omega t}, \quad \mathbf{e}_y = \hat{\mathbf{e}}_y e^{i\omega t}. \tag{11}$$

In order to use the method of harmonic linearisation we replace the only nonlinearity $v_y^{-1}\dot{\mathbf{e}}_y$ in Eq. (8) by a linear function as follows

$$v_y^{-1}\dot{\mathbf{e}}_y \approx K_y \dot{\mathbf{e}}_y, \quad K_y = \frac{\int\limits_0^{2\pi} v_y^{-1}\dot{\mathbf{e}}_y \cdot\cdot\dot{\mathbf{e}}_y d(\omega t)}{\int\limits_0^{2\pi} \dot{\mathbf{e}}_y \cdot\cdot\dot{\mathbf{e}}_y d(\omega t)} = \frac{\int\limits_0^{2\pi} v_y d(\omega t)}{\int\limits_0^{2\pi} v_y^2 d(\omega t)},$$

where K_y is a scalar linearisation factor. Inserting Eq. (11) into the latter equation yields

$$K_y = \frac{4}{\pi\omega\Gamma_y},$$

where $\Gamma_y = \sqrt{\frac{1}{2}\hat{\mathbf{e}}_y \cdot\cdot\hat{\mathbf{e}}_y^*}$ and asterisk denotes the complex conjugate. Therefore the system of the constitutive equations for the elastoplastic materials, Eqs. (7) and (8), takes now the form

$$\begin{cases} \mathbf{s} = 2G\left[\mathbf{e} - \int\limits_0^{\infty} \mathbf{e}_y p(y)dy\right] \\ \mathbf{e} = \mathbf{e}_y + \dfrac{4y}{\pi\omega\Gamma_y}\dot{\mathbf{e}}_y \end{cases}. \tag{12}$$

Substituting general complex exponential forms for the strain stresses, Eq. (11), into the second equation in (12), gives

$$\mathbf{e} = \mathbf{e}_y\left(1 + i\frac{4y}{\pi\Gamma_y}\right), \tag{13}$$

which implies that

$$\Gamma^2 = \Gamma_y^2\left[1 + \left(\frac{4y}{\pi\Gamma_y}\right)^2\right],$$

with $\Gamma = \sqrt{\frac{1}{2}\hat{\mathbf{e}} \cdot \cdot \hat{\mathbf{e}}^*}$ denoting the amplitude of the shear stress intensity[23]. This leads to the following equation for Γ_y

$$\Gamma_y = \Gamma\sqrt{1 - \left(\frac{4y}{\pi\Gamma}\right)^2}. \tag{14}$$

Inserting this into (13) yields the following expression for $\mathbf{e}_y$

$$\mathbf{e}_y = \mathbf{e}\left[1 - \left(\frac{4y}{\pi\Gamma}\right)^2 - i\frac{4y}{\pi\Gamma}\sqrt{1 - \left(\frac{4y}{\pi\Gamma}\right)^2}\right].$$

Substituting this into Eq. (7) results in the following constitutive equation

$$\mathbf{s} = 2G_c\mathbf{e},$$

where

$$G_c = G\left[1 - \int\limits_0^1 \left(1 - \eta^2 - i\eta\sqrt{1-\eta^2}\right) p\left(\frac{\pi\Gamma}{4}\eta\right)\frac{\pi\Gamma}{4}d\eta\right] \tag{15}$$

is referred to as a complex shear modulus and $\eta = 4y/\pi\Gamma$ is a nondimensional yield stress. As follows from Eq. (14) $0 \le \eta \le 1$, which determines the integration limits in the latter equation.

2.3 *Substructural energy*

As SEA operates with the energy flow between substructures, let us begin with deriving expressions for substructural energy. The virial theorem, e.g. Ref.[24], states that the total energy $E = 2\langle T\rangle$ only for elastic systems under small deformation. For this reason, it seems reasonable not to deal with the total energy, but with kinetic energy of substructures. Because of the localization of vibration within the structural members, the vector of the absolute displacement of the substructure $\mathbf{u}_n(\mathbf{r},t)$ may be sought in the form of an expansion in terms of the substructure's normal modes $\mathbf{u}_{nk}(\mathbf{r})$

$$\mathbf{r} \in V_n, \quad \mathbf{u}(\mathbf{r},t) = \sum_{k=1}^{\infty} \mathbf{u}_{nk}(\mathbf{r})q_{nk}(t), \tag{16}$$

where $q_{nk}(t)$ are referred to as the generalised coordinates. The normal modes of elastic vibrations are known to be orthonormal, i.e.

$$\int\limits_{V_n} \rho\mathbf{u}_{nk} \cdot \mathbf{u}_{nl}dV = \delta_{kl}, \quad \int\limits_{V_n} \left(k\vartheta_{nk}\vartheta_{nl} + 4G\mathbf{e}_{nk} \cdot \cdot\mathbf{e}_{nl}\right)dV = \Omega_{nk}^2\delta_{kl}, \tag{17}$$

where ϑ_{nk} and $\mathbf{e}_{nk}$ are dilatation and deviatoric part of the strain tensor corresponding to $\mathbf{u}_{nk}$, Ω_{nk} denotes the k-th eigenfrequency of the substructure V_n and δ_{kl} is the Kronecker delta. Applying Galerkin's approach, i.e. multiplying the equation for dynamics (1) by $\mathbf{u}_{nk}(\mathbf{r})$, integrating over the substructural volume V_n and taking into account the above property of the normal modes one arrives at the following equation for the generalised coordinate

$$\ddot{q}_{nk} + \Omega_{nk}^2 q_{nk} + \sum_{l=1}^{\infty} D_{kl} q_{nk} = \int_{V_n} \mathbf{h} \cdot \mathbf{u}_{nk} dV, \tag{18}$$

cf. [23]. Here the complex exponential form of the variables is used and

$$D_{kl} = \int_V 2\left(G - G_c\right) \mathbf{e}_{nk} \cdot \cdot \mathbf{e}_{nl} dV. \tag{19}$$

As mentioned in [23] the effect of plastic deformation is normally very small in application problems, so $|D_{kl}| << \Omega_{nk}^2$. The asymptotically leading part of the solution of (18) is governed by the following equation

$$\ddot{q}_{nk} + \left(\Omega_{nk}^2 + D_{kk}\right) q_{nk} = \int_{V_n} \mathbf{h} \cdot \mathbf{u}_{nk} dV. \tag{20}$$

The assumption of the proportional damping is one of the standard assumptions of SEA[11]. By means of spectral decomposition for the external forces and generalised coordinates

$$\mathbf{h}(\mathbf{r}, t) = \int_{-\infty}^{\infty} \hat{\mathbf{h}}(\mathbf{r}, \omega) e^{i\omega t} d\omega, \quad q_{nk}(t) = \int_{-\infty}^{\infty} \hat{q}_{nk}(\omega) e^{i\omega t} d\omega \tag{21}$$

we come to the following spectrum of the generalised coordinate

$$\hat{q}_{nk}(\omega) = \frac{1}{-\omega^2 + \Omega_{nk}^2 \left(1 + i\psi\right)} \int_{V_n} \mathbf{h} \cdot \mathbf{u}_{nk} dV, \tag{22}$$

where

$$\psi = \int_0^1 \eta \sqrt{1 - \eta^2} p\left(\frac{\pi \Gamma}{4} \eta\right) \frac{\pi \Gamma}{4} d\eta \tag{23}$$

describes internal friction effects.

Because of the inherent uncertainty of the structural properties and external loading at high frequencies, a statistical average is needed. Following

Ref.[9] we obtain and transform expression for the mean value of the kinetic energy of n-th subsystem

$$\langle T_n \rangle = \left\langle \frac{1}{2} \int_{V_n} \rho \dot{\mathbf{u}} \cdot \dot{\mathbf{u}} dV \right\rangle = \frac{1}{2} \sum_{k=1}^{\infty} \sum_{l=1}^{\infty} \langle \dot{q}_{nk} \dot{q}_{nl} \rangle \int_{V_n} \rho \mathbf{u}_{nk} \cdot \mathbf{u}_{nl} dV = \frac{1}{2} \sum_{k=1}^{\infty}$$

$$\int\limits_{-\infty}^{\infty}\!\!\int \int\limits_{V_n}\!\!\int \frac{\mathbf{u}_{nk}(\mathbf{r}) \cdot \langle \mathbf{h}(\mathbf{r},\omega) \mathbf{h}^*(\mathbf{r}_1,\omega_1) \rangle \cdot \mathbf{u}_{nk}(\mathbf{r}_1) \omega \omega_1 e^{i(\omega-\omega_1)t} d\omega d\omega_1}{\left[-\omega^2 + \Omega_{nk}^2 \left(1 + i\psi\right) \right] \left[-\omega_1^2 + \Omega_{nk}^2 \left(1 - i\psi\right) \right]} dV(\mathbf{r}) dV(\mathbf{r}_1)$$

As the stationary random processes are considered, the same notation is used for both temporal and statistical averages.

Another standard assumption of SEA is that the external distributed load is a stationary delta-correlated isotropic spatial white noise, spectral density of which is constant within each subsystem, i.e.

$$\mathbf{r} \in V_n, \ \langle \mathbf{h}(\mathbf{r},\omega) \mathbf{h}^*(\mathbf{r}_1,\omega_1) \rangle = S_n(\omega) \delta(\mathbf{r} - \mathbf{r}_1) \delta(\omega - \omega_1) \mathbf{E}, \qquad (24)$$

where $S_n(\omega)$ is the spectral density of external excitation. This allows one to simplify the equation for the kinetic energy

$$\langle T_n \rangle = \frac{1}{2} \sum_{k=1}^{\infty} \int_{-\infty}^{\infty} \frac{S_n(\omega) \omega^2 d\omega}{\left(-\omega^2 + \Omega_{nk}^2\right)^2 + \psi^2 \Omega_{nk}^4} \int_{V_n} \mathbf{u}_{nk} \cdot \mathbf{u}_{nk} dV. \qquad (25)$$

The further advance is possible under the next SEA assumption, namely that the mass density is constant over each substructure, then due to the orthogonality relation (17) we arrive at the following result

$$\langle T_n \rangle = \frac{1}{2} \frac{1}{\rho_n} \sum_{k=1}^{\infty} \int_{-\infty}^{\infty} \frac{S_n(\omega) \omega^2 d\omega}{\left(-\omega^2 + \Omega_{nk}^2\right)^2 + \psi^2 \Omega_{nk}^4}. \qquad (26)$$

The next assumptions of SEA are a broad band loading and a small damping. Then, due to Bolotin's method[25] one can place the spectral density at the resonance frequency beyond the integral and evaluate the integral obtained. The result is

$$\langle T_n \rangle = \frac{\pi}{2} \sum_{k=1}^{\infty} \frac{1}{\rho_n} \frac{S_n(\Omega_{nk})}{\psi \Omega_{nk}}. \qquad (27)$$

In the case of high modal density the sum above may be replaced by an integral over frequency[25]

$$\langle T_n \rangle = \frac{\pi}{2} \frac{1}{\rho_n} \int_0^{+\infty} \frac{S_n(\Omega)}{\psi \Omega} \nu_n(\Omega)\, d\Omega, \tag{28}$$

where ν_n is referred to as the asymptotic modal density of n-th subsystem. Introducing the spectral density of the averaged kinetic energy of n-th subsystem Σ_n, we obtain

$$\Sigma_n(\omega) = \frac{\pi}{4} \frac{1}{\rho_n} \frac{\nu_n(\omega)}{\psi \omega} S_n(\omega). \tag{29}$$

As mentioned above, SEA deals with a certain average modal energy of a subsystem having a centre frequency ω. The spectral density $\Sigma_n(\omega)$ may be viewed as this energy.

2.4   *Power of input, dissipation and transfer. The SEA equation*

Let us now transform the terms in the power balance equation (10). To begin with, we consider the average input power, i.e.

$$\left\langle \int_{V_n} \mathbf{h} \cdot \dot{\mathbf{u}}\, dV \right\rangle = \sum_{k=1}^{\infty} \int_{-\infty}^{\infty} \frac{i\omega e^{i(\omega - \omega_1)t}\, d\omega d\omega_1}{-\omega^2 + \Omega_{nk}^2(1 + i\psi)}$$

$$\iint_{V_n} \mathbf{u}_{nk}(\mathbf{r}) \cdot \langle \mathbf{h}(\mathbf{r}, \omega)\, \mathbf{h}^*(\mathbf{r}_1, \omega_1) \rangle \cdot \mathbf{u}_{nk}(\mathbf{r}_1)\, dV(\mathbf{r})\, dV(\mathbf{r}_1).$$

By virtue of the above assumptions we can simplify the latter expression, to obtain

$$\left\langle \int_{V_n} \mathbf{h} \cdot \dot{\mathbf{u}}\, dV \right\rangle = \frac{1}{\rho_n} \sum_{k=1}^{\infty} \int_{-\infty}^{\infty} \frac{S_n(\omega)\, i\omega d\omega}{-\omega^2 + \Omega_{nk}^2(1 + i\psi)}.$$

Transforming the integral in such a way that integration is performed only over the positive frequencies yields

$$\left\langle \int_{V_n} \mathbf{h} \cdot \dot{\mathbf{u}}\, dV \right\rangle = \frac{2}{\rho_n} \sum_{k=1}^{\infty} \Omega_{nk}^2 \int_0^{\infty} \frac{S_n(\omega)\, \psi \omega d\omega}{\left(-\omega^2 + \Omega_{nk}^2\right)^2 + \Omega_{nk}^4 \psi^2}.$$

Asymptotic evaluation of the integral gives

$$\left\langle \int_{V_n} \mathbf{h} \cdot \dot{\mathbf{u}} dV \right\rangle = \frac{\pi}{\rho_n} \sum_{k=1}^{\infty} S_n \left(\Omega_{nk} \right).$$

When the modal density is high one may replace the sum by an integral over the frequency

$$\left\langle \int_{V_n} \mathbf{h} \cdot \dot{\mathbf{u}} dV \right\rangle = \frac{\pi}{\rho_n} \int_{0}^{+\infty} S_n \left(\Omega \right) \nu_n \left(\Omega \right) d\Omega.$$

Introducing spectral density of the input power we obtain

$$\Sigma_n^{(input)} \left(\omega \right) = \frac{\pi}{2\rho_n} \nu_n \left(\omega \right) S_n \left(\omega \right). \tag{30}$$

The power dissipated in the substructure is given by Eq. (9)

$$\left\langle \int_{V_n} 2G\dot{\mathbf{e}} \cdot\cdot \int_{0}^{\infty} \mathbf{e}_y p(y) dy dV \right\rangle = -\left\langle \int_{V_n} \boldsymbol{\tau} \cdot\cdot \dot{\mathbf{e}} dV \right\rangle = -\left\langle \int_{V_n} \left(k\vartheta\dot{\vartheta} + \mathbf{s} \cdot\cdot \dot{\mathbf{e}} \right) dV \right\rangle.$$

Using Eq. (16) and introducing the complex shear modulus G_c, yields

$$\left\langle \int_{V_n} \boldsymbol{\tau} \cdot\cdot \dot{\mathbf{e}} dV \right\rangle = \int_{V_n} \sum_{k=1}^{\infty} \sum_{l=1}^{\infty} \left\langle \left(k\vartheta_{nk}\vartheta_{nl} + 2G_c \mathbf{e}_{nk} \cdot\cdot \mathbf{e}_{nl} \right) q_{nk}\dot{q}_{nl} \right\rangle dV.$$

By means of the spectral decompositions for generalised coordinates (21) and (22) we arrive at the following result

$$\left\langle \int_{V_n} \boldsymbol{\tau} \cdot\cdot \dot{\mathbf{e}} dV \right\rangle = \int_{V_n} dV \sum_{k=1}^{\infty} \sum_{l=1}^{\infty} \left(k\vartheta_{nk}\vartheta_{nl} + 2G_c \mathbf{e}_{nk} \cdot\cdot \mathbf{e}_{nl} \right) \int_{-\infty}^{\infty} \int\int_{V_n}$$

$$\frac{\mathbf{u}_{nk}(\mathbf{r}) \cdot \left\langle \mathbf{h} \left(\mathbf{r}, \omega \right) \mathbf{h}^* \left(\mathbf{r}_1, \omega_1 \right) \right\rangle \cdot \mathbf{u}_{nl}(\mathbf{r}_1) \left(-i\omega_1 \right) e^{i(\omega - \omega_1)t}}{\left[-\omega^2 + \Omega_{nk}^2 \left(1 + i\psi \right) \right] \left[-\omega_1^2 + \Omega_{nl}^2 \left(1 - i\psi \right) \right]} d\omega d\omega_1 dV(\mathbf{r}) dV(\mathbf{r}_1).$$

Simplifying the result under the above assumptions and taking info account that the correlation function of a stationary random function and its time derivative vanishes, i.e. $\langle q_{nk}\dot{q}_{nk} \rangle = 0$, one can cast the latter expression in the form

$$\left\langle \int_{V_n} \boldsymbol{\tau} \cdot\cdot \dot{\mathbf{e}} dV \right\rangle = I_1 + I_2,$$

236   *A. K. Belyaev*

where

$$I_1 = \frac{1}{\rho_n} \sum_{k=1}^{\infty} \sum_{\substack{l=1 \\ l \neq k}}^{\infty} \int_{-\infty}^{\infty} \int_{V_n} \frac{S_n(\omega)(k\vartheta_{nk}\vartheta_{nl} + 2G\mathbf{e}_{nk}\cdot\cdot\mathbf{e}_{nl})(-i\omega)\,d\omega dV}{[-\omega^2 + \Omega_{nk}^2(1+i\psi)][-\omega^2 + \Omega_{nl}^2(1-i\psi)]},$$

$$I_2 = \frac{1}{\rho_n} \sum_{k=1}^{\infty} \sum_{l=1}^{\infty} \int_{-\infty}^{\infty} \int_{V_n} \frac{S_n(\omega)\,2i\psi\mathbf{e}_{nk}\cdot\cdot\mathbf{e}_{nl}(-i\omega)\,d\omega dV}{[-\omega^2 + \Omega_{nk}^2(1+i\psi)][-\omega^2 + \Omega_{nl}^2(1-i\psi)]}.$$

Clearly $I_1 = 0$ because of the orthogonality condition (17). Evaluating I_2 yields the following equation for the power dissipated in the substructure

$$\left\langle \int_{V_n} \boldsymbol{\tau}\cdot\cdot\dot{\mathbf{e}}dV \right\rangle = 2G\frac{2}{\rho_n} \int_{V_n} \int_0^{\infty} \sum_{k=1}^{\infty} \frac{S_n(\omega)\,\psi\omega}{\left(-\omega^2 + \Omega_{nk}^2\right)^2 + \psi^2\Omega_{nk}^4}$$

$$\left\{ \left(\mathbf{e}_{nk}\cdot\cdot\mathbf{e}_{nk} + \sum_{l=1}^{\infty} \frac{\left(-\omega^2 + \Omega_{nk}^2\right)\left(-\omega^2 + \Omega_{nl}^2\right)\mathbf{e}_{nk}\cdot\cdot\mathbf{e}_{nl}}{\left(-\omega^2 + \Omega_{nl}^2\right)^2 + \psi^2\Omega_{nl}^4} \right) \right\}\,d\omega dV.$$

Introducing now the spectral density of the power dissipated in the substructure we obtain

$$\Sigma_n^{(diss)}(\omega) = S_n(\omega)\frac{2G\omega\psi}{\rho_n} \sum_{k=1}^{\infty} \frac{1}{\left(-\omega^2 + \Omega_{nk}^2\right)^2 + \psi^2\Omega_{nk}^4} \qquad (31)$$

$$\int_{V_n} \left\{ \mathbf{e}_{nk}\cdot\cdot\mathbf{e}_{nk} + \sum_{l=1}^{\infty} \frac{\left(-\omega^2 + \Omega_{nk}^2\right)\left(-\omega^2 + \Omega_{nl}^2\right)\mathbf{e}_{nk}\cdot\cdot\mathbf{e}_{nl}}{\left(-\omega^2 + \Omega_{nl}^2\right)^2 + \psi^2\Omega_{nl}^4} \right\}\,dV.$$

The power transferred to another substructures through the surface of the n-th substructure may be obtained in a number of ways. For example, Langley[26] suggested an approach based on a net power transfer. In what follows, equation for the energy transferred are derived under the assumption that all substructures are involved in the power exchange, i.e.e

$$\left\langle \int_{S_n} \mathbf{f}\cdot\dot{\mathbf{u}}dS \right\rangle = \left\langle \int_{S_n} \mathbf{N}\cdot\boldsymbol{\tau}\cdot\dot{\mathbf{u}}dS \right\rangle = \int_{S_n} \mathbf{N}\cdot\langle\boldsymbol{\tau}\cdot\dot{\mathbf{u}}\rangle\,dS$$

$$= \int_{S_n} \mathbf{N}\cdot\sum_{k=1}^{\infty}\sum_{l=1}^{\infty}(k\vartheta_{nk}\mathbf{u}_{nl} + 2G_c\mathbf{e}_{nk}\cdot\mathbf{u}_{nl})\sum_{j=1}^{N}\langle q_{jk}\dot{q}_{jl}\rangle\,dS,$$

where $\mathbf{N}$ is the unit vector of the outward normal to the surface. Again we represent the power as a sum of two integrals

$$J_1 = \int\limits_{S_n} \mathbf{N} \cdot \sum_{\substack{k=1 \\ l\neq k}}^{\infty} \sum_{l=1}^{\infty} \left(k\vartheta_{nk}\mathbf{u}_{nl} + 2G\mathbf{e}_{nk} \cdot \mathbf{u}_{nl}\right) \sum_{j=1}^{N} \langle q_{jk}\dot{q}_{jl}\rangle \, dS,$$

$$J_2 = \int\limits_{S_n} \mathbf{N} \cdot \sum_{k=1}^{\infty} \sum_{l=1}^{\infty} 2 \left\langle (G_c - G)\,\mathbf{e}_{nk} \cdot \mathbf{u}_{nl} \sum_{j=1}^{N} q_{jk}\dot{q}_{jl} \right\rangle \, dS,$$

with the condition $\langle q_k \dot{q}_k \rangle = 0$ being taken into account. After applying spectral decompositions, evaluating the integral under the above assumptions and transforming the integral obtained in such a way that integration is performed only over the positive frequencies the latter expression becomes

$$J_1 = \int\limits_{S_n} \mathbf{N} \cdot \sum_{\substack{k=1 \\ l\neq k}}^{\infty} \sum_{l=1}^{\infty} \left(k\vartheta_{nk}\mathbf{u}_{nl} + 2G\mathbf{e}_{nk} \cdot \mathbf{u}_{nl}\right) dS$$

$$\sum_{j=1}^{N} \frac{1}{\rho_j} \int_0^{\infty} \frac{2\left(\Omega_{jk}^2 - \Omega_{jl}^2\right) S_j\left(\omega\right)\omega^3\psi d\omega}{\left[\left(-\omega^2 + \Omega_{jk}^2\right)^2 + \psi^2\Omega_{jk}^4\right]\left[\left(-\omega^2 + \Omega_{jl}^2\right)^2 + \psi^2\Omega_{jl}^4\right]}.$$

Integral J_2 is evaluated by analogy, to give

$$J_2 = \int\limits_{S_n} \mathbf{N} \cdot \sum_{k=1}^{\infty} \sum_{l=1}^{\infty} 2G\mathbf{e}_{nk} \cdot \mathbf{u}_{nl} dS$$

$$\sum_{j=1}^{N} \frac{2}{\rho_j} \int_0^{\infty} \frac{\left(-\omega^2 + \Omega_{jk}^2\right)\left(-\omega^2 + \Omega_{jl}^2\right) S_j\left(\omega\right)\omega\psi d\omega}{\left[\left(-\omega^2 + \Omega_{jk}^2\right)^2 + \psi^2\Omega_{jk}^4\right]\left[\left(-\omega^2 + \Omega_{jl}^2\right)^2 + \psi^2\Omega_{jl}^4\right]}$$

and we finally arrive at the spectral density of the power transmitted

$$\Sigma_n^{(trans)}\left(\omega\right) = \sum_{j=1}^{N} S_j\left(\omega\right)\frac{2\omega\psi}{\rho_j} \sum_{k=1}^{\infty} \sum_{l=1}^{\infty} \int\limits_{S_n} dS\, \mathbf{N} \cdot \tag{32}$$

$$\frac{(k\vartheta_{nk}\mathbf{E} + 2G\mathbf{e}_{nk})\left(\Omega_{jk}^2 - \Omega_{jl}^2\right)\omega^2 + G\mathbf{e}_{nk}\left(-\omega^2 + \Omega_{jk}^2\right)\left(-\omega^2 + \Omega_{jl}^2\right)}{\left[\left(-\omega^2 + \Omega_{jk}^2\right)^2 + \psi^2\Omega_{jk}^4\right]\left[\left(-\omega^2 + \Omega_{jl}^2\right)^2 + \psi^2\Omega_{jl}^4\right]} \cdot \mathbf{u}_{nl}.$$

Using Eq. (29) one expresses $S_j(\omega)$ in terms of $\Sigma_j(\omega)$ and substitutes the result in Eqs. (31) and (32). Inserting the equations obtained into the power balance equation (10) one arrives at equation which can be conveniently written in the standard SEA form that was in general case first derived by means of Green's function by Langley[26]

$$\Sigma_n^{(input)} = \omega\eta_n\Sigma_n + \sum_{j\neq n} \omega\eta_{nj}\nu_j(\omega)\left(\frac{\Sigma_n}{\nu_n(\omega)} - \frac{\Sigma_j}{\nu_j(\omega)}\right). \tag{33}$$

Here due to the general terminology of SEA η_n and η_{nj} are referred to as the loss factors and the coupling loss factors. Further, $\Sigma_n/\nu_n(\omega)$ may be understood as the modal energy, since the modal density $\nu_n(\omega)$ gives actually a number of modes which are resonant over a unit frequency band with as the centre frequency ω. Another derivation was suggested by Pradlwarter and Schueller[27] who derived the SEA equation by applying stochastic modal analysis. As proven by Langley[26] the reciprocity condition for the coupling loss factors is met provided that the coupling between the subsystems is conservative.

3   Vibrational conductivity approach to high frequency dynamics

3.1   *Rationale for the description of high frequency dynamics by the methods of thermodynamics*

The aim of the forthcoming parts is to derive a boundary value problem for high frequency vibration by the methods of thermodynamics. The idea of such a description is based on the obvious analogy between the thermal motion of molecules and high frequency vibration. On the one hand the temperature is a thermodynamic macroparameter, while on the other it is a measure of the vibrational energy of molecules. However in thermodynamics, the concept of temperature can only be introduced once the assumption was made that the number of microstates (i.e. degrees of freedom) in the system is large. The basic idea of thermodynamics is to ignore the microscopic nature and to observe that the thermodynamic state of a system may be completely specified in terms of small numbers of macroscopic quantities. This is the approach of thermodynamics, one of the objectives of which is to find the relationships between the measurable quantities. It is well known from the equilibrium statistical thermodynamics that the relative fluctuation, i.e. the ratio of the root mean square fluctuation of the energy to the mean energy is of order $1/\sqrt{N}$ where N is the number of degrees of freedom. Since a cubic centimeter of a solid contains some 10^{20} atoms, the relative fluctuation is about 10^{-10}. For this reason, temperature alone is sufficient to describe the field of thermal vibration. It is very instructive to cite the thermal physicist's point of view[5]: "the concept of temperature is not required in mechanics and is only meaningful for systems containing many particles". This remark is absolutely correct. It can however be improved in such a way that the concept of temperature may be extended to mechanical systems with a number of degrees of freedom which is too large to be analysed by the conventional methods of vibration theory, but too small to apply the concept of energy equipartition. This is exactly the case of high frequency dynamics. The ideology of SEA may be considered as a rationale for the use of the methods of thermodynamics to high frequency dynamics.

According to the central result of SEA, Eq. (33) the power flow between two coupled substructures is proportional to the difference of the vibrational energy of each substructure. In other words, the fundamental SEA relation between power flow and modal energy difference is fully analogous to the Fourier law in thermodynamics which states that heat flow is proportional to temperature difference. This analogy has allowed Eichler[28] to reduce the vibration propagation in complex structures to a discrete transfer scheme, no differential equation of the heat conduction type, however, has been suggested. A boundary value problem of the heat conduction type was first applied by Palmov[29] to the analysis of random vibration in complex structures. Differential equations of the parabolic type were proposed later[30,31,32] for one-dimensional layered continua. A parabolic differential equation[33] has been reported to govern the propagation of a wave packet in a one-dimensional medium having a microstructure. Bending high frequency vibrations in heavily damped beams and plates were modelled with a parabolic equation by Pervozvansky[34]. The vibrational conductivity approach to high frequency dynamics of complex structures was proposed in the eighties by Belyaev and Palmov[35,9]. Nefske and Sung[36] have reported a power flow finite element analysis of dynamic systems. In their paper the boundary value problem of the heat conduction type was derived and applied to the analysis of beams. In the author's opinion, the derivation of the boundary value problem for the vibrational conductivity approach reported by Langley[37] should be considered as a general derivation for two-dimensional structural members. Kishimoto and Bernstein[38] have reported that the energy flow models based upon thermodynamic energy rather than stored energy can be used to predict the energy flow in low-dimensional systems. The fundamental thermodynamic energy of a system (i.e. "temperature") has been shown not to be its stored energy content, but rather its ability to shed "heat". Their approach permits one to overcome some shortcomings of the modal approach to SEA. An interesting application of the vibrational conductivity approach to granular media has been reported by Blekhman[39] when discussing vibrational rheology of granular media. In order to derive the governing equations of the granular media subjected to high frequency vibration, the granular medium has been supposed to be modelled as a compressible Newtonian fluid with parameters depending upon the local vibration while the imposed high frequency vibration has been computed using the vibrational conductivity approach. This model allows one to reveal and explain some "slow convective" flows of granular media in a vibrating vessel. Recently a number of discrete approaches leading to SEA equations and some continuous approaches resulting in differential equations of the parabolic type have been suggested for various structural components, namely, rods, single and coupled beams, plates and circular membranes at the IUTAM-Symposium on SEA[40].

The aforementioned observations and analogies can be viewed as a rationale for applying the methods of thermodynamics to the modelling of structures at high frequencies and for combining continuous and discrete approaches to high frequency dynamics.

3.2   Boundary-value problem of the vibrational conductivity approach to high frequency dynamics

Use is made here of a direct method which demands a given boundary value problem and deals only with the parameters' identification. Since high frequency vibrations in complex structures are believed to be described by some thermodynamic approach, the heat conduction equation is used. As shown in Sec. 2, SEA predicts not only vibration transfer between the substructures, but also a linear energy absorption within each substructure. For this reason, a "heat sink" term $-\alpha^2 S$ is included in the traditional heat conduction equation, to give the following vibrational conductivity equation

$$\mathbf{r} \in V, \ \nabla \cdot (\Gamma \nabla S) - \alpha^2 S = 0. \tag{34}$$

Here analogous to the heat conduction equation S is a "vibrational temperature", Γ is vibrational conductivity and α^2 is vibrational conductance. The boundary condition of the second kind is as follows

$$\mathbf{r} \in S, \ \mathbf{N} \cdot (\Gamma \nabla S) = F, \tag{35}$$

where F is the vibrational flux prescribed at the surface.

The main attraction of the vibrational conductivity approach is the scalar governing differential equation of the heat conduction type. In particular, the

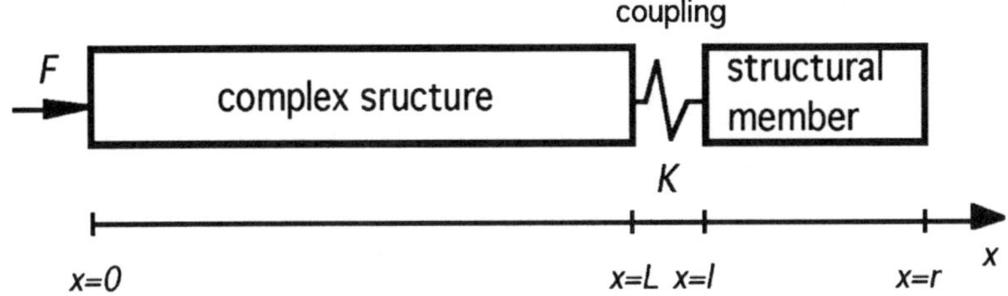

Figure 5. Complex structure and a structural member of interest

finite element formulation can be easily achieved by applying commercial finite element codes.

3.3 Local principle in the vibrational conductivity approach

The idea of the forthcoming analysis is to show that modelling local dynamics of structural members is feasible by combing the vibrational conductivity approach and SEA. Langley[37] has reported a variational principle and the associated boundary conditions for two-dimensional structural components and has shown that a one-term Rayleigh-Ritz solution leads to standard Statistical Energy Analysis. This idea permits us, in principle, to use SEA for local description of a structural member of interest and the vibrational conductivity approach for an overall description of the rest of the structure.

A different approach is suggested in what follows. Let us consider a system which consists of a small structural member attached to an extended complex structure, Fig. 5. When the longitudinal waves predominate, the analysis may be considered to be one-dimensional. The boundary-value problem of the vibrational conductivity approach in one dimension[9] is given by

$$\frac{d^2 S_1}{dx^2} - \alpha_1^2 S_1 = 0,\, 0 < x < L;\quad \frac{d^2 S_2}{dx^2} - \alpha_2^2 S_2 = 0,\, l < x < r, \qquad (36)$$

where S is the "vibrational temperature". The boundary conditions and the condition of interaction of two parts of the system are as follows

$$x = 0,\quad -\Gamma_1 \frac{dS_1}{dx} = F, \qquad (37)$$

$$x = r,\quad \frac{dS_2}{dx} = 0,\, \Gamma_1 \frac{dS_1}{dx} = H\left(S_1 - S_2\right) = \Gamma_2 \frac{dS_2}{dx},$$

where H is factor of the interface contact vibrational conductance. The solution of the boundary-value problem (36), (37) is

$$S_1 = \frac{F}{\alpha_1 \Gamma_1} \frac{\cosh\left(\alpha_1 \left(L - x\right)\right)}{\sinh\left(\alpha_1 L\right)} + Q_1 \cosh\left(\alpha_1 x\right), \ \ S_2 = Q_2 \cosh\left(\alpha_2 \left(r - x\right)\right),$$

where

$$Q_1 = -\frac{F}{\Delta} \frac{\alpha_2 \Gamma_2}{\alpha_1 \Gamma_1} \frac{\sinh \alpha_2 \left(r - l\right)}{\sinh \alpha_1 L}, \ Q_2 = \frac{F}{\Delta}, \ \Delta = \alpha_1 \Gamma_1 \sinh \alpha_1 L \cosh \alpha_2 \left(r - l\right) +$$

$$\alpha_2 \Gamma_2 \cosh \alpha_1 L \sinh \alpha_2 \left(r - l\right) + \delta \sinh \alpha_1 L \sinh \alpha_2 \left(r - l\right), \ \ \delta = \alpha_1 \Gamma_1 \alpha_2 \Gamma_2 / H.$$

As the substructure $l \leq x \leq r$ is assumed to be small compared to the whole structure, an averaged value of the vibrational temperature within the substructure is

$$\bar{S}_2 = \frac{1}{r - l} \int_l^r S_2\left(x\right) dx = \frac{Q_2}{\alpha_2 \left(r - l\right)} \sinh \alpha_2 \left(r - l\right).$$

Since at high frequencies $\sinh \alpha_2\left(r - l\right) \approx \cosh \alpha_2\left(r - l\right)$ and $\sinh \alpha_1 L \approx \cosh \alpha_1 L$, an asymptotic equation for the averaged vibrational temperature in the substructure is

$$\bar{S}_2 = \frac{F}{\alpha_2\left(r - l\right)\sinh\left(\alpha_1 L\right)} \frac{1}{\alpha_2 \Gamma_2 \left(1 + \alpha_1 \Gamma_1 \left[\dfrac{1}{H} + \dfrac{1}{\alpha_2 \Gamma_2}\right]\right)}. \tag{38}$$

Identification of the parameters of the vibrational conductivity approach and the local principle implies a comparison of the solution of the boundary value problem, Eqs. (36), (37), with a solution from another approach operating with mechanical parameters. As follows from Fig. 2 high frequency dynamics is a low frequency limit of thermodynamics, and structural vibrations are described by the equations governing the vibrational conductivity approach (34) and (35). On the other hand, the high frequency dynamics is a high frequency limit of structural dynamics. Thus, if the same vibrational field is modelled by means of these two approaches, this permits us to identify the parameters of the vibrational conductivity approach, provided that closed form solutions are obtained from both approaches. Therefore we proceed to deriving the governing equation for high frequency vibration from structural dynamics.

4 High frequency structural dynamics

4.1 Boundary value problem of high frequency structural dynamics

In this Section we derive the boundary problem of high frequency dynamics by means of modal analysis and the Hamilton variational principle of dynamics. Similar derivations have been performed by Palmov[41] and Belyaev and Palmov[9,42]. We consider an actual mechanical structure V depicted in Fig. 3. As vibration localises within the structural members, we can write for n-th substructure

$$\mathbf{r} \in V_n, \ \mathbf{u}_n\left(\mathbf{r}, t\right) = \sum_{k=1}^{\infty} \mathbf{u}_{nk}\left(\mathbf{r}\right) q_{nk}\left(t\right) + \mathbf{U}\left(\mathbf{r}, t\right). \tag{39}$$

The function $\mathbf{U}\left(\mathbf{r}, t\right)$ is usually introduced in problems of mathematical physics to improve the convergence in the vicinity of a boundary[43]. The reason for introducing $\mathbf{U}$ in equation (39) is quite different. The objective of the present study is to obtain a closed form solution describing the high frequency vibration in complex structures. To this end, the boundary value problem for $\mathbf{U}$ must be as simple as possible, that is, it is reasonable to require maximum smoothness of the function $\mathbf{U}\left(\mathbf{r}, t\right)$ with respect to the spatial coordinate $\mathbf{r}$ within the whole structure.

Now the question of how to specify the substructures' normal modes arises. The higher normal modes are known to be sensitive to the boundary conditions, the latter being vague for any substructure, see Preface. However, fortunately for structural dynamicists any set of normal modes is known to be complete and as shown by Hale and Meirovitch[44,45], the normal modes in energy space of substructures are not required to satisfy any boundary conditions in the internal boundaries of substructures. Hence, the normal modes may be chosen according to any suitable principle. For our analysis, the convenience of interpretation of the obtained results is of primary priority. For this reason, the normal modes are so specified that they vanish on the substructural boundaries S_n, that is, the substructural normal modes $\mathbf{u}_{nk}\left(\mathbf{r}\right)$ are nontrivial solutions of the following boundary-value problem

$$\begin{aligned}
&\mathbf{r} \in V_n, \ \nabla \cdot \left[\mathbf{C} \cdot\!\cdot \left(\nabla \mathbf{u}_{nk}\right)\right] + \rho \Omega_{nk}^2 \mathbf{u}_{nk} = 0, \\
&\mathbf{r} \in S_n, \ \mathbf{u}_{nk} = 0,
\end{aligned} \tag{40}$$

where $\mathbf{C}\left(\mathbf{r}\right)$ is the tensor of elastic moduli. In this case the function $\mathbf{U}\left(\mathbf{r}, t\right)$ coincides with the actual displacement on the substructural boundaries S_n and besides, it is supposed to be a smooth function of $\mathbf{r}$ within the entire structure. Hence $\mathbf{U}\left(\mathbf{r}, t\right)$ may be referred to as the displacement of the framework or the displacement of the primary structure.

The basic idea of the forthcoming analysis is as follows. Equation (39) ensures that the actual vibrational field is split. From a mechanical point of view, the function $\mathbf{U}$ describes the gross effects which are relevant to the overall dynamic behaviour while the modal sum represents the local effects, such as vibration localization, parameter uncertainties, weak coupling and heterogeneities. From a thermodynamical point of view, $\mathbf{U}$ is a macroparameter, while the modal sum describes the microstates. High frequency vibration exhibits some thermodynamic properties however the number of degrees of freedom (microstates) is not large enough to ignore the modal vibrations and to specify the "vibrational state" in terms of a few macroscopic parameters. From a mathematical point of view, the method of multiple scales[46] is applied where $\mathbf{U}$ is a slow variable while the modal sum represents the fast variables.

The kinetic energy of the structure with consideration to the expansion (39) and the normal modes' properties (17) is as follows

$$T = \frac{1}{2} \sum_{n=1}^{N} \int_{V_n} \rho \dot{\mathbf{u}}_n \cdot \dot{\mathbf{u}}_n dV = \tag{41}$$

$$\frac{1}{2} \int_V \rho \dot{\mathbf{U}} \cdot \dot{\mathbf{U}} dV + \frac{1}{2} \sum_{n=1}^{N} \sum_{k=1}^{\infty} \left[\dot{q}_{nk}^2 + 2\dot{q}_{nk} \int_{V_n} \rho \mathbf{u}_{nk} \cdot \dot{\mathbf{U}} dV \right].$$

In view of the smoothness of $\mathbf{U}(\mathbf{r}, t)$ and the essentially heterogeneous nature of the structure the following approximation appears to be valid

$$\int_V \rho \dot{\mathbf{U}} \cdot \dot{\mathbf{U}} dV = \sum_{n=1}^{N} (\dot{\mathbf{U}})_n \cdot (\dot{\mathbf{U}})_n \int_{V_n} \rho dV = \sum_{n=1}^{N} \rho^{(a)} (\dot{\mathbf{U}})_n \cdot (\dot{\mathbf{U}})_n V_n = \int_V \rho^{(a)} \dot{\mathbf{U}} \cdot \dot{\mathbf{U}} dV$$

where $\rho^{(a)} = V^{-1} \int_V \rho dV$ is an average density of the structure, i.e. an overall parameter. The latter equality corresponds to the standard transition from the Riemann-Stieltjes sum to the corresponding integral, which is admissible for large N. In the last sum appearing in equation (41) $\mathbf{U}(\mathbf{r}, t)$ may be placed beyond the integral since it is a smooth function of $\mathbf{r}$, while $\rho(\mathbf{r})$ is a rapidly changing function of $\mathbf{r}$. If we introduce the average displacement of the centre of mass of the substructure V_n when it vibrates due to the normal mode $\mathbf{u}_{nk}(\mathbf{r})$, i.e.

$$\mathbf{u}_{nk}^{(a)} = \frac{1}{\rho^{(a)} V_n} \int_{V_n} \rho \mathbf{u}_{nk} dV,$$

expression (41) can be rewritten in the following form

$$T = \int_V \rho^{(a)} \dot{\mathbf{U}} \cdot \dot{\mathbf{U}} dV + \frac{1}{2} \sum_{n=1}^{N} \sum_{k=1}^{\infty} \left[\dot{q}_{nk}^2 + 2\rho^{(a)} \mathbf{u}_{nk}^{(a)} \cdot \dot{\mathbf{U}} V_n \dot{q}_{nk} \right]. \qquad (42)$$

The equation for the potential energy is obtained similarly[9,42]

$$\Pi = \frac{1}{2} \int_V (\nabla \mathbf{U}) \cdot\cdot \mathbf{C}^{(a)} \cdot\cdot (\nabla \mathbf{U}) \, dV + \frac{1}{2} \sum_{n=1}^{N} \sum_{k=1}^{\infty} \Omega_{nk}^2 q_{nk}^2. \qquad (43)$$

Here $\mathbf{C}^{(a)}$ is an average tensor of elastic moduli, i.e. an overall parameter. The normal modes $\mathbf{u}_{nk}(\mathbf{r})$ form a complete set of functions. Hence, the function $\mathbf{U}(\mathbf{r}, t)$ in equation (39) is "redundant", which means that it is not feasible to obtain a unique boundary problem for $\mathbf{U}(\mathbf{r}, t)$ without an additional condition. This condition has actually been imposed by equation (43), where it has been assumed that

$$\sum_{n=1}^{N} \sum_{k=1}^{\infty} q_{nk} \int_{V_n} (\nabla \mathbf{U}) \cdot\cdot \mathbf{C}^{(a)} \cdot\cdot (\nabla \mathbf{u}_{nk}) \, dV = 0. \qquad (44)$$

Equation (43) implies that the primary structure and the internal degrees of freedom (i.e. secondary systems) are orthogonal in potential energy and exchange energy through the kinetic energy alone. Thus the primary structure and secondary systems excite each other kinematically rather than through the strength factors. The work of the external loads is as follows

$$W = \sum_{n=1}^{N} \sum_{k=1}^{\infty} p_{nk} q_{nk} + \int_V \mathbf{h} \cdot \mathbf{U} dV + \int_S \mathbf{f} \cdot \mathbf{U} dS, \; p_{nk} = \int_{V_n} \mathbf{h} \cdot \mathbf{u}_{nk} dV, \qquad (45)$$

where p_{nk} are the generalised forces.

Assuming δU and δq_{nk} be independent variations and applying the Hamilton variational principle yields the following boundary value problem

$$\mathbf{r} \in V_n, \; \nabla \cdot \left[\mathbf{C}^{(a)} \cdot\cdot (\nabla \mathbf{U}) \right] - \rho^{(a)} \left(\frac{\partial^2 \mathbf{U}}{\partial t^2} + \sum_{k=1}^{\infty} \mathbf{u}_{nk}^{(a)} \ddot{q}_{nk} \right) \mathbf{U} + \mathbf{h} = 0, \qquad (46)$$

$$\mathbf{r} \in V_n, \; \ddot{q}_{nk} + 2\psi_{nk} \Omega_{nk} \dot{q}_{nk} + \Omega_{nk}^2 q_{nk} = p_{nk} - \rho^{(a)} \mathbf{u}_{nk}^{(a)} \cdot \frac{\partial^2 \mathbf{U}}{\partial t^2} V_n, \qquad (47)$$

$$\mathbf{r} \in S, \ \mathbf{N} \cdot \left[\mathbf{C}^{(a)} \cdot\cdot (\nabla \mathbf{U}) \right] = \mathbf{f}. \tag{48}$$

The modal damping is introduced into equation (47) via a dimensionless damping factor ψ_{nk}. Eq. (46) governs the dynamics of the primary structure while Eq. (47) is in fact an infinite number of equations for the modal coordinates of the structural members, as $k = 1, 2, \dots \infty$. Eqs. (46) and (47) are seen to be coupled through inertial terms.

It is worth mentioning some alternative approaches leading to similar boundary value problems governing high frequency vibrations in engineering structures. Papers[47,48] model the primary structure as an elastic carrier medium while the secondary systems are represented by oscillators attached to the carrier medium. The general three-dimensional medium of complex structure has been reported by Palmov[47]. Modelling of extended complex structures by a Cosserat rod with a microstructure[48] permits the consideration of high frequency longitudinal, torsional and bending vibrations. Complex structures modelled by one-dimensional random media are studied in Ref.[49]. This approach has the merit that uncertainties in mechanical parameters and boundary conditions, as well as the heterogeneity of the complex structures are considered in the framework of the same approach.

4.2   Time-reduced boundary value problem

Since the main objective of the present study is to study the dynamic properties of structures at high frequencies, it is convenient to consider the analysis in the frequency domain. Assuming spectral representations for the external loads, Eq. (21), $\mathbf{U}(\mathbf{r}, t)$ and $q_{nk}(t)$, and substituting these into Eqs. (46)-(48) results in the following boundary problem for the spectra

$$\mathbf{r} \in V_n, \ \nabla \cdot \left[\mathbf{C}^{(a)} \cdot\cdot \left(\nabla \hat{\mathbf{U}} \right) \right] + \omega^2 \rho^{(a)} \left(\hat{\mathbf{U}} + \sum_{k=1}^{\infty} \mathbf{u}_{nk}^{(a)} \hat{q}_{nk} \right) + \hat{\mathbf{h}} = 0, \tag{49}$$

$$\mathbf{r} \in V_n, \ \left[-\omega^2 + 2i \psi_{nk} \Omega_{nk} \omega + \Omega_{nk}^2 \right] \hat{q}_{nk} = \hat{p}_{nk} + \omega^2 \rho^{(a)} \mathbf{u}_{nk}^{(a)} \cdot \hat{\mathbf{U}} V_n, \tag{50}$$

$$\mathbf{r} \in S, \ \mathbf{N} \cdot \left[\mathbf{C}^{(a)} \cdot\cdot \left(\nabla \hat{\mathbf{U}} \right) \right] = \hat{\mathbf{f}}, \tag{51}$$

where $\hat{\mathbf{U}}(r, \omega)$ is the spectrum of $\mathbf{U}(\mathbf{r}, t)$. Using equation (50) to obtain an expression for $\hat{q}_{nk}$ and substituting it into equation (49), the governing differential equation for high-frequency vibration of the primary structure is obtained as:

$$\nabla \cdot \left[\mathbf{C}^{(a)} \cdot \cdot \left(\nabla \hat{\mathbf{U}} \right) \right] + \omega^2 \mathbf{M}(\omega) \cdot \hat{\mathbf{U}} + \hat{\mathbf{h}}_e = 0. \tag{52}$$

Here

$$\hat{\mathbf{h}}_e = \hat{\mathbf{h}} + \rho^{(a)} \omega^2 \sum_{k=1}^{\infty} \frac{p_{nk} \mathbf{u}_{nk}^{(a)}}{-\omega^2 + 2i\psi_{nk}\Omega_{nk}\omega + \Omega_{nk}^2}, \tag{53}$$

$$\mathbf{M}(\omega) = \rho^{(a)} \left[\mathbf{E} + \rho^{(a)} \omega^2 \sum_{k=1}^{\infty} \frac{\mathbf{u}_{nk}^{(a)} \mathbf{u}_{nk}^{(a)} V_n}{-\omega^2 + 2i\psi_{nk}\Omega_{nk}\omega + \Omega_{nk}^2} \right], \tag{54}$$

where $\mathbf{u}_{nk}^{(a)} \mathbf{u}_{nk}^{(a)}$ denotes a vector dyad[50]. Eq. (52) differs from the conventional time-reduced vibration equation of an elastic body only in that the tensor $\mathbf{M}(\omega)$ replaces the mass density and the spectrum of effective body force $\hat{\mathbf{h}}_e$ appears instead of $\mathbf{h}$. The tensor $\mathbf{M}(\omega)$ is seen to be dependent upon both the overall (averaged mass density $\rho^{(a)}$) and the local (Ω_{nk}, ψ_{nk}, $\mathbf{u}_{nk}^{(a)}$, V_n) parameters, and is symmetric, cf. (54). In addition to this, real engineering structures have such complicated compositions that it is impossible to indicate the axis of anisotropy for the substructures' spectral characteristics. The tensor is thus assumed to be isotropic[9,42], i.e.

$$\mathbf{M}(\omega) = M(\omega)\mathbf{E}, \tag{55}$$

where

$$M(\omega) = \frac{1}{3} \mathbf{M}(\omega) \cdot \cdot \mathbf{E} = \rho^{(a)} \left[1 + \frac{1}{3} \rho^{(a)} \omega^2 \sum_{k=1}^{\infty} \frac{\mathbf{u}_{nk}^{(a)} \cdot \mathbf{u}_{nk}^{(a)} V_n}{-\omega^2 + 2i\psi_{nk}\Omega_{nk}\omega + \Omega_{nk}^2} \right]. \tag{56}$$

This parameter is crucial for further analysis. It reflects the inertial and spectral properties of the complex structure and for this reason it may be referred to as the generalised mass of the complex structure. As follows from equation (56) $\mathbf{M}(\omega)$ is the sum of an infinite number of modal resonance curves corresponding to the vibration of a single degree of freedom system; this is shown schematically in Fig. 6. Two frequency domains with distinct dynamic properties are observed there: (i) a low frequency domain with rather distant modal resonance curves, and (ii) high frequency domain with high modal overlap. The width of each resonance curve is known to be $2\psi_{nk}\Omega_{nk}$ at the

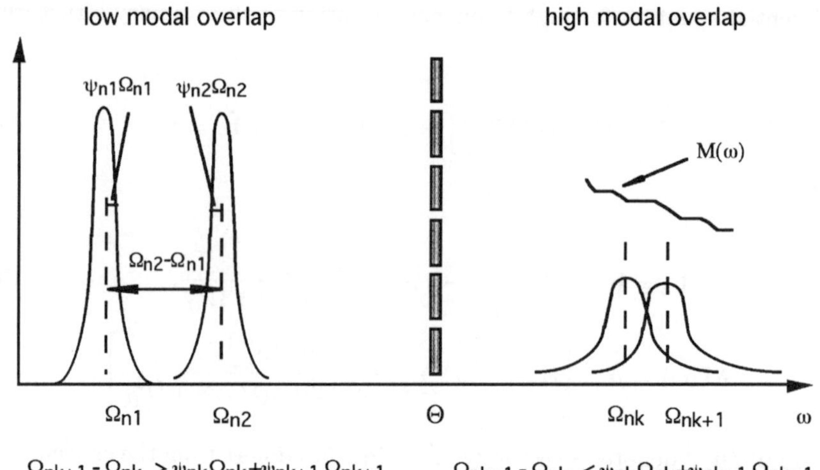

Figure 6. Frequency domains of low and high modal overlap

"half-power" level. Provided that the width of the resonance curve is large compared to the eigenfrequency separation (so-called high modal overlap), i.e.

$$\Delta\Omega_{nk} = \Omega_{nk+1} - \Omega_{nk} < \psi_{nk}\Omega_{nk} + \psi_{nk+1}\Omega_{nk+1} \tag{57}$$

the modal resonance curves merge. In this case $\mathbf{M}(\omega)$ is a smooth function of frequency. In order to demonstrate that there always exists a frequency domain of high modal overlap, let us consider, as an example, the longitudinal vibration of a rod. Its asymptotic natural frequency is known to be equal to $\omega_k = k\pi a/l$ [25], where a is the sound velocity and l is the length of the rod. The frequency separation is thus

$$\Delta\Omega_k = \Omega_{k+1} - \Omega_k = \pi a/l.$$

For this case, the inequality (57) takes the following form

$$2k + 1 > \psi^{-1}$$

or asymptotically for large k

$$2k > \psi^{-1}. \tag{58}$$

The material damping ψ is known to be frequency-independent, see Palmov[23]. Clearly, with growth of the ordinary number of the mode k the inequality

(58) will be satisfied. Hence there will always exist a frequency region with high modal overlap. Beams are well-known for having low modal density. The natural frequencies of beams are spaced well apart since the asymptotic natural frequency of bending vibration is[25]

$$\Omega_k = (k\pi/l)^2 \sqrt{EI/\rho A},$$

where ρ is the mass density, A is the cross-sectional area, EI is the bending rigidity, and l is the beam length. For this case, the inequality (57) is as follows

$$\frac{2k^2 + 2k + 1}{2k + 1} > \psi^{-1}$$

or asymptotically for large k

$$k > \psi^{-1}. \tag{59}$$

Once again, the inequality holds for large values of k.

In the domain of high frequency overlap, i.e.

$$\Omega_{nk+1} \approx \Omega_{nk},$$

inequality (57) may be rewritten as

$$\frac{\Omega_{nk}}{\Delta\Omega_{nk}} > \frac{1}{2\psi_{nk}}$$

or asymptotically for large k

$$\frac{\omega}{\Delta\Omega} > \frac{1}{2\psi}, \tag{60}$$

where $\Delta\omega$ is the eigenfrequency separation. Equation (60) is a more convenient form which permits us to simplify the analysis, e.g. for plates regardless of the boundary conditions $\Delta\Omega = \sqrt{4\pi\rho h/D}/A$, where h is the thickness, D is the flexural rigidity and A is the plate area. As the natural frequency increases, the left hand side of inequality (60) increases. The same conclusion is valid for three-dimensional bodies, see the asymptotic theory of eigenfrequency distribution by Bolotin[25]. The left hand side of equation (60) $(\omega/\Delta\Omega)$ increases as the frequency increases while the right hand side of the equation (60) $(1/2\psi)$ is frequency-independent. Therefore for each structural member there exists a frequency domain wherein the modal resonance curves are overlapped, cf. Fig. 7. This frequency domain of high modal overlap is referred to as the high frequency region[42]. An experimental study of honeycomb plates

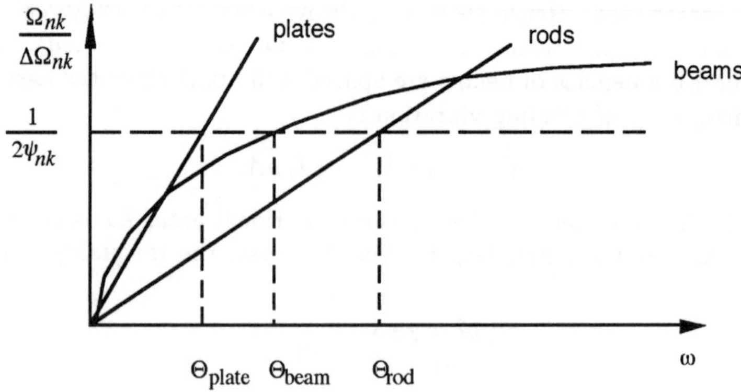

Figure 7. Critical frequencies for rods, beams and plates

by Clarkson and Ranky[51] has confirmed the existence of the frequency domain wherein the modal resonance curves are located so densely that one cannot indicate a particular modal resonance curve. In the domain of high modal overlap the modal curves merge, and their sum appears to be a locally smooth frequency function. In this case the modal sum in equation (56) can be treated as an integral over the high-frequency region rather than a sum

$$M(\omega) = \rho^{(a)} \left[1 + \omega^2 \int\limits_{\Theta}^{\infty} \frac{\Phi(\alpha)\, d\alpha}{-\omega^2 + 2i\psi\alpha\omega + \alpha^2} \right]. \tag{61}$$

Here a locally smooth function of the eigenfrequency distribution Φ is introduced as follows

$$\Phi(\Omega_{nk}) \Delta\Omega_{nk} = \frac{1}{3}\rho^{(a)} \mathbf{u}_{nk}^{(a)} \cdot \mathbf{u}_{nk}^{(a)} V_n. \tag{62}$$

Equation (61) indicates that in the high-frequency domain ($\omega > \Theta$) the structure behaves like a system with a continuous spectrum of eigenfrequencies.

In summary, for any structure with a modal density $d_{nk} = (\Delta\Omega_{nk})^{-1}$ and a damping ψ_{nk} there exists a critical frequency Θ, see Fig. 7. For frequencies greater than the critical frequency, i.e. $\omega > \Theta$, the structure exhibits the behaviour of a mechanical system with a continuous spectrum of natural frequencies. The value for the critical frequency Θ is not well defined and can be obtained only asymptotically from the following equation

$$\Theta = (2\psi_{nk}d_{nk})^{-1} \tag{63}$$

(cf. Eq. (60)). This parameter is specific for each structure and is very much dependent upon the size of the structure, namely, the larger the structure, the lower its critical frequency Θ. As follows from the tests by Clarkson and Ranky[51], the critical frequency is approximately 1200 Hz for the honeycomb plates with cross-sectional area of 1 m^2. The critical frequencies for rods, beams and plates can easily be derived from equation (63) as

$$
\begin{aligned}
rod: \quad & \Theta_{rod} = \frac{\pi}{2}\frac{1}{\psi}\frac{a}{l} \\[2mm]
beam: \quad & \Theta_{beam} = \pi^2\frac{1}{\psi^2}\frac{a}{l^2}\sqrt{\frac{I}{A}} \\[2mm]
plate: \quad & \Theta_{plate} = 2\pi\frac{1}{\psi}\frac{a}{A}\sqrt{\frac{h^2}{12\,(1-\nu^2)}}
\end{aligned}
\tag{64}
$$

where ν is Poisson's ratio. The latter equations show that for such widely used structural members as beams, rods and plates the critical frequency Θ increases as the velocity of sound (a) increases and the damping (ψ) and geometric sizes (l or A) decrease. The same tendency is expected for an arbitrary structural member. It is evident that for any structure there exists an intermediate frequency range wherein the properties of both low and high frequency vibrations are observable. This conclusion is confirmed by equation (63) which defines the critical frequency only asymptotically. For this reason, splitting the frequency region into low- and high-frequency domains may be performed from the point of view of efficiency and cost of numerical computations. For example, the papers by Fischer *et al*[52,53] have considered the broad-band random vibration in structural members and reported the use of FE-analysis at low frequencies and an integral description at high frequencies. The frequency that separates the low- and high-frequency domains, i.e. the critical frequency in our case, was chosen in these papers as the upper frequency bound of efficient numerical computations.

The spectral properties of the complex structures are assumed to be gross parameters since the spectral characteristics of the structure can be obtained only as a result of certain experiments involving excitation over narrow-bands or single frequency excitation of the whole structure. Equation (61) may be rewritten as follows

$$M(\omega) = \rho^{(a)}\left[\delta(\omega) - i\kappa(\omega)\right]^2, \tag{65}$$

where $\delta(\omega)$ and $\kappa(\omega)$ are nondimensional frequency-dependent parameters. Parameter $\delta(\omega)$ has been shown[9] to allow for the following estimation $\delta(\omega) \approx 1$. Comparison of Eqs. (61) and (65) yields

$$\kappa(\omega) = \omega^3 \int_{\Omega}^{\infty} \frac{\psi\alpha\Phi(\alpha)\,d\alpha}{\left(\alpha^2 - \omega^2\right)^2 + 4\psi^2\alpha^2\omega^2}. \tag{66}$$

When the damping ψ in the latter equation is finite, then $\kappa(\omega)$ is finite as well. The parameter $\kappa(\omega)$ however remains finite even for vanishingly small material damping. In order to prove this, the integral in equation (63) is estimated using methods of random vibration theory[25] for small damping ($\psi << 1$) and with a locally smooth function $\Phi(\alpha)$

$$\kappa(\omega) = \frac{1}{2}\pi\omega\Phi(\omega). \tag{67}$$

Palmov[23] has derived equation (67) by means of the Dirac-Plemelj relation[54]. From this last formula one can see that $\kappa(\omega)$ and consequently the absorption of high frequency vibration does not depend on the damping ψ. Actually, this coefficient is absent in Eq. (67). For vanishingly small damping the value of absorption is determined by the distribution function $\Phi(\alpha)$, that is, the internal degrees of freedom act as dynamic absorbers on the carrier structure. In the frequency domain of high modal overlap the resonance curves of the internal degrees of freedom merge yielding considerable spatial absorption of vibration over the whole high-frequency region. Dispersion and multiple scattering have been reported[49] to be additional reasons for the considerable spatial decay in complex structures at high frequencies. The effect of the local modes on the primary structures has been analysed by Der Kiureghian and Igusa[55] by means of numerical example and parametric study. Xu and Igusa[56] have considered a main structure supporting a large numbers of substructures with closely spaced frequencies. For the particular case of equidistant eigen-frequencies it was found that the vibrating multiple substructures have the same effect as viscous damping; this is in full agreement with equations (66) and (67). The paper by Saudi et al[57] discusses the effect of changes in the primary structure's properties as a result of the attachment of secondary systems. In the present analysis this effect is reflected in the generalised mass of the structure. The problem of closely spaced frequencies and some related aspects were recently revisited by Elishakoff[58] in his review paper.

In summary, the boundary value problem that governs three-dimensional high frequency vibration in complex structures takes the following form

$$\mathbf{r} \in V, \ \nabla \cdot \left[\mathbf{C}^{(a)} \cdot \cdot \left(\nabla \hat{\mathbf{U}} \right) \right] + \omega^2 \rho^{(a)} \left[1 - i\kappa \right]^2 \hat{\mathbf{U}} + \hat{\mathbf{h}}_e = 0,$$
$$\mathbf{r} \in S, \ \mathbf{N} \cdot \left[\mathbf{C}^{(a)} \cdot \cdot \left(\nabla \hat{\mathbf{U}} \right) \right] = \hat{\mathbf{f}}. \tag{68}$$

5 Local principle in the high frequency structural dynamics

5.1 *Boundary value problem in one dimension*

The local principle[59,60] implies combination of local dynamics of substructures with global dynamical properties of the structure in the framework of high frequency structural dynamics. By virtue of this principle, the parts of the system under consideration, Fig. 5, are modelled by means of two approaches. A complex structure $0 \le x \le L$ is described by the above method of high frequency structural dynamics, whereas a small substructure of interest $l \le x \le r$ is modelled by conventional methods from vibration theory. The coupling is commonly considered to be conservative[11,13], which allows the coupling to be modelled by a spring of rigidity K.

Applying the high frequency structural dynamics in one dimension one obtains the following differential equation in the frequency domain for the part $0 \le x \le L$

$$C_1 \frac{d^2 \hat{u}_1}{d^2 x} + M_1 \omega^2 \left(1 - i\kappa \right)^2 \hat{u}_1 = 0, \ 0 < x < L, \tag{69}$$

where M_1 and C_1 are the averaged mass per unit length and the averaged longitudinal rigidity of the structure, respectively, and $\hat{u}_1$ is the absolute displacement of the framework. The boundary condition at $x = 0$ is

$$x = 0, -C_1 \frac{d\hat{u}_1}{dx} = \hat{F}(\omega). \tag{70}$$

The dynamics of the substructure $l \le x \le r$ is governed by the equation

$$C_2 u_2'' - \mu_2 \ddot{u}_2 = 0; \ ' = \frac{\partial}{\partial x}, \tag{71}$$

where $C_2(x)$ and $\mu_2(x)$ are the longitudinal rigidity and mass density per unit length, respectively, and $u_2(x, t)$ is the absolute displacement. Let $v_n(x)$, $n = 1, 2, ..\infty$, be the normal modes of the free rod $l \le x \le r$, i.e.

$$\int\limits_{l}^{r} \mu_2 v_k v_n \, dx = \delta_{kn}; \quad \int\limits_{l}^{r} C_2 v'_k v'_n \, dx = \Omega_n^2 \delta_{kn},$$

where Ω_n is the n-th eigenfrequency. Accounting for the properties of the normal modes $v'_n(l) = v'_n(r) = 0$ and the boundary condition $u'_2(r) = 0$ yields

$$\ddot{q}_n + \Omega_n^2 q_n = -\left[C_2 u'_2(l,t)\right] v_n(l). \tag{72}$$

Balancing the coupling forces yields

$$C_1 u'_1(L) = K\left[u_2(l) - u_1(L)\right] = C_2 u'_2(l). \tag{73}$$

5.2   *Vibration in the substructure*

Spectral decomposition (21) allows one to go over into the frequency domain where Eq. (72) is given by

$$\hat{q}_n(\omega) = -\left[C_2 \hat{u}'_2(l,\omega)\right] v_n(l) / \left(-\omega^2 + \Omega_n^2\right). \tag{74}$$

The solution of Eq. (69) is

$$\hat{u}_1(x,\omega) = \hat{F} G(x,\omega), \quad G(x,\omega) = -\frac{1}{C_1 \lambda_1}\left(\sin \lambda_1 x + B \cos \lambda_1 x\right), \tag{75}$$

where $G(x,\omega)$ is the transfer function, $\lambda_1 = \omega\left(1 - i\kappa\right)/a_1$ is the wave number, $a_1 = \sqrt{C_1/M_1}$ is the group velocity in the structure and the integration constants A and B are as follows

$$B = \frac{\cos \lambda_1 L + \dfrac{K \sin \lambda_1 L}{C_1 \lambda_1}\left[1 + K \sum\limits_{n=1}^{\infty} \dfrac{v_n^2(l)}{-\omega^2 + \Omega_n^2}\right]^{-1}}{\sin \lambda_1 L - \dfrac{K \cos \lambda_1 L}{C_1 \lambda_1}\left[1 + K \sum\limits_{n=1}^{\infty} \dfrac{v_n^2(l)}{-\omega^2 + \Omega_n^2}\right]^{-1}}. \tag{76}$$

The spectrum of the modal coordinate is $\hat{q}_n(\omega) = \hat{F}(\omega) U_n(\omega)$, where the transfer function $U_n(\omega)$ is given by

$$U_n(\omega) = \frac{K v_n(l)}{-\omega^2 + \Omega_n^2}\left[K \cos \lambda_1 L - C_1 \lambda_1 \sin \lambda_1 L\left(1 + K \sum\limits_{n=1}^{\infty} \frac{v_n^2(l)}{-\omega^2 + \Omega_n^2}\right)\right]^{-1}.$$

Adopting the SEA assumption about a constant mass density per unit length over the substructure ($\mu(x) = const$) and repeating the derivation of Sec. 2, one obtains the following mean square of displacement averaged over the substructure

$$\bar{u}_2^2 = \frac{1}{r-l} \int_l^r \langle u_2^2(x,t) \rangle \, dx = \frac{1}{\mu(r-l)} \sum_{n=1}^{\infty} \int_{-\infty}^{\infty} S_F(\omega) |U_n(\omega)|^2 \, d\omega, \qquad (77)$$

where $\langle \rangle$ denotes the mathematic expectation which can be deemed as average over ensemble of substructures or realisations and $S_F(\omega)$ is the spectrum of external random or uncertain force $F(t)$.

In order to simplify the expressions for $|U_n(\omega)|^2$ we introduce first the modal damping by means of the damping factor ξ_n

$$|U_n(\omega)|^2 = \frac{K^2 v_n^2(l)}{\left| -\omega^2 + i\xi_n\omega + \Omega_n^2 \right|^2}$$

$$\left| K \cos \lambda_1 L - C_1 \lambda_1 \sin \lambda_1 L \left(1 + K \sum_{n=1}^{\infty} \frac{v_n^2(l)}{-\omega^2 + i\xi_n\omega + \Omega_n^2} \right) \right|^{-2}. \qquad (78)$$

In the vicinity of the natural frequency Ω_n the resonant term prevails in the sum of the latter equation. This allows us to neglect all non-resonant terms and recast Eq. (77) as follows

$$\bar{u}_2^2 = \frac{2K^2}{\mu(r-l)} \sum_{n=1}^{\infty} v_n^2(l) \int_0^{\infty} \frac{S_F(\omega)}{|K\cos\lambda_1 L - C_1\lambda_1 \sin\lambda_1 L|^2}$$

$$\frac{d\omega}{\left| -\omega^2 + i\xi_n\omega + \Omega_n^2 - \dfrac{K v_n^2(l) C_1 \lambda_1 \sin \lambda_1 L}{K \cos \lambda_1 L - C_1 \lambda_1 \sin \lambda_1 L} \right|^2}.$$

It is seen that the response function of a the n-th mode reflects the backward effect of the vibrating structure on vibration of its components.

For not very short structures at high frequencies one can take that asymptotically $\tan \lambda_1 L = \tan\left[\omega(1-i\kappa)L/a_1\right] = -i$. Thus the influence of the structure on vibration of the component can be taken into account by introducing an effective modal damping factor $\hat{\xi}_n$ and the effective natural frequency $\hat{\Omega}_n$ in the following form[61]

$$-\omega^2 + i\xi_n\omega + \Omega_n^2 + i\frac{K v_n^2(l) C_1 (1 - i\kappa)}{(K + iC_1\lambda_1) a_1}\omega = -\omega^2 + i\hat{\xi}_n\omega + \hat{\Omega}_n^2. \qquad (79)$$

Solving this equation for the new parameters yields

$$\hat{\xi}_n = \xi_n + \frac{K^2 v_n^2(l) C_1/a_1}{\left(K + \kappa C_1 \Omega_n/a_1\right)^2 + \left(C_1 \Omega_n/a_1\right)^2},$$

$$\hat{\Omega}_n^2 = \Omega_n^2 + \frac{K v_n^2(l) C_1 \left[C_1 \Omega_n \left(1 + \kappa^2\right)/a_1 + \kappa K\right]/a_1}{\left(K + \kappa C_1 \Omega_n/a_1\right)^2 + \left(C_1 \Omega_n/a_1\right)^2}.$$

Equation for $\bar{u}_2^2$ can now be cast as follows

$$\bar{u}_2^2 = \frac{2K^2}{\mu(r-l)} \sum_{n=1}^{\infty} v_n^2(l) \int_0^{\infty} \frac{S_F(\omega)\, d\omega}{\left|K \cos \lambda_1 L - C_1 \lambda_1 \sin \lambda_1 L\right|^2 \left|-\omega^2 + i\hat{\xi}_n \omega + \hat{\Omega}_n^2\right|^2}.$$

The integral over frequency is seen to have two terms. The first term in the denominator is a smooth function of frequency while the second term produces a resonant curve. In conformity with Sec. 2 we assume asymptotically that

$$\bar{u}_2^2 = \frac{2K^2}{\mu(r-l)} \sum_{n=1}^{\infty} \frac{S_F(\Omega_n) v_n^2(l)}{\left|K \cos \lambda_1(\Omega_n) L - C_1 \lambda_1(\Omega_n) \sin \lambda_1(\Omega_n) L\right|^2}$$

$$\int_0^{\infty} \frac{d\omega}{\left|-\omega^2 + i\hat{\xi}_n \omega + \hat{\Omega}_n^2\right|^2}.$$

Evaluating the latter integral and taking into account that asymptotically $|\cos \lambda_1(\Omega_n) L| = |\sin \lambda_1(\Omega_n) L|$ gives

$$\bar{u}_2^2 = \frac{\pi K^2}{\mu(r-l)} \sum_{n=1}^{\infty} \frac{1}{\hat{\Omega}_n^2 \hat{\xi}_n(\Omega_n)} \frac{S_F(\Omega_n) v_n^2(l)}{\left|\cos \lambda_1(\Omega_n) L\right|^2 \left[\left(K + \frac{\kappa C_1 \Omega}{a_1}\right)^2 + \left(\frac{C_1 \Omega}{a_1}\right)^2\right]}.$$

One replaces the sum in the latter equation by the integral over the frequency

$$\bar{u}_2^2 = \frac{\pi K^2}{\mu(r-l)} \int_0^{\infty} \frac{1}{\Omega^2 \hat{\xi}_n(\Omega) \left|\cos \lambda_1(\Omega) L\right|^2} \frac{S_F(\Omega) v^2(l) \nu(\Omega)\, d\Omega}{\left[\left(K + \frac{\kappa C_1 \Omega}{a_1}\right)^2 + \left(\frac{C_1 \Omega}{a_1}\right)^2\right]}.$$

It is assumed that $\hat{\xi}(\Omega)$ and $v^2(l)$ imply frequency-dependent functions which are some smooth approximations to discrete functions $\hat{\xi}_n$ and $v_n^2(l)$, respectively. The spectral density $S_{u2}(\omega)$ of the displacement averaged over the substructure is thus given by

$$S_{u2}(\omega) = \frac{\pi K^2 S_F(\omega)}{2\mu(r-l)} \frac{1}{\omega^2 \hat{\xi}(\omega) |\cos \lambda_1(\omega) L|^2} \frac{v^2(l)\nu(\omega)}{\left[\left(K + \dfrac{\kappa C_1 \Omega}{a_1}\right)^2 + \left(\dfrac{C_1 \Omega}{a_1}\right)^2\right]}.$$

Since $2|\cos \lambda_1(\omega) L|^2 = \sinh(2\kappa\omega L/a_1)$ at high frequencies one can adopt the following representation of the latter result

$$S_{u2} = \frac{\pi K^2 S_F(\omega)}{\mu(r-l)\,\omega^2 \hat{\xi}(\omega) \sinh\left(\dfrac{2\kappa\omega L}{a_1}\right)} \frac{v^2(l)\nu(\omega)}{\left[\left(K + \dfrac{\kappa C_1 \Omega}{a_1}\right)^2 + \left(\dfrac{C_1 \Omega}{a_1}\right)^2\right]}. \tag{80}$$

This equation is now applicable to identifying the approach parameters.

5.3 Vibrational field in the structure

The field of vibration is determined by transfer function $G(x,\omega)$, Eqs. (75) and (76), and can be cast in the form

$$G(x,\omega) = -\frac{1}{C_1 \lambda_1} \frac{\cos(\lambda_1(L-x) + \varphi)}{\sin(\lambda_1 L + \varphi)},$$

where phase φ depends upon the parameters of coupling and attached substructure. The limiting cases are: $\varphi = 0$ for $K = 0$ and $\varphi = \pi/2$ for $K = \infty$. The square of the absolute value for the transfer function

$$|G(x,\omega)|^2 = \frac{a^2}{C_1^2 \omega^2 (1 + \kappa^2)} \frac{\cosh \dfrac{2\kappa\omega(L-x)}{a} + \cos\left[\dfrac{2\omega(L-x)}{a} + \varphi\right]}{\cosh \dfrac{2\kappa\omega L}{a} - \cos\left[\dfrac{2\omega L}{a} + \varphi\right]} \tag{81}$$

is seen to be a highly oscillating function of frequency because of the circular functions. At high frequencies the circular functions are small compared to the hyperbolic functions and their presence results only in "unneeded" complication of an inherently smooth dependence. The circular functions originate from the boundary conditions which are known to be uncertain for any complex structure. Fig. 8 displays $|G(x,\omega)C_1\lambda_1|^2$ for five different values of φ. In order to obtain a reliable part of the transfer function, it appears reasonable to perform an ensemble average as is usual in SEA. The ensemble covers the different realisations of the structure, each realisation having slightly different mechanical properties. Here we perform an ensemble average by averaging over some frequency range using the technique suggested by Palmov[23] for

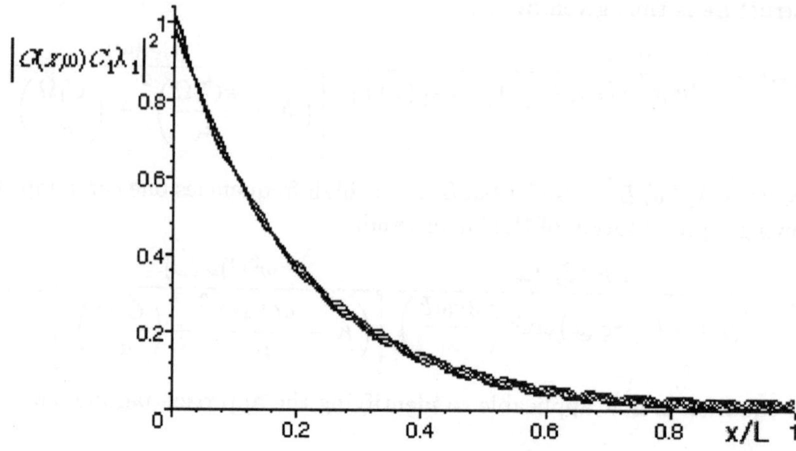

Figure 8. Square of the transfer function for five values of φ

a viscoelastic rod. Averaging the right hand side of equation (81) over the period of each trigonometric function yields

$$\left\langle |G(x,\omega)|^2 \right\rangle = \frac{a^2}{C_1^2 \omega^2 \left(1 + \kappa^2\right)}$$

$$\int\limits_0^{2\pi} \left[\cosh \frac{2\kappa\omega\left(L - x\right)}{a} + \cos\varphi_1\right] d\varphi_1 \int\limits_0^{2\pi} \frac{d\varphi_2}{\cosh \dfrac{2\kappa\omega L}{a} - \cos\varphi_2}.$$

The variables $\varphi_1 = 2\omega\left(L - x\right)/a + \varphi$ and $\varphi_2 = 2\omega L/a + \varphi$ may be considered to be independent since the values of x and L are generally incommensurable. Evaluating the integrals yields

$$\left\langle |\hat{u}_1(x,\omega)|^2 \right\rangle = \frac{a^2 \left\langle \hat{F}\left(\omega\right) \hat{F}^*\left(\omega\right) \right\rangle}{C_1^2 \omega^2 \left(1 + \kappa^2\right)} \frac{\cosh \dfrac{2\kappa\omega\left(L - x\right)}{a}}{\sinh \dfrac{2\kappa\omega L}{a}}. \tag{82}$$

This equation provides us with an expression for the vibrational field in the structure in which one may have confidence.

6   Parameters of the vibrational conductivity approach

6.1   Identification of the parameters

Considering uncoupled structure $0 \leq x \leq L$ implies $H = 0$ and $K = 0$. In this case, Eq. (38) of the vibrational conductivity approach reduces to

$$S_1 = \frac{F}{\alpha_1 \Gamma_1} \frac{\cosh\left(\alpha_1 \left(L - x\right)\right)}{\sinh\left(\alpha_1 L\right)}. \tag{83}$$

The confident part of the vibrational field in the structure (82) has the same character of functional dependence. Comparison of two latter equations allows one to identify nearly all parameters of the vibrational conductivity approach

$$S_1 = \left\langle \hat{u}_1\left(\omega\right) \hat{u}_1^*\left(\omega\right) \right\rangle, \; F = \omega^{-2} \left\langle \hat{F}\left(\omega\right) \hat{F}^*\left(\omega\right) \right\rangle, \tag{84}$$

$$\alpha_1 = 2\omega\kappa/a_1, \; \Gamma_1 = C_1^2 \left(2a_1\omega\kappa\right)^{-1}.$$

What remains to identify is parameter H describing properties of vibration conductance between the substructures. Equations (84) must hold also for the substructures, i.e.

$$\bar{S}_2 = S_{u2}\left(\omega\right), \; F = \omega^{-2} S_F\left(\omega\right), \; \alpha_2 = 2\omega\kappa_2/a_2, \; \Gamma_2 = C_2^2 \left(2a_2\omega\kappa_2\right)^{-1}. \tag{85}$$

The parameters α_2 and Γ_2 are actually not yet identified since one must relate the factor of the spatial decay κ_2 to the damping factor $\hat{\xi}$. To this end, we must adopt a rheological model for the substructure. We presume that the substructure in question is a rod with a Kelvin-Voigt rheological model[23], i.e.

$$E\left(1 + \beta \frac{\partial}{\partial t}\right) \frac{\partial^2 u}{\partial x^2} - \rho \frac{\partial^2 u}{\partial t^2} = 0, \tag{86}$$

where β is a parameter of the model. Multiplying this equation with the normal mode $v_n\left(x\right)$ and integrating over the length of the rod yields an ordinary differential equation for the generalised coordinate q_n

$$\ddot{q}_n + \beta \Omega_n^2 \dot{q}_n + \Omega_n^2 q_n = 0.$$

This equation says that

$$\hat{\xi}_n = \beta \Omega_n^2. \tag{87}$$

On the other hand, inserting $u\left(x, t\right) = \hat{u}\left(x\right) \exp\left(i\Omega_n t\right)$ into Eq. (86) yields

$$E\left(1+i\Omega_n\beta\right)\hat{u}'' + \rho\Omega_n^2\hat{u} = 0. \tag{88}$$

Comparing Eq. (88) with Eq. (69) indicates that qualitatively coincident vibration decay is obtained if one adopts that

$$2\kappa = \Omega_n\beta. \tag{89}$$

Removing β from Eqs. (87) and (89) enables one to relate spatial decay κ_2 to the damping factor $\hat{\xi}$

$$\hat{\xi}_n = 2\kappa\Omega_n.$$

Introducing a continuous approximation yields

$$\hat{\xi} = 2\kappa\omega. \tag{90}$$

Comparing expressions (38) and (80) shows that the identities (85) hold true if

$$\alpha_2\Gamma_2\left(1+\alpha_1\Gamma_1\left[\frac{1}{H}+\frac{1}{\alpha_2\Gamma_2}\right]\right) = \frac{\mu\kappa_2\omega}{\alpha_2}\frac{\left(K+\dfrac{\kappa C_1\omega}{a_1}\right)^2 + \left(\dfrac{C_1\omega}{a_1}\right)^2}{\pi v^2(l)K^2\nu(\omega)}. \tag{91}$$

The rod with a constant distributed mass density has the normalised vibration modes which satisfy the following condition[25]

$$v^2(l) = 2/\left[\mu\left(r-l\right)\right].$$

The asymptotic modal density for the rod is given by[25]

$$\nu(\omega) = \frac{r-l}{\pi a_2}.$$

Inserting these equations together with Eqs. (85) and (90) into Eq. (91) results after a little algebra in the following condition

$$1+\alpha_1\Gamma_1\left[\frac{1}{H}+\frac{1}{\alpha_2\Gamma_2}\right] = 1+\left(\frac{C_1}{a_1}\right)^2\left[\left(\frac{\omega}{K}\right)^2 + \frac{2\kappa\omega a_1}{KC_1}\right]. \tag{92}$$

As $\alpha_1\Gamma_1 = \left(C_1/a_1\right)^2$ this condition requires that the brace brackets on the left and right hand sides of the latter equation must coincide. This is however possible only under the assumption that

$$H << \alpha_2\Gamma_2,$$

otherwise the left hand side of this condition depends on the parameters of the substructure while the right hand side is independent of these. Also, the second term in the brace brackets on the right hand side must be neglected otherwise H which describes the local vibration conductance is dependent upon the global parameters of the system. Thus,

$$K << \frac{C_1}{a_1/\omega}. \tag{93}$$

Hence

$$H = (K/\omega)^2. \tag{94}$$

Substituting Eq. (94) into Eq. (92) leads to the following inequality imposed on K

$$K << \frac{C_2}{a_2/\omega}. \tag{95}$$

The restrictions obtained permit a very simple interpretation. The denominators in Eqs. (93) and (95) are wave lengths in the corresponding part of the system. Hence, Eqs. (93) and (95) imply that the modelling proposed is valid in the case of a weak coupling, i.e. if the stiffness of the coupling is much smaller than the stiffnesses of both coupled structures of the length coinciding with the wave length at the frequency under consideration. This condition is ordinarily satisfied since the contact rigidity is very low.

The equations developed for α, Γ and H are also valid for the three-dimensional case since they do not depend upon the dimension of the problem. The expressions for S and F are expected to be some linear combination of the corresponding one-dimensional ensemble averages since the original problem is linear. The only frame-indifferent linear combination of the one-dimensional solutions is known to be the first invariant of the tensor of ensemble average of displacement

$$S = E\langle U_x U_x^*\rangle + E\langle U_y U_y^*\rangle + E\langle U_z U_z^*\rangle = E\langle \mathbf{U}\cdot\mathbf{U}^*\rangle. \tag{96}$$

Pervozvansky[34] has shown that only radial vibrations are modeled in the framework of the vibrational conductivity approach which implies the following ensemble average

$$\Gamma = \omega^{-2} E \left\langle \tau_N \tau_N^* \right\rangle,$$

where τ_N is the normal stress caused by the external loading. As seen from equation (96), the vibrational temperature S may be viewed as the spectral density of random high frequency vibration for the random loading; in the case of deterministic loading S is the ensemble-averaged square of the vibration amplitude. The parameters of the vibrational conductivity approach are seen to be frequency-dependent, therefore, the vibrational temperature S turns out to be a frequency-dependent parameter. Thus, for each frequency ω there exists a field of vibrational temperature $S(\mathbf{r}, \omega)$. This is in contrast to thermodynamics wherein according to the law of equipartition of energy each degree of freedom has the same amount of energy, i.e. the thermal energy is uniformly distributed through the frequency band of thermal vibration. This difference determines a border between high frequency dynamics and thermodynamics only qualitatively. The problem of how to derive a closed form expression for the frequency separating high frequency dynamics and thermodynamics remains to be tackled.

6.2 *Numerical example*

As many assumptions have been made, it is reasonable to prove the assumptions on an example. The following parameters are taken for the numerical job: $L = 2$ m, $r - l = 0.8$ m, $\kappa_1 = 0.1$, $\kappa_1 = 0.025$, $C_1 = 10^9$ N, $C_2 = 10^8$ N, $K = 3 \cdot 10^8$ N, $\mu_1 = 500$ kg/m, $\mu_1 = 500$ kg/m. Figure 9 displays a smooth curve which corresponds to transfer function $\bar{S}_2/F$ from the vibrational conductivity approach, Eq. (38). The second curve represents square of absolute value of the transfer function for the substructure, Eq. (77), i.e.

$$[\mu \left(r - l\right)]^{-1} \sum_{n=1}^{\infty} |U_n(\omega)|^2 , \text{ where } U_n(\omega) \text{ is given by Eq. (78).}$$

One observes a discrepancy at low frequencies which is in a full conformity with the general idea of the above analysis. The present approach is developed for simulation of complex structures at high frequencies where it is seen to be sufficiently accurate. In addition to this, the conditions (93) and (95) are not fulfilled at low frequencies. The accuracy of the vibrational conductivity approach has been discussed in papers[36,62,63,37]. Without going into detail, we mention that the accuracy of the approach increases as the frequency increases.

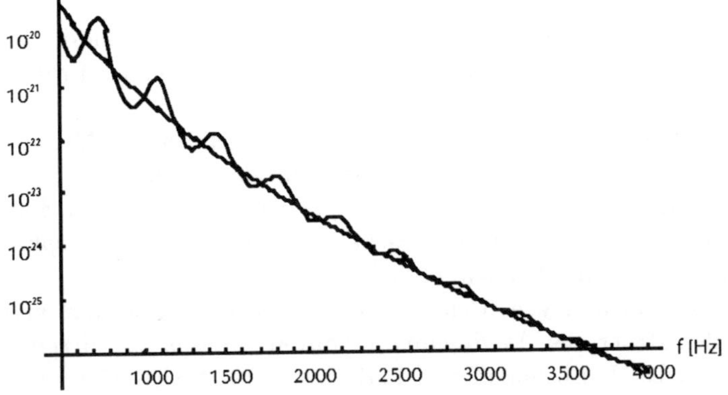

Figure 9. Comparison of the transfer functions from two approaches

7   Conclusions

The local principle in the vibrational conductivity approach to high frequency dynamics is suggested. It allows one to combine SEA for modelling local dynamics of a structural member and the vibrational conductivity approach for integral description of the entire complex structure. The study has shown that all parameters of the present method can be determined in terms of the parameters of high frequency structural dynamics. The latter are expressed in terms of the overall mechanical properties of the structure and local properties of the structural member attached to the complex structure. Only stiffness of the coupling is needed to model the vibration transfer via coupling. It was shown that the only restriction imposed on the parameters is the requirement of the coupling to be weak. This condition is ordinarily met at high frequencies since dynamic rigidity of the coupling is known to be very low.

References

1. I. Newton, *Philosophiae naturalis principia mathematica* (London, 1687).
2. L. Euler, *Mechanica sive motus scientia analytice exposita* (St. Petersburg, 1736).
3. S.P. Timoshenko, *History of strength of materials* (Dover, New York, 1983).
4. C.A. Truesdell, Historical introit. The origin of Rational Thermodynamics. In *Rational Thermodynamics,* ed. C.A. Truesdell, 1-48 (Springer-

Verlag, New York, 1984).

5. P.C. Riedl, *Thermal Physics* (The Macmillan Press, London, 1976).
6. F.J. Fahy, *Sound and Structural Vibration* (Academic Press, London, 1985).
7. C.A. Truesdell, *A First Course in Rational Continuum Mechanics* (The Johns Hopkins University, Baltimore, Maryland, 1972).
8. J. Glimm, Nonlinear and stochastic phenomena: the grand challenge for partial differential equations. *Society for Industrial and Applied Mathematics Review* **33**, 626-643 (1991).
9. A.K. Belyaev and V.A. Palmov, Integral theories of random vibration of complex structures. In *Random Vibration - Status and Recent Developments,* eds. I. Elishakoff and R.H. Lyon, 19-38 (Elsevier, Amsterdam, 1986).
10. R.A. Ibrahim, Structural dynamics with parameter uncertainties. *Applied Mechanics Review* **40**, 309-328 (1987).
11. F.J. Fahy, Statistical energy analysis: a critical overview. *Philosophical Transaction of the Royal Society London A* **346**, 431-447 (1994).
12. L.I. Mandelstamm, *Lectures on the theory of vibration* (In Russian) (GITTL, Moscow, 1929).
13. C.H. Hodges and J. Woodhouse, Theories of noise and vibration transmission in complex structures. *Reports on Progress in Physics* **49**, 107-170 (1986).
14. D. Li and H. Benaroya, Dynamics of periodic and near-periodic structures. *Applied Mechanics Review* **45**, 447-459 (1992).
15. H. Benaroya, ed., Localization and the effects of irregularities in structures, special issue. *Applied Mechanics Review* **49**, 56-135 (1996).
16. C. Pierre and E.H. Dowell, Localization of vibrations by structural irregularity. *Journal of Sound and Vibration* **114**, 549-564 (1987).
17. P.J. Cornwell and O.O. Bendiksen, Localization of vibration in large space reflectors. *American Institute of Aeronautics and Astronautics Journal* **27**, 219-226 (1989).
18. C. Pierre, Weak and strong vibration localization in disordered structures: a statistical investigation. *Journal of Sound and Vibration* **139**, 111-132 (1990).
19. P.D. Cha and C. Pierre, Vibration localization by disorder in assemblies of monocoupled, multimode component systems. *Transaction of the American Society of Mechanical Engineers, Journal of Applied Mechanics* **58**, 1072-1081 (1991).
20. S.T. Ariaratnam and W-C. Xie, Localization of stress wave propagation in disordered multi-wave structure. In *Structural Safety and Reliability,*

eds. G.I. Schuëller, M. Shinozuka and J.T.P. Yao, 77-83 (A.A. Balkema, Rotterdam, 1994).

21. R.S. Langley, Mode localization up to high frequencies in coupled 1D subsystems. *Journal of Sound and Vibration* **185**, 79-91 (1995).

22. P.W. Anderson, Absence of diffusion in certain random lattices. *Physical Review* **109**, 1492-1505 (1958).

23. V.A. Palmov, V*ibrations of Elastoplastic Bodies* (Springer, Berlin-Heidelberg, 1998).

24. H. Goldstein, *Classical Mechanics* (Addison-Wesley Publishing Company, Reading, Mass. 1950).

25. V.V. Bolotin, *Random Vibrations of Elastic Systems* (Nijhoff, The Hague, 1984).

26. R.S. Langley, A general derivation of the Statistical Energy analysis equations for coupled dynamic systems. *Journal of Sound and Vibration* **135**, 499-508 (1989).

27. H.J. Pradlwarter and G.I. Schuëller, Statistical Energy Analysis in view of stochastic modal analysis. In *IUTAM Symposium on Statistical Energy Analysis*, eds. F.J. Fahy and W.G. Price, 209-220 (Kluwer, Dodrecht, 1999).

28. E. Eichler, Thermal circuit approach to vibrations in coupled systems and noise reduction of a rectangular box. *Journal of the Acoustical Society of America* **37**, 995-1007 (1965).

29. V.A. Palmov, Description of high-frequency vibration of complex dynamic objects by methods of the theory of thermal conduction (in Russian). In *Selected Methods of Applied Mechanics*, ed. V.N. Chelomey, 241-251 (VINITI, Moscow, 1974).

30. V.D. Belov and S.A. Rybak, Applicability of the transport equation in one-dimensional wave-propagation problem. *Journal of Soviet Physics - Acoustics* **21**, 110-114 (1975).

31. V.D. Belov, S.A. Rybak and B.D. Tartakovskii, Propagation of vibrational energy in absorbing structures. *Journal of Soviet Physics - Acoustics* **23**, 115-119 (1977).

32. I.A. Bulitskaya, A.I. Vyalyshev and B.D. Tartakovskii, Propagation of vibrational and acoustic energy along a structure with losses. *Journal of Soviet Physics - Acoustics* **29**, 333-334 (1983).

33. A.K. Belyaev, Description of a one-dimensional vibrational state with a parabolic equation. *Soviet Applied Mechanics* **21**, 297-301 (1985).

34. A.A. Pervozvansky, High-frequency vibrations in case of considerable dissipation (in Russian). *Reports of the USSR Academy of Science* **288**, 1068-1072 (1986).

35. A.K. Belyaev and V.A. Palmov, Theory of vibroconductivity. In *Recent Advances in Structural Dynamics,* ed. M. Petyt, 157-168 (University of Southampton, 1980).

36. D.J. Nefske and S.H. Sung, Power flow finite element analysis of dynamic systems: basic theory and application to beam. *Journal of Vibration, Acoustics, Stress, and Reliability in Design* **111**, 94-100 (1989).

37. R.S. Langley, On the vibrational conductivity approach to high frequency dynamics for two-dimensional structural components. *Journal of Sound and Vibration* **182**, 637-657 (1995).

38. Y. Kishimoto and D.S. Bernstein, Thermodynamic modeling of interconnected systems, Part I: conservative coupling, Part II: Dissipative coupling. *Journal of Sound and Vibration* **182**, 23-76 (1995).

39. I.I. Blekhman, *Vibrational Mechanics* (in Russian) (Nauka, Moscow, 1994).

40. F.J. Fahy and W.G. Price, eds., *IUTAM Symposium on Statistical Energy Analysis* (Kluwer, Dodrecht, 1999).

41. V.A. Palmov, Integral methods for the analysis of vibration of dynamic structures (in Russian). *Advances in Mechanics* **2**, 3-24 (1979).

42. A.K. Belyaev, Vibrational state of complex mechanical structures under broad-band excitation. *International Journal of Solids and Structures* **27**, 811-823 (1991).

43. R.D. Mindlin and L.E. Goodman, Beam vibration with time-dependent boundary conditions. *Transaction of the American Society of Mechanical Engineers, Journal of Applied Mechanics* **17**, 377-380 (1950).

44. A.L. Hale and L. Meirovitch, A general substructure synthesis method for the dynamic simulation of complex structures. *Journal of Sound and Vibration* **69**, 309-326 (1980).

45. L. Meirovitch, *Computational Methods in Structural Dynamics* (Sijthoff-Noordhoff International Publishers, 1980).

46. A.H. Nayfeh, *Introduction to Perturbation Techniques* (Wiley, New York, 1981).

47. V.A. Palmov, On a model of medium of complex structure (in Russian). *Prikladnaya Matematika i Mekhanika (Journal of Applied Mathematics and Mechanics)* **33**, 768-773 (1969).

48. A.K. Belyaev, Dynamic simulation of high-frequency vibration of extended complex structures. *Mechanics of Structures and Machines* **20**, 155-168 (1992).

49. A.K. Belyaev, High-frequency vibration of extended complex structures. *Probabilistic Engineering Mechanics* **8**, 15-24 (1993).

50. A.I. Lurie, *Nonlinear Theory of Elasticity* (North-Holland, Amsterdam, 1990).

51. B.L. Clarkson and M.F. Ranky, Modal density of honeycomb plates. *Journal of Sound and Vibration* **91**, 103-118 (1983).

52. P. Fischer, A.K. Belyaev and H.J. Pradlwarter, Combined integral and FE analysis of broad-band random vibration in structural members. *Probabilistic Engineering Mechanics* **10**, 241-250 (1995).

53. P. Fischer, H.J. Pradlwarter and G.I. Schuëller, Evaluation of surface vibrations in the low and high frequency domain. In *Proceedings of the 1995 Design Engineering Technical Conferences,* ed. H.H. Cudney, **3A**, 1029-1036 (The American Society of Mechanical Engineering, 1995).

54. J.A. Desanto, *Scalar Wave Theory* (Springer-Verlag, Berlin, 1992).

55. A. Der Kiureghian and T. Igusa, Effect of local mode on equipment response. In *Structural Mechanics in Reactor Technology,* ed. F.H. Wittman, **K2**, 1087-1092 (A.A.Balkema, Rotterdam, 1987).

56. K. Xu and T. Igusa, Dynamic characteristics of multiple substructures with closely spaced frequencies. *Earthquake Engineering and Structural Dynamics* **21**, 1059-1070 (1992).

57. A. Saudi, T. Aziz and A. Ghobarah, A new stochastic analysis for multiple supported MDOF secondary systems: Part I: Dynamic interaction effects, Part II: Tuning and spatial coupling effects. *Nuclear Engineering and Design* **147**, 235-262 (1994).

58. I. Elishakoff, Random vibration of structures: A personal perspective. *Applied Mechanics Review* **48**, 809-825 (1995).

59. A.K. Belyaev and V.A. Palmov, Locality principle in structural dynamics. In *Recent Advances in Structural Dynamics,* eds. M. Petyt and H.F.Wolfe, **1**, 229-238 (University of Southampton, 1984).

60. A.K. Belyaev, On the application of the locality principle in structural dynamics. *Acta Mechanica* **83**, 213-222 (1990).

61. A.K. Belyaev and H.J. Pradlwarter, Wide-band random vibration in members of complex structures. *International Journal of Solids and Structures* **32**, 3629-3641 (1995).

62. R.S. Langley, Analysis of beam and plate vibrations by using the wave equation. Journal of Sound and Vibration **150**, 47-65 (1991).

63. A. Carcaterra and A. Sestieri, Energy trends in high frequency structural problems. In *Structural dynamics: Recent advances,* eds. N.S. Fergusson, H.W. Wolfe and C. Mei, 482-493 (The Institute of Sound and Vibrations Research, University of Southampton, 1994).

COMPUTATIONAL METHODS FOR ELASTO-PLASTIC MULTIBODY SYSTEMS

J. GERSTMAYR

*University of Linz, Institute for Mechanics and Machine Design, Altenbergerstr.
69, A-4040 Linz, AUSTRIA
E-mail: gerstmayr@mechatronik.uni-linz.ac.at*

Mechanical structures in engineering often undergo large rigid body motion together with small elastic deformation. Early observations concerning satellites
showed that the purely rigid body modeling is sometimes insufficient for the simulation of mechanical systems. The present work deals with the analysis and
simulation of flexible multibody systems with plastic deformation. The existing
theory of plane and beam-type multibody systems is extended with respect to
nonlinear material behavior, including moderately large deformation, as well. The
floating frame of reference formulation is used, which fits best to the small strain
plasticity formulation. The single bodies of a multibody system are connected via
nodes, which allow a very general formulation of the equations of motion. A short
section concerning numerics describes the numerical strategy, which is built up on
implicit Runge Kutta schemes. A computational program has been developed and
it is used to verify the implementability of the proposed theory. Several examples
and comparisons are performed to verify the accuracy of the proposed methods. A
damage model is included into the formulation. An example shows the accuracy
of the proposed method.

1 Introduction

Various mechanical principles and solution strategies are available for
multibody systems (MBSs). The present work deals with an extension of the
existing research with respect to elasto-plastic multibody systems, therefore
a brief overview on the literature concerning rigid and elastic multibody
systems is given first.

1.1 What is a 'Multibody System'?

Various definitions of 'multibody systems', which do not correspond exactly,
but roughly describe the notion of multibody systems, can be found in the
literature:

- *Schiehlen*[52]: 'Multibody systems are characterized by rigid and/or flexible bodies with inertia and springs, dampers and servo-motors without
  inertia, interconnected by rigid bearings or supports. Furthermore, friction and contact forces may be included. The dynamics of multibody

systems are primilarily characterized by their linear and/or nonlinear equations of motions. In addition the equations of reaction are needed for strength considerations.'

- *Bremer and Pfeiffer*[8]: 'A system (model) of several bodies which are connected with links and therefore constrained in the manifold of motions.' (*translated*)

- *Jalón and Bayo*[39]: 'The name multibody stands as a general term that encompasses a wide range of systems such as: mechanisms, automobiles and trucks (including steering systems, suspensions, etc.), robots, trains, industrial machinery (textile, packaging, etc.), space structures, antennas, satellites, the human body, and others.'

- *Shabana*[57]: 'Automobiles, space structures, robots, and machines are examples of mechanical and structural systems that consist of interconnected rigid and deformable components. The dynamics of these large-scale, multibody systems [...]' (preface to the second edition.)
  '[...] to develop methods for the dynamic analysis of multibody systems that consist of interconnected rigid and deformable components. [...] In general, a multibody system is defined to be a collection of subsystems called bodies, components, or substructures. The motion of the subsystems is kinematically constrained because of different types of joints, and each subsystem or component may undergo large translations and rotational displacements.'

- *Eich-Soellner and Führer*[16]: 'A multibody system (MBS) is a result of describing a mechanical system by simple components specially designed for mechanical systems. These consist of bodies, which have masses and torques and can be rigid or elastic. They are linked together by mass-less interconnections which can be force elements, like springs and dampers, or by joints reducing the degrees of freedom [...].'

- *Lennartsson*[43]: 'Multibody system is a device that consists of solid bodies, or links that are connected to each other by joints which restrict their relative motion. The study of multibody dynamics is the analysis of how such systems move under the influence of forces.'

The term multibody system in the definitions given above is sometimes restricted to the specific area of research. While most people use the term flexible bodies, Eich-Soellner and Führer[16] restrict their definition of a MBS to elastic bodies. Nevertheless, all definitions have in common that a multibody system represents a collection of bodies, which are connected by joints.

Coinciding with this definition, a first step into standardization of a MBS and a data-model for the representation of a MBS in a computer code is given by Otter et al.[47]. Although it is sometimes considered that the investigation of quasi-static motion in multi-rigid-bodies belongs to multibody systems, the challenging problem in the simulation is the dynamical (force-driven) and non-linear motion of interconnected bodies, the field of *multibody system dynamics* (MBD).

1.2 Examples

Examples and research areas for MBSs are listed in the literature mentioned above, e.g.

- vehicles (road+rail)

- space structures

- astronomy

- robotics

- mechanisms

- aircrafts

- bio-mechanics

Vehicle dynamics represents already an own field of research, because there are so many special parts, which are related to MBSs: suspension, steering assembly, gear unit, combustion engine, etc. In the field of space structures (satellites, space-roboters, ...), MBS simulation is very important, because no testing is possible in advance. The trend of constructing always faster and lighter robots, aircrafts and machines results in the need to perform simulations with flexible MBS. MBD also became very popular in connection with bio-mechanics, e.g. in the development of virtual operation simulators[14] and with simulations in athletics, see Schiehlen[53]. Investigations concerning bio-mechanics started already at the beginning of the last century by Fischer[18], who modeled the walking motion of human bodies by rigid bodies. Bio-mechanics as well as vehicle dynamics are research fields, in which the modeling of nonlinear material behavior will obviously play an increasingly important role in the future.

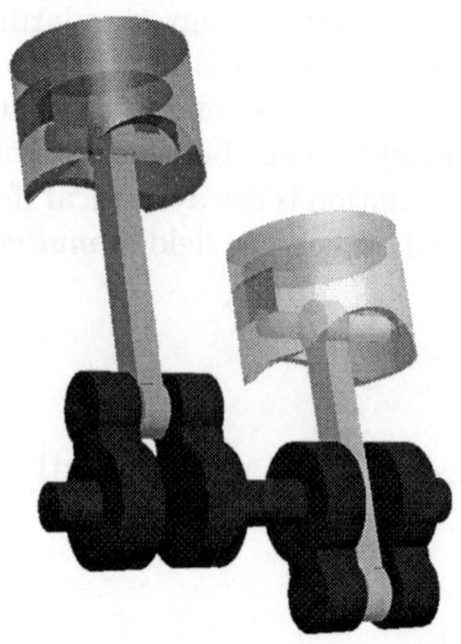

Figure 1. A multibody system.

1.3   *Multibody Systems and Plasticity*

Nowadays, large flexible deformation, friction and nonlinear material behavior is included into the computational algorithms. Nevertheless, the simulations are still sometimes inaccurate models of the real system. Consider for example the computer model of the inner parts of a combustion engine, shown in Figure 1, which is a typical MBS. A first approximation of this system clearly would be by means of rigid body dynamics, a time dependent force at the top of the pistons and a time dependent momentum at the driving shaft. An improvement would be to use flexible bodies, especially for the connecting rod and the crankshaft. Nevertheless, the complex distribution of pressure on the piston head as well as friction in the joints influence the motion of the system. Thermodynamics influences the material behavior of the piston. For an acoustic analysis, even the modeling of the of the slide bearings and the lubrication gap are of importance. For an analysis of the damping, the solid-fluid and the solid-air interaction as well as nonlinear material behavior have to be taken into account. In the case of an overload of the engine (e.g. overspeed), plastic deformation and/or nonlinear geometrical deformation have to be considered, as well as damage and crack-propagation would be important to be included into the simulation. Another type of multibody systems, coming from applications in astronautics, are systems with deployable mass. Consider for example a satellite which is deployed via a cable from a mother ship, where only the deployed cable but not the cable inside the ship is considered in the equations of motion, see e.g.[12] or[35]. Material nonlinearity has to be considered in order to model the damping of the cable.

1.4 State of the Art: Software

Up to now, only some material nonlinearities can be simulated by a MBS simulation (e.g. friction), but a general purpose program is still lacking, as well as the necessary computer power. A recent monograph including an overview of available multibody simulation software is given by von Schwerin[54]. A more detailed, but less recent overview of different available program packages is presented by Schiehlen[52]. One of the well-known commercial software tools is ADAMS, which has a graphical user interface, is capable for small deformation of the bodies by means of the component mode synthesis and has various possibilities for elastic joints and dampers. On the other hand there exist software packages developed at universities with open source code, e.g. MOBILE. These types of programs can be extended individually by the user but are difficult to handle without programming experience. Furthermore, multi-purpose symbolic and numerical computer tools should be mentioned, which are widely used in the field of MBS analysis, among them MAPLE[1] and MATHEMATICA, which are principally symbolic programs, but which have been extended by powerful numerical computation parts, and are therefore usable for the analysis of small MBS. In the field of control of MBS and in robotics, the numerical software tool MATLAB in connection with SIMULINK frequently is used. Many other tools in the field of MBS simulation developed by smaller companies and at universities shall not be forgotten.

1.5 State of the Art: Research

The following section shortly shall review the current research in multi-body systems, and it shall emphasize the importance of developing multibody systems with nonlinear material behavior. For the description of the system, the strategies are mainly divided by the use of independent coordinates (leads to ODEs) and the use of dependent coordinates with constraints (leads to DAEs). A comparison of both formulations can be found in Schwerin[54] and Jalón and Bayo[39].

For the derivation of the equations of motion of the MBD system, the following mechanical principles are frequently used:

- Principle of virtual displacements

- Balance of moment and moment of momentum

- Hamilton's principle

- Lagrange's equations

- Virtual power

- Gibbs-Appell method

- Kane's equations

- Lesser's formulation

Bremer and Pfeiffer[8] compare different notations and unify them by the so-called *projection method*. In a more recent work Bremer[7] presents a state of the art report of elastic multibody systems including a historical overview of the mechanical principles.

The equations of motion can be generated fully or partially by means of a symbolic computational tool (e.g. MATLAB, MAPLE), for an overview see Lennartsson[43], or the equations of motion can be written directly into the program-code, mostly with a dependent coordinates formulation. The fully symbolic methods are often applicable to 'small scale' systems[a] only, but they can generate a set of ODEs for a given MBS with arbitrary links and bodies. The multibody formalisms can be distinguished depending on which mechanical effects are included, as well. The main categories are rigid and flexible multibody systems, see e.g. Jalón and Bayo[39]. The literature on rigid body MBS usually includes the problem of initial positioning, real-time simulation, Coulomb friction, impact, collision and backlash. In the field of flexible multibody systems, several different approaches can be found in the literature. According to Shabana[56], the most used methods are *floating frame of reference, convected coordinate system, finite segment method, large rotation vector* and the *absolute nodal formulation*. The dynamical problem can be divided into several classes of problems, like the static equilibrium position problem, linearized dynamics, inverse dynamics, forward dynamics (dynamic simulation) and percussion and impacts. The present work only deals with forward dynamics. The forward simulation of MBSs must be solved by means of a numerical solution procedure except a few special example problems like the physical pendulum. From the numerical point of view, ODEs and in particular stiff ODEs are treatable by a wide range of numerical methods and are

---

[a]depending on the type of problem and on the used method up to several hundreds of degrees of freedom.

well studied in the literature, for a standard reference see Hairer et al.[28,27]. The numerical solution of DAEs is still under intensive investigation. DAEs arise whenever implicit differential equations have to be solved, as they appear in electric circuits, chemical processes, and of course in MBD. Standard textbooks are Hairer and Wanner[28], Brenan et al.[9] and Eich-Soellner et Führer[16]. For a recent work on implicit Runge-Kutta methods with DAEs in MBSs, see Haug et al.[29]. A state-space-based algorithm for DAEs is given by Haug et al.[30]. Recent contributions on numerical integration of MBS are given in the habilitation thesis of Simeon[61] and in Simeon[58,59,60].

Very little can be found in the open literature in the direction of elasto-plastic multibody systems. One very early approach, coming from an application in crash-test simulations, has been performed by Nikravesh et al.[46]. Nikravesh et al. use rigid bodies with plastic hinges (finite segment method) to simulate the plastic behavior of an experimental setup. The comparison between the experiment and the simulation shows reasonable accordance in the overall behavior, but the oscillations of the experiment can not be resolved with this type of modeling. A second approach, performed by Ambrosio and Nikravesh[4], uses an updated Lagrangian finite element formulation of the equations of motion for an elasto-plastic multibody system undergoing large displacements, rotations and deformations. There, deformation is related to a co-rotated configuration and a condensation technique is used to reduce the number of DOF. Although ride stability and crash simulations of vehicles were mentioned as examples for elasto-plastic multibody systems, no numerical solution techniques for the gained equations of motion and no computational examples are presented by the authors[4]. This shows again the necessity for the development of computational algorithms in the field of elasto-plastic MBSs. In a more recent work, Pan and Haug[48] write that they can treat nonlinear material deformation, nevertheless their examples include only elastic behavior.

The main goal of the present work is a synthesis of existing multibody dynamics knowledge with existing knowledge in plasticity and the output is a simulation tool for elasto-plastic multibody systems. Several famous monographs in the area of theory of plasticity exist, e.g. Hill[31] and Lubliner[45]. For a state of the art, see e.g. Khan and Huang[41]. A classical approach to the variational formulation of elasto-plasticity is given by Washizu[64].

In the present work, plasticity is treated as sources of eigenstrain similar to the eigenstrains in thermoelasticity. Vibrations of elasto-plastic beams with thermal shock loading are studied by Irschik and Ziegler[38]. An overview about dynamical thermo-viscoplasticity until 1995 is given by Irschik and Ziegler[37]. A dynamical study with visco-plastic beams, closely related to the dynamical formulation of plasticity in the present work, is given by Irschik and

Brunner[33]. A dynamical analysis of Timoshenko beams with eigenstrains is given by Adam[2]. The dynamic response of elastic-viscoplastic sandwich beams is given by Adam and Ziegler[3]. The biaxial dynamic bending of elasto-plastic beams is presented by Irschik[32]. For the problem of plasticity and damage in structural elements performing guided rigid-body motion, see Gerstmayr et al.[23]. The special case of the elasto-plastic pendulum has been studied by Gerstmayr and Irschik[24]. Some further details of the present exposition on elasto-plastic multibody dynamics systems, see Gerstmayr[22].

2   Elasto-Plastic Multibody System

2.1   Overview

In the present formulation beam-type bodies with an underlying rigid body motion are used to represent the multibody system. Only small deformations are permitted and only plane motion is considered. However, the second order theory of structures is included into the formulation, see section 4, in order to handle the effect of stiffening in fast rotating bodies and to account for moderately large displacements. The floating frame of reference formulation is used[55]. Axial displacement u and deflection w of the beam axis are used as characteristic displacements.

2.2   Hamilton's Principle

Hamilton's principle adapted to non-conservative systems is used to derive the equations of motion for the deflection and the axial deformation. Constraints and node forces will be introduced later in order to include the equations for the rigid body motion. In the following, we use Hamilton's principle extended to non-conservative systems,

$$\delta \int_{t1}^{t2} L \, dt + \int_{t1}^{t2} \delta W_{nc} \, dt = 0 \tag{1}$$

where L denotes Hamilton's functional and W_{nc} is the work of forces which may be non-conservative. Beams are considered where the mass is concentrated at the axis. Thus, the kinetic energy is defined by

$$T = \frac{1}{2}\rho A \int_0^L \dot{v}^2 \, dx \tag{2}$$

Two different Cartesian coordinate systems are used for a single beam: A global (fixed) coordinate system with lower case letters (x, z) and a local (co-rotated) coordinate system, identified by capital letters (X, Z), where X is the axial coordinate and Z is in direction transverse to the beam. The line

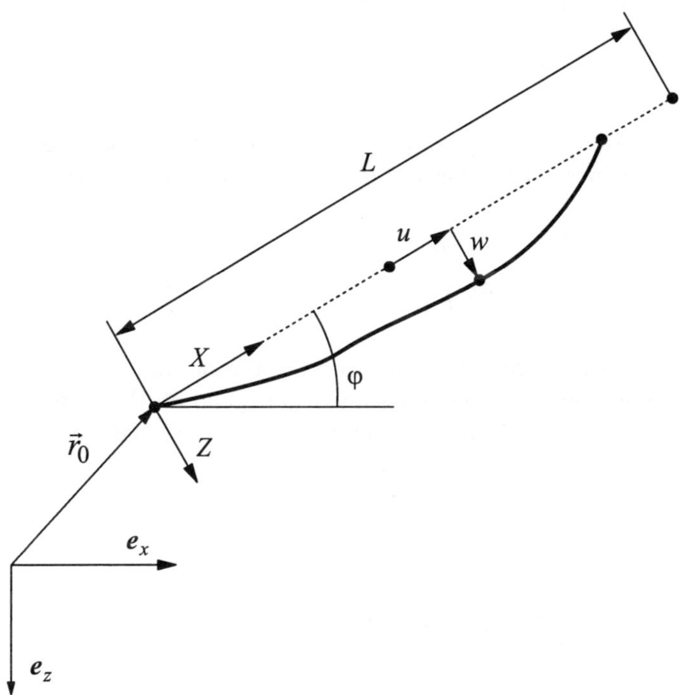

Figure 2. Reference Beam, Rigid Body Degree of Freedom φ, Translation Vector $\vec{r}_0$.

straight defined by $(X, Z = 0)$ denotes the undeformed axis of the beam. A point in the local coordinate system is transformed into the global coordinate system by rotation and translation, see Figure 2, with the rotation matrix $\mathbf{R}$ defined by

$$\mathbf{R} = \begin{bmatrix} \cos(\varphi) & \sin(\varphi) \\ -\sin(\varphi) & \cos(\varphi) \end{bmatrix} \tag{3}$$

The position vector of the axis point $(X, Z = 0)$ in global coordinates is composed of a translatory and a rotatory part, see Figure 2,

$$\vec{r} = \vec{r}_0 + \mathbf{R} \begin{pmatrix} X + u \\ w \end{pmatrix} \tag{4}$$

The velocity is gained by the derivative with respect to time of equ. (4)

$$\vec{v} = \dot{\vec{r}} = \dot{\vec{r}}_0 + \dot{\mathbf{R}} \begin{pmatrix} X + u \\ w \end{pmatrix} + \mathbf{R} \begin{pmatrix} \dot{u} \\ \dot{w} \end{pmatrix} \tag{5}$$

The abbreviations $\frac{\partial}{\partial t} = \dot{()}$ and $\frac{\partial}{\partial X} = ()'$ are used throughout this work. The potential energy only consists of the gravitational potential V_g, which is defined by

$$V = V_g = \rho A g \int\limits_0^L \left[(X + u) \sin \varphi + w \cos \varphi + r_{0y} \right] dx \tag{6}$$

Other forces, e.g. a moment due to an actuator, are included into node forces. The bending moment is assumed to depend on the deflection and on the nonlinear curvature $\overline{\kappa}$, the normal force is assumed to depend on the axial deformation and the mean nonlinear strain $\overline{\varepsilon}^m$

$$\begin{aligned} M &= -EI(w'' + \overline{\kappa}) \\ N &= EA(u' - \overline{\varepsilon}^m) \end{aligned} \tag{7}$$

The nonlinear parts will be treated in section 4. The virtual work of non-conservative forces δW_{nc} is defined by

$$\delta W_{nc} = \int\limits_A \left[-M \delta w'' + N \delta u' \right] dx \tag{8}$$

2.3   Equations of Motion

The kinetic and the potential energy as well as the virtual work of non-conservative forces are inserted into equ. (1). By use of

$$\begin{aligned} \mathbf{R}^T \ddot{\vec{r}} = \mathbf{R}^T &+ \ddot{\varphi} \begin{pmatrix} 0 & 1 \\ -1 & 0 \end{pmatrix} \begin{pmatrix} X + u \\ w \end{pmatrix} + \dot{\varphi}^2 \begin{pmatrix} -1 & 0 \\ 0 & -1 \end{pmatrix} \begin{pmatrix} X + u \\ w \end{pmatrix} + \\ 2\dot{\varphi} &\begin{pmatrix} 0 & 1 \\ -1 & 0 \end{pmatrix} \begin{pmatrix} \dot{u} \\ \dot{w} \end{pmatrix} + \begin{pmatrix} \ddot{u} \\ \ddot{w} \end{pmatrix} = \\ \mathbf{R}^T &+ \begin{pmatrix} \ddot{\varphi} w - \dot{\varphi}^2 (X + u) + 2 \dot{\varphi} \dot{w} + \ddot{u} \\ -\ddot{\varphi} (X + u) - \dot{\varphi}^2 w - 2 \dot{\varphi} \dot{u} + \ddot{w} \end{pmatrix} \end{aligned} \tag{9}$$

the weak formulation for the deformation of a single body follows to

$$\int_0^L \left\{ \begin{bmatrix} N' \\ -M'' \end{bmatrix} - \rho A \mathbf{R}^T (\ddot{\vec{r}}_0 - \vec{g}) - \right.$$

$$\left. \rho A \begin{bmatrix} -w\ddot{\varphi} + \ddot{u} + (-u-X)\dot{\varphi}^2 - 2\dot{w}\dot{\varphi} \\ (u+X)\ddot{\varphi} + \ddot{w} - w\dot{\varphi}^2 + 2\dot{u}\dot{\varphi} \end{bmatrix} \right\} \begin{bmatrix} \delta u \\ \delta w \end{bmatrix} dx = \vec{0} \tag{10}$$

and the according partial differential equations (PDEs) are

$$\begin{bmatrix} N' \\ -M'' \end{bmatrix} = \rho A \left(\mathbf{R}^T (\ddot{\vec{r}}_0 - \vec{g}) + \begin{bmatrix} -w\ddot{\varphi} + \ddot{u} + (-u-X)\dot{\varphi}^2 - 2\dot{w}\dot{\varphi} \\ (u+X)\ddot{\varphi} + \ddot{w} - w\dot{\varphi}^2 + 2\dot{u}\dot{\varphi} \end{bmatrix} \right) \tag{11}$$

These equations are independent of the constitutive equations, because the deformation is represented only in terms of the axial force N and the bending moment M.

2.4 Simplification of the Equations of Motion

While the example section uses the full equations of motion, certain aspects of simplifications of the equations of motion due to negligible changes in the results will be discussed here. For thin beams, the two field variables in equ. (11) may be very different in size and frequency. For example, with a height and width of $H = B = 0.1L$, the relation of cross-area A to the inertia term I is $A/I \approx 1000$. If local dynamical effects as well as the effect on global motion of the axial deformation u is not of special interest, $\ddot{u}$ can be neglected in equ. (11). In order to take into account the normal force, which has a linear part due to gravity and a quadratic part due to the centrifugal force, u should be approximated at least with a polynomial of order two. For a further simplification of the equations of motion, terms in equ. (11) can be neglected as small under the assumptions of a small axial deformation

$$u \ll X$$
$$\dot{u} \ll w\dot{\varphi} \tag{12}$$

and equ. (11) is simplified to

$$\begin{bmatrix} N' \\ -M'' \end{bmatrix} = \rho A \left(\mathbf{R}^T (\ddot{\vec{r}}_0 - \vec{g}) + \begin{bmatrix} -w\ddot{\varphi} - X\dot{\varphi}^2 - 2\dot{w}\dot{\varphi} \\ X\ddot{\varphi} + \ddot{w} - w\dot{\varphi}^2 \end{bmatrix} \right) \tag{13}$$

In several examples, where only very small deformations occurred, it turned out that various terms, like the Coriolis term are negligible. Equ. (11) can be therefore further simplified by assuming

$$w \ll X$$
$$\dot{w} \ll X\dot{\varphi} \tag{14}$$
$$w\dot{\varphi}^2 \ll \ddot{w}$$

This gives

$$\begin{bmatrix} EA(u'' - \bar{\varepsilon}^{m'}) \\ EI(w'''' + \bar{\kappa}'') \end{bmatrix} = \rho A \left(\mathbf{R}^T(\ddot{\vec{r}}_0 - \vec{g}) + \begin{bmatrix} -X\dot{\varphi}^2 \\ X\ddot{\varphi} + \ddot{w} \end{bmatrix} \right) \tag{15}$$

In most applications, the single bodies are connected to the ground or interconnected with different kinds of joints, e.g. with translational, revolute or cylindrical joints, see Shabana[55]. Therefore, boundary conditions are introduced in the following, in order to formulate the constraint equations.

2.5   Boundary Conditions

In the present study, deformations of the beam shall be approximated uniquely with a large translation $\vec{r}_0$, a large rotation φ, a small deflection w and a small axial deformation u according to the floating frame of reference formulation of Shabana[55]. Consider a deflection

$$w(X, t) = aX \tag{16}$$

with some small parameter $a \ll 1$. If the condition $w(X,t)|_{X=0, X=L} = 0$ (see Figure 2) is not used the deflection aX can be interpreted as rotation about the left end of the beam, as well. Under these aspects, and with the definitions of Figure 2, the boundary conditions are

$$w(X,t)|_{X=0} = 0, \; w(X,t)|_{X=L} = 0 \tag{17}$$

$$u(X,t)|_{X=0} = 0 \tag{18}$$

In consideration of the constraint equations, the axial deformation at the right end as well as the derivation of the deflection at both ends are of special interest. On the one hand 4 boundary conditions are needed for the deflection w and 2 boundary conditions are needed for the axial deformation u. On the other hand, the axial deformation at $X = L$ is equal to the elongation of the beam and the slopes of the deflections at $X = 0$ and $X = L$ are approximately the rotations at the ends of the beam for sufficiently small rotations of the cross-section, see also Figure 3. Therefore, the additional 3 boundary conditions introduce 3 new unknowns $u_L(t)$, $\phi_0(t)$ and $\phi_L(t)$, which are defined by

$$w'(X,t)|_0 = \phi_0(t), \ w'(X,t)|_L = \phi_L(t) \tag{19}$$

$$u(X,t)|_L = u_L(t) \tag{20}$$

This 3 unknowns will be used later in the constraint equations.

3 Space Discretization

3.1 Shape Functions

In order to solve the problem numerically by a time-integration method, the Ritz method is applied to discretize the field equations (11) in space. In the following, an asterisk (*) is used for the approximation of a field variable by the Ritz-projector, see e.g. Braess[6]. The deformation of the beam is expanded into a series of functions separated in space and time

$$u^*(X,t) = \sum_{i=0}^{m} b_i(t) P_i(X)$$

$$\tag{21}$$

$$w^*(X,t) = \sum_{j=0}^{n} a_j(t) P_j(X)$$

where the generalized coordinates are denoted by $a_j(t)$, $j = 0,\ldots,n$ for the deflection, and $b_i(t)$, $i = 0,\ldots,m$ for the axial displacement. Polynomial shape functions $(P_i(X) = P_j(X))$ are used. A different approach would be the component mode synthesis, see e.g. Craig[11], where a certain set of eigenforms is used as shape functions, While the eigenforms may show better convergence in certain problem cases, there is the problem of finding an optimal set of eigenforms. Different types of polynomials $P_i(x)$ have been studied. It turned out that orthogonality was not necessary but reasonable in order to avoid long expressions of the discretized equations. While Hermite-polynomials have been studied first, Legendre polynomials are used in the following. The special design of the Legendre polynomials leads to moderately large coefficients for the generalized coordinates and therefore numerical rounding errors are generally low.

3.2 Galerkin's Method

Galerkin's method is a special method of the weighted residual techniques. Assume $R^*(X)$ to be the residual of a differential equation due to a numerical approximation. In our case, we have

$$R^*(x) = \begin{bmatrix} -N^{*\prime} \\ M^{*\prime\prime} \end{bmatrix} + \rho A \left(\mathbf{R}^T(\ddot{\vec{r}}_0 - \vec{g}) + \begin{bmatrix} -w^*\ddot{\varphi} + \ddot{u}^* + (-u^* - X)\dot{\varphi}^2 - 2\dot{w}^*\dot{\varphi} \\ (u^* + X)\ddot{\varphi} + \ddot{w}^* - w^*\dot{\varphi}^2 + 2\dot{u}^*\dot{\varphi} \end{bmatrix} \right)$$

$$(22)$$

In the weighted residual technique, the weighted integral of the residual $R^*(X)$ is forced to vanish for every weight-function $W_i(X)$,

$$\int_0^L R^*(X)W_i(X)\,dx = 0 \tag{23}$$

where the number of weight-functions coincides with the number of unknown generalized coordinates. In Galerkin's method, the shape-functions itself are taken for the weight functions. In the case of polynomial shape functions of equ. (21) the principle leads to

$$\int_0^L R^*(X)P_i(X)\,dx = 0 \tag{24}$$

To get optimal results, the shape functions $P_i(X)$ shall satisfy kinematical boundary conditions, see equs. (17-20), and dynamical boundary conditions. If the dynamical boundary conditions of the approximation due not correspond with the exact ones, the deficient boundary forces generate additional virtual work, which can highly influence the accuracy of the solution of the problem. Adding the deviation of the boundary forces to the differential equations, the approximation can be improved. Appropriate force terms can be derived by means of partial integration. Although the Galerkin method is a practical way to interpret the discretized system, the same equations can be derived, if the shape functions equ. (21) are inserted into the weak formulation, equ. (10). For this mechanical interpretation of the Galerkin technique as the virtual work of the error forces, see Ziegler[65].

3.3 Boundary Forces

Collecting the right hand side of equ. (11) to an "external force term"

$$\vec{f}_{ext} = \rho A \left(\mathbf{R}^T(\ddot{\vec{r}}_0 - \vec{g}) + \begin{bmatrix} -w\ddot{\varphi} + \ddot{u} + (-u - X)\dot{\varphi}^2 - 2\dot{w}\dot{\varphi} \\ (u + X)\ddot{\varphi} + \ddot{w} - w\dot{\varphi}^2 + 2\dot{u}\dot{\varphi} \end{bmatrix} \right) \tag{25}$$

one arrives at the two equations

$$N^{*'} - f^*_{ext_X} = 0$$
$$-M^{*''} - f^*_{ext_Z} = 0 \tag{26}$$

where the relations

$$N^* = EA(u^{*'} - \bar{\varepsilon}^m)$$
$$M^* = -EI(w^{*''} + \bar{\kappa}) \tag{27}$$

are used according to equs. (7). Applying Galerkin's procedure, equ. (24), gives

$$\int_0^L \left(N^{*'} - f^*_{ext_X} \right) P_i \, dX = 0$$
$$\int_0^L \left(-M^{*''} - f^*_{ext_Z} \right) P_j \, dX = 0 \tag{28}$$

with $i = 1 \ldots m$ and $j = 1 \ldots n$. Equ. (28) also follow by inserting the shape functions into the variational formulation, equ. (10). After applying partial integration, these equations read

$$N^{*'} P_i \Big|_0^L - \int_0^L \left(N^* P_i' - f^*_{ext_X} P_i \right) dX = 0$$
$$-M^{*'} P_j \Big|_0^L + M^* P_j' \Big|_0^L + \int_0^L \left(-M^* P_j'' - f^*_{ext_Z} P_j \right) dX = 0 \tag{29}$$

The derivative of the bending moment is replaced by the shear force $Q^* = M^{*'}$ according to the Bernoulli-Euler beam theory. Denoting the assigned boundary forces by N, M and Q, the errors of the boundary forces are $(N - N^*)$, $(M - M^*)$ and $(Q - Q^*)$. Adding the virtual work of these terms to equ. (29), the equations are transformed to

$$N P_i \Big|_0^L - \int_0^L \left(N^* P_i' - f^*_{ext_X} P_i \right) dX = 0$$
$$-Q P_j \Big|_0^L + M P_j' \Big|_0^L + \int_0^L \left(-M^* P_j'' - f^*_{ext_Z} P_j \right) dX = 0 \tag{30}$$

Substituting equ. (27), this yields

$$NP_i\big|_0^L - \int_0^L \left(EA(u^{*\prime} - \overline{\varepsilon}^m)P_i' - f_{ext_X}^* P_i \right) dX = 0$$

$$-QP_j\big|_0^L + MP_j'\big|_0^L + \int_0^L \left(EI(w^{*\prime\prime} + \overline{\kappa})P_j'' - f_{ext_Z}^* P_j \right) dX = 0 \tag{31}$$

The forces N, M and Q can be considered as the Lagrange parameters of the constraints, as used e.g. by Shabana[55].

Assuming the rigid body angle and the nonlinear terms $(\overline{\kappa}, \overline{\varepsilon}^m)$ to be known, equs. (29) represent a set of $(n+m)$ ODEs and $(n+m)$ unknowns a_i and b_i, which can be solved by standard time-integration methods. In order to get reasonable results, the choice of the shape functions should be appropriate to the kinematical boundary conditions (17, 18). The work of the erroneous boundary forces produces an additional error in the solution of the discretized system.

By the way of contrast, equs. (30) are a set of $(n+m)$ ODEs and $(n+m+6)$ unknowns, as the real boundary forces are not known. Additionally, inserting the approximated field variables into the boundary conditions (17,18,19,20), the 6 lacking equations are found. Including these equations, both kinematical and dynamical boundary conditions appear in the equations. These equations are easy-to-use for a general MBS. Unfortunately, the resulting equations are a system of DAEs. The integral terms of equ. (31) are solved by symbolic computation and converted into C++ code by MAPLE. However, the driving terms $\overline{\kappa}$ and $\overline{\varepsilon}^m$ need to be computed from the constitutive equations and need a further treatment, see section 4.2.

3.4   *Simplification of the Discretized Field Equations*

Fortunately, the boundary forces appear linearly in equs. (30). By the use of appropriate shape functions for equ. (21), a linear system of equations can be found for the boundary forces. Therefore, the exact[b] boundary forces can be directly extracted from Galerkin's procedure. By the use of Hermite, Legendre or (x^i)-type[c] polynomials and by a number of shape functions $m \geq 3$

---

[b]exact in the sense of being more accurate by several orders of the approximations compared to the calculation from the field variables. If the approximation of the field variables is not enough often differentiable, the the calculated forces could be zero, e.g. for $w^*(X,t) = a_0(t) + a_1(t)X + a_2(t)X^2 \rightarrow Q|_{X=0} = EI\, w'''|_{X=0} = 0$, assuming $\overline{\kappa} = 0$.

[c]polynomials with $P_i(x) = x^i$

and $n \geq 1$ for the Ritz approximation, equ. (21), the above linear system of equations turned out to be solvable by means of symbolic computation. The formulas of the polynomials can by directly taken from MAPLE. As expected, rather long expressions are obtained for the boundary forces. Nevertheless, for Legendre polynomials the expressions are shorter than for the other tested polynomials. The system of equations e.g. for the normal forces $N_0 = N|_0$ and $N_L = N|_L$, which are not coupled to the bending moments and shear forces in the Bernoulli-Euler beam, reads

$$N_L P_0(L) - N_0 P_0(0) - \int_0^L \left(N^* P_0' - f^*_{ext_X} P_0 \right) dX = 0$$

$$N_L P_1(L) - N_0 P_1(0) - \int_0^L \left(N^* P_1' - f^*_{ext_X} P_1 \right) dX = 0 \tag{32}$$

If the determinant of the linear system is not zero, these equations can be solved for the normal forces. Appropriate shape functions therefore should satisfy the condition of a vanishing determinant, $det = P_0(L)P_1(0) - P_0(0)P_1(L) \neq 0$. For example in the case of Legendre polynomials, the first 2 terms read

$$\begin{aligned} P_0(\xi) &= 1 \\ P_1(\xi) &= \xi \end{aligned} \tag{33}$$

when using a coordinate transformation $\xi = X/L$. The determinant then is $det = 1 \cdot 0 - 1 \cdot 1 = -1$.

3.5 Boundary Conditions

In the case of $m \geq 3$ and $n \geq 1$ for the Ritz approximation (21), the algebraic equations stemming from the boundary conditions, equ. (17-20), and their derivatives with respect to time can be inserted into the Galerkin-type ODEs (31), as well. Because the ODEs equ. (31) and the ODEs given by the 2nd order theory, see equ. (48) subsequently, are only linear in u and linear in w, the unknowns $a_0(t), a_1(t), a_2(t), a_3(t)$ and $b_0(t), b_1(t)$ can be explicitly solved for a given set of shape functions and written as function of the remaining unknowns.

3.6   Example

For the shape functions $P_i(X) = (X/L)^i$ in the Ritz approximation (21), the coefficients $a_0, \dots, a_3, b_0, b_1$ can be expressed as

$$
\begin{aligned}
a_0 &= 0, \ a_1 = \phi_0 L, \\
a_2 &= -\left(2\phi_0 + \phi_L\right) L + \sum_{i=4}^{n} (i - 3) a_i, \\
a_3 &= \left(\phi_0 + \phi_L\right) L - \sum_{i=4}^{n} (i - 2) a_i, \\
b_0 &= 0, \ b_1 = -\sum_{i=2}^{m} b_i
\end{aligned}
\tag{34}
$$

Furthermore equs. (34) are differentiated with respect to time and inserted into the ODEs(30)

4   Further Nonlinearities

4.1   Restrictions

One of the aims of this work was to study different kinds of nonlinearities of MBSs. The highly nonlinear inertia terms of equs. (11) lead to nonlinear ordinary differential equations in time. Due to the use of symbolic computation and the automatic generation of C++ code, further nonlinearities in the equations of motion (29) do not lead to further implementational efforts, if the chosen solution technique is convergent.

4.2   Sources of Self-Stress

The motivation for the use of the floating frame of reference approach was that the equations of motion can be easily related to the constitutive equations in case of small deformations. The following derivation deals with two-dimensional problems, where only the axial strain ε_{xx} is considered as predominant strain and where comparatively small deformations are considered. With regard to the computational implementation, sources of self-stress are introduced, which represent the physically nonlinear part of strain, compare with[36,34,33]. For problems which are related e.g. to plasticity, damage or material damping, the total strain ε_{xx} can be thus split into two parts

$$
\varepsilon_{xx} = \varepsilon_{xx}^e + \bar{\varepsilon}_{xx}.
\tag{35}
$$

where ε^e_{xx} represents the elastic part of strain and $\bar{\varepsilon}_{xx}$ represents the nonlinear part of strain. In the case of pure elastic bodies,

$$\bar{\varepsilon}_{xx} = 0 \tag{36}$$

follows, whereas in the case of elasto-plastic bodies, the nonlinear part of strain is set equal to the plastic part of strain,

$$\bar{\varepsilon}_{xx} = \varepsilon^p \tag{37}$$

The constitutive equation for the axial stress can be written, using the elastic part of strain

$$\sigma_{xx} = E\varepsilon^e_{xx} = E(\varepsilon_{xx} - \bar{\varepsilon}_{xx}) \tag{38}$$

The total strain is calculated from the Bernoulli-Euler kinematics of the beam, which is

$$\varepsilon_{xx} = u' - Zw'' \tag{39}$$

Inserting the constitutive relation (38) into the reaction forces, one gets

$$N = \int_A \sigma_{xx} dA = EA(u' - \bar{\varepsilon}_m)$$
$$M = \int_A \sigma_{xx} Z dA = -EI(w'' + \bar{\kappa}) \tag{40}$$

which verifies the assumption (7). The nonlinear behavior of the beam is collected in the nonlinear curvature $\bar{\kappa}$ and in the mean nonlinear strain $\bar{\varepsilon}_m$, which are related to the nonlinear strain $\bar{\varepsilon}_{xx}$ via cross-sectional integration

$$\bar{\kappa} = \frac{1}{I_y} \int_A \bar{\varepsilon}_{xx} Z \, dA$$
$$\bar{\varepsilon}_m = \frac{1}{A} \int_A \bar{\varepsilon}_{xx} \, dA \tag{41}$$

The total strain can be written as function of normal force N, nonlinear strain $\bar{\varepsilon}_{xx}$ and deflection w

$$\varepsilon_{xx} = \frac{N}{EA} + \bar{\varepsilon}_m - Zw''. \tag{42}$$

Equ. (42) is used to iteratively solve the nonlinear part of strain $\bar{\varepsilon}_m$.

4.3 Moderately Large Strains

Several approaches for a fully nonlinear theory of the deformation of beams can be found in the literature, e.g. a new approach from Li[44], or several well known approaches from Simo[62] or Bathe[5]. Because the approaches lead to a coupled system of nonlinear (partial) differential equations, the solution is gained by a numerical solution technique, mainly by the finite element method. Most of the structures in mechanical engineering however do not undergo large deformation and are therefore either calculated by the linear theory of structures. As mentioned e.g. by Bremer and Pfeiffer[8], the linearization of the equations of a structural element (beam, plate, ...) with superimposed rigid body rotation may lead to errors with are of the same order as the linear terms. This is due to forces (like centrifugal forces) which are of order zero. In 2 dimensional beam kinematics the 2nd order theory of structures must be included to get the stiffening terms, which are necessary e.g. for fast rotating beams or for beams with large axial load. The 2nd order theory is also used in statics, see e.g. Rubin and Schneider[50]. We therefore include this theory into our formulation.

4.4 Second Order Theory of Structures

In the 2nd order theory, geometrically linearized relations are used, but the dynamical principle is formulated in deformed configuration, compared to the first order theory. The axial force is assumed to be constant over the length of the beam, and it may be large compared to transverse forces. Therefore the problem becomes nonlinear if the normal force is not known. In the following derivation the normal force is assumed to be not constant over length. Although not the full geometrical nonlinearity is included into this theory, much better results are achieved for moderately large deformations, compared to the linear case. This is shown in a numerical example in the doctoral thesis of J. Gerstmayr[22]. There, the additional nonlinear part causes a stiffening due to centrifugal force. The following derivations refer to elastic beams. Starting with the Bernoulli-Euler beam theory, the following relations hold:

$$M = -EIw''$$
$$Q = M'$$

(43)

According to Rubin and Schneider[50] the equilibrium in the deformed configuration needs to distinguish between transversal force R and shear force Q (normal to the deformed axis). Both forces are in relation by

$$R = Q + Nw'$$

(44)

The equation of balance of linear momentum for the transversal force in the 2nd order theory of structures reads

$$-R' = f_{ext\,z} \tag{45}$$

where $f_{ext\,z}$ is the external force, used in equs. (25). Inserting equs. (44) and (43) into equ. (45) gives

$$EIw'''' - (Nw')' = f_{ext\,z} \tag{46}$$

Calculating the normal force from the axial deformation

$$N = EAu', \tag{47}$$

and inserting into equ. (46), one obtains

$$EIw'''' - EA\,(u'w'' + u''w') = f_{ext\,z} \tag{48}$$

For the second order theory of structures, equ. (48) must be used instead of equs. (11b). Furthermore, the node-equilibrium has to be written in terms of R and N instead of Q and N. In the elasto-plastic case, equs. (7) are used instead of equs. (43a, 47). In the case of plasticity, the transverse component of equ. (11) becomes

$$EI(w'''' + \overline{\kappa}'') - EA\,((u' - \overline{\varepsilon}_m)w')' = f_{ext\,z} \tag{49}$$

In the Galerkin formulation, partial integration is used, to gain the boundary forces and to avoid the numerical problematic derivative of $\overline{\varepsilon}_m$. In the case of the second order theory, equ. (30b) is replaced by

$$-RP_j\big|_0^L + MP_j'\big|_0^L + \int_0^L \left(-M^*P_j'' - N^*w^{*'}P_j' - f^*_{ext\,z}P_j\right)\,dX = 0 \tag{50}$$

The integrals for the plastic part of strain in equ. (50), compare with equ. (27), are calculated by means of symbolic computation for given polynomials P_j and given space-wise distribution of $\overline{\kappa}$ and $\overline{\varepsilon}_m$, see equs. (86,87).

5   Constrained Motion

In order to link the flexible bodies, joints are introduced. In engineering applications, joints occur in the form of supports and bearings, which cause special restrictions to the motion of the connected bodies. Joints thus can be split into interconnecting joints, which link the free ends of two bodies, and ground-joints, which connect a body to the ground. In two dimensions, revolute and translational joints are of special importance. In the exact modeling of a flexible MBS, the DOF are infinite, while in a discrete approximation the DOF are depending on the number of shape functions (=number of generalized coordinates), rigid body DOF and constraints.

6   Algebraic equations

Similar to a finite element formulation, nodes are introduced additionally to the elements. Every node has three DOF, namely two coordinates (x_N, z_N) and a rotation ϕ_N, see Figure 3. Every beam-element is attached to two nodes. Two elements are linked via an interconnecting node. Every DOF of a node can be fixed, which means that this DOF is attached to the ground. Relations between the degrees of freedom of one element and its two nodes are formulated subsequently.

6.1   *Rigid Body Angle* φ

The rigid body rotation φ of a single beam-type body can be calculated from the two nodes attached to this body

$$\varphi = \begin{cases} \mathrm{atan}\left(\frac{z_{N0}-z_{NL}}{x_{N0}-x_{NL}}\right) & x_{N0} \neq x_{NL} \\ \pm\frac{\pi}{2} & x_{N0} = x_{NL}, z_{N0} \lessgtr z_{NL} \end{cases} \tag{51}$$

Note that in equ. (51) the rigid body angle φ can only be uniquely defined in a range between $-\pi$ to $+\pi$. Recall that φ only appears in trigonometric functions in equ. (11), which are periodic in 2π for φ, or derivatives of φ with respect to time come into play. In the case of e.g. a linear torque-spring attached to the rigid body angle, special techniques have to be applied to determine the necessary offset of $\pm i\pi, i \in \mathbb{N}$. However, the rigid body angle is only an internal variable and not necessarily coupled to measurable quantities in a practical problem. Node angles which also include the small rotation of the beam end will be used in the following e.g. for the implementation of controllers, see section 9. Furthermore, the constraints are transformed into terms of accelerations, therefore node-angles will include the whole revolutions, as well, compared to the rigid body angle φ.

Introducing the two distances l_x and l_z which are depending on the nodal coordinates,

$$l_x = x_{NL} - x_{N0}, \; l_z = z_{NL} - z_{N0} \tag{52}$$

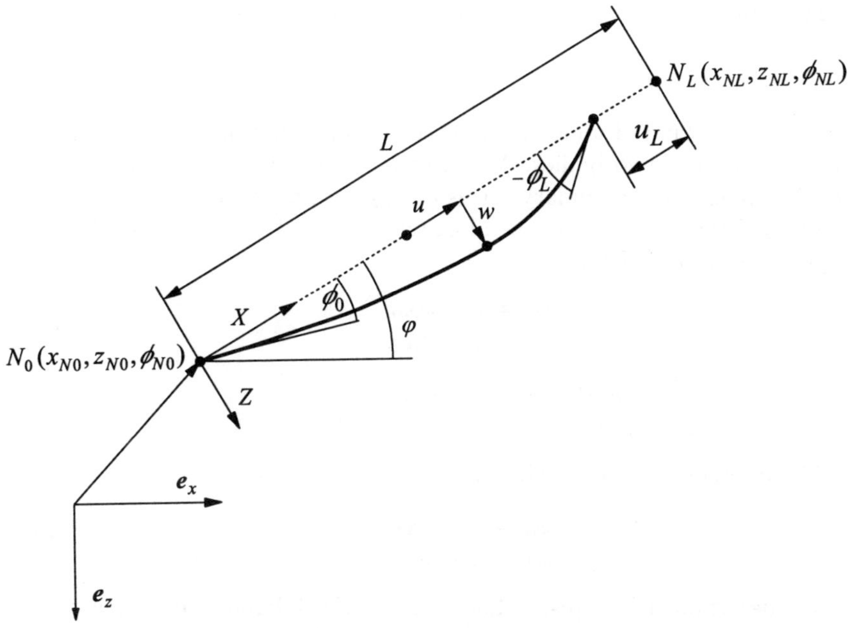

Figure 3. Beam Element Connected to 2 Nodes.

the angular velocity $\dot{\varphi}$ and the angular acceleration $\ddot{\varphi}$ can be determined for $l_x \neq 0$ by differentiation of equ. (51),

$$\dot{\varphi} = \frac{-l_x \dot{l_y} + l_y \dot{l_x}}{l_x^2 + l_y^2} \tag{53}$$

and

$$\ddot{\varphi} = \frac{-l_x \ddot{l_z} + 2\dot{l_x}\dot{l_z} - 2\frac{\dot{l_x}}{l_x}\dot{l_x}^2 + l_z \ddot{l_x}}{l_x^2 + l_z^2} - \frac{2(-l_x \dot{l_z} + l_z \dot{l_x})(l_z \dot{l_z} - \frac{\dot{l_x}}{l_x} l_z^2)}{(l_x^2 + l_z^2)^2} \tag{54}$$

and for $l_x = 0$

$$\dot{\varphi} = \frac{\dot{l_x}}{l_z} \tag{55}$$

$$\ddot{\varphi} = -\frac{2\dot{l_x}\dot{l_z} - l_z \ddot{l_x}}{l_z^2} \tag{56}$$

Note that bodies with $l_x = 0$ and $l_z = 0$ are not taken into account.

6.2   Interconnecting Joints

Depending on the kind of joint, different constraint equations can be defined for the connection of two bodies. When connecting a beam's end rigidly to a node, the kinematical boundary terms ϕ_0, ϕ_L, u_L, introduced in equs. (19, 20), can be written as function depending on the node DOF only - recall that φ is depending on node DOF only

$$\phi_0 = \varphi - \phi_{N0}$$
$$\phi_L = \varphi - \phi_{NL} \tag{57}$$

$$u_L = \sqrt{(x_{NL} - x_{N0})^2 + (z_{NL} - z_{N0})^2} - L \tag{58}$$

Equs. (57) differentiated twice with respect to time gives

$$\ddot{\phi}_0 = \ddot{\varphi} - \ddot{\phi}_{N0}$$
$$\ddot{\phi}_L = \ddot{\varphi} - \ddot{\phi}_{NL} \tag{59}$$

where $\ddot{\varphi}$ is determined by equs. (54,56). Equ. (58) differentiated twice with respect to time by use of l_x and l_z from equ. (52) gives

$$\ddot{u}_L = -\frac{\left(l_x \dot{l}_x + l_z \dot{l}_z\right)^2}{(l_x^2 + l_z^2)^{3/2}} + \frac{\dot{l}_x^2 + l_x \ddot{l}_x + \dot{l}_z^2 + l_z \ddot{l}_z}{\sqrt{l_x^2 + l_z^2}} \tag{60}$$

When connecting the beam's end $x = 0$ or $x = L$ to a node with free rotation, one of equs. (57) is replaced by the free-end condition, taking M from equ. (30)

$$M|_{x=0} = 0 \text{ or } M|_{x=L} = 0 \tag{61}$$

while equ. (58) remains the same.

6.3   Node Equilibrium and Ground Joints

Equs. (57, 58) are the equations for the newly introduced unknowns in equs. (19, 20). The DOF for every node, x_N, z_N and ϕ_N, need additional equations for the system to be solvable. In the case of a node, which is not attached to the ground, additional equilibrium equations are gained. Assuming the nodes to have mass m_N and rotatory inertia I_N, the balance of momentum and angular momentum gives

$$\sum F_x = m_N \ddot{x}_N, \qquad \sum F_z = m_N \ddot{z}_N, \qquad \sum M_y = I_N \ddot{\phi}_N \tag{62}$$

where F_x denotes a force, which acts at the node in the global x-direction and F_z in z-direction respectively. M_y denotes a moment about the y-axis. For an ideal joint, mass and inertia of the node vanish, $m_N = 0$ and $I_m = 0$, and equs. (62) become algebraic equations. If any of the node-DOF is attached rigidly to the ground, some of the following relations can be used

$$\{x_N = const., z_N = const., \phi_N = const.\} \tag{63}$$

The corresponding equilibrium equation of equs. (62) is replaced by a suitable condition from equ. (63). Furthermore, if two elements are linked by a revolute joint, one element is attached rigidly to the node in the form of free rotation. The various possibilities of the constraints shall be illustrated in the following example.

6.4 Example

Consider a plain flexible pendulum, shown in Figure 4. In the sense of the above definitions, we get 3 equations for either of the two nodes

$$x_{N1} = 0, \qquad z_{N1} = 0, \qquad \sum M_{y_{N1}} = 0$$

$$\sum F_{x_{N2}} = 0, \sum F_{z_{N2}} = 0, \sum M_{y_{N2}} = 0 \tag{64}$$

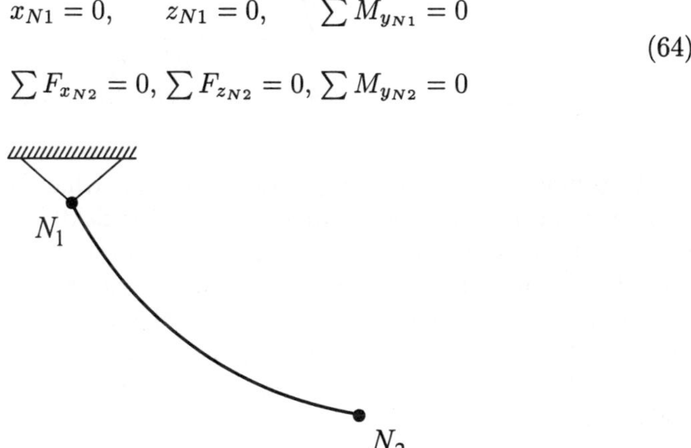

Figure 4. Pendulum: Node definitions and boundary conditions.

7 Plasticity and Damage

The main topic of the present work is the inclusion of plasticity into a flexible multibody system formulation extendable to other types of material nonlinearities. The behavior of the elastic solution, equs. (31) with $\bar{\varepsilon}_m = 0$ and

$\overline{\kappa} = 0$, is very different from solutions with plasticity. While the energy is conserved for the elastic equations[d], dissipative effects are present in the case of plasticity.

In the following, the focus is set on plastic strain, plastic hysteresis and damage. Only one-dimensional plasticity is treated, therefore a simple relation can be given for the actual nonlinear strain, depending on previous strain and actual loading.

7.1   Ideal Elastic-Plastic Materials

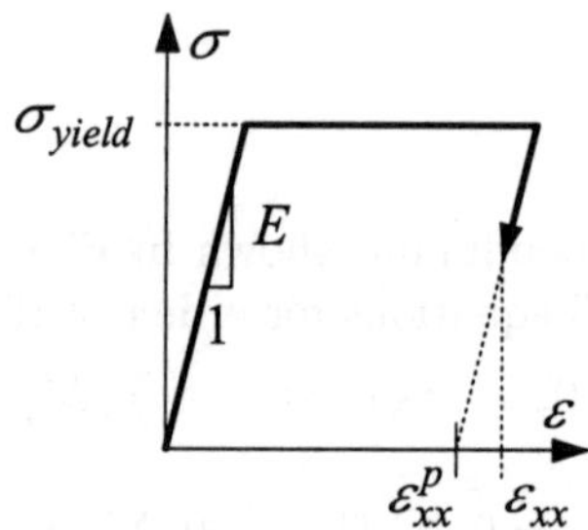

Figure 5. σ-ε Relation for Ideal Elastic-Plastic Materials.

According to the theory of plasticity, see e.g. Khan and Huang[41] and to the numerical treatment of plasticity, see e.g. Crisfield[13], the yield function is used to determine the mechanism of deformation (elastic/plastic) at a certain strain-increment. The yield function f can be written depending on the one-dimensional strain σ and some internal variable ξ. For a given stress-strain relation with linear isotropic hardening, the yield function is depending on the stress σ, internal hardening variable ξ, yield stress σ_Y and isotropic hardening module K

$$f(\sigma, \xi) = |\sigma| - (\sigma_Y + K\xi) \tag{65}$$

Figure 5 sketches the case without hardening. In the case of $f < 0$ only elastic deformation takes place, while for $f = 0$ plastic deformation occurs. The plastic strain rate is defined by

$$\dot{\varepsilon}^p = \dot{\xi}\frac{\partial f}{\partial \sigma} = \dot{\xi}\,\mathrm{sgn}(\sigma) \tag{66}$$

---

[d]only in the case of conservative forces

A dot denotes an infinitesimal increase per time, equ. (66) is written in rate from. The hardening variable can be related to the plastic strain by

$$\dot{\xi} = |\dot{\varepsilon}^p| \tag{67}$$

In the case of pure elasticity $(f < 0)$ with Young's modulus E, the rates are

$$\dot{\varepsilon}^p = 0, \quad \dot{\sigma} = E\dot{\varepsilon} \tag{68}$$

while for $f = 0$, plasticity occurs. This leads to

$$\dot{\varepsilon}^p = \dot{\varepsilon}\frac{E}{E+K}, \quad \dot{\sigma} = \dot{\varepsilon}\frac{EK}{E+K} \tag{69}$$

In the following, hardening is omitted for the sake of simplicity, nevertheless equ. (69) shall show, that an extension to isotropic hardening or to a piecewise linear stress-strain relation is easily performed.

7.2 Numerical Treatment

Within the computation of the plastic strains, the following scheme is used for the sake of efficiency. A stress-strain relation is given by the Young's modulus E for the elastic part and the yield-stress from an ideal elasto-plastic relation. Equ. (69) is written in increment form, as suggested by Khan[41], using the time-step h of the used time-integration scheme, which is assumed to be small. In the following, the index t denotes the actual (new) time-step and the index $t-h$ denotes quantities which have been calculated in the previous (old) time-step.

Depending on the yield function equ. (65) and with $\varepsilon_Y = \frac{\sigma_Y}{E}$ the plastic strain becomes

$$
\begin{aligned}
\varepsilon_t^p &= \varepsilon_{t-h}^p && \text{for } f < 0 \\
\varepsilon_t^p &= \varepsilon_{t-h}^p + \operatorname{sgn}(\sigma)\left(\left|\varepsilon_t - \varepsilon_{t-h}^p\right| - \varepsilon_Y\right) && \text{for } f = 0
\end{aligned}
\tag{70}
$$

The yield function is given by

$$f = |\sigma| - \sigma_Y \tag{71}$$

The total strain ε_t of equ. (70) is given in equ. (42). However, the total strain of equ. (42) depends on the actual plastic strain ε_t^p. Using the plastic strain of the previous step ε_{t-h}^p and inserting into equ. (42) gives a good approximation of the total strain, which is used for the estimation of the plastic part of strain of the actual step, ε_t^p in equ. (70). An iterative approach is used to solve the nonlinear part of strain in the actual step with arbitrary accuracy. In case that the plastic strains are not much larger than the elastic strains, only few iterations (< 20) are necessary to calculate the plastic strain with a relative accuracy of $1e - 6$, as used in the subsequent numerical examples. The error of the plastic strains is estimated via

$$Err_{plast} = \frac{\max_{i,j}\left|\varepsilon_{t,k}^{p,ij} - \varepsilon_{t,k-1}^{p,ij}\right|}{\max_{i,j}\left|\varepsilon_{t,k}^{p,ij}\right|} \quad \text{if} \quad \max_{i,j}\left|\varepsilon_{t,k}^{p,ij}\right| \neq 0$$

$$\text{or} \quad Err_{plast} = 0 \quad \text{else} \tag{72}$$

where k denotes the number of iterations performed, the initial value for $k = 0$ is taken from the previous step. The body is subdivided over its length and height into $k_x \times k_z$ rectangles - the plastic cells, in which the plastic part of strain is assumed to be space-wise constant. The plastic part of strain of one cell (i,j) with the geometry $\frac{i-1}{k_x}L \leq X \leq \frac{i}{k_x}L$ and $\frac{i-1}{k_z}H \leq Z \leq \frac{j}{k_z}H$, $i = 1 \ldots k_x$, $j = 1 \ldots k_z$ is denoted by $\varepsilon_{xx}^{p,ij}$. Thus, the plastic part of strain is given by

$$\varepsilon_{xx}^{p}(X, Z) =$$

$$\sum_{i=1}^{k_x}\sum_{j=1}^{k_z}\left(\mathcal{H}(X-\tfrac{i-1}{k_x}L)-\mathcal{H}(X-\tfrac{i}{k_x}L)\right)\left(\mathcal{H}(Z-\tfrac{i-1}{k_z}H)-\mathcal{H}(Z-\tfrac{j}{k_z}H)\right)\varepsilon_{xx}^{p,ij} \tag{73}$$

where $\mathcal{H}()$ denotes the Heaviside-function.
The proposed solving technique, described in section 8, uses the assumption that time-steps are small enough in order to resolve the change of plastic deformation. An according time-step control could be used to determine the switching time between the states $f = 0$ and $f < 0$. Nevertheless, with larger number of plastic zones, the number of switching instants increases and therefore the effect of state-switching is neglected by using small time-steps. Numerical experiments showed that this method is computationally efficient and accurate enough. For a comparison of the proposed method with the FE-program ABAQUS concerning an oscillating beam with plastic deformation, see Gerstmayr[22].

7.3 Damage

Highly stressed machine elements often undergo damage, which results in a decrease of stiffness. "Damage of materials is defined as the progressive physical process by which they break", Lemaitre[42]. At the microscale, damage occurs due to the mechanism of debonding. At the mesoscale, damage manifests itself in various ways, like brittle damage, ductile damage, creep damage, low cycle fatigue and high cycle fatigue. In the present formulation, only the mesoscale behavior of damage is considered, where ductile damage in combination with low cycle fatigue is treated exemplarily. Ductile damage occurs with high values for the plastic strain over a certain threshold which is set to zero for simplicity. Low cycle fatigue occurs with high values for stress/strain and cyclic plastic strain.

The one-dimensional surface damage-variable-model, which traces back to L.M. Kachanov[40], is used here, see Lemaitre[42]. The damage is interpreted at the microscale as the creation of micro-surfaces of discontinuities, e.g. plastic enlargement of micro-cavities. At the mesoscale, the pattern of micro-cavities can be approximated in any plane by the area of intersections of flaws with that plane. For one-dimensional problems, as applicable for the beam theory, a dimensionless damage parameter $D(X, Z)$ is introduced according to the above assumptions

$$D = \frac{A_D}{A} \tag{74}$$

where A is the original cross-area at some (small) representative volume, and A_D is the damaged area corresponding to the originally undamaged area. Thus, the damage parameter is bounded by

$$\begin{aligned}
D = 0 &\rightarrow \text{undamaged material} \\
D = 1 &\rightarrow \text{fully damaged material}
\end{aligned} \tag{75}$$

The uniaxial stress σ in a small cross-section A of a specimen, loaded axially by force F, is usually defined by $\sigma = F/A$. According to the damage parameter D, a effective stress concept has been introduced by Y.N. Rabotnov, where the effective stress σ_{eff} in a damaged area loaded by a force F is given by

$$\sigma_{eff} = \frac{F}{A - A_D} \tag{76}$$

which can be written in terms of the damage parameter and the mean stress σ as

$$\sigma_{eff} = \frac{\sigma}{1 - D} \tag{77}$$

Lemaitre's strain equivalence principle is used, in order to model each type of damage at the mesoscale. For example, the elasticity law in the plastic and damaged range is extended from $\varepsilon^e = \sigma/E$ to

$$\varepsilon^e = \frac{\sigma}{E(1 - D)} \tag{78}$$

and equ. (35) is therefore extended to

$$\frac{\sigma_{xx}}{1-D} = E\left(\varepsilon_{xx} - \bar{\varepsilon}_{xx}\right) \tag{79}$$

In the following, the classical linear elastic, perfectly plastic constitutive

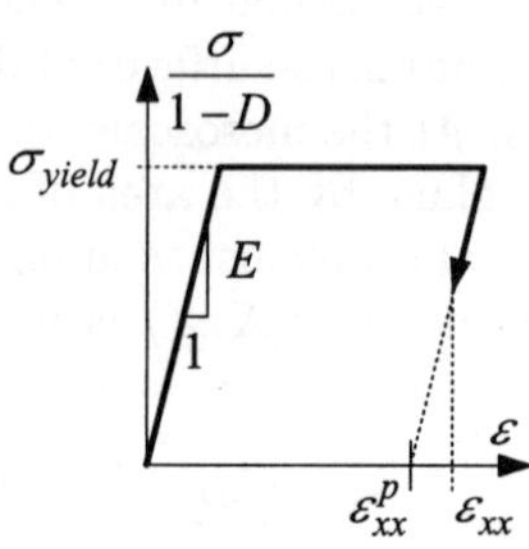

Figure 6. σ-ε Relation for ideal elastic-plastic materials and damage.

relation is used. Applying the effective stress concept, the constitutive relation can be given in the damaged case, as depicted in Figure 6. For a yield stress σ_Y in the undamaged case, yielding occurs in a damaged material when

$$\frac{|\sigma|}{1-D} = \sigma_Y \tag{80}$$

Equ. (70) still can be applied in the computational algorithm, but with some modified value for ε_Y. According to equ. (80) one gets

$$\varepsilon_Y = \frac{\sigma_{YD}}{E} = \frac{\sigma_Y(1-D)}{E} \tag{81}$$

The difference to the case without damage becomes apparent by comparing equ. (38) with equ. (79). The nonlinear part of strain, introduced into the present formulation as eigenstrain, is connected to total strain, plastic part of strain and damage by

$$\bar{\varepsilon}_{xx} = (1-D)\varepsilon^p_{xx} + D\varepsilon_{xx} \tag{82}$$

Note that for the initial value $D \doteq 0$ of the undamaged material the relation $\bar{\varepsilon}_{xx} = \varepsilon^p_{xx}$ follows. The nonlinear curvature $\bar{\kappa}$ is defined by

$$\bar{\kappa} = \frac{1}{I} \int_A \bar{\varepsilon} Z \, dA \tag{83}$$

The mean nonlinear part of strain is defined by

$$\bar{\varepsilon}_m = \frac{1}{H} \int\limits_{-H/2}^{H/2} \bar{\varepsilon}\, dZ \tag{84}$$

According to the space-wise discretization of the plastic part of strain, equ. (73), the damage variable D is approximated by a space-wise constant value D^{ij} for every plastic cell (i,j). The nonlinear part of strain, equ. (82), is calculated for one cell (i,j) with the relation

$$\bar{\varepsilon}_{xx}^{ij} = (1 - D^{ij})\varepsilon_{xx}^{p,ij} + D^{ij}\varepsilon_{xx}^{ij} \tag{85}$$

The total strain ε_{xx}^{ij} is calculated from ε_{xx}, see equ. (42), with $X = \frac{i-0.5}{k_x}L$ and $Z = \frac{j-0.5}{k_z}H$. The nonlinear curvature, see equ. (83), is approximated piecewise constant over the length L, where $\bar{\kappa}(X) = \bar{\kappa}^i$ for $\frac{i-1}{k_x}L \leq X \leq \frac{i}{k_x}L, i = 1 \ldots k_x$. By use of equ. (83), the nonlinear curvature is calculated as

$$\bar{\kappa}^i = \frac{1}{I}\sum_{j=1}^{k_z}\bar{\varepsilon}^{ij}\left(\frac{j-0.5}{k_z} - \frac{1}{2}\right)\frac{WH^2}{k_z} \tag{86}$$

The mean nonlinear part of strain, see equ. (84), is approximated piecewise constant over the length L, where $\bar{\varepsilon}_m(X) = \bar{\varepsilon}_m^i$ for $\frac{i-1}{k_x}L \leq X \leq \frac{i}{k_x}L, i = 1 \ldots k_x$. By use of equ. (83), the nonlinear curvature can be calculated as

$$\bar{\varepsilon}_m^i = \frac{1}{k_z}\sum_{j=1}^{k_z}\bar{\varepsilon}^{ij} \tag{87}$$

7.4 *Evolution of the Damage Parameter*

In the time-integration scheme, a time-evolution model for the damage variable has to be chosen appropriately. Differing from the theory of micro-cavities, which reduce the effective cross-area for the case of tension but do not or only slightly influence the compression region, in the present work, no difference is made between damage in the compression and in the tension case for the sake of simplicity. An evolution equation is used at the mesoscale, similar to that of Frantziskonis and Desai[21,20]. They have derived an exponential decay of $(1-D)$ for geo-technical materials, based on micro-mechanical considerations. It is assumed that the evolution of damage in metals can be modeled similar to geo-technical materials. As suggested by Fotiu et al.[19] the dissipated plastic work per unit volume W_d is introduced in the argument of the exponential function

$$D = D_u \left[1 - e^{-\alpha W_d^2}\right] \tag{88}$$

where D_u and α denote material parameters. According to equ. (79), the dissipated plastic work per unit volume at the end of a time-step $W_d(t + h)$ is connected with the initial value of the time-step $W_d(t)$ by

$$W_d(t + h) = W_d(t) \text{ if } |\varepsilon_{xx}(t + h) - \varepsilon_{xx}^t(t)| \leq \varepsilon_Y$$

$$W_d(t + h) = W_d(t) + \sigma_Y \left(1 - D(t + h)\right) \left[|\varepsilon_{xx}(t + h) - \varepsilon_{xx}^t(t)| - \varepsilon_Y\right] \text{ else} \tag{89}$$

where the actual stress is used during yielding. The dissipated plastic work in structural elements performing guided rigid body motions is shown in Gerstmayr et al.[23]. The additional (always positive) amount of dissipated energy is approximated by a constant increase in a time-step with the assumption, that time-steps are small compared to the change of the strain and the damage parameter. It shall be emphasized, that this is only an example for the constitutive modeling. Furthermore, no matter which time-integration scheme is used for the dynamical equations equ. (31), the plastic dissipated energy is only approximated constant over time, which requires a proportional decrease of the time-step to the desired accuracy of the simulation. Nevertheless, the essential effects of microstructural changes with a minimum of material parameters, which are always difficult to determine, can be brought into play.

8   Numerical Time-Integration

8.1   *Implicit Runge Kutta Methods*

Many analytical techniques for analyzing (stability, existence, uniqueness of solution) and for solving differential equations are only valid for linear differential equations. Some partial differential equations can be solved by means of separation of variables in the form of infinite series of eigenfunctions. Analytical techniques need mostly differentiability and smoothness of the solution. Thus, for most multibody systems numerical solution methods are inevitable. According to the textbook of Hairer et al.[27], a large field of numerical methods do exist. The methods can be separated into "one-step" and multi-step methods. Multi-step methods are very popular as they are computationally very efficient. Some of them are suitable for DAEs, e.g. the BDF formulas.

However, there is the problem of calculating the starting values and the extension to variable step sizes is difficult. One-step methods are less efficient, but easier to implement, especially several explicit formulas like the explicit Euler method. Several classes of methods exist, most of them can be described with the Runge-Kutta tableaus. These methods can be classified with respect to numerical stability, order of consistence and computational and implementational effort.

In the case of mechanical systems, especially when combining longitudinal and transversal deformable degrees of freedom in structural mechanics, stiff differential equations must be expected. Therefore either the time-step has to be adapted to the problem, which is inefficient, or stiffly stable methods must be used for stiff ODEs, either stiffly stable multi-step methods or stiffly stable Runge Kutta methods, e.g. stiffly stable Implicit Runge Kutta (IRK) methods for arbitrary stiffness.

As it has been shown in the literature, see e.g. Hairer and Wanner[28], IRK methods cannot be used for arbitrary DAEs. As mentioned above, the constraint equations of the elasto-plastic multibody system have been differentiated twice with respect to time, in order to lead to an index 1 DAE system, where some of the standard time-integration methods can be used as well. Knowing the fact, that Runge Kutta methods of the Gauss-type are generally not suitable for DAEs, especially for DAEs of higher index, a compromise has been made in the present work, where the midpoint rule is used for the ODE part of the system and the constraints are solved exactly with the nonlinear solver at the midpoint and at the end of the time-step. This leads to a very small numerical drift off in the constraint equations. The numerical accuracy which has only been verified by examples is satisfying.

Several classes of IRK methods for higher index DAEs exist, especially stiffly accurate methods shall be mentioned, e.g. the so-called *Radau IIA* method. An alternative approach to the solution of the multibody problem would therefore be of course to take only the first derivative of the constraint equations or the constraint equations itself and to use a stiffly accurate IRK methods. In order to lower the implementational effort, the initial value problem is avoided by using zero initial-velocities. Further details can be found in the literature cited above.

Depending on the index, numerical problems may arise with the nonlinear equations following from the Runge-Kutta tableau. The Jacobian of the non-linear equations systems may have bad condition numbers depending on the complexity of the system. Nevertheless, regularization of the equations can be performed by means of choosing the time-step and the numerical differentiation appropriate, see Engl[17].

8.2   Index of the DAE-System

[e] In the DAE-literature many definitions of the index exist. In the following, the differentiation index defined by Hairer and Wanner[28] is used. It is assumed to have the special case of a DAE system in semi-explicit form, consisting of a system of first order ODEs

$$\dot{\vec{x}} = \vec{F}(\vec{x}, \vec{z}, t) \tag{90}$$

and of algebraic equations

$$\vec{G}(\vec{x}, \vec{z}, t) = 0 \tag{91}$$

with the vector of generalized coordinates $\vec{x}$, the algebraic variables $\vec{z}$ and time t. According to[9], a property known as the *index* plays a key role in the behavior of DAEs. If the constraint equation (91) is differentiated with respect to time t, one gets

$$\vec{G}_x(\vec{x}, \vec{z}, t)\dot{\vec{x}} + \vec{G}_z(\vec{x}, \vec{z}, t)\dot{\vec{z}} = -\vec{G}_t(\vec{x}, \vec{z}, t) \tag{92}$$

If $\vec{G}_z$ is non-singular, the system (90,92) is an implicit ODE and the system (90,91) is said to have index one. If this is not the case, the system (90,92) can be rewritten in the semi-explicit form of equ. (90,91) and the process can be repeated. The number of differentiations required to obtain a formulation with non-singular $\vec{G}_z$ is called the differentiation index. Mechanical systems mostly have indices between one and three. DAEs with an index larger than one are called DAEs of *higher index* and are much more difficult to solve than DAEs with index one.

The constraints equ. (59,60, 61,62) are used for the realization of a multibody system in the implemented program. The matrix $\vec{G}_z(\vec{x}, \vec{z}, t)$ is different for every multibody system and also time-dependent. Nevertheless, it is assumed that the system has index 1 in most cases except to e.g. bifurcation points, beams with zero length, etc. which are not taken into account. The numerical solution of the constraint equation $\vec{G}_z(\vec{x}, \vec{z}, t) = 0$ for given $\vec{x}$ and unknown $\vec{z}$, see equ. (96) subsequently, requires $\vec{G}_z(\vec{x}, \vec{z}, t)$ to be regular, otherwise Newton's method can not be performed. In such a case the algebraic variables could be approximated with the solution of the half-step, see equ. (94), nevertheless this method is, in general, not numerical stable.

---

[e] Some symbols are used in this section, which have already been introduced above in another context. The present section uses the notification common in the mathematical literature

8.3   Object Oriented Nonlinear Solver

The idea of the implemented solver for the equations of motion was to use the object oriented programming[51] in the sense of a hierarchical working scheme. Each part of the solver is represented by an object, which can "communicate" with his parent and child-objects[67]. Appropriate interfaces and communication flags are used for exchanging data and for distribution of each task. The basic tasks are solved by a linear algebra package, which handles vectors and matrices, and which performs basic calculations like addition, multiplication, calculation of the inverse of a matrix and solving a system of linear equations. Standard libraries in many programming languages can be obtained from the internet, e.g. the object-oriented linear algebra package LAPACK++[15]. In the program developed together with the previous theory, a nonlinear solver is implemented with the capability of inheritance, which is a main feature of object oriented languages. For a given nonlinear system of equations

$$\vec{f}(\vec{x}) = \vec{0} \tag{93}$$

a base object (in C++ a "class") is developed, containing a pure virtual function call $\vec{f}(\vec{x})$, which evaluates the vector function $\vec{f}$ of equ. (93). A sub-procedure evaluates the Jacobian $\mathbf{J}$ of $\vec{f}(\vec{x})$. The main solving procedure, which is called with some parameters like accuracy, initial value and the name of the desired solution strategy, uses the Jacobian to iteratively approximate the solution of the nonlinear system of equations until the demanded accuracy is reached. Depending on the parameters, strategies like standard or modified Newton's method and "trust region" are used. For an overview of methods for the solution of nonlinear equations systems, see e.g. Pozridikidis[49]. The implemented method reports the success of the procedure, in order to choose another strategy or a smaller time-stepsize if the procedure does not converge. A specific object with a certain nonlinear function $\vec{f}$, which can be calculated from several sub-procedures and sub-objects, is "derived" from this pure virtual nonlinear-solver class. The main advantage of this technique is that in every area - the basic mathematical functions, the nonlinear solver and the nonlinear problem - improvements can be made independently. Furthermore, code-doubling is avoided, which leads to less possibilities for implementation-errors. E.g. the nonlinear solver can be applied directly to a system of DAEs to solve the nonlinear algebraic equations for initial values.
Implicit time-integration methods generally lead to a system of nonlinear equations compared to explicit methods. Although some ODEs, e.g. that of a linear mechanical oscillator, can be transformed into a system of linear equations, and despite the fact, that some equations of the nonlinear system

of equations are linear and could be eliminated, the nonlinear solver is used to solve all of the equations at once.

8.4   Object Oriented Time-Integrator

The necessary symbolic preparatory work to the problem is to generate a system of first order ODEs (90) and corresponding algebraic equations (91), which is performed by the symbolic preprocessor, see section 8.5. For a given time-step and a given IRK-tableau, the nonlinear system of equations can be formally written down, e.g. for the implicit midpoint rule the nonlinear equations read

$$\vec{f}_{impl} \begin{pmatrix} \vec{x}_{t+h/2} \\ \vec{z}_{t+h/2} \end{pmatrix} = \begin{pmatrix} \vec{x}_{t+h/2} - \vec{x}_t - \frac{h}{2}\vec{F}\left(\vec{x}_{t+h/2}, \vec{z}_{t+h/2}\right) \\ \vec{G}\left(x_{t+h/2}, \vec{z}_{t+h/2}\right) \end{pmatrix} = \vec{0} \qquad (94)$$

The explicit formula for the full step is

$$\vec{x}_{t+h} = \vec{x}_t + h\,\vec{F}\left(\vec{x}_{t+h/2}, \vec{z}_{t+h/2}\right) \qquad (95)$$

The algebraic variables for the full step can be calculated from the algebraic constraints

$$\vec{f}_{alg} \begin{pmatrix} \vec{x}_{t+h} \\ \vec{z}_{t+h} \end{pmatrix} = \vec{G}\left(x_{t+h}, \vec{z}_{t+h}\right) = \vec{0} \qquad (96)$$

Equs.(94) or equs.(96) can be written as a system of nonlinear equations $\vec{f}$, with unknowns $\vec{y}$

$$\vec{f}\left(\vec{y}\right) = \vec{0} \qquad (97)$$

Newton's method is applied using the Jacobian $\mathbf{J} = \dfrac{\partial \vec{f}\left(\vec{y}\right)}{\partial \vec{y}}$

$$\vec{y}_{i+1} = \vec{y}_i - \left[\mathbf{J}\big|_{\vec{y}=\vec{y}_i}\right]^{-1}\vec{f}\left(\vec{y}_i\right) \qquad (98)$$

where in the case of convergence, $\vec{y}_{i+1}$ is a better approximation than $\vec{y}_i$. When choosing the time-stepsize small enough, the convergence criteria for Newton's method (start-vector is near enough to solution) is always fulfilled. Nevertheless, very small time-steps can cause large condition numbers for the Jacobian, which leads to an ill-posed problem when solving the linear equations system equ. (98).

The Jacobian $\mathbf{J}$ is computed by means of the finite differences method. The element (i, j) in the Jacobian is approximated by

$$\mathbf{J}_{ij} = \frac{\partial \vec{f}_i\left(\vec{y}\right)}{\partial y_j} \approx \frac{f_i\left(u_1, ..., u_j + \delta/2, ..., u_n\right) - f_i\left(u_1, ..., u_j - \delta/2, ..., u_n\right)}{\delta} \qquad (99)$$

The choice of the differentiation parameter δ is critical. For smaller values of δ the approximation is getting better, taking into account a sufficient number of digits in the computation. It turned out in the numerical experiments, that for moderately large values of δ, the Jacobian is "smoothed" and the Newton method showed better convergence than with smaller values[f]. As the optimal convergence speed was not necessary for the computations, rather large values between $\delta = 10^{-5} \div 10^{-1}$ were used. For optimization of the finite differences, extrapolation methods[g] and automatic differentiation[26] shall be mentioned. The solution of equ. (98) can be computed by solving a system of linear equations, e.g. by Gaussian elimination. Because the number of unknowns was lower than 1000 in the examples, numerical solution methods would not be much more efficient. For a stable solution of the system of equations, Pivot-search is used. A high speed-up is gained by use of modified Newton's method, where the Jacobian is inverted once (computationally costly), but used in equation equ. (98) as long, as the convergence rate is better than a certain criterion (in the examples: not more than 50 iterations). If the modified method does not converge, the standard Newton method is used, starting with the original starting vector. Therefore from the point of convergence radius, the method is at least as good as the standard Newton's method.

8.5 Special Notes to the Implementation

A program, called `mbsplast`, which can perform a forward simulation of beam-type elasto-plastic multibody systems and uses the above described theory has been implemented. The symbolic computation program MAPLE is used as symbolic "pre-processor". The derivation of the equations of motions in the form of DAEs is completely generated in MAPLE. The symbolic program derives the kinematical relations, equ. (5), and the equations of motion, equ. (11). The field variables are approximated by a variable number of shape functions, which are provided by the MAPLE libraries. These shape functions are inserted into the Ritz-Galerkin formulation, equ. (29). Symbolic integration is used to generate the set of ODEs. The modification of the equations of motion, described in Section 3.4 is performed fully in MAPLE. The resulting equations have a length of up to 1500 lines of program code, with the number of shape functions used in the examples. The constraints are symbolically

---

[f]Theoretically, the best value for numerical differentiation would be around $\sqrt{\epsilon}$, where ϵ is the accuracy of the floating point operation in the computer, which was $\epsilon \approx$ 1e-16 in the performed calculations

[g]calculate the differentiation quotient e.g. $\delta = 0.1, 0.01, 0.001, 0.0001, \ldots$ and extrapolate for $\delta \to 0$

differentiated with respect to time. A set of interfaces is defined, which has to be the same in the MAPLE program and in the C-program. The resulting, mostly explicit equations and the DAEs are written automatically with the MAPLE function C() into C++ header files, which can be included in a smart way into the developed C-program.

9   Simulation of Controlled Motion

As described in section 8.3, additional algebraic equations are easily implemented into the present computational code. Thus, equations from actors and sensors, as they appear often in a real problem of mechanical engineering, can be easily added via the use of controllers. A controller is a function which measures various quantities of the system (e.g. rotation ϕ_N, displacement in nodes x_N, z_N) and calculates actuating forces (e.g. F_x, F_z, M_y) by means of a certain law of control. The forces are added to the node equilibria, see equ. (62). P, PD and PID control elements have been implemented and have been tested for various problems. Collocated sensors and actors can be used, e.g. if the moment at a node is acting and the node angle and its derivatives with respect to time are measured. For example, in case of a PD controller, which shall control the angle ϕ_N (controlled variable) at a node to a certain reference angle ϕ_{end} by means of a certain actuating moment M_c, the control equation for the moment is

$$M_c = -R_p\left(\phi_N - \phi_{end}\right) - R_d\left(\dot{\phi}_N - \dot{\phi}_{end}\right) \tag{100}$$

where R_p and R_d denote the corresponding proportional and differential parameters for the control element. While a P or a PD control element changes only the behavior of the mechanical system, a PID controller leads to an additional differential equation, which changes the size of the DAE system, as well. Compared to equation equ. (100), an additional integral term appears in the PID controller for the angle ϕ_N,

$$M_c = -R_p\left(\phi_N - \phi_{end}\right) - R_d\left(\dot{\phi}_N - \dot{\phi}_{end}\right) - R_i\left(\int_0^t \phi_N(\tau)\,d\tau - \phi_{i\_end}\right) \tag{101}$$

The integral can be eliminated via a new variable $\overline{\phi}_N = \int_0^t \phi_N(\tau)\,d\tau$ or $\dot{\overline{\phi}}_N = \phi_N$, which is inserted into equ. (101) and via the new differential equation

$$\dot{\overline{\phi}}_N = \overline{\phi}_N \tag{102}$$

Equ. (102) is added to the system of differential equations for the original multibody system. Note that the initial conditions can be influenced by the controller. One way to overcome this is to activate the controller not before the end of the first time-step.

10 Slider-Crank Mechanism

The slider-crank mechanism has been studied frequently in the literature with rigid and with flexible components, see e.g.[10] or[59]. It appears in applications like a gas-engine, a pump and many other applications, where rotational motion shall be transformed into translational motion. In the following example, consider a simple 2 dimensional model of a slider crank mechanism as shown in Figure 7. An (electric) engine is acting on the driving beam 1 with a moment

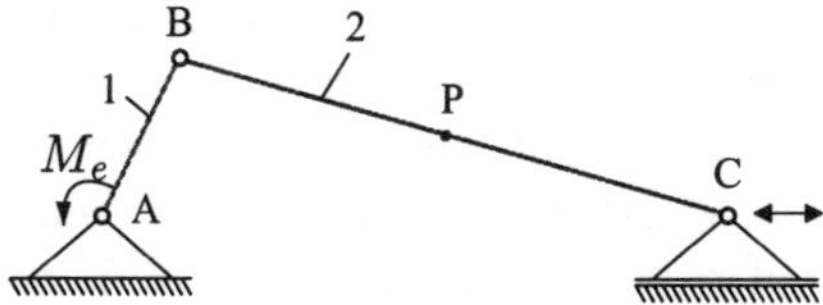

Figure 7. Sketch of a 2 dimensional slider crank mechanism.

M_e. Beam 1 starts to rotate and the right support C moves horizontally. Link B moves on a circle, therefore oscillating inertial forces act transversely to the beam 2. Consider a comparatively thin beam 2 and an increasing rotation speed of the driving arm 1. At a certain speed, the inertial forces are large enough to induce plastic deformation in beam 2. This is simulated in the following example. Furthermore reverse plasticity repeatedly occurs during the motion. Only with the use of a fine discretization the sharp edges between different plastic zones produced by reversed plasticity can be identified, see Figure 12. Without damage, plastic shakedown occurs, and a decreasing area remains, where plastic strain changes its magnitude in every cycle. Nevertheless, when a Kachanov damage model is coupled with the elasto-plastic multibody system, beam 2 weakens with increasing dissipated plastic energy. Thus, the simulated slider-crank mechanism becomes unstable, and the deformation becomes so large that the simulation stops. The real mechanism then must be considered as broken.

The parameters for the slider-crank mechanism are the following: The geometrical parameters for beam 1 are $H = 0.04$m, $B = 0.04$m and $L = 0.3$m. The beam is quite high for the Bernoulli Euler theory, but this beam mainly acts as a flywheel and therefore undergoes only very small deformation. Beam 2 is slender, with $H = 0.01$m, $B = 0.01$m and $L = 1$m. Gravity is not taken into account. The material parameters are the same for both beams, and they are taken from steel, with density $\rho = 7800\frac{\text{kg}}{\text{m}^3}$, and Young's modulus $E = 210000\frac{\text{N}}{\text{mm}^2}$. An ideal elasto-plastic material is used with a yield stress of $\sigma_{\text{yield}} = 350\frac{\text{N}}{\text{mm}^2}$. The time and space-wise discretization has been chosen under the aspects of accuracy and computation time. The time-evolution of several variables, expecially the time-evolution of the critical value of the damage changed still by a few percent (towards the end of the simulation) compared to a solution with half the resolution in time and space. One reason for that is, that the time-evolution of these variables is approximately exponentially. Nevertheless, important results, like the shape of the plastic zones, do not change when the discretization is further refined. Under those aspects, the time-step size is chosen by $t = 0,25$ms, the thick beam 1, which undergoes no plastic deformation, is modeled by one beam-element and $20 \times 200 (x \times z)$ plastic cells over height and length. The second beam is separated into 5 beam-elements, where every beam-element has $70 \times 200 (x \times z)$ plastic cells . The number of polynomials is $N = 7$ and $M = 3$. The damage parameters, see equ. (88), are $D_u = 1$ and $\alpha = \dfrac{0,525e12}{\sigma_{\text{yield}}^2}$.

The time-evolution of the driving moment M_e is shown in Figure 8. Due

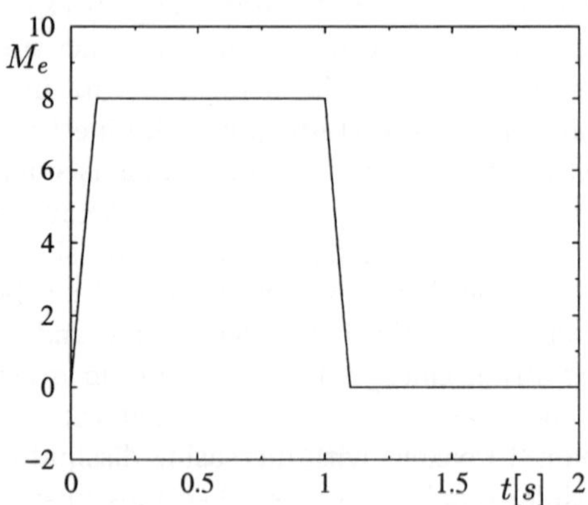

Figure 8. Time-evolution of driving moment.

to the driving moment beam 1 accelerates until an average angular velocity of approximately 7 rpm is reached, afterwards the mechanism undergoes no driving forces. The speed is of special interest, because with a slower speed, the machine would deform purely elastic and with a higher speed, it would break immediately due to inertial forces. Figures 9-11 show the simulation window for different instants. This visualization shall sketch the behavior of the mechanism, which can be studied in detail in the animation file generated by the computer code[h]. Figure 11 clearly shows the development of damage and plasticity in a moment, where the mechanism is almost broken. At the position of maximum damage - near the middle of the driven beam - a plastic hinge develops. The calculation is stopped at this point, because the deformation and strain is too large for the used theory. Figure 12 shows the plastic part of strain in a 3 dimensional view at $t = 2.5$s. This view shows the cliffs of the plastic strain which develop through reverse plasticity. Due to the printers resolution, only every 2nd cell is plotted (165×200 cells instead of 350×200 cells). Figure 13 shows the mid-span deflection at point P, see Figure 7, of the driven beam with and without damage. The slow drift of both solutions can be clearly seen. When the beam is weakened up to a certain level of damage (compare simulation window), the deformation rapidly increases. This moment in time is assumed to be the time when the part breaks. Figure 14 shows the moment-curvature diagram, where the normalized bending moment $\dfrac{M}{EI}$ at the middle of the driven beam is plotted over the nonlinear curvature, see equ. (83) with and without damage. Figure 15 shows the time-evolution of the plastic part of strain at point P at $z = H/2$. It shall be mentioned again that the used damage model uses the assumption that the material is damaged under pressure as well. Figure 15 would have - besides the influence of the axial force N - the opposite sign for $z = -H/2$. Figure 16 shows the time-evolution of the damage parameter D at point P, at $z = H/2$. Figure 17 shows the performed rotations of the driving beam over time and the time-evolution of the rotation speed of the driving beam. After a short period of acceleration, compare with Figure 8, the rotation speed - and the kinetic energy - decreases due to the dissipated plastic energy. Without damage, the beam shakes down, as shown in Figure 15, and the rotation speed keeps almost constant. With damage, the plastic activity becomes bigger with increasing damage and therefore the plastic dissipated energy still increases. The computational time on a MIPS R12000 processor with 300 MHz used for the simulation with damage needed 6 hours 50 minutes, where approx. 20% of the computational time was needed for graphical purposes. The average time needed for the calculation of one time-step is therefore about 1.5

---

[h]The animation file can be downloaded with a web browser (e.g. Netscape) under
http://www.mechatronik.uni-linz.ac.at/institute/TMGM/tmech/gerstmayr.html

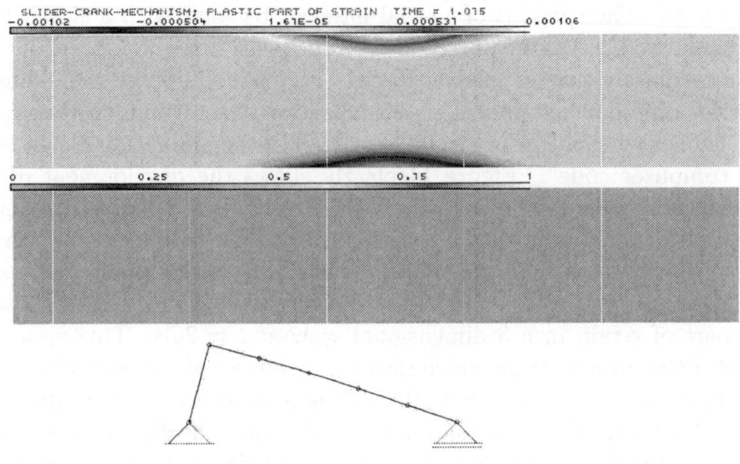

Figure 9. Simulation window, plastic strain and damage at $t = 1.075$s.

seconds, which is assumed to be very fast compared to standard dynamic FE-calculations with elasto-plasticity and comparable discretization.

Conclusions

A computational algorithm for studying the development of plasticity and damage in flexible multibody systems is presented. The present formulation uses the floating frame of reference formulation of beam-type multibody systems with an extension to elasto-plastic strains. A high resolution of the plastic zones is gained by a separation of the problem into an elastic and a plastic part. Examples demonstrate the functionality and show the high potential for applications.

The extension to 3 dimensional elasto-plastic multibody systems, which has important applications like crash simulations for vehicles, is still under investigation. A preliminary purely elastic 3 dimensional formulation is currently in publication.

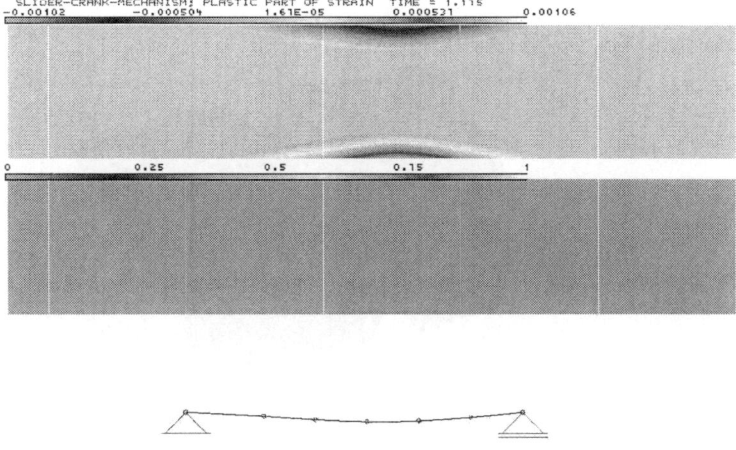

Figure 10. Simulation window, plastic strain and damage at $t = 1.175$s.

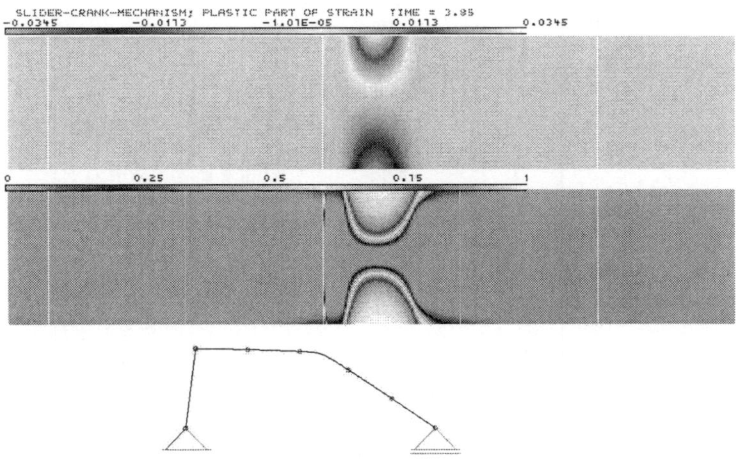

Figure 11. Simulation window, plastic strain and damage at $t = 3.85$s.

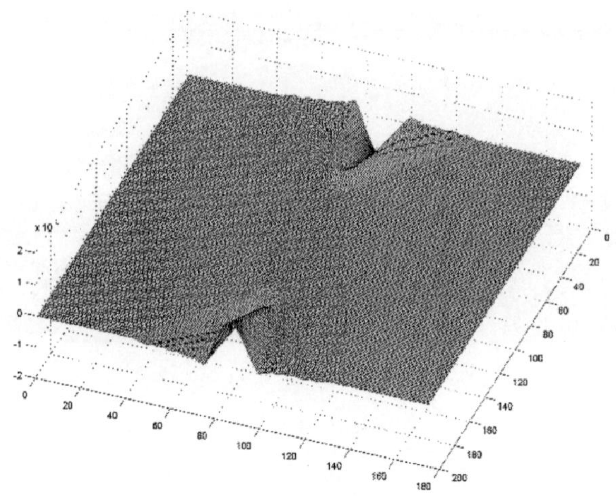

Figure 12. Plastic part of strain over length and height of driven beam at $t = 2.5$s, 3 dimensional view.

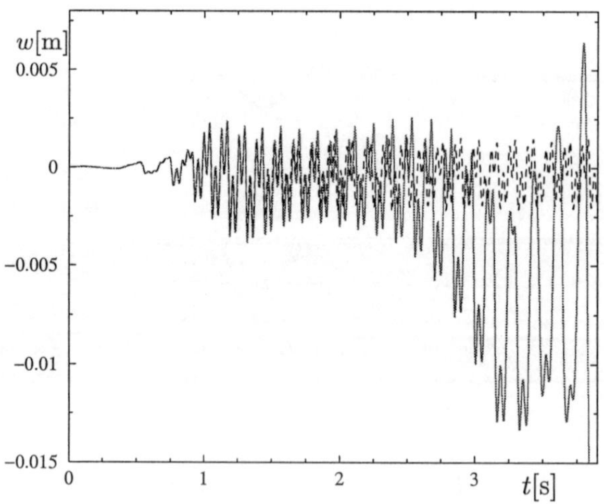

Figure 13. Mid-span deflection at point P, see Figure 7. Comparison of elasto-plastic simulation (*dotted black*) and elasto-plastic simulation with damage (*solid red*).

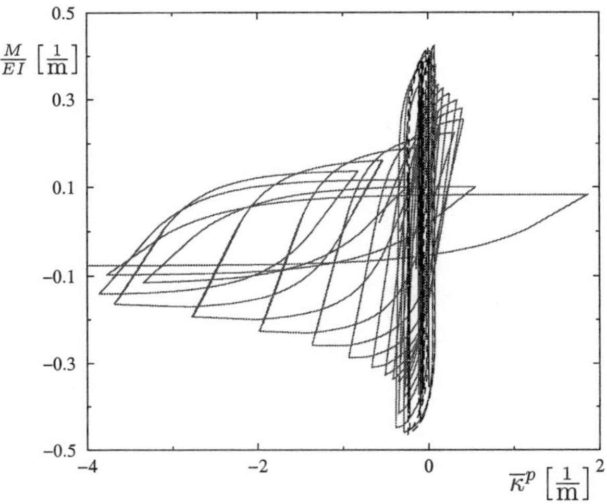

Figure 14. Normalized bending moment at point P, see Figure 7, of driven beam. Comparison of simulation with damage (*solid red*) and without damage (*dotted black*).

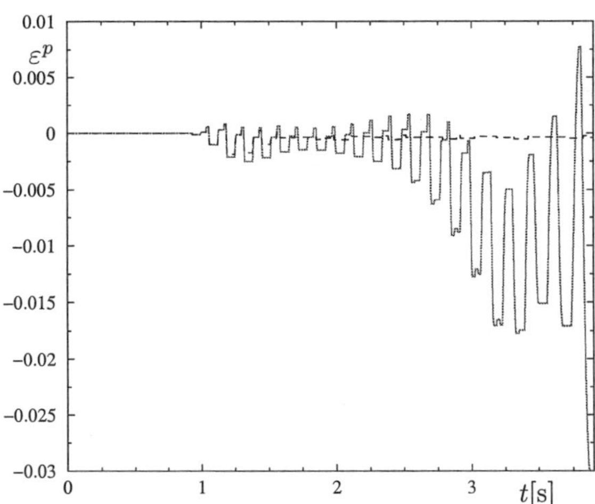

Figure 15. Plastic part of strain at point P, see Figure 7, at $z = H/2$. Comparison of simulation with damage (*solid red*) and without damage (*dotted black*).

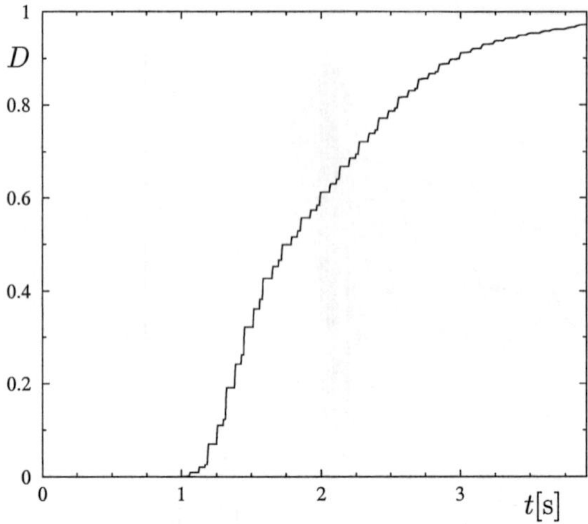

Figure 16. Damage parameter in slider-crank mechanism at point P, $z = H/2$.

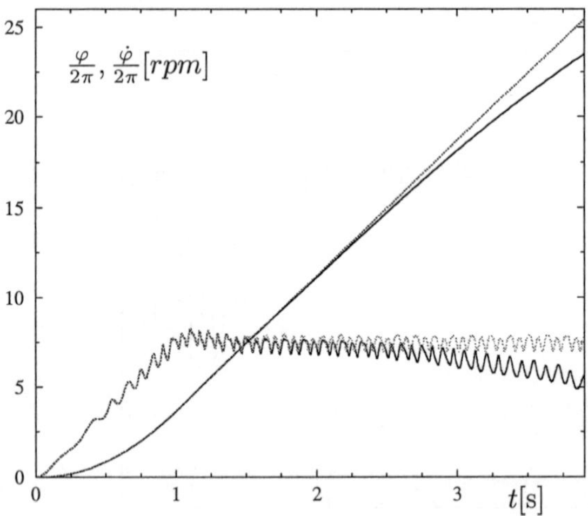

Figure 17. Number of revolutions of the driving beam over time with damage (*solid black*) and without damage (*dotted red*). Comparison of rotational speed with damage (*solid blue*) and without damage (*dotted green*).

Acknowledgments

The author especially wishes to thank Prof. H. Irschik at the University of Linz, Austria, for his valuable comments and discussions.
This work has been supported by the Austrian Science Found - 'Fonds zur Förderung der wissenschaftlichen Forschung' (FWF) - within the subproject F1311 of the Special Research Program SFB F013 'Numerical and Symbolic Scientific Computing'

References

1. *MAPLE 6, Waterloo Maple Inc., 57 Erb Street West, Waterloo, ON, Canada N2L 6C2, Jan 2000. www.maplesoft.com.*

2. C. Adam, *Dynamics of linear elastic Timoshenko beams with eigenstrains*, Asian J. of Struct. Engrg. **2** (1996), no. 1 & 2, 15–32.

3. C. Adam and F. Ziegler, *Dynamic response of elasto-viscoplastic sandwich beams with asymmetrically arranged thick layers*, IUTAM Symposium on Transformation Problems in Composite and Active Materials (Kluwer Academic Publishers, Netherlands, 1998), 221–232.

4. J.A.C. Ambrosio and P.E. Nikravesh, *Elasto-plastic deformations in multibody dynamics*, Nonlinear Dynamics **3** (1992), 85–104.

5. K.J. Bathe, *Large displacement analysis of three dimensional beam structures*, Int. J. Numer. Meth. Engrg. **4** (1979), 961–986.

6. D. Braess, *Finite elements. theory, fast solvers, and applications in solid mechanics*, Cambridge University Press, 1997.

7. H. Bremer, *On the dynamics of elastic multibody systems*, Appl. Mech. Rev. **52** (1999), no. 9, 275–306.

8. H. Bremer and F. Pfeiffer, *Elastische Mehrkörpersysteme*, B. G. Teubner Stuttgart, 1992.

9. K. E. Brenan, S. L. Campbell, and L. R. Petzold, *Numerical solution of initial-value problems in differential-algebraic equations*, SIAM, Philadelphia, 1996.

10. S. C. Chu, K. C. Pan, *Dynamic response of a high-speed slider crank mechanism with an elastic connecting rod*, ASME J. Eng. Industry, no. 92, 542–550. (1975).

11. R.R. Craig, *Structural dynamics, an introduction to computer methods*, John Wiley & Sons, New York, 1981.

12. E. B. Crellin, F. Janssens, D. Poelaert, W. Steiner, and H. Troger, *On balance and variational formulations of the equation of motion of a body deploying along a cable*, J. Appl. Mech. (1997).

13. M.A. Chrisfield, *Non-linear finite element analysis of solids and structures, Volume I and II*, John Wiley & Sons, Chichester, 1991.

14. S. von Dombrowski, *Modellierung von Balken bei großen Verformungen für ein kraftreflektierendes Eingabegerät*, Master's thesis, Universität Stuttgart, Institut für Flugmechanik und Flugregelung, 1997.

15. J. Dongarra, R. Pozo, and D. Walker, *Lapack++: A design overview of object-oriented extensions for high performance linear algebra*, Proceedings of Supercomputing '93, IEEE Computer Society Press, 1993, http: // math.nist.gov/ lapack++, pp. 162–171.

16. E. Eich-Soellner and C. Führer, *Numerical methods in multibody dynamics*, B.G. Teubner Stuttgart, 1998.

17. H.W. Engl, M. Hanke, and A. Neubauer, *Tikhonov regularisation of nonlinear differential-algebraic equations*, Institutsbericht Nr. 385, Institut für Mathematik, 1989.

18. O. Fischer, *Einführung in die Mechanik lebender Mechanismen*, Leipzig, 1906.

19. P. Fotiu, H. Irschik, and F. Ziegler, *Dynamic plasticity: Structural drift and modal projections*, Nonlinear Dynamics in Engineering Systems, Proc. IUTAM-Symp. Stuttgart 1989 (W. Schiehlen, ed.), Springer-Verlag Berlin, 1990, 75–82.

20. G. Frantziskonis and C.S. Desai, *Constitutive model with strain softening*, Int. J. Sol. Struct. **6** (1987), 733–750.

21. G. Frantziskonis and C.S. Desai, *Elastoplastic model with damage for strain softening geomaterials*, Acta Mechanica **68** (1987), 151–170.

22. J. Gerstmayr, *A solution strategy for elasto-plastic multibody systems and related problems*, Ph.D. thesis, Institute of Mechanics and Machine Design, Division of Technical Mechanics, Linz, 2001.

23. J. Gerstmayr, H. J. Holl, and H. Irschik, *Development of plasticity and damage in vibrating structural elements performing guided rigid-body motions*, Arch. Appl. Mech. **71** (2001), 135–145.

24. J. Gerstmayr, H. Irschik, *Vibrations of the elasto-plastic pendulum*, Int. J. Nonlin. Mech., accepted for publication.

25. J. Gerstmayr, H. Irschik, *Dynamic analysis of machine elements exposed to plasticity and damage*, In Proceedings of the STAMM 2000, P. E. O'Donoghue and J. N. Flavin, eds. , 86–92, Elsevier Paris, 2000.

26. A. Griewank, D. Juedes, and J. Srinivasan, *ADOL-C: A package for the automatic differentiation of algorithms written in C/C++*, Preprint MCS-P180-1190, Mathematics and Computer Science Division, Argonne National Laboratory, 1991.

27. E. Hairer, S. P. Nørsett, and G. Wanner, *Solving ordinary differential*

equations I, nonstiff problems, Springer Verlag Berlin Heidelberg, 1987.

28. E. Hairer and G. Wanner, *Solving ordinary differential equations II, stiff and differential-algebraic problems*, Springer Verlag Berlin Heidelberg, 1991.

29. E. J. Haug, D. Negrut, and C. Engstler, *Implicit Runge-Kutta integration of the equations of multibody dynamics in descriptor form*, Mech. Struct. & Mach. **27** (1999), 337–364.

30. E. J. Haug, D. Negrut, and M. Iancu, *A state-space-based implicit integration algorithm for differential-algebraic equations of multibody dynamics*, Mech. Struct. & Mach. **25** (1997), 311–334.

31. R. Hill, *The mathematical theory of plasticity*, Oxford University Press, New York, 1950.

32. H. Irschik, *Biaxial dynamic bending of elastoplastic beams*, Acta Mechanica **62** (1986), 155–167.

33. H. Irschik and W. Brunner, *Vibrations of multi-layered composite elasto-viscoplastic beams using second order theory*, Nonlin. Dyn. **6** (1994), 37–48.

34. H. Irschik, P. Fotiu, and F. Ziegler, *Extension of maysel's formula to the dynamic eigenstrain problem*, J. Mech. Behav. of Mat. **5** (1993), 59–66.

35. H. Irschik and H. J. Holl, *The equations of Lagrange written for a non-material volume*, Accepted for Publication in *Acta Mechanica*.

36. H. Irschik and F. Ziegler, *Dynamics of linear elastic structures with self-stress: A unified treatment for linear and nonlinear problems*, ZAMM, vol. 68, 1988, 199–205.

37. H. Irschik and F. Ziegler, *Dynamic processes in structural thermo-viscoplasticity*, Appl. Mech. Rev. (1995), 301–315.

38. H. Irschik and F. Ziegler, *Thermal shock loading of elastoplastic beams*, Journal of Thermal Stresses, vol. 8, 1985, 53–69.

39. J. G. de Jalón and E. Bayo, *Kinematic and dynamic simulation of multi-body systems: The real time challenge*, Springer, New York, 1994.

40. L.M. Kachanov, *Introduction to continuum damage mechanics*, Martinus Nijhoff Dordrecht, The Netherlands, 1986.

41. A.S. Khan and S. Huang, *Continuum theory of plasticity*, John Wiley & Sons, New York, 1995.

42. J. Lemaitre, *A course on damage mechanics*, 2nd ed., Springer-Verlag Berlin Heidelberg, 1996.

43. A. Lennartsson, *Efficient multibody dynamics*, Ph.D. thesis, Royal Institute of Technology, Department of Mechanics, Stockholm, 1999.

44. Mingrui Li, *The finite deformation theory for beam, plate and shell part I. the two-dimensional beam theory*, Comp. Meth. Appl. Mech. Engrg.

146 (1997), 53–63.

45. J. Lubliner, *Plasticity theory*, Macmillan Publishing Company, New York/London, 1990.

46. P. Nikravesh, I. Chung, and R.L. Benedict, *Plastic hinge approach to vehicle crash simulation*, Computers and Structures **16** (1983), 395–400.

47. M. Otter, M. Hocke, A. Daberkow, and G. Leister, *An object oriented data model for multibody systems*, Advanced Multibody System Dynamics, ed. W. Schiehlen, Kluwer Academic Publishers, Dordrecht, 1993, 19–48.

48. W. Pan and E.J. Haug, *Dynamic simulation of general flexible multibody systems*, Mech. Struct. & Mach. **27** (1999), 217–251.

49. C. Pozrikidis, *Numerical computation in science and engineering*, Oxford University Press, 1998.

50. H. Rubin and K.J. Schneider, *Baustatik. Theorie I. und II. Ordnung*, 3rd ed., Werner-Verlag, 1996.

51. J. Rumbaugh, M. Blaha, W. Premerlani, F. Eddy, and W. Lorensen, *Object-oriented modeling and design*, Prentice Hall, Englewood Cliffs, New Jersey, 1991.

52. W. Schiehlen, *Multibody systems handbook*, Springer-Verlag, Berlin, 1990.

53. W. Schiehlen, *Multibody system dynamics: Roots and perspectives*, Multibody System Dynamics (1997).

54. R. von Schwerin, *Multibody system simulation; numerical methods, algorithms, and software*, Springer Verlag Berlin, 1999.

55. A.A. Shabana, *Dynamics of multibody systems*, John Wiley & Sons, 1989.

56. A.A. Shabana, *Flexible multibody dynamics: Review of past and recent developments*, Multibody System Dynamics **1** (1997), 189–222.

57. A.A. Shabana, *Dynamics of multibody systems*, 2nd ed., Cambridge University Press, 1998.

58. B. Simeon, *Numerische Integration mechanischer Mehrkörpersysteme: Projizierende Deskriptorformen, Algorithmen und Rechenprogramme*, Tech. report, Fortschritt-Berichte VDI, Reihe 20, Nr. 130, Düsseldorf, VDI Verlag, 1994.

59. B. Simeon, *Modelling a flexible slider crank mechanism by a mixed system of DAEs and PDEs*, Math. Modelling of Systems **2** (1996), 1–18.

60. B. Simeon, *DAE's and PDE's in elastic multibody systems*, Numer. Algorithms **19** (1998), 235–246.

61. B. Simeon, *Numerische Simulation gekoppelter Systeme von partiellen und differential-algebraischen Gleichungen in der Mehrkörperdynamik*, Habilitation Thesis, University of Karlsruhe (TH), 1999.

62. J.C. Simo, *A finite strain beam formulation. the three dimensional dynamic problem, Part I*, Comp. Meth. Appl. Mech. Engrg. **58** (1986),

79–116.

63. W.J. Stronge and T.X. Yu, *Dynamic models for structural plasticity*, Springer Verlag London, 1993.

64. K. Washizu, *Variational methods in elasticity and plasticity*, 2nd ed., Pergamon Press, Oxford, 1974.

65. F. Ziegler, *Mechanics of solids and fluids*, Springer-Verlag New York, 1991.

66. F. Ziegler and H. Irschik, *Thermal stress analysis based on maysel's formula*, Thermal Stresses II (London - New York) (R. B. Hetnarski, ed.), Elsevier, 1987, 120–188.

67. T. Zimmermann, Y. Dubois-Pèlerin, and P. Bomme, *Object-oriented finite element programming: Governing principles*, Comp. Meth. Appl. Mech. Engrg. **98** (1992), 291–303.

New Trends in Optimal Structural Control

K.G.Arvanitis[1], E.C.Zacharenakis[2], A.G.Soldatos[3] and G.E.Stavroulakis[4]

[1] *Aristotle University of Thessaloniki, Dept. of Agriculture,P.O.Box 275, 54006 Thessaloniki, Greece (email: karvan@agro.auth.gr)*

[2] *Technological Educational Institute of Crete, Dept. of Civil Engineering, Stavromenos, Heraklion, 71500, Greece (email:zacharen@stef.teiher.gr)*

[3] *National Technical University of Athens, Dept. of Electrical and Computer Engineering, Division of Computer Science, Zographou 15773, Athens, Greece (email: asoldat@cc.ece.ntua.gr)*

[4] *Carolo Wilhelmina Technical University, Dept. of Civil Engineering, Institute of Applied Mechanics, D-38106 Braunschweig, Germany and Univeristy of Ioannina, Greece (email:g.stavroulakis@tu-bs.de).*

1. Introduction

A building during an earthquake, an aircraft flying in a turbulent weather, a slender bridge excited from winds or a slender spray boom attached on a tractor in agricultural applications are all typical examples of dynamically excited structures. The structural analysis of these systems, under the usual linear elasticity assumption, can be performed at various accuracy levels using available analytical or computational dynamics' techniques. Notions line direct integration of the dynamical system, or modal analysis which appear in this context are certainly known. The task of structural analysis is the calculation of the structural vibration (waveforms) and of the corresponding stresses for appropriate, relevant dynamical excitations. During structural design one uses this information in connection with strength criteria (i.e., restrictions on the arising stresses), serviceability criteria (i.e., restrictions on the vibration amplitudes or on the arising velocities and accelerations), fatigue criteria etc. Since the dynamical response of a linear elastic structure depends on the mass and elasticity properties, the classical passive design is based on appropriate choise of the materials and of the dimensions of the structure. Usually, a conservative design is based on the peak values of the corresponding quantities and leads, unavoidably, to certain waste of resourses.

In order to avoid this waste of resourses one may apply active or passive control systems which modify internal stresses or the mass or the damping of the dynamical system and keep the response of the resulting 'intelligent' structure within the required limits. The resulting system is a smart, intelligent or adaptive structure. The science which deals with these systems is known as adaptronics, mechatronics or structronics. All these names can be found today in the currivula of several Universities and attract the interest of industrial users. Nevertheless, the basic ideas of controlled structures are quite old. The fact that electronic components and devises for the structural monitoring, and the realisation of the control are becoming cheaper, allows for the application of control concepts in an increasing number of products and structures.

The potential of active and passive control to provide us with more elegant dynamical structures is well known in the mechanical engineering community. Applications in civil engineering are rather limited. They concern, mainly, vibration reduction for wind loadings in slender structures (bridges and high-raise buldings). The number of applications in the much more interesting area of earthquake aseismic design is not very large. One reason is certainly that the control power which is required in this case is higher than the one used for the reduction of wind-induced vibrations. Further reasons are the inherent uncertainties and the required robustness and maybe nonlinearity of the corresponding control systems. In fact, early attempts to use classical linear-quadratic control schemes in civil engineering were not very successful. Nevertheless, several decades before, the required knowledge was restricted to the community of control engineers and the required software was not user-friendly. The picture has radically changed. Today software packages (like SIMULINK/MATLAB) allow for the consideration of quite complicated control schemes for the design of structures.

This chapter tries to introduce some recent models of optimal structural control for the design of civil and mechanical engineering structures. Notions like robust design, nonlinear and H-infinity controllers and multirate control schemes, which are of importance for the implementation with electronic circuits, are discussed. The methods are compared with more classical optimal control tools for the preliminary aseismic design of a shear-type frame and of vibration suppression of a spraying boom in agricultural applications. Relevant references are given in the text, where the interested reader may find more mathematical details (including proofs of the methods and algorithms used here) as well as additional material. Neither the list of references nor the theories and applications discussed in this chapter are complete. The purpose of this text is to motivate the interested engineer to use structural control concepts and to show him that the available tools are powerfull and efficient for the everyday design.

2. Some Basic Notions of Dynamical Systems

Throughout this work, our interest is focused on linear time-invariant finite dimensional dynamical system having the standard state space form

$$\dot{x}(t) = Ax(t) + B\zeta(t) + Dq(t) \quad \text{(2.1a)}$$

or

$$\dot{x}(t) = Ax(t) + B\zeta(t) + w(t) \quad \text{(2.1b)}$$

$$y_m(t) = Cx(t) + J_1\zeta(t) \quad \text{(2.1c)}$$

$$y_c(t) = Ex(t) + J_2\zeta(t) \quad \text{(2.1d)}$$

where, $x(t) \in R^n$ is the state vector, $\zeta(t) \in R^m$ is the input or control vector, $q(t) \in R^d$ is the external loading vector, $w(t) \in R^n$ is the external disturbance vector, $y_m(t) \in R^{p_1}$ is the vector of measurement (observed) outputs, $y_c(t) \in R^{p_2}$ is the controlled output vector, and where all the matrices have real entries and appropriate dimensions.

For a continuous-time, linear time-invariant system, with a state space model of the general form

$$\dot{x}(t) = Ax(t) + B\xi(t) \ , \ y(t) = Cx(t) + J\zeta(t) \quad \text{(2.2)}$$

its frequency domain description is given by its matrix transfer function, from generalized inputs $\xi(t) \in R^m$ (controls and/or disturbances) to generalized outputs $y(t) \in R^p$ (measured and/or controlled outputs), which has the form

$$H_{\xi y}(s) \triangleq J + C(sI_n - A)^{-1} B$$

System (2.2) is said to be *left invertible* if and only if $q_r = m$, where r is the smaller integer for which, $q_{r+i} = q_r, \ \forall i \geq 1$, and where, in general,

$$q_j = \text{rank}(N_j) - \text{rank}(N_{j-1})$$

while the sequence of matrices N_j is defined as

$$N_0 = J \ , \ N1 = CB \ , \ N_j = \begin{bmatrix} N_0 & \\ N_1 & \\ CAB & \mathbf{0} \\ \vdots & N_{j-1} \\ CA^{j-1}B & \end{bmatrix}, j \geq 2$$

System (2.2) is said to be *internally stable* if all the eigenvalues λ_i, i=1,...,n of matrix $\mathbf{A}$ are located in the open left complex half-plane, $\mathbf{C}^-$, i.e. if $Re(\lambda_i) < 0, \forall i = 1, 2, \cdots, n$. We then say that $\mathbf{A}$ is an *asymptotically stable* matrix.

System (2.2) or the pair $(\mathbf{A}, \mathbf{B})$ is called *controllable*, if and only if [1]-[3]

$$\text{rank} S = n \; , \; S = \begin{bmatrix} \mathbf{B} & \mathbf{AB} & \cdots & \mathbf{A}^{n-1}\mathbf{B} \end{bmatrix}$$

System (2.2) or the pair $(\mathbf{C}, \mathbf{A})$ is called *observable*, if and only if [1]-[3]

$$\text{rank} R = n \; , \; S = \begin{bmatrix} \mathbf{C}^T & \mathbf{A}^T\mathbf{C}^T & \cdots & \left(\mathbf{A}^T\right)^{n-1}\mathbf{C}^T \end{bmatrix}$$

System (2.2) or the triplet $(\mathbf{A}, \mathbf{B}, \mathbf{C})$ is called *minimal,* if $(\mathbf{A}, \mathbf{B})$ is controllable and $(\mathbf{C}, \mathbf{A})$ is observable.

Suppose now that a given matrix pair $(\mathbf{A}, \mathbf{B})$ is not controllable (i.e. $\text{rank} S = r < n$). Then, there exist a nonsingular transformation matrix $\mathbf{T} \in \mathbf{R}^{n \times n}$, such that

$$\widetilde{\mathbf{A}} \triangleq \mathbf{T}^{-1}\mathbf{AT} \equiv \begin{bmatrix} \widetilde{\mathbf{A}}_1 & \widetilde{\mathbf{A}}_2 \\ 0 & \widetilde{\mathbf{A}}_3 \end{bmatrix} , \; \widetilde{\mathbf{B}} \triangleq \mathbf{T}^{-1}\mathbf{B} \equiv \begin{bmatrix} \widetilde{\mathbf{B}}_1 \\ 0 \end{bmatrix}$$

where $\widetilde{\mathbf{A}}_1 \in \mathbf{R}^{r \times r}$ and $\widetilde{\mathbf{B}}_1 \in \mathbf{R}^{r \times m}$ are such that $\left(\widetilde{\mathbf{A}}_1, \widetilde{\mathbf{B}}_1\right)$ is controllable. Then, the eigenvalues of $\widetilde{\mathbf{A}}_1$ is called the set of *controllable modes* of $(\mathbf{A}, \mathbf{B})$ and the eigenvalues of $\widetilde{\mathbf{A}}_3$ is called the set of *uncontrollable modes* of $(\mathbf{A}, \mathbf{B})$ [3]. By duality, if a given matrix pair $(\mathbf{C}, \mathbf{A})$ is not observable (i.e. $\text{rank} R = q < n$), then, there exist a nonsingular transformation matrix $\mathbf{U} \in \mathbf{R}^{n \times n}$, such that

$$\overline{\mathbf{A}} \triangleq \mathbf{U}^{-1}\mathbf{AU} \equiv \begin{bmatrix} \overline{\mathbf{A}}_1 & 0 \\ \overline{\mathbf{A}}_2 & \overline{\mathbf{A}}_3 \end{bmatrix} , \; \overline{\mathbf{C}} \triangleq \mathbf{CU} \equiv \begin{bmatrix} \overline{\mathbf{C}}_1 & 0 \end{bmatrix}$$

Then, the eigenvalues of $\overline{\mathbf{A}}_1$ is called the set of *observable modes* of $(\mathbf{C}, \mathbf{A})$ while the eigenvalues of $\overline{\mathbf{A}}_3$ is called the set of *unobservable modes* of $(\mathbf{C}, \mathbf{A})$ [3].

The set of the uncontrollable modes of $(\mathbf{A}, \mathbf{B})$ and the unobservable modes of $(\mathbf{C}, \mathbf{A})$ is called the set of *invariant zeros* of system $(\mathbf{A}, \mathbf{B}, \mathbf{C}, \mathbf{J})$ [4].

System (2.2) or the matrix pair $(\mathbf{A}, \mathbf{B})$ is called *stabilizable,* if there is a constant matrix $\mathbf{F} \in \mathbf{R}^{m \times n}$ such that $\mathbf{A}_{cl} = \mathbf{A} + \mathbf{BF}$ is stable. Of course, $(\mathbf{A}, \mathbf{B})$ is stabilizable if it is controllable [1], [2].

System (2.2) or the matrix pair $(\mathbf{C}, \mathbf{A})$ is called *detectable,* if there is a cons-tant matrix $\mathbf{K} \in \mathbf{R}^{m \times n}$ such that matrix $\mathbf{A}^T + \mathbf{C}^T\mathbf{F}$ is stable. Of course, $(\mathbf{C}, \mathbf{A})$ is detectable if it is observable [1], [2].

The duality between controllability (resp. stabilizability) and observability (resp. detectability) is obvious.

A collection of m integers $\{\delta_1, \delta_2, \cdots, \delta_m\}$ is called a *controllability index vector* (or a *set of locally minimum controllability indices*) of the pair $(\mathbf{A}, \mathbf{B})$, if the following relationships simultaneously hold

$$\sum_{i=1}^{m} \delta_i = n, \, rank[\mathbf{b}_1 \quad \cdots \quad \mathbf{A}^{\delta_1 - 1}\mathbf{b}_1 \quad \cdots \quad \mathbf{b}_m \quad \cdots \quad \mathbf{A}^{\delta_m - 1}\mathbf{b}_m] = n$$

where $\mathbf{b}_i, i = 1, 2, \cdots, m$ is the ith column of matrix $\mathbf{B}$.

A collection of p integers $\{\beta_1, \beta_2, \cdots, \beta_p\}$ is called an *observability index vector* (or a *set of locally minimum observability indices*) of the pair $(\mathbf{C}, \mathbf{A})$, if the following relationships simultaneously hold

$$\sum_{i=1}^{p} \beta_i = n, \, rank\left[\mathbf{c}_1^T \quad \cdots \quad \left(\mathbf{A}^T\right)^{\beta_1 - 1}\mathbf{c}_1^T \quad \cdots \quad \mathbf{c}_p^T \quad \cdots \quad \left(\mathbf{A}^T\right)^{\beta_p - 1}\mathbf{c}_p^T\right] = n$$

where $\mathbf{c}_i^T, i = 1, 2, \cdots, p$ is the ith column of matrix $\mathbf{C}^T$.

System (2.2) is said to be *strongly connected* if

$$\left(\mathbf{H}_{\xi y}\right)_{ij}(s) = (\mathbf{J})_{ij} + \mathbf{c}_i\left(s\mathbf{I}_n - \mathbf{A}\right)^{-1}\mathbf{b}_j \neq 0$$

where $\left(\mathbf{H}_{\xi y}\right)_{ij}(s)$ and $(\mathbf{J})_{ij}$ denote the i-j elements of the matrix transfer function $\mathbf{H}_{\xi y}(s)$ and of the matrix $\mathbf{J}$, respectively and $\mathbf{c}_i$ is the ith row of matrix $\mathbf{C}$.

An eigenvalue λ_i of system (2.2) is said to be a *fixed mode* with respect to a certain controller if its location in the s-plane cannot be influenced by this controller.

Consider now connecting, to each of the inputs of system (2.2), an ideal sampler with sampling period $T_0 \in \mathbf{R}^+$ and a so-called zero order hold circuit (i.e. a D/A converter), such that

$$\xi(t) = \xi(kT_0), \, \forall t \in [kT_0, (k+1)T_0), k = 0, 1, 2, \cdots$$

where, in general, $\phi(kT_0)$, is the value of $\phi(t)$ at t=kT0, $k = 0, 1, 2, \cdots$, obtained by sampling $\phi(t)$ with period T_0, while detecting each system output at every kT_0. In this case, it is not difficult to show that (see [5] fore details)

$$\mathbf{x}[(k+1)T_0] = \Phi\mathbf{x}(kT_0) + \hat{\mathbf{B}}\xi(kT_0) \, , \, \mathbf{y}(kT_0) = \mathbf{C}\mathbf{x}(kT_0) + \mathbf{J}\xi(kT_0), k \geq 0 \quad (2.3)$$

where

$$\Phi = \exp(\mathbf{A}T_0) \ , \ \ \hat{\mathbf{B}} = \int_0^{T_0} \exp(\mathbf{A}\lambda)\mathbf{B}d\lambda \qquad (2.4)$$

System (2.3) is called the *step invariant transformation* of system (2.2), or the sampled-data system corresponding to (2.2). Obviously, the step-invariant transformation maps the state matrices as follows

$$(\mathbf{A}, \mathbf{B}, \mathbf{C}, \mathbf{J}) \rightarrow (\Phi, \hat{\mathbf{B}}, \mathbf{C}, \mathbf{J})$$

The discrete transfer function matrix of system (2.3) is given by

$$\mathbf{T}_{\xi y}(z) \triangleq \mathbf{J} + \mathbf{C}(z\mathbf{I}_n - \Phi)^{-1}\hat{\mathbf{B}}$$

where the indeterminate z refers to the usual $\mathbf{Z}$-transform [5].

System (2.3) is said to be internally stable if $\rho(\Phi) < 1$, where $\rho(\Phi) = v_{\max}(\Phi)$ denotes the *spectral radius* of matrix Φ and $v_{\max}(\Phi)$ is the largest absolute eigenvalue of Φ. We then say that Φ is asymptotically stable.

The sampling period T_0 is called *pathological* (relative to $\mathbf{A}$ in (2.2)) if $\mathbf{A}$ has two eigenvalues with equal real parts and imaginary parts that differ by an integral multiple of $\dfrac{2\pi}{T_0}$. Otherwise, the sampling period is called non-pathological. If the sampling period is non-pathological, then controllability of $(\mathbf{A}, \mathbf{B})$ implies controllability of $(\Phi, \hat{\mathbf{B}})$ and observability of $(\mathbf{C}, \mathbf{A})$ implies observability of $(\mathbf{C}, \Phi)$. Throughout this paper, we assume that sampling periods are non-pathological.

A primary objective of feedback is to reject disturbances (which always act to dynamical systems) to certain amount and to provide good tracking facilities. Additional requirements for an efficient control system are the properties of stability (rapidly decaying modes), noise filtering (suppression of sensor noise) and moderate effort or actuator signal magnitude. We address most of these objectives in subsequent Sections.

3. Disturbance Rejection Using State Feedback

When designing a control system, one often begins with a plant, which is subject to external disturbances. Disturbances have a detrimental effect on the system performance. Therefore, it is a reasonable and common design objective, at least to try, to reduce the effect of these disturbances to an acceptable level. In the present Section, our purpose is to review a method for complete rejection of external disturbances, and for the solution of the so-called *state feedback disturbance rejection problem*.

Consider that we are given a system of the form (2.1a), (2.1c), (2.1d) with

$y_m(t) \equiv y_c(t)$ and $J_1 = J_2 = 0$. Assume that the above system is left invertible and consider applying to it the state feedback control law of the form

$$\zeta(t) = Fx(t) + G\omega(t) \ , \ \det G \neq 0 \qquad (3.1)$$

where, $\omega(t) \in \mathbf{R}^m$ is a new input vector.

The disturbance rejection problem (also called, disturbance decoupling) considered in this Section, is stated as follows: Find, if possible, matrices $\mathbf{F}$ and $\mathbf{G}$ (with $\det \mathbf{G} \neq 0$) such that the following relation holds

$$\left[C(sI_n - A_{cl})^{-1} B_{cl} \quad C(sI_n - A_{cl})^{-1} D \right] = \left[H(s) \quad 0_{p_1 \times d} \right] \quad (3.2)$$

where $\mathbf{A}_{cl} = \mathbf{A} + \mathbf{BF}$ and $\mathbf{B}_{cl} = \mathbf{BG}$. Obviously, since

$$Y_m(s) = C(sI_n - A_{cl})^{-1} B_{cl} \Omega(s) + C(sI_n - A_{cl}F)^{-1} DQ(s)$$

where, $Y_m(s), \Omega(s), Q(s)$, are the Laplace transforms of $y_m(t), \omega(t), q(t)$, respectively, relation (3.2) means that the impact of the disturbances on the system outputs is eliminated. Next, define

$$V = \begin{bmatrix} B & D \end{bmatrix} \ , \ E_j = \begin{bmatrix} H_j & 0 \end{bmatrix} \ , \ H_j = CA_{cl}^{j-1} B_{cl}$$

$$\widetilde{G} = \begin{bmatrix} \hat{G} & 0 \\ 0 & I \end{bmatrix} \equiv \begin{bmatrix} G^{-1} & 0 \\ 0 & I \end{bmatrix} \ , \ \widetilde{L} = \begin{bmatrix} L \\ 0 \end{bmatrix} \ , \ L = \hat{G}F$$

With these definitions, one can easily put (3.2) in the form of the following set of equations

$$CV = E_1 \widetilde{G} \qquad (3.3a)$$

$$CA^{j-1}V = E_j \widetilde{G} - \sum_{\rho=1}^{j-1} E_{j-\rho} \widetilde{L} A^{\rho-1} V \ , \ \forall j \geq 2 \qquad (3.3b)$$

We next manipulate (3.3) on the basis of a time-invariant version of the Periodic Structure Algorithm [6]. To this end, let $\beta = \min\{j : CA^j V \neq 0\}$ and

$$\Gamma_0 = CA^\beta \qquad (3.4)$$

Using the above definitions, relations (3.3a) and (3.3b) reduce to

$$\Gamma_0 V = \begin{bmatrix} H_{\beta+1} \hat{G} & 0 \end{bmatrix} \qquad (3.5a)$$

$$\Gamma_0 A^{j-1} V = \begin{bmatrix} H_{\beta+j} \hat{G} & 0 \end{bmatrix} - \sum_{\rho=1}^{j-1} H_{\beta+j-\rho} L A^{\rho-1} V \qquad (3.5b)$$

$\forall j \geq 2$. Now, let S_i be invertible matrices such that

$$S_i \Gamma_i V = \begin{bmatrix} \hat{\Gamma}_i \\ \tilde{\Gamma}_i \end{bmatrix} V \equiv \begin{bmatrix} \hat{\Gamma}_i V \\ 0 \end{bmatrix} \ , \quad \forall i \geq 0 \qquad (3.6)$$

where

$$\Gamma_i = \begin{bmatrix} \hat{\Gamma}_{i-1} \\ \tilde{\Gamma}_{i-1} A \end{bmatrix} \qquad (3.7)$$

with Γ_0 defined as in (3.4). Now, pre-multiplying (3.5a) and (3.5b) by S_0, we obtain the following two systems of equations

$$\hat{\Gamma}_0 V = \begin{bmatrix} \hat{H}_{\beta+1,0} \hat{G} & 0 \end{bmatrix} \qquad (3.8a)$$

$$\hat{\Gamma}_0 A^{j-1} V = \begin{bmatrix} \hat{H}_{\beta+j,0} \hat{G} & 0 \end{bmatrix} - \sum_{\rho=1}^{j-1} \hat{H}_{\beta+j-\rho,0} L A^{\rho-1} V \qquad (3.8b)$$

$\forall j \geq 2$, and

$$\tilde{\Gamma}_0 A V = \begin{bmatrix} \tilde{H}_{\beta+2,0} \hat{G} & 0 \end{bmatrix} \qquad (3.9a)$$

$$\tilde{\Gamma}_0 A^{j-1} V = \begin{bmatrix} \tilde{H}_{\beta+j,0} \hat{G} & 0 \end{bmatrix} - \sum_{\rho=1}^{j-1} \tilde{H}_{\beta+j-\rho,0} L A^{\rho-1} V \qquad (3.9b)$$

$\forall j \geq 3$, where $\hat{H}_{\beta+j,0}$, $\forall j \geq 1$, is the $q_0 \times m$ matrix containing the first q_0 rows of the matrix $S_0 H_{\beta+j}$, whereas $\tilde{H}_{\beta+j}$, $\forall j \geq 1$ is the $(p - q_0) \times m$ matrix containing the last $p - q_0$ rows of the matrix $S_0 H_{\beta+j}$, where $q_0 = \text{rank}\{\Gamma_0 V\}$.

Grouping appropriately equations (3.8), (3.9a), (3.9b), we arrive at the following system of equations

$$\Gamma_1 V = \begin{bmatrix} H_{\beta+1,1} \hat{G} & 0 \end{bmatrix} \qquad (3.10a)$$

$$\Gamma_1 A^{j-1} V = \begin{bmatrix} H_{\beta+j,1} \hat{G} & 0 \end{bmatrix} - \sum_{\rho=1}^{j-1} H_{\beta+j-\rho,1} L A^{\rho-1} V \qquad (3.10b)$$

$\forall j \geq 2$, where matrix Γ_1 is defined by (3.7), for i=1 and matrices $H_{\beta+j,1}$, $\forall j \geq 1$ is defined by

$$H_{\beta+j,i} = \begin{bmatrix} \hat{H}_{\beta+j,i-1} \\ \tilde{H}_{\beta+j+1,i-1} \end{bmatrix} \qquad (3.11)$$

for i=1. The rows of the block matrices $\hat{\mathbf{H}}_{\beta+j,i-1}$ and $\widetilde{\mathbf{H}}_{\beta+j+1,i-1}$ consist of the first q_i and the last $p-q_i$ rows of the matrices $\mathbf{S}_i\mathbf{H}_{\beta+j,i-1}$ and $\mathbf{S}_i\mathbf{H}_{\beta+j+1,i-1}$, respectively. Pre-multiplying (3.10a) and (3.10b) with $\mathbf{S}_1$, as defined by (3.6), and following the same procedure as above, we obtain

$$\Gamma_2\mathbf{V} = \begin{bmatrix} \mathbf{H}_{\beta+1,2}\hat{\mathbf{G}} & \mathbf{0} \end{bmatrix}$$

$$\Gamma_2\mathbf{A}^{j-1}\mathbf{V} = \begin{bmatrix} \mathbf{H}_{\beta+j,2}\hat{\mathbf{G}} & \mathbf{0} \end{bmatrix} - \sum_{\rho=1}^{j-1}\mathbf{H}_{\beta+j-\rho,2}\mathbf{L}\mathbf{A}^{\rho-1}\mathbf{V}$$

$\forall j\geq 2$, where matrices $\mathbf{H}_{\beta+j,2}$, $\forall j\geq 1$, is the expression (3.11) for i=2.

In the ith step of the proposed algorithm, we obtain the following system of equations

$$\Gamma_i\mathbf{V} = \begin{bmatrix} \mathbf{H}_{\beta+1,i}\hat{\mathbf{G}} & \mathbf{0} \end{bmatrix}$$

$$\Gamma_i\mathbf{A}^{j-1}\mathbf{V} = \begin{bmatrix} \mathbf{H}_{\beta+j,i}\hat{\mathbf{G}} & \mathbf{0} \end{bmatrix} - \sum_{\rho=1}^{j-1}\mathbf{H}_{\beta+j-\rho,i}\mathbf{L}\mathbf{A}^{\rho-1}\mathbf{V}$$

$\forall j\geq 2$, where matrices $\mathbf{H}_{\beta+j,i}$, $\forall j\geq 1$ are given by (3.11).

Observe now that $\widetilde{\Gamma}_r \equiv \mathbf{0}$. This is due to the construction of matrices $\mathbf{S}_i$ and Γ_i. Consequently, we obtain the following system of equations

$$\hat{\Gamma}_r\mathbf{V} = \begin{bmatrix} \hat{\mathbf{H}}_{\beta+1,r}\hat{\mathbf{G}} & \mathbf{0} \end{bmatrix}$$

$$\hat{\Gamma}_r\mathbf{A}^{j-1}\mathbf{V} = \begin{bmatrix} \hat{\mathbf{H}}_{\beta+j,r}\hat{\mathbf{G}} & \mathbf{0} \end{bmatrix} - \sum_{\rho=1}^{j-1}\hat{\mathbf{H}}_{\beta+j-\rho,r}\mathbf{L}\mathbf{A}^{\rho-1}\mathbf{V}$$

$\forall j\geq 2$, which, according to the results reported in [7], is equivalent to the following system of equations

$$\mathbf{P}_0\hat{\Gamma}_r\mathbf{V} = \begin{bmatrix} \mathbf{I} & \mathbf{0} \end{bmatrix} \quad (3.12a)$$

$$\sum_{i=0}^{k}\mathbf{P}_i\hat{\Gamma}_r\mathbf{A}^{k-i}\mathbf{V} + \mathbf{F}\mathbf{A}^{k-1}\mathbf{V} = 0 \quad (3.12b)$$

where,

$$\mathbf{P}_0 = \mathbf{G}\mathbf{Z}_{\beta+1,r} \ , \ \ \mathbf{Z}_{\beta+1,r} = \hat{\mathbf{H}}_{\beta+1,r}^{-1} \ , \ \ \mathbf{P}_j = -\sum_{\rho=0}^{j-1}\mathbf{P}_\rho\hat{\mathbf{H}}_{\beta+j-\rho+1}\mathbf{Z}_{\beta+1,r} \ , \ \ \forall j\geq 1 \quad (3.13)$$

It is now clear that, on the basis of the proposed time-invariant version of the Periodic Structure Algorithm, the disturbance rejection problem has been reduced to the problem of solving the system of equations (3.12), with respect to P_i, F. Note that, matrix G is not considered as unknown, since, on the basis of (3.13), it follows that G may be any arbitrary non-singular matrix.

Now, define $s_r = \text{rank}(T_r) - \text{rank}(T_{r-1})$, where, in general,

$$T_1 = CV \;\; , \;\; T_j = \begin{bmatrix} T_1 & \\ CAV & \mathbf{0} \\ \vdots & T_{j-1} \\ CA^{j-1}V & \end{bmatrix}$$

The necessary and sufficient solvability conditions for disturbance rejection problem using state feedback are [7]

$$\text{(i)} \; s_r = m \qquad \text{(3.14a)}$$

$$\text{(ii)} \; \hat{\Gamma}_r D = 0 \qquad \text{(3.14b)}$$

Moreover, a parameterization of all admissible controllers is given by [7]

$$F = -\left(\hat{\Gamma}_r B\right)^{-1} \hat{\Gamma}_r A + \hat{R}\Xi$$

where $\hat{R}$ is an $m \times (n - \lambda)$ arbitrary matrix, Ξ is a basis for Ker Π, $\lambda = \text{rank}\Pi$ and

$$\Pi = \begin{bmatrix} D & A_s D & \cdots & A_s^{n-1} D \end{bmatrix} \;\; , \;\; A_s = A - B\left(\hat{\Gamma}_r B\right)^{-1} \hat{\Gamma}_r A$$

Although, the disturbance rejection problem can be solved using state feedback, under conditions (3.14a), (3.14b), it is of paramount importance to know when, either stability or the ability of freely assigning the closed-loop system eigenvalues is guaranteed, additionally to disturbance rejection. According to the results in [7], the necessary conditions for simultaneous disturbance rejection and internal stability of the closed-loop system, under a state feedback law of the form (3.1), are:

(i)   Conditions (3.14a) and (3.14b) are satisfied.

(ii)   The state-space system

$$\dot{x}(t) = A_s x(t) + B\tilde{u}(t) \;\; , \;\; \tilde{y}(t) = \Xi x(t) \qquad \text{(3.15)}$$

       is strongly connected.

(iii) The set of the fixed modes of system (3.15), under the output feedback law of the form

$$\tilde{u}(t) = \hat{R}\tilde{y}(t) \qquad \text{(3.16)}$$

denoted by $\mathbf{S}_P$, fulfils $\mathbf{S}_P \subset \mathbf{C}^-$.

The sufficient conditions for system (2.1a), (2.1c) (with $\mathbf{J}_1 = \mathbf{0}$) to be simultaneously disturbance decoupled and stabilizable by a state feedback of the form (3.1), are:

(i)   Conditions (3.14a) and (3.14b) are satisfied.

(ii)   System (3.15) is strongly connected.

(iii) $\mathbf{S}_P \subset \mathbf{C}^-$

(iv) $\lambda + 1 \le m + \gamma$, where γ is the number of the fixed modes of system (3.15) under the feedback law (3.16) (multiplicities included).

The necessary conditions for simultaneous disturbance rejection and pole placement in the closed-loop system, under a state feedback law of the form (3.1), are:

(i)   Conditions (3.14a) and (3.14b) are satisfied.

(ii)   System (3.15) is strongly connected.

(iii) $\mathbf{S}_P = \varnothing$

Finally, the sufficient conditions for simultaneous disturbance rejection and pole placement in the closed-loop system, under a state feedback law of the form (3.1), are:

(i)   Conditions (3.14a) and (3.14b) are satisfied.

(ii)   System (3.15) is strongly connected.

(iii) $\mathbf{S}_P = \varnothing$

(iv) $\lambda + 1 \le m$

The results of this Section provide a formal framework for the facing of the effect of external disturbances in linear systems. However, for some systems, it may be impossible (cf. conditions (3.14a), (3.14b)) to reduce the effect of the disturbances below a certain threshold value. The disturbance rejection problem would have no solution in this case. We are, therefor, forced to find other ways, in order to design stabilizing controllers, which will reduce the effect of the disturbances to a prespecified level. Such a way, is the LQ (LQG) regulation approach, which is reviewed in the following Section.

4. Linear Quadratic Control and Optimal Noise Rejection

The linear quadratic (LQ) optimization is a fundamental part of control theory. Combining elegant mathematical methods and application considerations, the LQ-based design offers clear-cut solutions to MIMO control problems.

The central problem in LQ optimization is related to the determination of a feedback control strategy, for a linear system influenced by external disturbances, which, after its application, reduces the impact of the disturbances on the system states (or outputs), to a certain amount, described by a quadratic index of performance. In this section, the basic theory and design methods of LQ control, as well as some of its important variations, will be presented.

4.1. Continuous-Time Linear Quadratic Regulation (LQR)

In mathematical terms, the problem addressed by the linear quadratic regulation (LQR) theory is the following: Suppose that we have a plant model in the state space form (2.1a), (2.1c), (2.1d) with $\mathbf{y}_m(t) \equiv \mathbf{y}_c(t)$ and $\mathbf{J}_1 = \mathbf{J}_2 = \mathbf{0}$. Then, find optimal control forces $\zeta(t)$ such that to minimize the following quadratic performance index (also called, the cost functional)

$$J = \frac{1}{2} \int_0^{t_f} \left(\mathbf{x}^T(t)\mathbf{Q}\mathbf{x}(t) + \zeta^T(t)\mathbf{R}\zeta(t) \right) dt \qquad (4.1)$$

with $\mathbf{Q} = \mathbf{Q}^T \geq \mathbf{0}$ and $\mathbf{R} = \mathbf{R}^T > \mathbf{0}$. Note that, this problem is usually called the *finite horizon LQ state regulation problem*.

Using optimality conditions for the previously formulated optimization problem (see, e.g. [8], [9]), one gets the following optimal control law

$$\zeta(t) = \mathbf{F}(t)\mathbf{x}(t) = -\mathbf{R}^{-1}\mathbf{B}^T\mathbf{P}(t)\mathbf{x}(t) \qquad (4.2)$$

where $\mathbf{P}(t)$ is the solution of the following differential matrix Riccati equation (DMRE)

$$\dot{\mathbf{P}}(t) + \mathbf{P}(t)\mathbf{A} + \mathbf{A}^T\mathbf{P}(t) - \mathbf{P}(t)\mathbf{B}\mathbf{R}^{-1}\mathbf{B}^T\mathbf{P}(t) + \mathbf{Q} = \mathbf{0} \qquad (4.3)$$

with the terminal condition $\mathbf{P}(t_f) = \mathbf{0}$.

The solution $\mathbf{P}(t)$ of (4.3) reaches steady-state conditions far enough away from t_f. When $t_f \to \infty$ (in which case, the problem is called *the infinite horizon LQR problem*) the time-dependent solution of (4.3) becomes time-invariant. That is $\mathbf{P}(t) = \mathbf{P}$ and $\dot{\mathbf{P}}(t) = \mathbf{0}$. In this case, one gets the control law

$$\zeta(t) = \mathbf{F}\mathbf{x}(t) = -\mathbf{R}^{-1}\mathbf{B}^T\mathbf{P}\mathbf{x}(t)$$

where $\mathbf{P}$ can be found as the positive definite solution of the algebraic Riccati equation (ARE)

$$\mathbf{P}\mathbf{A} + \mathbf{A}^T\mathbf{P} - \mathbf{P}\mathbf{B}\mathbf{R}^{-1}\mathbf{B}^T\mathbf{P} + \mathbf{Q} = \mathbf{0} \qquad (4.4)$$

Note that, equation (4.4) has a positive definite solution and the closed-loop system with system matrix $\mathbf{A}_{cl} = \mathbf{A} - \mathbf{B}\mathbf{R}^{-1}\mathbf{B}^T\mathbf{P}$ is asymptotically stable if $(\mathbf{A}, \mathbf{B})$ is stabilizable and $(\mathbf{W}_1, \mathbf{A})$ is detectable, where $\mathbf{W}_1$ is defined as $\mathbf{W}_1\mathbf{W}_1^T = \mathbf{Q}$. A way to find the solution $\mathbf{P}$ of (4.4) is the following: Form the matrix

$$\mathbf{H}_1 = \begin{bmatrix} \mathbf{A} & -\mathbf{B}\mathbf{R}^{-1}\mathbf{B}^T \\ \mathbf{Q} & -\mathbf{A}^T \end{bmatrix}$$

Since matrix $\mathbf{H}$ has the Hamiltonian property $\mathbf{SH} = (\mathbf{SH})^T$, where $\mathbf{S} = \begin{bmatrix} \mathbf{0} & -\mathbf{I}_n \\ \mathbf{I}_n & \mathbf{0} \end{bmatrix}$, there exists an eigen-vector matrix $\mathbf{X} = \begin{bmatrix} \mathbf{X}_{11} & \mathbf{X}_{12} \\ \mathbf{X}_{21} & \mathbf{X}_{22} \end{bmatrix}$ satisfying

$$\mathbf{H}_1 \begin{bmatrix} \mathbf{X}_{11} & \mathbf{X}_{12} \\ \mathbf{X}_{21} & \mathbf{X}_{22} \end{bmatrix} = \begin{bmatrix} \mathbf{X}_{11} & \mathbf{X}_{12} \\ \mathbf{X}_{21} & \mathbf{X}_{22} \end{bmatrix} \begin{bmatrix} \Lambda & \mathbf{0} \\ \mathbf{0} & -\Lambda \end{bmatrix}$$

where $\Lambda = \mathop{\mathrm{diag}}\limits_{i=1,\cdots,n} \{\lambda_i\}$ where λ_i are the eigenvalues of $\mathbf{H}1$ satisfying $\mathrm{Re}(\lambda_i) \leq 0, \forall i = 1, \cdots, n$. Then,

$$\mathbf{P} = \mathbf{X}_{21}\mathbf{X}_{11}^{-1}$$

In the case where, the cost functional (4.1) is of the form

$$J = \frac{1}{2} \int_0^\infty \left\{ \begin{bmatrix} \mathbf{x}(t) \\ \zeta(t) \end{bmatrix}^T \begin{bmatrix} \mathbf{Q} & \mathbf{N} \\ \mathbf{N}^T & \mathbf{R} \end{bmatrix} \begin{bmatrix} \mathbf{x}(t) \\ \zeta(t) \end{bmatrix} \right\} dt \ , \ \mathbf{R} > 0 \ , \ \mathbf{Q} - \mathbf{N}\mathbf{R}^{-1}\mathbf{N}^T \geq 0$$

the above problem is referred to as the *LQR problem with cross-product terms*. Under the assumption that $(\mathbf{A}, \mathbf{B})$ is stabilizable and $(\mathbf{W}_2, \mathbf{A} - \mathbf{B}\mathbf{R}^{-1}\mathbf{N}^T\mathbf{C})$ is detectable, with $\mathbf{W}_2\mathbf{W}_2^T = \mathbf{Q} - \mathbf{N}\mathbf{R}^{-1}\mathbf{N}^T$, the solution of the above problem is

$$\zeta(t) = \mathbf{F}\mathbf{x}(t) = -\mathbf{R}^{-1}\left(\mathbf{B}^T\Pi + \mathbf{N}^T\right)\mathbf{x}(t)$$

where Π is the positive definite solution of the modified algebraic Riccati equation (MARE)

$$\Pi\left(\mathbf{A} - \mathbf{B}\mathbf{R}^{-1}\mathbf{N}^T\right) + \left(\mathbf{A} - \mathbf{B}\mathbf{R}^{-1}\mathbf{N}^T\right)^T\Pi - \Pi\mathbf{B}\mathbf{R}^{-1}\mathbf{B}^T\Pi + \left(\mathbf{Q} - \mathbf{N}\mathbf{R}^{-1}\mathbf{N}^T\right) = 0 \quad (4.5)$$

The solution of (4.5) can be obtained by using an algorithm similar to that proposed above for the solution of (4.4). To this end, one must simply replace $\mathbf{H}_1$ by the Hamiltonian matrix

$$\mathbf{H}_2 = \begin{bmatrix} \mathbf{A} - \mathbf{B}\mathbf{R}^{-1}\mathbf{N}^T & -\mathbf{B}\mathbf{R}^{-1}\mathbf{B}^T \\ \mathbf{Q} - \mathbf{N}\mathbf{R}^{-1}\mathbf{N}^T & -\left(\mathbf{A} - \mathbf{B}\mathbf{R}^{-1}\mathbf{N}^T\right)^T \end{bmatrix}$$

It has been shown in [10], that LQR exhibits very good robustness properties with a gain margin in the range $\left[\frac{1}{2}, \infty\right]$ and a phase margin of at least $60°$ at the system input. Here, by robustness we mean that any unstructured system perturbation within an ε bound does not destabilize the closed-loop system. Note also that, the minimu inward and upward gain margins of the optimal regulator are defined, in general, to be positive scalars, GM_{in} and GM_{up}, for whish a simultaneous insertion of the gains g_i, $i = 1,2,\cdots,m$ in the feedback loop, defined by the connection of the system under control and the regulator, will not destabilize the closed-loop system if $GM_{in} \leq g_i \leq GM_{up}$. Similarly, the phase margin of the regulator is defined to be the scalar PM, for which a simultaneous insertion of the phase factors $\exp(j\varphi_i)$, $i = 1,2,\cdots,m$ in the above ith feedback loop will keep the closed-loop stable if $|\varphi_i| \leq PM$. For a detailed interpretation of gain and phase margins, see [10], [11].

Thus, provided that the states are available, the LQ regulator is preferable to state feedback pole assignment technique from the standpoint of robustness, since in the latter case no robustness guarantees are available. However, it has been shown in [12] that, LQ optimal regulators for the performance index with cross-product terms do not guarantee the stability margins of the standard LQ optimal regulators. In this case, a serious loss of gain and phase margins is expected, depending on the selection of cost matrices $\mathbf{Q}$, $\mathbf{R}$ and $\mathbf{N}$.

4.2. Linear Quadratic Gaussian Regulation (LQG) and Kalman Filtering

The LQR design assumes that all the states of the system are available for feedback, disturbances are deterministic and measurements are not affected by noise. Obviously, this is not the case for most practical situations. A more realistic system model is obtained if $\mathbf{q}(t)$ in (2.1a) is a zero-mean white noise (namely zero-mean Gaussian stochastic) process with corresponding covariance matrix $\mathbf{R_q} \geq 0$, i.e.

$$E\{\mathbf{q}(t)\} = \mathbf{0} \ , \ E\{\mathbf{q}(t)\mathbf{q}^T(\tau)\} = \mathbf{R_q}\delta(t-\tau) \ , \ t,\tau \in \mathbf{R}$$

(with $E\{\bullet\}$ denoting the expected value of the probabilistic variable in the brackets and $\delta(t)$ representing the Dirac function in the continuous-time case) and if (2.1c) is replaced by

$$\mathbf{y}_m(t) = \mathbf{C}\mathbf{x}(t) + \mathbf{v}(t) \qquad (4.6)$$

where $\mathbf{v}(t)$ is also a zero-mean white-noise process with covariance kernel $\mathbf{R_v} > 0$, i.e.

$$E\{\mathbf{v}(t)\} = \mathbf{0} \ , \ E\{\mathbf{v}(t)\mathbf{v}^T(\tau)\} = \mathbf{R_v}\delta(t-\tau) \ , \ t,\tau \in \mathbf{R}$$

We also assume that $\mathbf{q}(t)$ and $\mathbf{v}(t)$ are in fact uncorrelated with each other, namely that

$$E\{q(t)v^T(t)\} = 0$$

The problem is then to devise a feedback-control law that minimizes the cost functional

$$J = E\left\{\frac{1}{2}\int_0^{t_f}\left(x^T(t)Qx(t) + \zeta^T(t)R\zeta(t)\right)dt\right\} \quad (4.7)$$

The above problem can be solved through the application of the well known *separation principle* (also called the *certainty equivalence principle*) [8], [9], which consists of solving separately an LQR problem and a Kalman filtering problem for generating state estimates, in order to implement full-state feedback designs, in the case where the state vector is not accessible. A Kalman filter is the optimal linear state estimator under the existence of an additive process noise and measurement noise. It permits the entire sensor signals to be used for estimation of all the state variables. The purpose of the Kalman filter is to provide an estimate $\hat{x}(t)$ of $x(t)$ according to the filter equation

$$\dot{\hat{x}}(t) = A\hat{x}(t) + B\zeta(t) + G(t)\left[y_m(t) - C\hat{x}(t)\right]$$

where $G(t) = \Sigma(t)C^T R_v^{-1}$ and where $\Sigma(t)$ is the positive semi-definite symmetric solution of the DMRE

$$\dot{\Sigma}(t) = A\Sigma(t) + \Sigma(t)A^T + DR_q D^T - \Sigma(t)C^T R_v^{-1}C\Sigma(t)$$

$$\Sigma(0) = E\left\{(\hat{x}(0) - x(0))(\hat{x}(0) - x(0))^T\right\}$$

This result is essentially the dual of that of the linear quadratic state feedback control system design. The filter is stable because the system is assumed observable and matrix $\Sigma(t)$ is positive semi-definite. The covariance matrices R_q and R_v can be regarded as design parameters in tuning the bandwidth and the characteristics of the Kalman filter, which is equivalent to the weighting matrices Q and R in the LQR design.

The resulting optimal feedback controller combines a Kalman state estimator and the state feedback law

$$\zeta(t) = F(t)\hat{x}(t)$$

where $F(t)$ is given by (4.2), (4.3). In the steady state the control law becomes time-invariant and, in this case, Σ is obtained through the solution of the ARE

$$A\Sigma + \Sigma A^T + DR_q D^T - \Sigma C^T R_v^{-1}C\Sigma = 0$$

while P is obtained through (4.4).

It is clear that the Kalman state estimator design is the dual problem of LQR design, where matrices $\mathbf{B}$, $\mathbf{C}$, $\mathbf{Q}$ and $\mathbf{R}$ in the LQR design are equivalent to matrices $\mathbf{C}^{\mathrm{T}}$, $\mathbf{D}^{\mathrm{T}}$, $\mathbf{R}_q$ and $\mathbf{R}_v$ in the Kalman state estimator design, respectively.

It has been demonstrated in [13] that, unlike LQ regulators, the LQG regulators have no intrinsic robustness properties and can at times exhibit poor stability margins. In other words, inserting a Kalman filter into the loop can sometimes cause the stability margins to shrink drastically. This is one of the reasons why the H^∞-optimal regulator, with its known property of robustness, is preferred to LQG regulator in certain control applications.

4.3. Singlerate Sampled-Data LQG Control

In the case where the control is realized by a digital computer, the control action is considered to be taken at the sampling instant. Let the sampling interval be T_0 and $\zeta(t)$ be constant for $kT_0 \le t < (k+1)T_0$. In this case equations (2.1a) and (4.6) takes the form

$$\mathbf{x}\big[(k+1)T_0\big] = \Phi\mathbf{x}(kT_0) + \hat{\mathbf{B}}\zeta(kT_0) + \mathbf{w}_d(kT_0) \qquad \text{(4.8a)}$$

$$\mathbf{y}(kT_0) = \mathbf{C}\mathbf{x}(kT_0) + \mathbf{v}(kT_0) \quad k \ge 0 \qquad \text{(4.8b)}$$

where Φ and $\hat{\mathbf{B}}$ are given by (2.4) and

$$\mathbf{w}_d(kT_0) = \int_0^{T_0} \exp(\mathbf{A}\lambda)\mathbf{D}q(kT_0 + T_0 - \lambda)d\lambda$$

Note that $\mathbf{w}_d(kT_0)$ and $\mathbf{v}(kT_0)$ are uncorrelated zero-mean white-noise processes with corresponding covariance kernels $\hat{\mathbf{R}}_{w_d} \ge 0$, $\hat{\mathbf{R}}_v > 0$, respectively, i.e.

$$E\big\{\mathbf{w}_d(kT_0)\mathbf{w}_d^{\mathrm{T}}(\ell T_0)\big\} = \hat{\mathbf{R}}_{w_d}\delta(k-\ell) \;,\; E\big\{\mathbf{v}(kT_0)\mathbf{v}^{\mathrm{T}}(\ell T_0)\big\} = \hat{\mathbf{R}}_v\delta(k-\ell),\, k,\ell \in \mathbf{N}$$

where here $\delta(k)$ denotes the Dirac function in the discrete-time case.

In the infinite horizon case (i.e. $t_f = \infty$), the performance index (4.7) is transformed into

$$J_d = E\left\{\frac{1}{2}\sum_{k=0}^{\infty}\begin{bmatrix}\mathbf{x}((kT_0))\\ \zeta(kT_0)\end{bmatrix}^{\mathrm{T}}\begin{bmatrix}\hat{\mathbf{Q}} & \hat{\mathbf{N}}\\ \hat{\mathbf{N}}^{\mathrm{T}} & \hat{\mathbf{R}}\end{bmatrix}\begin{bmatrix}\mathbf{x}((kT_0))\\ \zeta(kT_0)\end{bmatrix}\right\}$$

where [5], [14]

$$\begin{bmatrix} \hat{\mathbf{Q}} & \hat{\mathbf{N}} \\ \hat{\mathbf{N}}^{\mathrm{T}} & \hat{\mathbf{R}} \end{bmatrix} = \int_{0}^{T_0} \exp\left(\begin{bmatrix} \mathbf{A}^{\mathrm{T}} & 0 \\ \mathbf{B}^{\mathrm{T}} & 0 \end{bmatrix}\lambda\right)\begin{bmatrix} \mathbf{Q} & 0 \\ 0 & \mathbf{R} \end{bmatrix}\exp\left(\begin{bmatrix} \mathbf{A} & \mathbf{B} \\ 0 & 0 \end{bmatrix}\lambda\right) d\lambda \qquad (4.9)$$

Then, the sampled-data control of a continuous system for a given performance index is obtained by the optimal control of the discrete system (4.8) for the corresponding performance index (4.9). We call this design, the *single-rate sampled data LQ regulation design*, since in this design both control and state signals are sampled at a uniform rate (i.e. with same sampling period).

Therefore, when direct state feedback is available, the optimal control is given by

$$\zeta(kT_0) = -\hat{\mathbf{F}}\mathbf{x}(kT_0) \qquad (4.10)$$

where

$$\hat{\mathbf{F}} = \left(\hat{\mathbf{R}} + \hat{\mathbf{B}}^{\mathrm{T}}\hat{\mathbf{P}}\hat{\mathbf{B}}\right)^{-1}\left(\hat{\mathbf{B}}^{\mathrm{T}}\hat{\mathbf{P}}\Phi + \hat{\mathbf{N}}^{\mathrm{T}}\right) \qquad (4.11)$$

and $\hat{\mathbf{P}}$ is the positive definite solution of the modified discrete algebraic Riccati equation (DARE)

$$\hat{\mathbf{P}} = \Phi^{\mathrm{T}}\hat{\mathbf{P}}\Phi + \hat{\mathbf{Q}} - \left(\Phi^{\mathrm{T}}\hat{\mathbf{P}}\hat{\mathbf{B}} + \hat{\mathbf{N}}\right)\left(\hat{\mathbf{R}} + \hat{\mathbf{B}}^{\mathrm{T}}\hat{\mathbf{P}}\hat{\mathbf{B}}\right)^{-1}\left(\hat{\mathbf{B}}^{\mathrm{T}}\hat{\mathbf{P}}\Phi + \hat{\mathbf{N}}^{\mathrm{T}}\right) \qquad (4.12)$$

Note that, equation (4.12) has a positive definite solution and the closed-loop system with system matrix $\Phi_{\mathrm{cl}} = \Phi - \hat{\mathbf{B}}\left(\hat{\mathbf{R}} + \hat{\mathbf{B}}^{\mathrm{T}}\hat{\mathbf{P}}\hat{\mathbf{B}}\right)^{-1}\left(\hat{\mathbf{B}}^{\mathrm{T}}\hat{\mathbf{P}}\Phi + \hat{\mathbf{N}}^{\mathrm{T}}\right)$ is asymptotically stable if $\left(\Phi, \hat{\mathbf{B}}\right)$ is stabilizable and $\left(\mathbf{W}_3, \Phi\right)$ is detectable, where $\mathbf{W}_3$ is defined as $\mathbf{W}_3\mathbf{W}_3^{\mathrm{T}} = \hat{\mathbf{Q}} - \hat{\mathbf{N}}\hat{\mathbf{R}}^{-1}\hat{\mathbf{N}}^{\mathrm{T}}$. A way to find the solution $\hat{\mathbf{P}}$ of (4.12) is the following [15]: Form the Hamiltonian matrix

$$\mathbf{H}_3 = \begin{bmatrix} \left(\Phi - \hat{\mathbf{B}}\hat{\mathbf{R}}^{-1}\hat{\mathbf{N}}^{\mathrm{T}}\right) + \hat{\mathbf{B}}\hat{\mathbf{R}}^{-1}\hat{\mathbf{B}}^{\mathrm{T}}\left(\Phi^{\mathrm{T}} - \hat{\mathbf{N}}\hat{\mathbf{R}}^{-1}\hat{\mathbf{B}}^{\mathrm{T}}\right)^{-1}\left(\hat{\mathbf{Q}} - \hat{\mathbf{N}}\hat{\mathbf{R}}^{-1}\hat{\mathbf{N}}^{\mathrm{T}}\right) & -\hat{\mathbf{B}}\hat{\mathbf{R}}^{-1}\hat{\mathbf{B}}^{\mathrm{T}}\left(\Phi^{\mathrm{T}} - \hat{\mathbf{N}}\hat{\mathbf{R}}^{-1}\hat{\mathbf{B}}^{\mathrm{T}}\right)^{-1} \\ -\left(\Phi^{\mathrm{T}} - \hat{\mathbf{N}}\hat{\mathbf{R}}^{-1}\hat{\mathbf{B}}^{\mathrm{T}}\right)^{-1}\left(\hat{\mathbf{Q}} - \hat{\mathbf{N}}\hat{\mathbf{R}}^{-1}\hat{\mathbf{N}}^{\mathrm{T}}\right) & \left(\Phi^{\mathrm{T}} - \hat{\mathbf{N}}\hat{\mathbf{R}}^{-1}\hat{\mathbf{B}}^{\mathrm{T}}\right)^{-1} \end{bmatrix}$$

Then, there exists an eigenvector matrix $\mathbf{Y} = \begin{bmatrix} \mathbf{Y}_{11} & \mathbf{Y}_{12} \\ \mathbf{Y}_{21} & \mathbf{Y}_{22} \end{bmatrix}$ satisfying

$$\mathbf{H}_3\begin{bmatrix} \mathbf{Y}_{11} & \mathbf{Y}_{12} \\ \mathbf{Y}_{21} & \mathbf{Y}_{22} \end{bmatrix} = \begin{bmatrix} \mathbf{Y}_{11} & \mathbf{Y}_{12} \\ \mathbf{Y}_{21} & \mathbf{Y}_{22} \end{bmatrix}\begin{bmatrix} \Lambda & 0 \\ 0 & \Lambda^{-1} \end{bmatrix}$$

where $\Lambda = \operatorname*{diag}_{i=1,\cdots,n}\{\lambda_i\}$ and λ_i are the eigenvalues of H3 satisfying $|\lambda_i| \leq 1$, i=1,2,...,n. Then, $\hat{\mathbf{P}} = \mathbf{Y}_{21}\mathbf{Y}_{11}^{-1}$.

It has been shown in [16], [17] that, the stability margins of discrete-time LQ regulators are inferior to the stability margins of the continuous LQR, and the robustness of the discrete LQ could be jeopardized. Therefore, matrices $\mathbf{Q}$, $\mathbf{R}$ as well as the sampling period T_0 must selected very carefully in order to ensure good robustness properties of the discrete LQ regulator.

In the case where the state variables are not accessible a discrete time Kalman state estimator can be used instead. Then, the optimal feedback strategy is given by

$$\zeta(kT_0) = -\hat{\mathbf{F}}\hat{\mathbf{x}}(kT_0)$$

where $\hat{\mathbf{F}}$ is given by (4.11), (4.12), and $\hat{\mathbf{x}}(kT_0)$ is an estimate of $\mathbf{x}(kT_0)$, obtained by the discrete-time Kalman filter

$$\hat{\mathbf{x}}[(k+1)T_0] = \Phi\mathbf{x}(kT_0) + \hat{\mathbf{B}}\zeta(kT_0) + \hat{\mathbf{G}}[\mathbf{y}_m(kT_0) - \mathbf{C}\hat{\mathbf{x}}(kT_0)] \qquad (4.13)$$

where

$$\hat{\mathbf{G}} = \Phi\hat{\Sigma}\mathbf{C}^T\left(\mathbf{C}\hat{\Sigma}\mathbf{C}^T + \hat{\mathbf{R}}_v\right)^{-1} \qquad (4.14)$$

and $\hat{\Sigma}$ is the positive definite solution of the DARE

$$\hat{\Sigma} = \Phi\hat{\Sigma}\Phi^T + \hat{\mathbf{R}}_{w_d} - \Phi\hat{\Sigma}\mathbf{C}^T\left(\mathbf{C}\hat{\Sigma}\mathbf{C}^T + \hat{\mathbf{R}}_v\right)^{-1}\mathbf{C}\hat{\Sigma}\Phi^T \qquad (4.15)$$

4.4. Multirate Sampled-Data LQG Control

Crudely speaking, a multirate control system is a system, in which sampling is performed with different rates at different locations. Multirate sampling schemes have long been the focus of interest by many control designers (see [18], [19] and the references cited therein). There are several reasons to use such a sampling scheme in digital control systems. First of all, in complex, multivariable control systems, often it is unrealistic, or sometimes impossible, to sample all physical signals uniformly at one single rate [19]. In such situations, one is forced to use multirate sampling. Furthermore, in general, one gets better performance if one can sample and hold faster. But faster A/D and D/A conversions mean higher cost in implementation. For signals with different bandwidths, better trade-offs between performance and implementation cost can be obtained using A/D and D/A converters at different rates. On the other hand, multirate controllers are in general time-varying. Thus multirate control can achieve what singlerate cannot; e.g. gain improvement, simultaneous stabilization and decentralized control.

We next review a multirate version of the sampled-data LQG problem. To this end, consider applying to system (2.1a), (4.6) (with $\mathbf{q}(t)$ and $\mathbf{v}(t)$ being as in the previous section), the multirate control strategy of Figure 1, in which, it is assumed that all samplers start simultaneously at $t=0$, and that the hold circuits H_0 and H_N are zero order holds with holding times T_0 and $T_N = \dfrac{T_0}{N}$, $N \in \mathbf{Z}^+$,

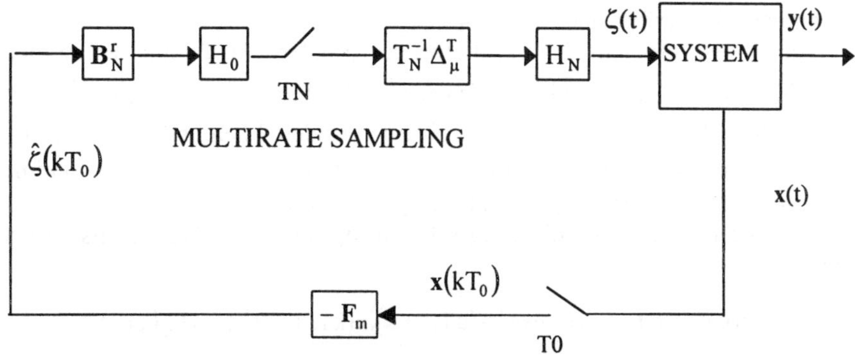

Figure 1. Multirate sampled-data control of linear systems

respectively. The inputs of the plant are constrained to the following piecewise constant controls

$$\zeta(kT_0 + \mu T_N + s) = T_N^{-1}\Delta_\mu^T B_N^r \hat\zeta(kT_0) \ , \ \hat\zeta(kT_0) \in R^{\,p_N} \qquad (4.16a)$$

$$\hat\zeta(kT_0) = -F_m x(kT_0) \ , \ k \ge 0 \qquad (4.16b)$$

for $t = kT_0 + \mu T_N$, $\mu = 0,1,\cdots,N-1$, $k \ge 0$ and for $s \in [0, T_0)$. Note that, in (4.16a),

$$\Delta_\mu \stackrel{\wedge}{=} \hat{A}_N^{N-\mu-1}\hat{B}_{T_N} \ , \ \hat{A}_N = \exp(AT_N) \ , \ \hat{B}_{T_N} = \int_0^{T_N}\exp(A\lambda)Bd\lambda$$

and B_N^r is the $n \times p_N$ full rank matrix defined by

$$B_N^r = B_N\left(B_N^T B_N\right)^{-1}$$

where B_N is the $n \times p_N$ full rank matrix satisfying

$$W_N(T_0,0) = B_N B_N^T$$

with $W_N(T_0,0) \ge 0$ being the *generalized reachability Grammian of order N* on the interval $[0, T_0]$, which is defined by

$$W_N(T_0,0) = T_N^{-1}\sum_{\mu=0}^{N-1}\Delta_\mu\Delta_\mu^T$$

Note also that $p_N = \text{rank}\{W_N(T_0,0)\}$.

The multirate LQ regulation problem is then stated as follows: Find a control law of the form (4.16a), (4.16b) which when applied to system (2.1a), (4.6), minimizes the performance index (4.7).

The above problem is essentially the problem of finding an optimal $\hat{\zeta}_{opt}(kT_0) \in \mathbf{R}^{PN}$, $k \geq 0$, which minimizes (4.7). Moreover, since $\hat{\zeta}(kT_0)$ obeys (4.16b), the LQ optimal regulation problem considered here, can be reduced to the determination of the optimal gain $\mathbf{F}_m$ which minimizes (4.7). This optimal gain can be determined using the following procedure:

Observe first that the following relationship holds (see [18] for its derivation)

$$\mathbf{x}(kT_0 + \mu T_N + s) = \exp\{\mathbf{A}(\mu T_N + s)\}\mathbf{x}(kT_0) + \mathbf{B}_N^*(\mu, s)\,\hat{\zeta}(kT_0)$$

where

$$\mathbf{B}_N^*(\mu, s) = T_N^{-1}\left\{\exp(\mathbf{A}s)\mathbf{V}_\mu^T + \hat{\mathbf{B}}_s \Delta_\mu^T\right\}\mathbf{B}_N^r$$

In above relation, the matrices $\mathbf{V}_\mu$ and $\hat{\mathbf{B}}_s$ are defined as follows

$$\mathbf{V}_\mu = \hat{\mathbf{A}}_N^{N-\mu}\Theta_\mu(T_N)\Theta_\mu^T(T_N) \;,\; \hat{\mathbf{B}}_s = \int_0^s \exp(\mathbf{A}\lambda)\mathbf{B}d\lambda$$

where, matrix $\Theta_\mu(T_N)$ is defined as follows

$$\Theta_\mu(T_N) = \left[\hat{\mathbf{B}}_{T_N} \quad \hat{\mathbf{A}}_N\hat{\mathbf{B}}_{T_N} \quad \cdots \quad \hat{\mathbf{A}}_N^{\mu-1}\hat{\mathbf{B}}_{T_N}\right]$$

It is pointed out that

$$\mathbf{B}_N^*(N-1, T_N) = \mathbf{B}_N$$

Therefore, at the sampling instants $t = kT_0$, we can easily obtain

$$\mathbf{x}[(k+1)T_0] = \Phi\mathbf{x}(kT_0) + \mathbf{B}_N\hat{\zeta}(kT_0) + \mathbf{w}_d(kT_0) \quad (4.17)$$

Now, define the following matrices

$$\tilde{\mathbf{Q}}_N = \int_0^{T_0} \exp(\mathbf{A}^T\lambda)\mathbf{Q}\exp(\mathbf{A}\lambda)d\lambda \equiv \sum_{\mu=0}^{N-1}\left(\hat{\mathbf{A}}_N^\mu\right)^T\hat{\mathbf{Q}}(T_N)\hat{\mathbf{A}}_N^\mu$$

$$\tilde{\mathbf{G}}_N = T_N^{-1}\left\{\sum_{\mu=0}^{N-1}\left(\hat{\mathbf{A}}_N^\mu\right)^T\left[\hat{\mathbf{Q}}(T_N)\mathbf{V}_\mu^T + \hat{\mathbf{N}}(T_N)\Delta_\mu^T\right]\right\}\mathbf{B}_N^r$$

$$\tilde{\mathbf{R}}_N = T_N^{-2}(\mathbf{B}_N^r)^T\left\{\sum_{\mu=0}^{N-1}[\mathbf{V}_\mu \quad \Delta_\mu]\begin{bmatrix}\hat{\mathbf{Q}}(T_N) & \hat{\mathbf{N}}(T_N) \\ \hat{\mathbf{N}}^T(T_N) & \hat{\mathbf{R}}(T_N) + T_N\mathbf{R}\end{bmatrix}\begin{bmatrix}\mathbf{V}_\mu^T \\ \Delta_\mu^T\end{bmatrix}\right\}\mathbf{B}_N^r$$

where

$$\hat{Q}(T_N) = \int_0^{T_N} \exp(A^T\lambda)Q\exp(A\lambda)d\lambda \quad , \quad \hat{N}(T_N) = \int_0^{T_N} \exp(A^T\lambda)Q\hat{B}_\lambda\, d\lambda$$

$$\hat{R}(T_N) = \int_0^{T_N} \hat{B}_\lambda^T Q\hat{B}_\lambda\, d\lambda$$

It is pointed out that matrices $\hat{Q}(T_N)$, $\hat{N}(T_N)$ and $\hat{R}(T_N)$ (and consequently matrices $\tilde{Q}_N$, $\tilde{G}_N$ and $\tilde{R}_N$), can easily be computed on the basis of the algorithm reported in [20].

The equivalent discrete LQR problem is defined by the performance index

$$J_m = \frac{1}{2}\sum_{k=0}^{\infty}\begin{bmatrix} x(kT_0) \\ \hat{u}(kT_0) \end{bmatrix}^T \begin{bmatrix} \tilde{Q}_N & \tilde{G}_N \\ \tilde{G}_N^T & \tilde{R}_N \end{bmatrix}\begin{bmatrix} x(kT_0) \\ \hat{u}(kT_0) \end{bmatrix}$$

The optimal control minimizing J_m is given by (4.16a), (4.16b) with

$$\mathbf{F}_m = \left(\tilde{R}_N + B_N^T\Pi_N B_N\right)^{-1}\left(\tilde{G}_N + B_N^T\Pi_N\Phi\right) \qquad (4.18)$$

where Π_N is the symmetric positive definite solution of the following modified DARE

$$\Pi_N = \Phi^T\Pi_N\Phi + \tilde{Q}_N$$
$$-\left(\tilde{G}_N + \Phi^T\Pi_N B_N\right)\left(\tilde{R}_N + B_N^T\Pi_N B_N\right)^{-1}\left(\tilde{G}_N^T + B_N^T\Pi_N\Phi\right) \qquad (4.19)$$

Note that there exists a unique positive definite solution Π_N of (4.19) and the corresponding closed-loop system with closed-loop system matrix $\Phi_{cl} = \Phi - B_N F_m$ is asymptotically stable if (and only if) (A, B) is controllable (stabilizable) and (W_1, A) is observable (detectable), where W_1 is such that $W_1 W_1^T = Q$.

When the state variables are not accessible a Kalman state estimator of the form (4.13)-(4.15) can be used, and the solution of the problem studied can be summarized by equations (4.16a), (4.16b), (4.18), (4.19) and (4.13)-(4.15).

There are several specific reasons to prefer the present multirate LQG design instead of the singlerate one presented in the previous Section: First of all, the multirate control scheme can produce better response characteristics and a less LQR cost than that of singlerate sampling, if the states are sampled at the same rate in both schemes. Moreover, if disturbances are present in the dynamics, a less optimal average cost and a smaller covariance of the state vector might be possible. On the other hand, in the multirate scheme presented above, the state vector can even be sampled at pathological rates, which can cause a singlerate system to lose controllability. In general, such a multirate control law provides a great flexibility in sampling rate selection. In particular, it is extremely useful, compared to singlerate sampling, if the measurement sampling rate is predetermined by some mechanical constraints, which cause sensor limitations.

4.5. Sampled-Data LQG Control Based on Multirate-Output Controllers

From the analysis presented in Section 4.3, it becomes clear that, for the solution of the singlerate sampled-data LQG control problem it is necessary to solve two Riccati equations: one associated with the Kalman estimator and the other associated with the LQ regulator. Obviously, this fact has a detrimental influence on the numerical complexity of the problem. On the other hand, the number of states of the estimator and of the system must be the same, and the estimator must run on-line. When the controlled system is of high order, this implies high computation rates in the controller. Because of these reasons, estimator based LQ regulators seem to be quite complicated as design tools.

Nevertheless, in some cases, the aforementioned complexity can be reduced drastically. Such a case, is the case where

$$\mathbf{q}(t) = \mathbf{q}(kT_0) \ , \ t \in \left[kT_0, (k+1)T_0\right) \qquad (4.20)$$

In this case, the first of relations (4.8) becomes

$$\mathbf{x}\left[(k+1)T_0\right] = \Phi\mathbf{x}(kT_0) + \hat{\mathbf{B}}\zeta(kT_0) + \hat{\mathbf{D}}\mathbf{q}(kT_0) \qquad (4.21)$$

Consider now detecting the ith plant measured output $y_{m,i}(t)$, $i=1,2,\ldots,p_1$, at every T_i, such that to obtain the values $y_{m,i}(kT_0 + \rho T_i)$, $\rho = 0,1,\ldots,N_i - 1$, where, c_i is the ith row of the matrix $\mathbf{C}$. Here, $N_i \in \mathbf{Z}^+$ are the so-called output multiplicities of the sampling and $T_i \in \mathbf{R}^+$ are the so-called output sampling periods having rational ratio, i.e. $T_i = \dfrac{T_0}{N_i}$, $i=1,2,\ldots,p_1$. The sampled values of the plant measured outputs obtained over $\left[kT_0, (k+1)T_0\right)$, are stored in the N^*-dimensional column vector $\hat{\gamma}(kT_0)$ of the form

$$\hat{\gamma}(kT_0) = \left[y_{m,1}(kT_0) \quad \cdots \quad y_{m,1}(kT_0 + (N_1 - 1)T_1) \quad \cdots \right.$$
$$\left. y_{m,p_1}(kT_0) \quad \cdots \quad y_{m,p_1}(kT_0 + (N_{p_1} - 1)T_{p_1})\right]^T$$

where $N^* = \sum\limits_{i=1}^{p_1} N_i$. The vector $\hat{\gamma}(kT_0)$ is used in the control law of the form

$$\zeta\left[(k+1)T_0\right] = \mathbf{L}_\zeta \zeta(kT_0) - \mathbf{L}_\gamma \hat{\gamma}(kT_0) \qquad (4.22)$$

where $\mathbf{L}_\zeta \in R^{m \times m}$, $\mathbf{L}_\gamma \in R^{m \times N^*}$. Controllers of the form (4.22), are known as *multirate-output controllers* (abbreviated here as MROCs) [21] and its configuration is depicted in Figure 2.

Concerning the properties of the multirate-output sampling mechanism proposed above, we point out that the following multirate formula holds

$$\mathbf{Hx}\big[(k+1)T_0\big] = \hat{\gamma}(kT_0) - \Theta_\zeta \zeta(kT_0) - \Theta_q \mathbf{q}(kT_0)$$

where the matrices $\mathbf{H} \in R^{N^* \times n}$, $\Theta_\zeta \in R^{N^* \times m}$, $\Theta_q \in R^{N^* \times d}$ have the forms

$$\mathbf{H} = \begin{bmatrix} \mathbf{H}_1 \\ \mathbf{H}_2 \\ \vdots \\ \mathbf{H}_{p_1} \end{bmatrix}, \quad \Theta_\zeta = \begin{bmatrix} (\Theta_\zeta)_1 \\ (\Theta_\zeta)_2 \\ \vdots \\ (\Theta_\zeta)_{p_1} \end{bmatrix}, \quad \Theta_q = \begin{bmatrix} (\Theta_q)_1 \\ (\Theta_q)_2 \\ \vdots \\ (\Theta_q)_{p_1} \end{bmatrix} \qquad \text{(4.23a)}$$

$$\mathbf{H}_i = \begin{bmatrix} \mathbf{c}_i \big(\hat{\mathbf{A}}_i^{N_i}\big)^{-1} \\ \vdots \\ \mathbf{c}_i \hat{\mathbf{A}}_i^{-1} \end{bmatrix}, \quad (\Theta_\zeta)_i = \begin{bmatrix} \mathbf{c}_i \hat{\mathbf{B}}_{i,N_i} \\ \vdots \\ \mathbf{c}_i \hat{\mathbf{B}}_{i,1} \end{bmatrix}, \quad (\Theta_q)_i = \begin{bmatrix} \mathbf{c}_i \hat{\mathbf{D}}_{i,N_i} \\ \vdots \\ \mathbf{c}_i \hat{\mathbf{D}}_{i,1} \end{bmatrix}, \quad i = 1,2,...,p_1 \quad \text{(4.23b)}$$

and where, for $i = 1,2,...,p_1$ and for $\rho = 1,2,...,N_i$,

$$\hat{\mathbf{A}}_i = \exp(\mathbf{A}T_i) \,, \quad \hat{\mathbf{B}}_{i,\rho} = \int_0^{-\rho T_i} \exp(\mathbf{A}\lambda)\mathbf{B}d\lambda \,, \quad \hat{\mathbf{D}}_{i,\rho} = \int_0^{-\rho T_i} \exp(\mathbf{A}\lambda)\mathbf{D}d\lambda \quad \text{(4.23c)}$$

Suppose now that $(\mathbf{C}, \mathbf{A})$ is an observable pair. Suppose also that

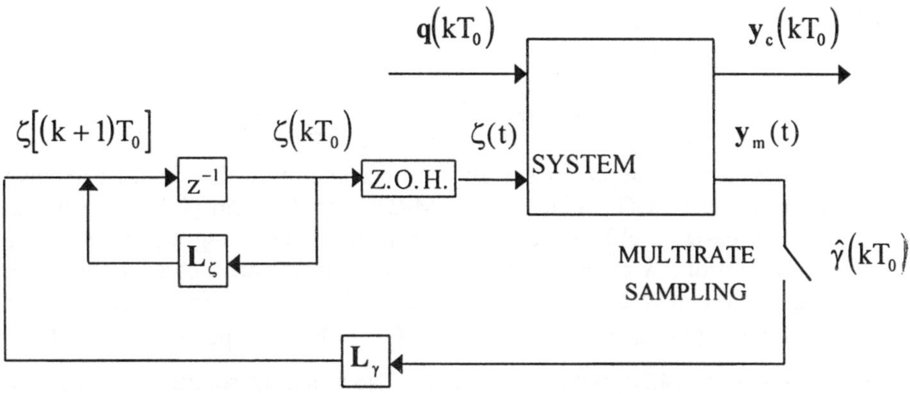

Figure 2. Control of linear systems using MROCs

$$\text{rank}\begin{bmatrix} A & D \\ C & 0_{p_1 \times d} \end{bmatrix} = n + d \qquad (4.24)$$

In this case, if $N_i \geq n_i$, where n_i comprise an observability index vector of the matrix pair (C^*, A^*), with

$$C^* = \begin{bmatrix} C & 0_{p_1 \times d} \end{bmatrix} \; , \; A^* = \begin{bmatrix} A & D \\ 0_{p_1 \times n} & 0_{p_1 \times d} \end{bmatrix} \qquad (4.25)$$

then, matrix $\begin{bmatrix} H & \Theta_q \end{bmatrix}$ has full column rank for almost every sampling period T_0 and the control law (4.22) is equivalent to the control law (4.10), for $k \geq 1$, by choosing properly the MROC matrices L_γ and L_ζ such that

$$L_\gamma \begin{bmatrix} H & \Theta_q \end{bmatrix} = \begin{bmatrix} \hat{F} & 0_{m \times d} \end{bmatrix} \; , \; L_\zeta = L_\gamma \Theta_\zeta \quad (4.26)$$

Moreover, in the case where matrix L_ζ has a prespecified desired form, say $L_{\zeta,sp}$, if

$$\text{rank}\begin{bmatrix} A & B & D \\ C & 0_{p_1 \times m} & 0_{p_1 \times d} \end{bmatrix} = n + m + d$$

and if $N_i \geq n_i^*$, where n_i^* comprise an observability index vector of the matrix pair $(\tilde{C}, \tilde{A})$, with

$$\tilde{C} = \begin{bmatrix} C & 0_{p_1 \times m} & 0_{p_1 \times d} \end{bmatrix} , \; \tilde{A} = \begin{bmatrix} A & B & D \\ 0_{p_1 \times n} & 0_{p_1 \times m} & 0_{p_1 \times d} \end{bmatrix} \qquad (4.27)$$

then, matrix $\begin{bmatrix} H & \Theta_\zeta & \Theta_q \end{bmatrix}$ has full column rank and there always exists a matrix L_γ such that

$$L_\gamma \begin{bmatrix} H & \Theta_\zeta & \Theta_q \end{bmatrix} = \begin{bmatrix} \hat{F} & L_{\zeta,sp} & 0_{m \times d} \end{bmatrix} \quad (4.28)$$

On the basis of the previous analysis, it becomes clear that, in order to solve the sampled-data LQG problem using MROCs of the form (4.22), one has to refer to an easier problem, i.e. to the problem of designing a *fictitious* state feedback control law of the form (4.10) which equivalently solves the problem studied for system (4.21), and then calculating the actual controller gains, either by solving (4.26) with respect to L_γ and L_ζ, or by first specifying $L_\zeta = L_{\zeta,sp}$ and then solving (4.28) with respect to L_γ. Note that the solution of (4.26) and (4.28) is guaranteed if matrices $\begin{bmatrix} H & \Theta_q \end{bmatrix}$ and $\begin{bmatrix} H & \Theta_\zeta & \Theta_q \end{bmatrix}$, respectively, have full column rank. Note also that, the case $\mathbf{L}_{\zeta,\mathrm{sp}} \equiv \mathbf{0}$ is of course permissible leading to a static MROC of the form

$$\zeta[(k+1)T_0] = -\mathbf{L}_\gamma \hat{\gamma}(kT_0)$$

In any case, the solution of the problem is reduced to the solution of a single DARE of the form (4.12), instead of the two DARE needed in the case of estimator based controllers. Moreover, since the closed-loop system poles are the eigenvalues of $\Phi_{cl} = \Phi - \hat{\mathbf{B}}\hat{\mathbf{F}} = \Phi - \hat{\mathbf{B}}\mathbf{KH}$ and m additional eigenvalues at the origin [21] (with m=0, if $\mathbf{L}_{\zeta,\mathrm{sp}} = \mathbf{0}$), the dynamics introduced in the closed-loop system by a MROC is, in any case, less than the dynamics introduced by estimator based controllers. This has beneficial influence on the overall complexity of the problem. Finally, it is pointed out that the stability robustness properties of MROC based LQ regulators have been investigated in [22]-[24].

Although, the analysis presented in this Section, has been restricted to the case where the disturbance are piecewise constant functions in intervals of the form $[kT_0, (k+1)T_0)$, more general classes of disturbances can be treated in a similar manner using the approach presented in [25].

4.6. Sampled-Data LQG Control Using Two-Point Multirate Controllers

Consider once again the case where (4.20) holds, and apply to the system under control the feedback strategy of Figure 3, which is a combination of the control strategies presented above in Sections 4.3 and 4.5. In particular, we assume that the control signal has the form (4.16a), while in the present case $\hat{\zeta}(kT_0)$ (i.e., the input or control vector) is updated through the relation

$$\hat{\zeta}[(k+1)T_0] = \mathbf{L}_\zeta \hat{\zeta}(kT_0) - \mathbf{L}_\gamma \hat{\gamma}(kT_0) \qquad (4.29)$$

In the present case, the ith plant output $y_i(t)$, is detected at every $T_i = T_0/M_i$, in order to obtain the values $y_i(kT_0 + \rho T_i)$, $\rho = 0,1,...,M_i - 1$, where, $M_i \in Z^+$, $i = 1,2,...,p$, are the so-called output multiplicities of the sampling. The sampled values of the plant outputs obtained over $[kT_0, (k+1)T_0)$, are stored in the M^*-dimensional column vector $\hat{\gamma}(kT_0)$ of the form

$$\hat{\gamma}(kT_0) = [y_1(kT_0) \quad \cdots \quad y_1(kT_0 + (M_1 - 1)T_1) \quad \cdots$$
$$y_p(kT_0) \quad \cdots \quad y_p(kT_0 + (M_p - 1)T_p)]^{\mathrm{T}}$$

where $M^* = \sum_{i=1}^{p} M_i$. It is worth noticing that, in general, $M_i \neq N$. That is, multirate sampling of the plant inputs and outputs may be performed at a diffe-

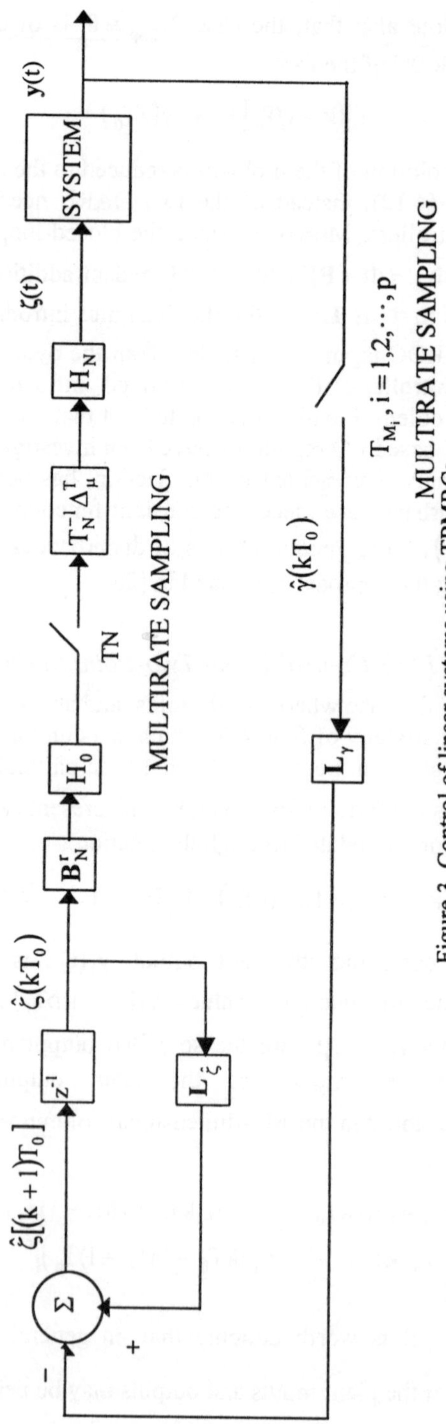

Figure 3. Control of linear systems using TPMRCs

rent rate. Controllers of the form (4.16a), (4.27) are usually called *Two-Point Multirate Controllers (TPMRCs)* [26].

Since (4.20) is satisfied, relation (4.17) becomes

$$\mathbf{x}[(k+1)T_0] = \Phi\mathbf{x}(kT_0) + \mathbf{B}_N\hat{\zeta}(kT_0) + \hat{\mathbf{D}}\mathbf{w}(kT_0) \quad \text{(4.30)}$$

Note also that

$$\mathbf{x}(kT_0 + \rho T_i) = \hat{\mathbf{A}}_i^\rho\mathbf{x}(kT_0) + \mathbf{B}_{M_i}^*(\rho)\,\hat{\zeta}(kT_0) + \tilde{\mathbf{D}}_{i,\rho}\mathbf{q}(kT_0)$$

$$\hat{\mathbf{A}}_i = \exp(\mathbf{A}T_i) \ , \ \ \mathbf{B}_{M_i}^*(\rho) = T_N^{-1}E_{M_i}(\rho)\mathbf{B}_N^r \ , \ \ \tilde{\mathbf{D}}_{i,\rho} = \int_0^{\rho T_i}\exp(\mathbf{A}\lambda)\mathbf{D}d\lambda$$

In (4.30), matrix $E_{M_i}(\rho)$ is defined as follows

$$E_{M_i}(\rho) = \sum_{j=0}^{a(i,\rho)-1}\int_{jT_N}^{(j+1)T_N}\exp\left\{\mathbf{A}\left(\rho\frac{N}{M_i}T_N - \xi\right)\right\}\mathbf{B}d\xi\Delta_j^T$$

$$+ \int_{a(i,\rho)T_N}^{\rho\frac{N}{M_i}T_N}\exp\left\{\mathbf{A}\left(\rho\frac{N}{M_i}T_N - \xi\right)\right\}\mathbf{B}d\xi\Delta_{a(i,\rho)}^T \quad \text{(4.31)}$$

$$a(i,\rho) = \text{INT}_S\left(\rho\frac{N}{M_i}\right)$$

and where $\text{INT}_S(\nu)$ is the greatest integer that is less than or equal to $\nu \in \mathbf{R}^+$.

The following basic formula of the multirate sampling mechanism holds (see [26] for its derivation)

$$\mathbf{H}\mathbf{x}[(k+1)T_0] = \hat{\gamma}(kT_0) - \Theta_\zeta\hat{\zeta}(kT_0) - \Theta_q\mathbf{q}(kT_0) \ , \ k \geq 0$$

where, matrices $\mathbf{H} \in \mathbf{R}^{M^* \times n}$ and $\Theta_\zeta \in \mathbf{R}^{M^* \times p N_0}$ are defined as follows

$$\mathbf{H} = \begin{bmatrix} \mathbf{c}_1^T\left(\hat{\mathbf{A}}_1^{M_1}\right)^{-1} \\ \vdots \\ \mathbf{c}_1^T\hat{\mathbf{A}}_1^{-1} \\ \vdots \\ \mathbf{c}_p^T\left(\hat{\mathbf{A}}_p^{M_p}\right)^{-1} \\ \vdots \\ \mathbf{c}_p^T\hat{\mathbf{A}}_p^{-1} \end{bmatrix} \ , \ \Theta_{\hat{\zeta}} = \begin{bmatrix} \mathbf{c}_1^T\hat{\mathbf{B}}_{1,0} \\ \vdots \\ \mathbf{c}_1^T\hat{\mathbf{B}}_{1,M_1-1} \\ \vdots \\ \mathbf{c}_p^T\hat{\mathbf{B}}_{p,0} \\ \vdots \\ \mathbf{c}_p^T\hat{\mathbf{B}}_{p,M_p-1} \end{bmatrix}$$

while Θ_q is defined as in (4.23). In (4.31),

$$\hat{\mathbf{B}}_{i,\rho} = \mathbf{B}^*_{M_i}(\rho) - \hat{\mathbf{A}}_i^{\rho - M_i}\mathbf{B}_N \ , \ i = 1,2,...,p \ , \ \rho = 0,1,...,M_i - 1$$

Suppose now that $(\mathbf{C}, \mathbf{A})$ is an observable pair and that (4.24) holds. Then, if $N_i \geq n_i$, where n_i comprise an observability index vector of the matrix pair (4.25), the control law (4.29) is equivalent to the control law (4.16b), by choosing properly the gains $\mathbf{L}_\gamma$ and $\mathbf{L}_\zeta$ such that

$$\mathbf{L}_\gamma\begin{bmatrix}\mathbf{H} & \Theta_q\end{bmatrix} = \begin{bmatrix}\mathbf{F}_m & 0_{m\times d}\end{bmatrix} \ , \ \mathbf{L}_\zeta = \mathbf{L}_\gamma\Theta_\zeta \qquad (4.32)$$

Moreover, if for some integers $M_i = \sigma_i^*, i = 1,2,...,p$, such that $M^* \geq n + p_N + d$, matrix $\begin{bmatrix}\mathbf{H} & \Theta_\zeta & \Theta_q\end{bmatrix}$ has full column rank, then for a prespecified $\mathbf{L}_\zeta = \mathbf{L}_{\zeta,sp}$ and for almost every sampling period T_0, there exists a matrix $\mathbf{L}_\gamma$ such that

$$\mathbf{L}_\gamma\begin{bmatrix}\mathbf{H} & \Theta_\zeta & \Theta_q\end{bmatrix} = \begin{bmatrix}\mathbf{F} & \mathbf{L}_{\zeta,sp} & 0_{m\times d}\end{bmatrix} \qquad (4.33)$$

Therefore, in order to solve the sampled-data LQG problem using TPMRCs, one has to design a *fictitious* state feedback control law of the form (4.16b) which equivalently solves the problem studied for system (4.30), and subsequently to calculate the actual controller gains, either by solving (4.32) with respect to $\mathbf{L}_\gamma$ and $\mathbf{L}_\zeta$, or by first specifying $\mathbf{L}_\zeta = \mathbf{L}_{\zeta,sp}$ and then solving (4.33) with respect to $\mathbf{L}_\gamma$. As in the case of MROCs, the solution of this equivalent problem reduces to the solution of a single DARE. Furthermore the dynamics introduced in the closed-loop by TPMRCs is of low order. Finally, it is pointed out that the stability robustness properties of TPMRC based LQ regulators have been investigated in [26].

4.7. Noise Rejection Using Generalized Sampled-Data Hold Functions

Dynamic multirate controllers, such as MROCs and TPMRCs, with its nice property of reconstructing the action of the state feedback, provide us two alternative approaches to solve the sampled-data LQG problem without resorting to the hard task of the state estimation. In the present Section, we will review a slightly different approach to the problem, with the same nice feature. This approach is based on periodic sampled-data output feedback controllers, often called *generalized sampled-data hold functions* (GSHF) [27].

To briefly present the method, we will consider, in the sequel, controllable and observable linear state space systems of the form (2.1b), (4.6), which will be acted upon by controls of the form

$$\zeta(t) = \mathbf{F}(t)\mathbf{y}_m(kT_0) \;\; , \;\; t \in [kT_0, (k+1)T_0), k \geq 0 \qquad \text{(4.34a)}$$

$$\mathbf{F}(t+T_0) = \mathbf{F}(t) \;\; , \;\; \text{for } t \in [0, T_0) \qquad \text{(4.34b)}$$

where $\mathbf{F}(t)$ is a T_0-periodic integrable and bounded matrix of appropriate dimension representing a generalized sampled-data hold function. Then, the closed loop system takes the form

$$\mathbf{x}[(k+1)T_0] = \Phi \mathbf{x}(kT_0) + \mathbf{F}_\phi \zeta(kT_0) + \omega(kT_0) \;\; , \;\; \mathbf{y}_m(kT_0) = \mathbf{C}\mathbf{x}(kT_0) + \mathbf{v}(kT_0)$$

or equivalently,

$$\mathbf{x}[(k+1)T_0] = (\Phi + \mathbf{F}_\phi \mathbf{C})\mathbf{x}(kT_0) + \mathbf{F}_\phi \mathbf{v}(kT_0) + \omega(kT_0)$$

$$\mathbf{F}_\phi = \int_0^{T_0} \exp[\mathbf{A}(T_0 - \lambda)]\mathbf{B}\mathbf{F}(\lambda)d\lambda \quad \text{(4.35a)}$$

$$\omega(kT_0) = \int_{kT_0}^{(k+1)T_0} \exp\{\mathbf{A}[(k+1)T_0 - \lambda]\}\mathbf{w}(\lambda)d\lambda \text{ (4.35b)}$$

It is supposed that $\mathbf{w}(t)$ and $\mathbf{v}(kT_0)$ are stationary zero-mean white Gaussian process with covariance kernels

$$\mathsf{E}[\mathbf{w}(t)\mathbf{w}^T(\tau)] = \mathbf{R}_w \delta(t - \tau), t, \tau \in \mathbf{R}$$

$$\mathsf{E}[\mathbf{v}(kT_0)\mathbf{v}^T(\ell T_0)] = \mathbf{R}_v \delta(k - \ell) \;\; , \;\; k, \ell \in \mathbf{N}, \mathbf{R}_v > 0$$

where by abuse of notation, $\delta(\bullet)$ represents the Dirac function in both the discrete-time and continuous-time case. Relation (4.35) implies that $\omega(kT_0)$ is a stationary zero-mean white Gaussian process with covariance kernel

$$\mathsf{E}[\omega(kT_0)\omega^T(\ell T_0)] = \mathbf{R}_\omega \delta(k - \ell) \;\; , \;\; k, \ell \in \mathbf{N}$$

$$\mathbf{R}_\omega = \int_0^{T_0} \exp[\mathbf{A}(T_0 - \lambda)]\mathbf{R}_w \exp[\mathbf{A}^T(T_0 - \lambda)]d\lambda$$

The optimal noise rejection problem via GSHF control, treated in this Section, is as follows: Given a symmetric positive definite weighting matrix $\mathbf{Q} \in \mathbf{R}^{n \times n}$ and a sampling period T_0, find a GSHF based optimal regulator of the form (4.34), in order to minimize the following cost functional

$$J_G = \lim_{k \to \infty} \mathsf{E}\left[\frac{1}{2}\mathbf{x}^T(kT_0)\mathbf{Q}\mathbf{x}(kT_0)\right]$$

The solution of the above problem has been established in [27], and it can be expressed in terms of a discrete Riccati equation, which resembles the Riccati equation appearing in discrete Kalman filtering. More precisely, it has been proven in [27] that, if the triplet $(\mathbf{A}, \mathbf{B}, \mathbf{C})$ is minimal, then for almost all $T_0 > 0$, the optimal noise rejection problem is solvable. Its solution is

$$\mathbf{F}(t) = \mathbf{B}^T \exp\left[\mathbf{A}^T (T_0 - t)\right] \mathbf{W}^{-1}(\mathbf{A}, \mathbf{B}, T_0)\mathbf{F}_\phi \qquad (4.36)$$

where

$$\mathbf{W}(\mathbf{A}, \mathbf{B}, T_0) = \int_0^{T_0} \exp\left[\mathbf{A}(T_0 - \lambda)\right]\mathbf{B}\mathbf{B}^T \exp\left[\mathbf{A}^T(T_0 - \lambda)\right]d\lambda$$

$$\mathbf{F}_\phi = -\mathbf{\Phi}\mathbf{X}\mathbf{C}^T(\mathbf{C}\mathbf{X}\mathbf{C}^T + \mathbf{R}_v)^{-1}$$

and $\mathbf{X} \in \mathbf{R}^{n \times n}$ is the unique positive definite solution of the DARE

$$\mathbf{\Phi}\mathbf{X}\mathbf{\Phi}^T - \mathbf{\Phi}\mathbf{X}\mathbf{C}^T(\mathbf{C}\mathbf{X}\mathbf{C}^T + \mathbf{R}_v)^{-1}\mathbf{C}\mathbf{X}\mathbf{\Phi}^T + \mathbf{R}_\omega - \mathbf{X} = 0$$

The minimum of the cost function J is then calculated by (see [27] for details)

$$J_{min} = \text{tr}\{\mathbf{Q}\mathbf{X}\}$$

Clearly, relation (4.36) provides us a solution to the optimal noise rejection problem via GSHF control, in the case where the hold function $\mathbf{F}(t)$ does not have any prespecified structure. Our attention is next focused on the special class of the time-varying T_0-periodic matrix functions $\mathbf{F}(t)$, for which every element of $\mathbf{F}(t)$, denoted by $f_{ij}(t)$, is piecewise constant over intervals of length $T_i = T_0 / N_i$, with $N_i \in \mathbf{Z}^+$, i.e.

$$f_{ij}(t) = f_{ij,\mu}, \forall t \in \left[\mu T_i, (\mu + 1)T_i\right)$$

for $\mu = 0,..., N_i - 1$. In this case, as it has been shown in [28], the following relation holds

$$\mathbf{F}_\phi = \widetilde{\mathbf{B}}\hat{\mathbf{F}} \qquad (4.37)$$

where, defining by **bi** the ith column of **B**,

$$\widetilde{\mathbf{B}} = \left[\hat{\mathbf{b}}_1 \quad \cdots \quad \hat{\mathbf{A}}_1^{N_1-1}\hat{\mathbf{b}}_1 \quad \cdots \quad \hat{\mathbf{b}}_m \quad \cdots \quad \hat{\mathbf{A}}_m^{N_m-1}\hat{\mathbf{b}}_m\right]$$

$$\hat{\mathbf{A}}_i \triangleq \exp(\mathbf{A}T_i) \ , \ \hat{\mathbf{b}}_i \triangleq \int_0^{T_i}\exp(\mathbf{A}\lambda)\mathbf{b}_i d\lambda$$

and where the $m \times p$ block matrix $\hat{\mathbf{F}}$ has the form

$$\hat{\mathbf{F}} = \begin{bmatrix} \hat{\mathbf{f}}_{11} & \cdots & \hat{\mathbf{f}}_{1p} \\ \vdots & \ddots & \vdots \\ \hat{\mathbf{f}}_{m1} & \cdots & \hat{\mathbf{f}}_{mp} \end{bmatrix}, \quad \hat{\mathbf{f}}_{ij} = \begin{bmatrix} f_{ij,N_i-1} \\ \vdots \\ f_{ij,0} \end{bmatrix}$$

In this case, ith row $\mathbf{f}_i^T(t)$ of the matrix $\mathbf{F}(t)$ and the ith block row of the matrix $\hat{\mathbf{F}}$ are interrelated as

$$\mathbf{f}_i^T(t) = \begin{bmatrix} f_{i1}(t) & \cdots & f_{ip}(t) \end{bmatrix} = e_{N_i-\mu} \begin{bmatrix} \hat{\mathbf{f}}_{i1} & \cdots & \hat{\mathbf{f}}_{ip} \end{bmatrix}, \quad \forall \mu T_i \le t < (\mu+1)T_i$$

for i=1,…,m and for $\mu=0$,…, N_i-1, where $e_{N_i-\mu} \in R^{N_i}$ is the row vector, whose elements are zero except for a unity appearing in the ($N_i - \mu$)th position.

It is clear that, in the case where the elements of $\mathbf{F}(t)$ are piecewise constant, the admissible hold function can be obtained on the basis of $\hat{\mathbf{F}}$. To find matrix $\hat{\mathbf{F}}$ one must solve (4.37). Note that (4.37) is always solvable if $N_i \ge n_i$, i=1,2,…,m, where n_i comprise a set of locally minimum controllability indices of the pair $(\mathbf{A},\mathbf{B})$ (for details see [28]).

From the above analysis, it becomes clear that GSHF based control has the efficacy of state feedback, in solving the LQG control problem without the requirement of state estimation. Moreover, unlike estimator based or dynamic multirate controller based LQG regulators, the GSHF LQG regulator does not introduce any exogenous dynamics in the closed-loop, thus further reducing the overall complexity of the problem. Finally, it is pointed out that, the stability robustness properties of GSHF based LQG regulators have been investigated in [29].

5. Robust Nonlinear Control

The fact of uncertainty is inherent in real-world systems. The facing of model uncertainty that mainly arises from the behavior of a plant under control that changes with time and from the fact that its parameters are imprecisely known, is a common task in control systems. The designer must ultimately insure stability and performance of the actual closed-loop system and the designed controller must be robust to the model uncertainty. In recent years, much effort has been devoted to the problem of obtaining stabilizing controllers for uncertain systems [30], [31]. In much of this research, the uncertainties are modeled in deterministic rather than stochastic manner, and they are characterized by certain structural conditions and known bounds (see [32] and the references cited therein). Such uncertainties could be due to uncertain disturbance inputs, uncertain parameters or nonlinear elements that are difficult to characterize.

In this Section, we consider the problem of stabilizing an uncertain system, in the case where the norm of the control input is bounded by a prespecified constant. We treat continuous-time dynamical systems whose nominal part is linear and whose uncertain part is norm-bounded. In particular, we consider dynamical systems of the form

$$\dot{x}(t) = \left[A + \Delta A(r(t))\right]x(t) + \left[B + \Delta B(s(t))\right]\zeta(t) + Dq(t) \ , \ x(t_0) = x_0 \qquad (5.1)$$

where $x(t) \in R^n$, $\zeta(t) \in R^m$, $q(t) \in R^d$, $r(t) \in R^\ell$, $s(t) \in R^q$, and where A, B, D, are known real constant matrices with appropriate dimensions, while

$$\Delta A(\bullet): \ R^\ell \to R^{n \times n} \ , \ \Delta B(\bullet): \ R^q \to R^{n \times m} \qquad (5.2)$$

are known continuous functions.

Uncertainties in the system matrix, the input matrix, and system input, respectively, are modeled by the unknown Lebesgue measurable functions

$$r(\bullet): R \to R \ , \ s(\bullet): R \to S \ , \ q(\bullet): R \to Q \qquad (5.3)$$

where R, S and Q are known compact subsets of the appropriate spaces. Thus, the only information concerning the unknown elements $r(t)$, $s(t)$ and $q(t)$ resides in their bounding sets R, S and Q, respectively.

Concerning system (5.1)–(5.3), we assume that the uncertainties belong to the range space of the input matrix B or, more precisely, that there exist two continuous functions $\Gamma(\bullet): R^\ell \to R^{m \times n}$, $\Lambda(\bullet): R^q \to R^{m \times n}$ and a constant matrix $W \in R^{m \times d}$, such that the so-called matching conditions are met; namely,

$$\Delta A(r) = B\,\Gamma(r) \ , \ \forall r \in R \qquad (5.4a)$$

$$\Delta B(s) = B\,\Lambda(s) \ , \ \forall s \in S \qquad (5.4b)$$

$$D = BW \qquad (5.4c)$$

and

$$\max_{s \in S} \lVert \Lambda(s) \rVert < 1 \qquad (5.4d)$$

The control purpose, in this Section is to find a control input $\zeta(t)$, such that the state $x(t)$ of the system (5.1)-(5.3) is restricted to a sufficiently small neighborhood of its initial, prior to the application of the disturbance, value. In mathematical terms, it is desired to find an appropriate control input $\zeta(t)$, such that the following two desirable properties to be achieved, for the uncertain system (5.1)–(5.3):

Property P1: (*Uniform Boundedness*). Given $x_0 \in R^n$, there is a positive

$d(\mathbf{x}_0) < \infty$ such that, for all solutions $\mathbf{x}(\bullet) : [t_0, t_1) \to \mathbf{R}^n, \mathbf{x}(t_0) = \mathbf{x}_0$,

$$\|\mathbf{x}(t)\| \le d(\mathbf{x}_0), \qquad \forall\, t \in [t_0, t_1)$$

Property P2: (*Uniform Ultimate Boundedness*). Given $\mathbf{x}_0 \in \mathbf{R}^n$ and $\mathbf{X} = \{\mathbf{x} \in \mathbf{R}^n : \|\mathbf{x}\| \le \delta > 0\}$, there is a nonnegative $T(\mathbf{x}_0, \mathbf{X}) < \infty$, such that, for all solutions $\mathbf{x}(\bullet) : [t_0, \infty) \to \mathbf{R}^n, \mathbf{x}(t_0) = \mathbf{x}_0$,

$$\mathbf{x}(t) \in \mathbf{X}, \qquad \forall\, t \ge t_0 + T(\mathbf{x}_0, \mathbf{X})$$

Loosely speaking, uniform boundedness implies that every solution emanating from initial state $\mathbf{x}_0$ remains within a bounded neighborhood whose radius may depend on $\mathbf{x}_0$. Uniform ultimate boundedness implies that every solution starting at $\mathbf{x}_0$ will enter and remain within a neighborhood of prescribed radius δ, after a finite time which may depend on $\mathbf{x}_0$ and δ. These two properties, sometimes stated in a slightly different but equivalent form, are the main ingredients of *practical stability* [33], [34].

Consider now a nonlinear full state feedback control law of the form

$$\zeta(t) = \mathbf{F}\mathbf{x}(t) + \mathrm{p}_\varepsilon(\mathbf{x}(t)) \quad \text{(5.5)}$$

such that, given $\varepsilon > 0$, there holds

$$\mathrm{p}_\varepsilon(\mathbf{x}(t)) = \begin{cases} -\dfrac{\mathbf{B}^T \mathbf{P}\mathbf{x}(t)}{\|\mathbf{B}^T \mathbf{P}\mathbf{x}(t)\|}\rho(\mathbf{x}(t)) \,, & \text{if } \|\mathbf{B}^T \mathbf{P}\mathbf{x}(t)\| \ge \varepsilon \\[2ex] -\dfrac{\mathbf{B}^T \mathbf{P}\mathbf{x}(t)}{\varepsilon}\rho(\mathbf{x}(t)) \,, & \text{if } \|\mathbf{B}^T \mathbf{P}\mathbf{x}(t)\| < \varepsilon \end{cases} \qquad \text{(5.6)}$$

where $\mathbf{P} \in \mathbf{R}^{n \times n}$ is the unique symmetric, positive-definite solution of the Lyapunov equation

$$\mathbf{P}\mathbf{A}_{cl} + \mathbf{A}_{cl}^T \mathbf{P} + \mathbf{Q} = 0 \quad \text{(5.7)}$$

for given symmetric, positive definite $\mathbf{Q} \in \mathbf{R}^{n \times n}$, where

$$\mathbf{A}_{cl} = \mathbf{A} + \mathbf{B}\mathbf{F}$$

and

$$\rho(\mathbf{x}(t)) = \left\{1 - \max_{s \in S}\|\Lambda(s)\|\right\}^{-1}\left\{\max_{r \in R}\|\Gamma(r)\mathbf{x}(t)\| + \max_{s \in S}\|\Lambda(s)\mathbf{F}\mathbf{x}(t)\| + \max_{q \in Q}\|\mathbf{W}\mathbf{q}(t)\|\right\} \quad \text{(5.8)}$$

A simple way to choose $\mathbf{F}$ is as follows: If $\mathbf{A}$ is stable set $\mathbf{F}=\mathbf{0}$. Otherwise, let $\mathbf{Q}_R \in \mathbf{R}^{n \times n}$ be any symmetric positive definite matrix. Then, the Riccati equation

$$\mathbf{P}^*\mathbf{A} + \mathbf{A}^T\mathbf{P}^* - \mathbf{P}^*\mathbf{B}\mathbf{B}^T\mathbf{P}^* + \mathbf{Q}_R = 0 \qquad (5.9)$$

has a unique positive definite solution $\mathbf{P}^* \in \mathbf{R}^{n \times n}$. Setting

$$\mathbf{F} = -\mathbf{B}^T\mathbf{P}^*$$

the matrix $\mathbf{A}_{cl}$ is asymptotically stable. Note that the solution $\mathbf{P}^*$ of equation (5.9) satisfies equation (5.7) with the positive definite matrix

$$\mathbf{Q} = \mathbf{Q}_R + \mathbf{F}^T\mathbf{F}$$

Thus, the controller under consideration is given by

$$\zeta(t) = -\mathbf{B}^T\mathbf{P}^*\mathbf{x}(t) + p_\varepsilon\big(\mathbf{x}(t)\big)$$

with $p_\varepsilon\big(\mathbf{x}(t)\big)$ given by (5.6).

Now, provided the stated assumptions are met, the control (5.5)-(5.8) guarantees practical stability for uncertain system (5.1)–(5.3) and, in particular, Properties P1 and P2 for every possible realization of uncertain elements $\mathbf{r}(\bullet)$, $\mathbf{s}(\bullet)$, $\mathbf{q}(\bullet)$, (for proof see [32]-[34]). Here, it should be noted that $\delta = \delta(\varepsilon)$ and can be made arbitrarily small by choice of ε; namely, decreasing ε results in a decrease of the radius of the ultimate boundedness set. Furthermore, if the initial state belongs to the Lyapunov ellipsoid, which defines the ultimate boundedness set, the whole solution remains within it.

Consider now the special case where in (5.1) $\Delta\mathbf{A}(\bullet) = \Delta\mathbf{B}(\bullet) = 0$ and the system states and matrices can be transformed and/or partitioned as follows

$$\mathbf{z}(t) = \begin{bmatrix} \mathbf{z}_1(t) \\ \mathbf{z}_2(t) \end{bmatrix} \;,\quad \mathbf{A} = \begin{bmatrix} \mathbf{A}_{11} & \mathbf{A}_{12} \\ \mathbf{A}_{21} & \mathbf{A}_{22} \end{bmatrix} \;,\quad \mathbf{B} = \begin{bmatrix} \mathbf{B}_1 \\ 0 \end{bmatrix} \;,\quad \mathbf{D} = \begin{bmatrix} \mathbf{D}_1 \\ 0 \end{bmatrix}$$

where $\mathbf{z}_1(t) \in \mathbf{R}^{n_1}$, $\mathbf{z}_2(t) \in \mathbf{R}^{n_2}$, $n_1 + n_2 = n$, and the block matrices are of appropriate dimensions. Obviously in this case, system (5.1) can be divided into two subsystems as follows

$$\dot{\mathbf{z}}_1(t) = \mathbf{A}_{11}\mathbf{z}_1(t) + \mathbf{A}_{12}\mathbf{z}_2(t) + \mathbf{B}_1\zeta(t) + \mathbf{D}_1\mathbf{q}(t) \;,\quad \mathbf{z}_1(t_0) = \mathbf{z}_1^0 \qquad (5.10a)$$

$$\dot{\mathbf{z}}_2(t) = \mathbf{A}_{21}\mathbf{z}_1(t) + \mathbf{A}_{22}\mathbf{z}_2(t) \;,\quad \mathbf{z}_2(t_0) = \mathbf{z}_2^0 \qquad (5.10b)$$

Obviously, the state of each of the two subsystems affects the dynamic behavior of the other. In particular, the inputs of the first subsystem are induced from the disturbance signal and the state of the second subsystem. The only input of the second subsystem comes from the state of the first one.

Here, it is desired to find a control law of the form

$$\zeta(t) = \mathbf{F}_1\mathbf{z}_1(t) + p_1\big(\mathbf{z}_1(t)\big) \qquad (5.11)$$

with $p_1(z_1(t))$ a nonlinear function of the state of the first subsystem, such that the overall state $[z_1(t) \quad z_2(t)]^T$ of the system is restricted to a sufficiently small neighborhood of its initial, prior to the application of the disturbance signal, value. The control law (5.11) is called *robust nonlinear partial state feedback*.

According to the results reported in [35], there exists a control law of the form (5.11) that renders the system practically stable, if and only if the following assertions simultaneously hold

a) There exists a matrix $W_1 \in R^{m \times d}$ satisfying the matching condition

$$D_1 = B_1 W_1$$

b) The pair (A_{11}, B_1) is stabilizable (controllable), i.e. there exists a matrix $F_1 \in R^{m \times n_1}$ such that the matrix $A_{cl,1} = A_{11} + B_1 F_1$ is stable and the Lyapunov equation

$$A_{cl,1}^T P_1 + P_1 A_{cl,1} = -Q_1$$

has a positive definite solution P_1 for some positive definite matrix Q_1.

c) $\lambda_{min}(Q_1)\lambda_{min}(Q_2) - (\|P_1 A_{12}\| + \|P_2 A_{21}\|)^2 > 0$, with P_2 the positive definite solution of the Lyapunov equation

$$A_{22}^T P_2 + P_1 A_{22} = -Q_2 \quad , \quad \text{for some } Q_2 > 0$$

where $\lambda_{min}(\bullet)$ the smallest eigenvalue of a square matrix.
Moreover, in this case, we obtain

$$p_1(z_1(t)) = \begin{cases} -\dfrac{B_1^T P_1 z_1(t)}{\|B_1^T P_1 z_1(t)\|} \rho_1(z_1(t)) \, , & \text{if } \|B_1^T P_1 z_1(t)\| \geq \varepsilon \\[2ex] -\dfrac{B_1^T P_1 z_1(t)}{\varepsilon} \rho_1(z_1(t)) \, , & \text{if } \|B_1^T P_1 z_1(t)\| < \varepsilon \end{cases}$$

where

$$\rho_1(z_1(t)) = \max_{q \in Q} \|W_1 q(t)\|$$

Note that, an admissible matrix F_1, can be obtained by following an algorithm analogous to that proposed above for the calculation of matrix F, in the full state feedback case.

6. H^∞-Robust Control Design

The H^∞-control problem was originally formulated in [36], as the problem of finding an appropriate feedback control law which when applied to a continuous or discrete-time system influenced by external disturbances, renders the corresponding closed-loop system internally stable while keeping the H_∞-norm of the map from external disturbances to controlled outputs of the system, less than a prespecified desired bound.

Similarly to LQG optimization techniques, H^∞-control design methods provide optimal controllers based on the minimization of a certain cost index. In both types of cost function, the dynamic cost weights have a similar effect. Moreover, closed-loop stability can be guaranteed whether the plant under control is non-minimum phase or unstable. However, the fundamental conceptual idea behind H^∞-designs involves the minimization of transfer function magnitudes which is quite different to the LQG requirement to minimize a complex-domain integral representing state and control power spectra. Moreover, the improvements in the robustness of LQG designs to parameter variations or model inaccuracies are achieved in a rather artificial manner, whereas the H^∞-control approach is much more suited for the improvement of robustness, since it is a design procedure developed specifically to allow for the modeling errors.

In this Section, the basic theory and design methods of H^∞-control, as well as some of its important variations, will be presented.

6.1. Continuous-Time State Feedback H^∞-Disturbance Attenuation

Consider the state-space dynamical system of the form (2.1a), (2.1c), (2.1d) with $\mathbf{x}(0) = \mathbf{0}$ and suppose for instance that $\mathbf{J}_1 = \mathbf{J}_2 = \mathbf{0}$. Assume that $(\mathbf{A}, \mathbf{B})$ is a stabilizable matrix pair and that, without loss of generality, the pair $(\mathbf{E}, \mathbf{A})$ has no unobservable modes in the imaginary axis.

Suppose now that the state variables are available for feedback through the control law of the form

$$\zeta(t) = \mathbf{F}\mathbf{x}(t) \qquad (6.1)$$

In the case where (6.1) is applied to system (2.1a), (2.1c), the closed-loop system becomes

$$\dot{\mathbf{x}}(t) = (\mathbf{A} + \mathbf{BF})\mathbf{x}(t) + \mathbf{Dq} \ , \ \mathbf{y}_m(t) = \mathbf{Cx}(t) \ , \ \mathbf{y}_c(t) = \mathbf{Ex}(t)$$

or, in the frequency domain,

$$\mathbf{Y}_c(s) = \mathbf{E}(s\mathbf{I}_n - \mathbf{A} - \mathbf{BF})^{-1}\mathbf{DQ}(s) = \mathbf{H}_{qy_c}(s)\mathbf{Q}(s)$$

where

$$\mathbf{H}_{qy_c}(s) \triangleq \mathbf{E}(s\mathbf{I}_n - \mathbf{A} - \mathbf{BF})^{-1}\mathbf{D}$$

denotes the transfer function between the external disturbances and the controlled outputs.

The H^∞ – norm of $\mathbf{H}_{qy_c}(s)$ is defined by the following chain of relations

$$\left\|\mathbf{H}_{qy_c}(s)\right\|_\infty = \sup_{s=j\omega}\left\{\left\|\mathbf{H}_{qy_c}(s)\mathbf{Q}(s)\right\|_2 : \mathbf{Q}(s) \in \mathbf{H}_2, \left\|\mathbf{Q}(s)\right\|_2 = 1\right\} = \sup_{s=j\omega}\left\{\left\|\mathbf{Y}_c(s)\right\|_2 : \mathbf{Q}(s) \in \mathbf{H}_2, \left\|\mathbf{Q}(s)\right\|_2 = 1\right\}$$

$$= \sup_\omega \sigma_{\max}\left[\mathbf{H}_{qy_c}(j\omega)\right] \quad (6.2)$$

where, in general, $\left\|\mathbf{F}(s)\right\|_2$ denotes the *2-norm* of a signal $\mathbf{f}(t)$ in the frequency domain, which is defined by

$$\left\|\mathbf{F}(s)\right\|_2 = \left\{\frac{1}{2\pi}\int_{-\infty}^{+\infty}\mathbf{F}^*(j\omega)\mathbf{F}(j\omega)d\omega\right\}^{\frac{1}{2}}$$

where, $\mathbf{F}(j\omega)$ is the frequency domain signal, ω is the angular frequency and $(\bullet)^*$ refers to the complex conjugate transpose. Moreover, in (6.2), $\mathbf{H}_2$ is the Hardy space referring to the class of functions, which are analytic and bounded in the open right half plane, i.e.

$$\mathbf{H}_2 = \left\{\mathbf{F} : \mathbf{F}(s)\text{ is analytic in }\mathrm{Re}(s) > 0\text{ and }\left\|\mathbf{F}\right\|_2 < \infty\right.$$

and $\sigma_{\max}(\bullet)$ denotes to the largest singular value of a matrix.

Relation (6.2) means that $\left\|\mathbf{H}_{qy_c}(s)\right\|_\infty$ is the maximum energy of the output $\mathbf{y}_c(t)$ for every disturbance $\mathbf{q}(t)$ with unit energy, i.e. $\left\|\mathbf{Q}(s)\right\|_2 = 1$.

The continuous-time state feedback H^∞-disturbance attenuation problem (also called the continuous-time minimum H^∞-norm regulation problem), can then be stated as follows: Given a constant $\gamma \in \mathbf{R}^+$, find, a suitable control of the form (6.1) such that, simultaneously,

(i)   The closed-loop system is internally stable.

(ii)   For a prespecified $\gamma > 0$, the H^∞-norm of $\mathbf{H}_{qy_c}(s)$ satisfies the bound

$$\left\|\mathbf{H}_{qy_c}(s)\right\| < \gamma$$

Note that the above control objective can be written in the equivalent form

$$\left\|\mathbf{y}_c\right\|_2^2 - \gamma^2\left\|\mathbf{q}\right\|_2^2 \leq -\varepsilon\left\|\mathbf{q}\right\|_2^2$$

for all $\mathbf{q}(t) \in \mathbf{L}_2[0,\infty)$ and for some $\varepsilon \in \mathbf{R}^+$, where $\mathbf{L}_2[0,\infty) = \mathbf{S}_+ \cap \mathbf{L}_2(-\infty,+\infty)$

and where

$$\mathbf{S}_+ = \left\{ \mathbf{d}(t) : \mathbf{R} \to \mathbf{R}^d, \text{ such that } q(t) \text{ is Lebesgue measurable and } q(t) = 0, \forall t < 0 \right\}$$

$$\mathbf{L}_2(-\infty, +\infty) = \left\{ q(t) : \mathbf{R} \to \mathbf{R}^d, \text{ such that } q(t) \text{ is Lebesgue measurable and } \int_{-\infty}^{+\infty} \|q(t)\|^2 \, dt < \infty \right\}$$

The solution of the above problem can be summarized in the following Proposition (for proof see [37])

Proposition 6.1. Under the stated assumptions of the system under control, the state feedback H^∞-disturbance attenuation problem is solvable if and only if, for some $\varepsilon > 0$ and positive definite matrices $\widetilde{\mathbf{R}} \in \mathbf{R}^{m \times m}$, $\widetilde{\mathbf{Q}} \in \mathbf{R}^{n \times n}$, there exists a positive definite solution $\mathbf{P}$ of the algebraic Riccati equation

$$\mathbf{PA} + \mathbf{A}^T \mathbf{P} - \frac{1}{\varepsilon} \mathbf{PB} \widetilde{\mathbf{R}}^{-1} \mathbf{B}^T \mathbf{P} + \frac{1}{\gamma} \mathbf{PDD}^T \mathbf{P} + \frac{1}{\gamma} \mathbf{E}^T \mathbf{E} + \varepsilon \widetilde{\mathbf{Q}} = 0 \qquad (6.3)$$

Moreover, in this case, the required feedback gain matrix is given by

$$\mathbf{F} = -\frac{1}{2\varepsilon} \widetilde{\mathbf{R}}^{-1} \mathbf{B}^T \mathbf{P}$$

It is worth noticing at this point that, a way to find the solution $\mathbf{P}$ of (6.3) can be obtained by using an algorithm similar to that proposed above for the solution of (4.4). To this end, one must simply replace $\mathbf{H}_1$ by the following Hamiltonian matrix

$$\mathbf{H}_4 = \begin{bmatrix} \mathbf{A} & \frac{1}{\gamma} \mathbf{DD}^T - \frac{1}{\varepsilon} \mathbf{B} \widetilde{\mathbf{R}}^{-1} \mathbf{B}^T \\ -\frac{1}{\gamma} \mathbf{E}^T \mathbf{E} - \varepsilon \widetilde{\mathbf{Q}} & -\mathbf{A}^T \end{bmatrix}$$

Proposition 6.1 states that there exists an internally stabilizing state feedback which makes the H_∞-norm less than some a priori given bound $\gamma > 0$ if and only if a positive definite matrix $\mathbf{P}$ and a positive constant ε exist such that $\mathbf{P}$ is a stabilizing solution of the Riccati equation (6.3) which is parameterized by ε. The main drawback of this result is that the Riccati equation is parameterized. However, under certain assumptions, it turned out that this parameter could be removed. In particular, we have the following stronger results (see [38]-[41] for details).

Proposition 6.2. Consider system (2.1a), (2.1d) with $\mathbf{x}(0) = 0$, $\mathbf{J}_2 \neq 0$ and let $\gamma > 0$ given. Assume that the system $(\mathbf{A}, \mathbf{B}, \mathbf{E}, \mathbf{J}_2)$ has no invariant zeros on the imaginary axis and that $\mathbf{J}_2$ is injective (i.e. matrix $\mathbf{R}_J = \mathbf{J}_2^T \mathbf{J}_2$ is nonsingular). Then, there exists an internally stabilizing state feedback law of the form (6.1), which makes the H_∞-norm of the transfer fnction matrix $H_{qy_c}(s) = (E + J_2 F)(sI - A - BF)^{-1}D$ less than γ, if and only if there is a positive semi-definite solution of the ARE

$$A^T P + PA + E^T E - \begin{bmatrix} B^T P + J_2^T E \\ D^T P \end{bmatrix}^T \begin{bmatrix} J_2^T J_2 & 0 \\ 0 & -\gamma^2 I \end{bmatrix}^{-1} \begin{bmatrix} B^T P + J_2^T E \\ D^T P \end{bmatrix} = 0$$

such that the matrix

$$A_{cl} = A - \begin{bmatrix} B & D \end{bmatrix} \begin{bmatrix} J_2^T J_2 & 0 \\ 0 & -\gamma^2 I \end{bmatrix}^{-1} \begin{bmatrix} B^T P + J_2^T E \\ D^T P \end{bmatrix}$$

is asymptotically stable. Furthermore, in this case, the required feedback gain matrix is given by

$$F = -R_J^{-1} \begin{bmatrix} J_2^T E + B^T P \end{bmatrix}$$

Proposition 6.3. Consider system (2.1a), (2.1d) with $x(0) = 0$, $J_2 \neq 0$ and let $\gamma > 0$ given. Assume that (A, B) is stabilizable and (E, A) is observable. Assume also that $J_2^T \begin{bmatrix} J_2 & E \end{bmatrix} = \begin{bmatrix} I & 0 \end{bmatrix}$. Then, there exists an internally stabilizing state feedback law of the form (6.1), which makes the H_∞-norm of $H_{qy_c}(s) = (E + J_2 F)(sI - A - BF)^{-1}D$ less than γ, if and only if there is a positive semi-definite solution of the ARE

$$A^T P + PA + P\left(\frac{DD^T}{\gamma^2} - BB^T\right)P + E^T E = 0$$

Furthermore, in this case, the required feedback gain matrix is given by

$$F = -B^T P$$

It is not difficult to see that if the weightings $Q_1 = E^T E$ and $R_1 = R_J$ or $R_1 = I$ denote the state and control signal weighting in an LQ or LQG synthesis problem, the solution will be very similar to the state feedback controllers defined in Propositions 6.2 and 6.3. In fact when $\gamma \to \infty$, certain terms in the algebraic Riccati equations tend to zero and the H^∞ and LQ or LQG controllers coincide.

Clearly Propositions 6.2 and 6.3, provide solution to the state feedback H^∞-control problem in the case where $J_2 \neq 0$. In the case where $J_2 = 0$, one may write

$$J_2 = \theta J \ , \ \theta > 0 \ , \ \theta \to 0 \ , \ J^T \begin{bmatrix} J & E \end{bmatrix} = \begin{bmatrix} I & 0 \end{bmatrix} \quad (6.4)$$

for some θ. Then we can establish the following Corollary.

Corollary 6.1. Consider system (2.1a), (2.1d) with $\mathbf{J}_2 = \varepsilon \mathbf{J}$, for some $\theta > 0$, $\theta \to 0$ and let $\gamma > 0$ given. Assume that $(\mathbf{A}, \mathbf{B})$ is stabilizable and $(\mathbf{E}, \mathbf{A})$ is observable. Assume also that $\mathbf{J}^T [\mathbf{J} \ \ \mathbf{E}] = [\mathbf{I} \ \ \mathbf{0}]$. Then, there exists an internally stabilizing state feedback law of the form (6.1), which makes the H_∞-norm of $\mathbf{H}_{qy_c}(s) = \mathbf{E}(s\mathbf{I} - \mathbf{A} - \mathbf{BF})^{-1}\mathbf{D}$ less than γ, if and only if there is a positive semi-definite solution of the ARE

$$\mathbf{A}^T\mathbf{P} + \mathbf{PA} + \mathbf{P}\left(\frac{\mathbf{DD}^T}{\gamma^2} - \frac{\mathbf{BB}^T}{\theta^2}\right)\mathbf{P} + \mathbf{E}^T\mathbf{E} = 0 \qquad (6.5)$$

Furthermore, in this case, the required feedback gain matrix is given by

$$\mathbf{F} = -\frac{1}{\theta^2}\mathbf{B}^T\mathbf{P}$$

The solution of (6.5) can be easily obtained by using an algorithm similar to that proposed above for the solution of (4.4). To this end, one must simply replace $\mathbf{H}_1$ by the following Hamiltonian matrix

$$\mathbf{H}_5 = \begin{bmatrix} \mathbf{A} & \dfrac{\mathbf{DD}^T}{\gamma^2} - \dfrac{\mathbf{BB}^T}{\theta^2} \\ -\mathbf{E}^T\mathbf{E} & -\mathbf{A}^T \end{bmatrix}$$

6.2. Discrete-Time State Feedback Minimum H^∞-Norm Regulation

Consider the discrete analogous of system (2.1a), (2.1d), obtained by sampling with period T_0 and holding with zero order hold circuits, and having the form

$$\mathbf{x}[(k+1)T_0] = \Phi\mathbf{x}(kT_0) + \hat{\mathbf{B}}\zeta(kT_0) + \hat{\mathbf{D}}\mathbf{q}(kT_0) \qquad (6.6a)$$

$$\mathbf{y}_c(kT_0) = \mathbf{E}(kT_0) + \mathbf{J}_2\zeta(kT_0) \ , \ \ \mathbf{x}(0) = 0 \ , \ \ k \geq 0 \qquad (6.6b)$$

It is assumed that $(\Phi, \hat{\mathbf{B}})$ is controllable (stabilizable), $(\mathbf{E}, \Phi)$ is observable (detectable) and that $\mathbf{J}_2^T [\mathbf{J}_2 \ \ \mathbf{E}] = [\mathbf{I} \ \ \mathbf{0}]$.

The discrete-time state feedback minimum H^∞-norm regulation problem is stated as follows: Find a feedback control strategy of the form

$$\zeta(kT_0) = \mathbf{Fx}(kT_0) \qquad (6.7)$$

in $\ell_2[0, \infty)$ which interanlly stabilizes the closed loop system and makes the H_∞-norm of the transfer function matrix $\mathbf{H}_{qy_c}(z) = (\mathbf{E} + \mathbf{J}_2\mathbf{F})(z\mathbf{I} - \mathbf{A} - \mathbf{BF})^{-1}\mathbf{D}$, from $\mathbf{q}(kT_0)$ to $\mathbf{y}_c(kT_0)$, less than a prespecified bound $\gamma > 0$, i.e.

$$\left\| \mathbf{H}_{qy_c}(z) \right\|_\infty < \gamma$$

where $\ell_2[0,\infty)$ is the space of sequences defined by

$$\ell_2[0,\infty) = \left\{ \mathbf{f} \triangleq \{\mathbf{f}_k\}_0^\infty : \|\mathbf{f}\|_2 \triangleq \left\{ \sum_0^\infty \mathbf{f}_k^T \mathbf{f}_k \right\}^{1/2} < \infty \right\}$$

and

$$\left\| \mathbf{H}_{qy_c}(z) \right\|_\infty = \sup_{\mathbf{q}(kT_0) \in \ell_2[0,\infty)} \frac{\left\| \mathbf{y}_c(kT_0) \right\|_2}{\left\| \mathbf{q}(kT_0) \right\|_2} = \sup_{\theta \in [0,2\pi]} \sigma_{\max}\left[\mathbf{H}_{qy_c}\left(e^{j\theta}\right) \right] = \sup_{|z|=1} \sigma_{\max}\left[\mathbf{H}_{qy_c}(z) \right]$$

where, $\sigma_{\max}\left[\mathbf{H}_{qy_c}(z) \right]$ is the largest singular value of $\mathbf{H}_{qy_c}(z)$, and where we have used the standard definition of the ℓ_2 norm of a discrete signal $s(kT_0)$, given by

$$\left\| s(kT_0) \right\|_2^2 = \sum_{k=0}^\infty s^T(kT_0) s(kT_0)$$

The solution of this problem can be obtained as in the following Proposition (for proof see [40]-[44])

Proposition 6.4. Under the stated assumptions, there exists an internally stabilizing state feedback law of the form (6.7), which makes the H_∞-norm of $\mathbf{H}_{qy_c}(z) = (\mathbf{E} + \mathbf{J}_2\mathbf{F})(z\mathbf{I} - \mathbf{A} - \mathbf{BF})^{-1}\mathbf{D}$ less than γ, if and only if there is a positive semi-definite solution of the DARE

$$\mathbf{X} = \Phi^T\mathbf{X}\Phi + \mathbf{E}^T\mathbf{E} - \Phi^T\mathbf{X}\hat{\mathbf{B}}\left(\mathbf{I} + \hat{\mathbf{B}}^T\mathbf{X}\hat{\mathbf{B}}\right)^{-1}\hat{\mathbf{B}}^T\mathbf{X}\Phi + \mathbf{X}\hat{\mathbf{D}}\left(\gamma^2\mathbf{I} + \hat{\mathbf{D}}^T\mathbf{X}\hat{\mathbf{D}}\right)^{-1}\hat{\mathbf{D}}^T\mathbf{X}$$

Furthermore, in this case, the required feedback gain matrix is given by

$$\mathbf{F} = -\left(\mathbf{I} + \hat{\mathbf{B}}^T\mathbf{X}\hat{\mathbf{B}}\right)^{-1}\hat{\mathbf{B}}^T\mathbf{X}\Phi$$

The close relationship between the discrete-time LQ or LQG and H^∞-control problems is once again obvious. Indeed, if $\gamma \to \infty$, $\mathbf{Q} = \mathbf{E}^T\mathbf{E}$ and $\mathbf{R} = \mathbf{I}$, one obtains the standard result for discrete-time LQ or LQG control.

In the case where $\mathbf{J}_2 = \mathbf{0}$, we may use (6.4) to obtain the following result

Corollary 6.2. Consider system (6.6), (2.1d) with $\mathbf{J}_2 = \varepsilon\mathbf{J}$, for some $\eta > 0$, $\eta \to 0$ and let $\gamma > 0$ given. Assume that $\left(\Phi, \hat{\mathbf{B}}\right)$ is controllable (stabilizable) and $(\mathbf{E}, \Phi)$ is observable (detectable). Assume also that $\mathbf{J}^T[\mathbf{J} \quad \mathbf{E}] = [\mathbf{I} \quad \mathbf{0}]$. Then, there exists an internally stabilizing state feedback law of the form (6.7), which makes the H_∞-norm of $\mathbf{H}_{qy_c}(z) = \mathbf{E}(z\mathbf{I} - \mathbf{A} - \mathbf{BF})^{-1}\mathbf{D}$ less than γ, if and only if there is a positive semi-definite solution of the DARE

$$\mathbf{X} = \mathbf{\Phi}^T\mathbf{X}\mathbf{\Phi} + \mathbf{E}^T\mathbf{E} - \mathbf{\Phi}^T\mathbf{X}\hat{\mathbf{B}}\left(\eta^2\mathbf{I} + \hat{\mathbf{B}}^T\mathbf{X}\hat{\mathbf{B}}\right)^{-1}\hat{\mathbf{B}}^T\mathbf{X}\mathbf{\Phi} + \mathbf{X}\hat{\mathbf{D}}\left(\gamma^2\mathbf{I} + \hat{\mathbf{D}}^T\mathbf{X}\hat{\mathbf{D}}\right)^{-1}\hat{\mathbf{D}}^T\mathbf{X} \quad (6.8)$$

Furthermore, in this case, the required feedback gain matrix is given by

$$\mathbf{F} = -\left(\eta^2\mathbf{I} + \hat{\mathbf{B}}^T\mathbf{X}\hat{\mathbf{B}}\right)^{-1}\hat{\mathbf{B}}^T\mathbf{X}\mathbf{\Phi}$$

Note that the solution of (6.8) can be easily obtained by using an algorithm similar to that proposed above for the solution of (4.12). To this end, one must simply replace $\mathbf{H}_3$ by the following Hamiltonian matrix

$$\mathbf{H}_6 = \begin{bmatrix} \mathbf{\Phi} + \dfrac{\hat{\mathbf{B}}\hat{\mathbf{B}}^T}{\eta^2}\left(\mathbf{\Phi}^T\right)^{-1}\mathbf{E}^T\mathbf{E} & \mathbf{\Phi}\dfrac{\hat{\mathbf{D}}\hat{\mathbf{D}}^T}{\gamma^2} - \dfrac{\hat{\mathbf{B}}\hat{\mathbf{B}}^T}{\eta^2}\left(\mathbf{\Phi}^T\right)^{-1}\left(\mathbf{I} - \mathbf{E}^T\mathbf{E}\dfrac{\hat{\mathbf{D}}\hat{\mathbf{D}}^T}{\gamma^2}\right) \\ -\left(\mathbf{\Phi}^T\right)^{-1}\mathbf{E}^T\mathbf{E} & \left(\mathbf{\Phi}^T\right)^{-1}\left(\mathbf{I} - \mathbf{E}^T\mathbf{E}\dfrac{\hat{\mathbf{D}}\hat{\mathbf{D}}^T}{\gamma^2}\right) \end{bmatrix}$$

6.3. Alternative Discrete-Time H^∞-Control Strategies: Dynamic Output Feedback vs. Multirate Output Controllers

In the case where the states of system (6.6) are not available for feedback, the solution of H^∞-control problem can be obtained on the basis of dynamic output (estimator based) feedback, along the lines of the H^∞-filtering theory reported in [43], [45], [46]. More precisely, consider system (6.6) and assume that discrete-time output measurements of the form

$$\mathbf{y}_m(kT_0) = \mathbf{C}\mathbf{x}(kT_0) + \mathbf{J}_1\zeta(kT_0) \quad (6.9)$$

are available for feedback. In this case, our aim is to find a dynamic output feedback control law of the form

$$\hat{\mathbf{x}}[(k+1)T_0] = \mathbf{\Phi}_c\hat{\mathbf{x}}(kT_0) + \hat{\mathbf{B}}_c\mathbf{y}_m(kT_0) \ , \quad \zeta(kT_0) = \mathbf{F}_c\hat{\mathbf{x}}(kT_0) \quad (6.10)$$

which internally stabilizes the closed-loop system and makes the H_∞-norm of the closed-loop transfer function less than a prespecified bound $\gamma > 0$. The solution of the problem can be obtained as in the following Proposition.

Proposition 6.5. Assume that $\left(\mathbf{\Phi}, \hat{\mathbf{B}}\right)$ is controllable (stabilizable), $(\mathbf{E}, \mathbf{\Phi})$ is observable (detectable) and that $\hat{\mathbf{D}}$ is left invertible. Define, $\hat{\mathbf{D}}_\gamma = \gamma^{-1}\hat{\mathbf{D}}$. Then, the dynamic output feedback minimum H^∞-norm regulation problem is solvable if and only if there exist positive semidefinite matrices $\mathbf{X}$ and $\mathbf{Y}$, satisfying:

$$X = \Phi^T X \Phi + E^T E - \Phi^T X \hat{B}\left(I + \hat{B}^T X \hat{B}\right)^{-1} \hat{B}^T X \Phi + X \hat{D}_\gamma \left(I + \hat{D}_\gamma^T X \hat{D}_\gamma\right)^{-1} \hat{D}_\gamma^T X$$

$$\left(\tilde{\Phi}, \tilde{D}\right) \text{ is stabilizable}$$

$$\left(C, \tilde{\Phi}\right) \text{ is detectable}$$

$$Y = \tilde{\Phi} Y \tilde{\Phi}^T + \tilde{D}\tilde{D}^T - \tilde{\Phi} Y C^T \left(I + C Y C^T\right)^{-1} C Y \tilde{\Phi}^T + Y \tilde{E}^T \left(I + \tilde{E} Y \tilde{E}^T\right)^{-1} \tilde{E} Y$$

where

$$\tilde{\Phi} = \left(I + \hat{D}_\gamma \hat{D}_\gamma^T X\right)\Phi$$

$$\tilde{D} = \left(I + \hat{D}_\gamma \hat{D}_\gamma^T X\right)\hat{D}_\gamma W_1^T$$

$$\tilde{E} = W_2 \left(I + \hat{B}^T X \hat{B}\right)^{-1} \hat{B}^T X \Phi$$

and where W_1 and W_2 are defined by

$$W_1^T W_1 = \left(I + \hat{D}_\gamma^T X \hat{D}_\gamma\right)^{-1}$$

$$W_2^T W_2 = I + \hat{B}^T X \hat{B}$$

If such X and Y exist, then one controller of the form (6.10) that satisfies the design requirements is

$$\Phi_c = \tilde{\Phi} - KC - \left(I + \hat{D}_\gamma \hat{D}_\gamma^T X\right)\hat{B}\left(I + \hat{B}^T X \hat{B}\right)^{-1} \hat{B}^T X \Phi$$

$$\hat{B}_c = K$$

$$F_c = -\left(I + \hat{B}^T X \hat{B}\right)^{-1} \hat{B}^T X \Phi$$

where

$$K = \tilde{\Phi} Y C^T \left(I + C^T Y C\right)^{-1}$$

From Proposition 6.5, it becomes clear that, for the solution of the minimum H^∞-norm regulation problem using dynamic output feedback controllers, it is necessary to solve (similarly to the LQG design) two discrete algebraic Riccati equations. Therefore, the solution of the problem is quite complicated, lengthy and numerically involved, particularly in cases where the system under control is of high order.

Fortunately, we can still apply the technique based on MROCs, in order to attack the minimum H^∞-norm regulation problem, in the case where the states are inaccessible, and to reduce drastically the overall complexity. Following the results presented in Section 4.5, the technique for H^∞-control using MROCs consists in finding a fictitious state feedback control law of the form (6.7) which equivalently solves the problem for the system (6.6), (6.9), according to Proposition 6.4 or Corollary 6.2, and then calculating the gains of the actual controller of the form (4.22), either by solving (4.26) with respect to L_γ and L_ζ, or by first specifying $L_\zeta = L_{\zeta,sp}$ and then solving (4.28) with respect to L_γ. The solvability of (4.26) and (4.28) is guaranteed by appropriate choices of the output multiplicities of the sampling. Therefore, the solution of the minimum H^∞-norm regulation problem, essentially consists on the solution of a single DARE and a linear matrix equation, instead of the two DARE needed in the case of output feedback controllers. Finally, more general classes of disturbances (beyond the class of piecewise constant disturbance treated here) can be treated using a similar analysis, which is presented in [25].

7. Structural Modelling: Principles and Applications

Elastic structures with an m-dimensional control vector $\zeta(t)$, which are assumed to be assembled from finite elements obeying linear material laws, can easily be put in the form of equations (2.1a)-(2.1d).

For example, in the case of linearly elastic structures in civil engineering with an m-dimensional control vector $\zeta(t)$ the dynamical model reads

$$\mathbf{M}\ddot{u}(t) + \check{\mathbf{C}}\dot{u}(t) + \mathbf{K}u(t) = \mathbf{M}_0\ddot{u}_g + \mathbf{B}_0\zeta(t) \qquad (7.1)$$

where $\mathbf{M} \in \mathbf{R}^{k\times k}$ is the mass matrix, $\check{\mathbf{C}} \in \mathbf{R}^{k\times k}$ is the damping matrix, $\mathbf{K} \in \mathbf{R}^{k\times k}$ is the stiffness matrix, $\ddot{u}_g \in \mathbf{R}^g$ is the ground earthquake acceleration vector (the load), $u(t) \in \mathbf{R}^k$ is the nodal displacement vector, $\dot{u}(t) \in \mathbf{R}^k$ is the velocity vector and $\ddot{u}(t) \in \mathbf{R}^k$ is the acceleration vector. Moreover, $\mathbf{M}_0 \in \mathbf{R}^{k\times g}$ and $\mathbf{B}_0 \in \mathbf{R}^{k\times m}$, are the loading and control forces arrangement matrices, respectively. Here, the method of additional masses is used for the approximate modeling of structures with the ground support earthquake acceleration $\ddot{u}_g$ (see e.g. [47]). The additional mass of the ground is used for the construction of the matrices $\mathbf{M}$ and $\mathbf{M}_0$ in (7.1). By using the substitution $\dot{u}(t) = v(t)$, one gets from (7.1) the state space model

$$\begin{bmatrix} \dot{u}(t) \\ \dot{v}(t) \end{bmatrix} = \begin{bmatrix} 0_{k\times k} & I_{k\times k} \\ -M^{-1}K & -M^{-1}\check{C} \end{bmatrix} \begin{bmatrix} u(t) \\ v(t) \end{bmatrix} + \begin{bmatrix} 0_{k\times m} \\ M^{-1}B_0 \end{bmatrix} \zeta(t) + \begin{bmatrix} 0_{k\times k} \\ M^{-1}M_0 \end{bmatrix} \ddot{u}_g(t) \quad (7.2)$$

Relation (7.2), may further be written in the standard matrix-vector state space forms (2.1a) or (2.1b), where, for n=2k, d=g, $A \in R^{n\times n}$, $B \in R^{n\times m}$, $D \in R^{n\times p}$, $x(t) \in R^n$, $q(t) \in R^d$ and $w(t) \in R^n$, have the forms

$$A = \begin{bmatrix} 0_{k\times k} & I_{k\times k} \\ -M^{-1}K & -M^{-1}\check{C} \end{bmatrix}, B = \begin{bmatrix} 0_{k\times m} \\ M^{-1}B_0 \end{bmatrix} \quad (7.3a)$$

$$D = \begin{bmatrix} 0_{k\times g} \\ M^{-1}M_0 \end{bmatrix}, \; x(t) = \begin{bmatrix} u(t) \\ v(t) \end{bmatrix} = \begin{bmatrix} u(t) \\ \dot{u}(t) \end{bmatrix}, \; q(t)=\ddot{u}_g, \; w(t)=Dq(t) \quad (7.3b)$$

The above general formulation for active control systems can be further specified for certain structural systems. For example, a shear-type structural frame may represent with sufficient accuracy the dynamic behaviour of a multistory building during an earthquake. In this case one accepts a a diagonal structure of the mass matrix M. To this end, one assumes as outputs the displacements of the nodes at the story over the foundation where the control force inputs are placed (control floor), i.e.

$$C = \begin{bmatrix} 0_{m\times g} & I_{m\times m} & 0_{m\times r} & 0_{m\times g} & 0_{m\times m} & 0_{m\times r} \end{bmatrix} \quad (7.4)$$

Let the following partition of displacement, velocity and acceleration degrees be considered

$$u = \begin{bmatrix} u_1 & u_2 & u_3 \end{bmatrix}^T, \; \dot{u} = \begin{bmatrix} \dot{u}_1 & \dot{u}_2 & \dot{u}_3 \end{bmatrix}^T, \; \ddot{u} = \begin{bmatrix} \ddot{u}_1 & \ddot{u}_2 & \ddot{u}_3 \end{bmatrix}^T \quad (7.5)$$

where u_1 is the g-dimensional vector of basement displacements, u_2 is the m-dimensional vector of control floor displacements and u_3 is the r-dimensional vector of the rest (upper stories) displacements. Note that k=g+m+r. Since a localized ground story control configuration is usually studied, a particular form of (7.1) is often more relevant. It reads

$$\begin{bmatrix} M_1 + M_g & 0 & 0 \\ 0 & M_2 & 0 \\ 0 & 0 & M_3 \end{bmatrix} \begin{bmatrix} \ddot{u}_1 \\ \ddot{u}_2 \\ \ddot{u}_3 \end{bmatrix} + \begin{bmatrix} \check{C}_{11} & \check{C}_{12} & \check{C}_{13} \\ \check{C}_{21} & \check{C}_{22} & \check{C}_{23} \\ \check{C}_{31} & \check{C}_{32} & \check{C}_{33} \end{bmatrix} \begin{bmatrix} \dot{u}_1 \\ \dot{u}_2 \\ \dot{u}_3 \end{bmatrix}$$
$$+ \begin{bmatrix} K_{11} & K_{12} & K_{13} \\ K_{21} & K_{22} & K_{23} \\ K_{31} & K_{32} & K_{33} \end{bmatrix} \begin{bmatrix} u_1 \\ u_2 \\ u_3 \end{bmatrix} = \begin{bmatrix} 0 \\ B_m \\ 0 \end{bmatrix} \zeta + \begin{bmatrix} M_g \\ 0 \\ 0 \end{bmatrix} \ddot{u}_g \quad (7.6)$$

since $\mathbf{B}_0 = \begin{bmatrix} \mathbf{0}_{g \times m} \\ \mathbf{B}_m \\ \mathbf{0}_{r \times m} \end{bmatrix}$, $\mathbf{M}_0 = \begin{bmatrix} \mathbf{M}_g \\ \mathbf{0}_{m \times g} \\ \mathbf{0}_{r \times g} \end{bmatrix}$, where $\mathbf{B}_m \in \mathbf{R}^{m \times m}$ is an invertible matrix to ensure the linear independence of the control inputs $\zeta(t)$ and where, in general, $\mathbf{M}_g \in \mathbf{R}^{g \times \ell}$ is an appropriate fictitious additional mass matrix. This is a general formulation, which includes, for example, rocking motions, where one has ℓ elements of measured ground accelerations and g degrees of freedom at the ground nodes. In the following, we set $\overline{\mathbf{M}}_1 = \mathbf{M}_1 + \mathbf{M}_g$. Thus, equations (7.3) take on the form

$$\mathbf{A} = \begin{bmatrix} \mathbf{0}_{k \times k} & \mathbf{I}_{k \times k} \\ -\mathbf{A}_1 & -\mathbf{A}_2 \end{bmatrix} \qquad (7.7a)$$

$$\mathbf{A}_1 = \begin{bmatrix} \overline{\mathbf{M}}_1^{-1}\mathbf{K}_{11} & \overline{\mathbf{M}}_1^{-1}\mathbf{K}_{12} & \overline{\mathbf{M}}_1^{-1}\mathbf{K}_{13} \\ \mathbf{M}_2^{-1}\mathbf{K}_{21} & \mathbf{M}_2^{-1}\mathbf{K}_{22} & \mathbf{M}_2^{-1}\mathbf{K}_{23} \\ \mathbf{M}_3^{-1}\mathbf{K}_{31} & \mathbf{M}_3^{-1}\mathbf{K}_{32} & \mathbf{M}_3^{-1}\mathbf{K}_{33} \end{bmatrix}, \quad \mathbf{A}_{22} = \begin{bmatrix} \overline{\mathbf{M}}_1^{-1}\check{\mathbf{C}}_{11} & \overline{\mathbf{M}}_1^{-1}\check{\mathbf{C}}_{12} & \overline{\mathbf{M}}_1^{-1}\check{\mathbf{C}}_{13} \\ \mathbf{M}_2^{-1}\check{\mathbf{C}}_{21} & \mathbf{M}_2^{-1}\check{\mathbf{C}}_{22} & \mathbf{M}_2^{-1}\check{\mathbf{C}}_{23} \\ \mathbf{M}_3^{-1}\check{\mathbf{C}}_{31} & \mathbf{M}_3^{-1}\check{\mathbf{C}}_{32} & \mathbf{M}_3^{-1}\check{\mathbf{C}}_{33} \end{bmatrix} \qquad (7.7b)$$

$$\mathbf{B} = \begin{bmatrix} \mathbf{0}_{k \times m} \\ \mathbf{0}_{g \times m} \\ \mathbf{M}_2^{-1}\mathbf{B}_m \\ \mathbf{0}_{r \times m} \end{bmatrix}, \quad \mathbf{D} = \begin{bmatrix} \mathbf{0}_{k \times g} \\ \overline{\mathbf{M}}_1^{-1}\mathbf{M}_g \\ \mathbf{0}_{m \times g} \\ \mathbf{0}_{r \times g} \end{bmatrix} \qquad (7.7c)$$

It is worth noticing that, more complicated modeling techniques are also plausible (see [48], [49], and the references cited therein).

In order to fix the ideas let us assume a base-isolated n-story building which is modelled by a so-called shear-type frame. The building is viewed as an interconnection of physical components. Moreover, a "lumped mass parameter" model, i.e., an assemblage of masses, each representing a combined floor mass, and linear springs and dampers for the inter-story connections is used. Motion is assumed to take place in one horizontal direction. Torsional and external damping effects, although present generally, are not considered here. In this example the horizontal displacements are denoted by y. It is obvious that y is composed of the horizontal translational degrees of freedom (d.o.f.s) of the displacement vector u used in relation (7.1).

Let $y_0(t)$ be the displacement of the ground and $y_i(t)$ be the displacement of the ith floor at time t relative to an inertial frame of reference; see Figure 4. Let m_i be the mass of the ith floor, $\check{c}_{i-1}$ be the damping coefficient and κ_{i-1} be the spring constant in the connection of the ith floor to the floor, or ground, below it.

Subsequently, the time argument, t, is omitted when no confusion is likely to arise. Then,

$$m_1 \ddot{y}_1 = -\breve{c}_0(\dot{y}_1 - \dot{y}_0) - \kappa_0(y_1 - y_0) + \breve{c}_1(\dot{y}_2 - \dot{y}_1) + \kappa_1(y_2 - y_1)$$
$$m_2 \ddot{y}_2 = -\breve{c}_1(\dot{y}_2 - \dot{y}_1) - \kappa_1(y_2 - y_1) + \breve{c}_2(\dot{y}_3 - \dot{y}_2) + \kappa_2(y_3 - y_2)$$

$$\cdot \quad \cdot \quad \cdot$$

$$m_i \ddot{y}_i = -\breve{c}_{i-1}(\dot{y}_i - \dot{y}_{i-1}) - \kappa_{i-1}(y_i - y_{i-1}) + \breve{c}_i(\dot{y}_{i+1} - \dot{y}_i) + \kappa_i(y_{i+1} - y_i) \qquad \text{(7.8a)}$$

$$\cdot \quad \cdot \quad \cdot$$

$$m_N \ddot{y}_N = -\breve{c}_{N-1}(\dot{y}_N - \dot{y}_{N-1}) - \kappa_{N-1}(y_N - y_{N-1})$$

$$y_i(t_0) = y_i^0 \; , \quad \dot{y}_i(t_0) = \dot{y}_i^0 \; , \text{ for } i = 1,2,...,N \qquad \text{(7.8b)}$$

are some initial conditions at time t=t0.

A state space model may be obtained by considering $\mathbf{x} = (x_1, x_2,..., x_{2N})^T$ where $x_i = \dot{y}_i$, $x_{i+N} = y_i$, $i = 1,2,...,N$. Equations (7.8a), (7.8b) become

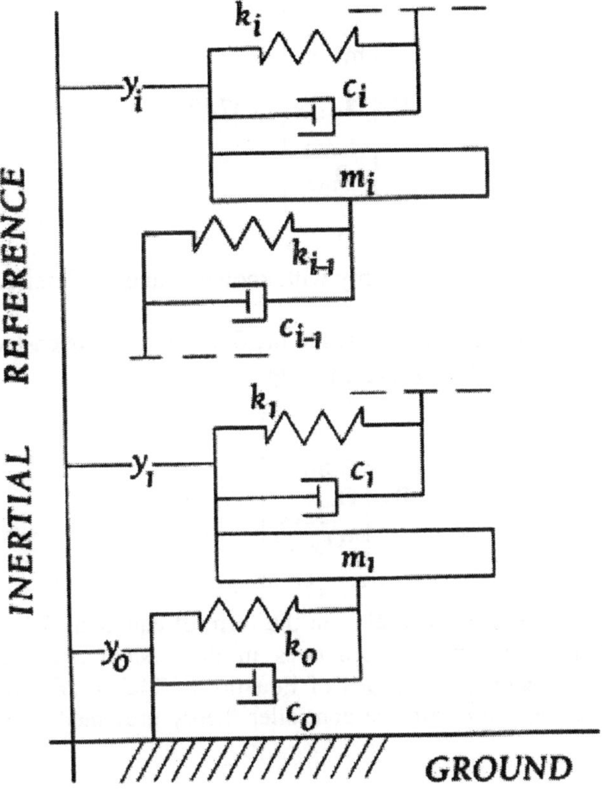

Figure 4. Lumped parameter model of a building in absolute coordinates

$$\dot{\mathbf{x}}(t) = \mathbf{A}\mathbf{x}(t) + \mathbf{D}\mathbf{q}(t) \ , \ \mathbf{x}(t_0) = \mathbf{x}_0 \qquad (7.9)$$

where

$$
\mathbf{A} = \left[
\begin{array}{ccccccc|ccccccc}
\frac{(-\breve{c}_0 - \breve{c}_1)}{m_1} & \frac{\breve{c}_1}{m_1} & 0 & \cdots & \cdots & \cdots & 0 & \frac{(-\kappa_0 - \kappa_1)}{m_1} & \frac{\kappa_1}{m_1} & 0 & \cdots & \cdots & \cdots & 0 \\[2mm]
\frac{\breve{c}_1}{m_2} & \frac{(-\breve{c}_1 - \breve{c}_2)}{m_2} & \frac{\breve{c}_2}{m_2} & 0 & \cdots & \cdots & 0 & \frac{\kappa_1}{m_2} & \frac{(-\kappa_1 - \kappa_2)}{m_2} & \frac{\kappa_2}{m_2} & 0 & \cdots & \cdots & 0 \\[2mm]
\cdots & \cdots & \cdots & \cdots & \cdots & \cdots & \cdots & \cdots & \cdots & \cdots & \cdots & \cdots & \cdots & \cdots \\[2mm]
0 & \cdots & \frac{\breve{c}_{i-1}}{m_i} & \frac{(-\breve{c}_{i-1} - \breve{c}_i)}{m_i} & \frac{\breve{c}_i}{m_i} & \cdots & 0 & 0 & \cdots & \frac{\kappa_{i-1}}{m_i} & \frac{(-\kappa_{i-1} - \kappa_i)}{m_i} & \frac{\kappa_i}{m_i} & \cdots & 0 \\[2mm]
\cdots & \cdots & \cdots & \cdots & \cdots & \cdots & \cdots & \cdots & \cdots & \cdots & \cdots & \cdots & \cdots & \cdots \\[2mm]
0 & \cdots & \cdots & \cdots & \cdots & \frac{\breve{c}_{N-1}}{m_N} & \frac{-\breve{c}_{N-1}}{m_N} & 0 & \cdots & \cdots & \cdots & \cdots & \frac{\kappa_{N-1}}{m_N} & \frac{-\kappa_{N-1}}{m_N} \\[2mm]
\hline
\multicolumn{7}{c|}{\mathbf{I}_{N\times N}} & \multicolumn{7}{c}{\mathbf{O}_{N\times N}}
\end{array}
\right]
$$

$$(7.10\text{a})$$

$$
\mathbf{D} = \left[
\begin{array}{cc}
\dfrac{\breve{c}_0}{m_1} & \dfrac{\kappa_0}{m_1} \\[3mm]
\multicolumn{2}{c}{\mathbf{O}_{(N-1)\times 2}} \\[1mm]
\hline
\multicolumn{2}{c}{\mathbf{O}_{N\times 2}}
\end{array}
\right] \qquad (7.10\text{b})
$$

and $\mathbf{q}(t) = \begin{bmatrix} \dot{y}_0 \\ y_0 \end{bmatrix}$ is the uncertainty with known bound, as described in Section
5. Let a control force $\zeta(t)$ be applied to the first floor. The open-loop controlled
system equations take the form (2.1a) with

$$
\mathbf{B} = \left[
\begin{array}{c}
\dfrac{1}{m_1} \\[3mm]
\mathbf{O}_{(N-1)\times 1} \\[1mm]
\hline
\mathbf{O}_{N\times 1}
\end{array}
\right] \qquad (7.10\text{c})
$$

The above state space model is in the form of equation (5.1) in Section 5,
with $\Delta\mathbf{A}(r(t)) = \Delta\mathbf{B}(s(t)) = 0$. Uncertainties in the system and input matrices,
subject to the matching conditions of Equations (5.4a), (5.4b), may be easily
included. In conformity with the controller theory presented in Section 5, we

assume $\mathbf{q}(t) \in \mathbf{Q}$, a known compact set. This assumption is reasonable, since maximum values of ground displacement y_0^{max} and velocity $\dot{y}_0^{max}$ are known for the worst earthquakes on record. Thus,

$$\mathbf{Q} = \left\{ \mathbf{q} \in \mathbf{R}^2 : \left| q_1 \right| \leq y_0^{max}, \left| q_2 \right| \leq \dot{y}_0^{max} \right\} \qquad (7.11)$$

It is readily verified that the matching condition of equation (5.4c) is met with $\mathbf{W} = [\,\breve{c}_0, \kappa_0\,]$, and $\mathbf{A}_{cl} = \mathbf{A}$ is an appropriate choice, since $\mathbf{A}$ is stable. Thus, the control that assures practical stability of the system of Equations (2.1a), (7.10a)-(7.10c), that is, the control that keeps the structure arbitrarily close to its initial configuration, is given by equation (5.6).

It is worth noticing, at this point, that system (7.8a), (7.8b) can be easily put in the form (5.10a), (5.10b) by defining

$$\mathbf{z}_1(t) = (\dot{y}_1, y_1)^T \quad, \quad \mathbf{q}(t) = (\dot{y}_0, y_0)^T \quad, \quad \mathbf{z}_2(t) = (\dot{y}_2, ..., \dot{y}_N, y_2, ..., y_N)^T$$

$$\mathbf{A}_{11} = \begin{bmatrix} (-\breve{c}_0 - \breve{c}_1)/m_1 & (-\kappa_0 - \kappa_1)/m_1 \\ 1 & 0 \end{bmatrix} \quad, \quad \mathbf{A}_{21} = \begin{bmatrix} \breve{c}_1/m_2 & \\ k_1/m_2 & \mathbf{0}_{2\times(2N-3)} \end{bmatrix}^T$$

$$\mathbf{A}_{12} = \begin{bmatrix} \breve{c}_1/m_1 & \mathbf{0}_{2\times(N-2)} & \kappa_1/m_1 & \mathbf{0}_{2\times(N-2)} \\ 0 & & 0 & \end{bmatrix}$$

$$\mathbf{A}_{22}\begin{bmatrix} (-\breve{c}_1-\breve{c}_2)/m_2 & \breve{c}_2/m_2 & 0 & \vdots & (-\kappa_1-\kappa_2)/m_2 & \kappa_2/m_2 & 0 \\ \vdots & \vdots & \vdots & \vdots & \vdots & \vdots & \vdots \\ c_{i-1}/m_i & (-\breve{c}_{i-1}-\breve{c}_i)/m_i & \breve{c}_i/m_i & \vdots & \kappa_{i-1}/m_i & (-\kappa_{i-1}-\kappa_i)/m_i & \kappa_i/m_i \\ \vdots & \vdots & \vdots & \vdots & \vdots & \vdots & \vdots \\ 0 & \breve{c}_{N-1}/m_N & -\breve{c}_{N-1}/m_N & \vdots & 0 & \kappa_{N-1}/m_N & -\kappa_{N-1}/m_N \\ \hline & \mathbf{I}_{(N-1)\times(N-1)} & & \vdots & & \mathbf{0}_{(N-1)\times(N-1)} & \end{bmatrix}$$

$$\mathbf{B}_1 = \begin{bmatrix} 1/m_1 & 0 \end{bmatrix}^T \quad, \quad \mathbf{D}_1 = \begin{bmatrix} \breve{c}_0/m_1 & \kappa_0/m_1 \\ 0 & 0 \end{bmatrix}$$

Elastic structures have also widespread application in agricultural engineering and related structures. One mentions here the lightweight structures of greenhouses, which suffer extended vibrations and eventually collapse under wind loading. These structural analysis and control problems follow the general formulation (7.1) and are not discussed further here.

Another field of applications concern mechanical apparati and devices which are used in agricultural engineering (e.g. spray booms attached on tractors) or in a more general context the mechanical analysis of the tractor itshelf, including design and vibration or noise isolation of the operator's cabin. Automatic control enhances the effectiveness of the system, and/or the confort of the operator (thus, its productivity). In a long term and in view of applications on automatic steering of the tractors, which have already been proposed and tested using available GPS (global positioning system) signals, the system theoretic modeling and control is, of course, indispensable. Note, in passing, that relevant work has been done on applications in robotics and other smart controlled structures. Detailed discussion of all these aspects are beyond the scope of this article.

Let us consider spray books attached on tractors which are used for the distribution of chemical sprays over field crops. The spray booms are flexible structures, sometimes weakly damped. Sprayer performance is measured by the uniformity of spray deposit distribution which can be affected by many factors. Among them, vertical movements of the nozzles with respect to the soil and horizontal motions of nozzles relative to the tractor, play a fundamental role. In particular, vertical nozzle movements are caused by a combination of rolling and vertical translations of tractor and spray boom. These are called "rigid body motions" and are direct consequence of soil surface roughness. In addition, rolling angular accelerations and vertical translational accelerations of the tractor induce vertical flexible deformations of the spray boom, which are partly attenuated by the boom framework. Continuous changes of vertical distance between crop and nozzle create local under- and overapplication of spray. Boom vibrations are magnified by the use of longer more flexible spray booms and the use of powerful agricultural machines, operating at higher field speeds. That is why disturbance sensitivity of the vehicle, boom suspension, and spray boom, is important in the overall performance of the sprayer (see [50]-[54] and the references cited therein).

Most tractor movements and spray boom vibrations under field conditions are small enough to be described by a linearized vector second order system of the form (7.1), where, in the present case, $\mathbf{u}(t)$ is a vector of independent generalized Lagrangian coordinates,. In the present case, equation (7.1) can be obtained analytically by finite element techniques or by applying multibody dynamics codes. A model used in [53]-[54] is outlined here, in order to demonstrate that the problem can be formulated and studied using the general theories of the previous sections. More technical details can be found in the original purlications. Because vertical translations and rolling of the tractor produce the most detrimental effect on the spray application, the equations of motion can be restricted to two dimensions The sprayer can then be described by three independent generalized Lagrangian coordinates u_{11}, u_{12} and u_{21}, which represent three degrees of freedom. The coordinates u_{11} and u_{12} describe the vertical translations and the rolling of the tractor with respect to the soil. The spray boom is suspended on the tractor with a pendulum system maintaining a rolling degree of freedom represented by u_{21}. When $\mathbf{u}(t) = \begin{bmatrix} u_{11} & u_{12} & u_{21} \end{bmatrix}^{\mathrm{T}}$, then [53]

$$\mathbf{M} = \begin{bmatrix} m_1 + m_2 & 0 & 0 \\ 0 & m_2(R-\lambda)^2 + I_1 + I_2 & -m_2(R-\lambda)\lambda + I_2 \\ 0 & -m_2(R-\lambda)\lambda + I_2 & m_2\lambda^2 + I_2 \end{bmatrix} \qquad (7.12)$$

$$\breve{\mathbf{C}} = \operatorname{diag}\{2c_1, 2c_1\lambda_1^2, c\} \qquad (7.13)$$

where m_1 is the mass of the tractor body, m_2 is the mass of the spray boom, R is the distance between center of gravity of the tractor body and the pivot of the pendulum, λ is the pendulum length of the pendulum suspension, I_1 and I_2 are the mass moments of inertia of the tractor body and of the spray boom, respectively, with regard to the x-axis of the inertia coordinate system, c_1 is the summed damping coefficient of two tires, λ_1 is half the tractor gauge and c is the damping coefficient of the torsional damper. The stiffness matrix is [53]

$$\mathbf{K} = \begin{bmatrix} 2\kappa_1 & 0 & 0 \\ 0 & 2\kappa_1\lambda_1^2 + m_2g(R+\kappa) & m_2g\lambda \\ 0 & m_2g\lambda & m_2g\lambda + \kappa \end{bmatrix} \qquad (7.14)$$

where, κ_1 is the summed stiffness of two tires, g is the gravitational constant and κ is the stiffness of torsional spring. Since the disturbances of the spray boom are due to vertical translations and rolling of the tractor with respect to soil, the disturbance distribution matrix has the form [53]

$$\mathbf{M}_0 = \begin{bmatrix} \kappa_1 & \kappa_1 & c_1 & c_1 \\ -\kappa_1\lambda_1 & \kappa_1\lambda_1 & -c_1\lambda_1 & c_1\lambda_1 \\ 0 & 0 & 0 & 0 \end{bmatrix}$$

The active control distribution matrix is $\mathbf{B}_0 = 10^3 \mathbf{I}_3$. Defining $\mathbf{x}(t) = \begin{bmatrix} \mathbf{u}(t) \\ \dot{\mathbf{u}}(t) \end{bmatrix}$ as the state vector, the equations of motion can be easily transformed into its equivalent state space form (2.1a), with system matrices given by (7.3). The measured as well as the controlled outputs of the system are the vertical distances between the two boom tips and the soil. We then obtain [53]

$$\mathbf{C} = \mathbf{E} = \begin{bmatrix} 1 & \lambda_s & \lambda_s & 0 & 0 & 0 \\ 1 & -\lambda_s & -\lambda_s & 0 & 0 & 0 \end{bmatrix}, \quad \mathbf{J}_1 = \mathbf{J}_2 = 0 \qquad (7.15)$$

where λ_s is half the total boom length.

Flexible hollow beams have dynamic characteristics and properties which are representative of common small spray booms. In real applications they are usually equipped by a cradle on which the beam is clamped. Longitudinal accelerations and yawing angular accelerations of the tractor, caused by soil unevenness, induce horizontal flexible deformations of the beam. An evaluation model of the structure is obtained by executing a finite element analysis on the beam, for which purpose, it is usually divided in a certain number, say r, of equal parts. The longitudinal acceleration of the node of the element coincident with the cradle serves as the disturbance input $\mathbf{q}(t)$ of the system. The application of finite elements to describe the dynamic behavior of the flexible beam, converts the beam from a distributed or continuous system, represented by a r-th order linear partial differential equation, into a lumped or discrete system, represented by linear vector second-order system consisting of k=2r coupled ordinary differential equations of the form [54]

$$\mathbf{M\ddot{u}}(t) + \mathbf{Ku}(t) = \mathbf{M}_0\mathbf{q}(t) + \mathbf{B}_0\boldsymbol{\zeta}(t) \qquad (7.16)$$

Relation (7.16) gives the equations of motion of an undamped mechanism in physical coordinates. Each element in $\mathbf{u}(t)$ represents a rotation or translation of a prespecified location or node of the beam. The number of coordinates in (7.16) can be increased by using a larger number of shorter elements to represent the beam. For weakly damped beams, a small modal damping is introduced into the equations of motion. To this end, the finite element model, expressed in nodal coordinates, is transformed into a modal model by executing the coordinate transformation $\mathbf{u}(t) = \Phi_{sm}\eta(t)$. Then, we have

$$\ddot{\eta}(t) + 2\mathbf{Z}\Omega\dot{\eta}(t) + \Omega^2\eta(t) = \mathbf{B}_{sm}\boldsymbol{\zeta}(t) + \mathbf{M}_{sm}\mathbf{q}(t) \quad (7.17)$$

where, $\eta = \begin{bmatrix} \eta_1 & \eta_2 & \cdots & \eta_k \end{bmatrix}^T$, $\Omega = \operatorname*{diag}_{i=1,2,\cdots,k}\{\omega_i\}$, $\Phi_{sm} = \begin{bmatrix} \phi_{sm}^1 & \phi_{sm}^2 & \cdots & \phi_{sm}^k \end{bmatrix}$,

$\mathbf{Z} = \operatorname*{diag}_{i=1,2,\cdots,k}\{\zeta_i\}$, ϕ_{sm}^i is the ith mode shape vector associated with the ith natural

frequency ω_i of the beam and $\phi_{sm}^i\eta_i$ is the ith mode of the beam. Each mode shape vector has n components and is generally normalized. Commonly, the first mode $\phi_{sm}^1\eta_1$ is called the fundamental mode or first harmonic, while the higher-order modes $\phi_{sm}^2\eta_2$, $\phi_{sm}^3\eta_3$,..., are called the second, third, ..., harmonic or the first, second, ..., overtone. Equation (7.17) describes the dynamic behavior of the beam as a system of k uncoupled equations of motion in $\eta(t)$. Because the beam is weakly damped the damping coefficients ζ_i, $i = 1,2,\cdots,k$ take small values for all modes. Each differential equation in (7.17) is related with one mode shape and one natural frequency of the beam and states the vibration of the beam in the related mode. All modes contribute with a different (modal participation) factor to the total elastic deformation of the beam according to the coordinate transformation $\mathbf{u}(t) = \Phi_{sm}\eta(t)$. A state-space representation of the beam is easily obtained from the modal model, and has the form of equation (2.1a) with

$$\mathbf{x}(t) = \begin{bmatrix} \eta(t) \\ \dot{\eta}(t) \end{bmatrix}, \quad \mathbf{A} = \begin{bmatrix} 0 & \mathbf{I} \\ -\Omega^2 & -2Z\Omega \end{bmatrix}, \quad \mathbf{B} = \begin{bmatrix} 0 \\ \mathbf{B}_{sm} \end{bmatrix}, \quad \mathbf{D} = \begin{bmatrix} 0 \\ \mathbf{M}_{sm} \end{bmatrix} \qquad (7.18)$$

The output of the above system is the beam tip deformation. Since the vector second-order system (7.17) contains k modal coordinates, the state-space model will have n=2k states. Strictly speaking, the elastic beam, as a continuous system, has an infinite number of flexible modes. However, high-order modes can be neglected for several reasons. They cannot be approximated well by a finite element model, they do not contribute to the total deformation of the structure because they are not activated under normal conditions and it is even impossible to excite or to detect them with common apparatus. Increasing the number of states in (7.18) by increasing the number of modes in (7.17) renders the model computationally less attractive without adding any benefit. Design models can, however, be derived from the evaluation model in a consistent way. This implies that the order of the design model must be small as possible with preservation of as much information as possible especially in the frequency band where the disturbance signals are active. Since the evaluation model is described in modal form, the desired design models are directly derived from the evaluation model by performing a model reduction technique which is known as Guyan reduction (see, e.g., [55]).

8. Application Examples

In this Section, the different control techniques presented above, will be examined, simulated and tested through a variety of illustrative application examples.

8.1. Example I (Seismic disturbance rejection of shear-type frame)

In this example, the general theory outlined in Section 3 is applied on the specific model of civil engineering structures described by equations (7.4), (7.7a), (7.7b). In the case where we use the partitioned form of the structure described by equation (7.6) it will be shown that the problem of disturbance rejection is solvable for the above civil engineering structure. Indeed, the examined system is left invertible since

$$q_1 = \operatorname{rank}(\mathbf{N}_1) - \operatorname{rank}(\mathbf{N}_0) = \operatorname{rank}(\mathbf{CB}) - \operatorname{rank}(0_{m\times m}) = \operatorname{rank}(0_{m\times m}) - 0 = 0$$

$$q_2 = \operatorname{rank}(\mathbf{N}_2) - \operatorname{rank}(\mathbf{N}_1) = \operatorname{rank}(\mathbf{CAB}) - \operatorname{rank}(\mathbf{CB})$$
$$= \operatorname{rank}(\mathbf{M}^{-1}\mathbf{B}_0) - 0 = \operatorname{rank}(\mathbf{M}_2^{-1}\mathbf{B}_m) = m$$

and the smaller integer for which, qr+i=qr, $\forall i \geq 1$, is r=2. Furthermore, relation (3.14a) always hold, since

$$s_2 = \text{rank}(\mathbf{T}_2) - \text{rank}(\mathbf{T}_1) = \text{rank}(\mathbf{CAV}) - \text{rank}(\mathbf{CV})$$

$$= \text{rank}\!\left(\!\left[\mathbf{M}_2^{-1}\mathbf{B}_m \quad \mathbf{0}_{m\times q}\right]\!\right) - \text{rank}\!\left(\mathbf{0}_{(m+q)\times(m+q)}\right) = 2$$

In order to check the validity of condition (3.14b), on can apply the time-invariant version of the Periodic Structure Algorithm presented in Section 3, and exploit the structure of the state space model (7.4), (7.7a), (7.7b). For this system one has $\beta = 1$, since $\mathbf{CV} = 0$ and $\mathbf{CAV} \neq 0$. Therefore, the algorithm presented in Section 3, stops at its first step and we have $\hat{\Gamma}_r = \Gamma_0 = \mathbf{CA}$. Thus to satisfy (3.14b) it must be $\mathbf{CAD} = 0$. But the latter always holds true in the system configuration considered here. Thus for the localized ground story configuration the state feedback disturbance rejection problem is always solvable if the control floor displacements are assumed to be the system outputs. In particular, $\mathbf{G}$ must be any arbitrary invertible matrix and a special solution for $\mathbf{F}$ is given by

$$\mathbf{F}_s = -\left(\hat{\Gamma}_r\mathbf{B}\right)^{-1}\hat{\Gamma}_r\mathbf{A} = \mathbf{B}_m^{-1}\left[\mathbf{K}_{21} \quad \mathbf{K}_{22} \quad \mathbf{K}_{23} \quad \check{\mathbf{C}}_{21} \quad \check{\mathbf{C}}_{22} \quad \check{\mathbf{C}}_{23}\right]$$

Observe also that

$$\mathbf{A}_s = \begin{bmatrix} \mathbf{0}_{k\times k} & \mathbf{I}_{k\times k} \\ -\mathbf{A}_{s,1} & -\mathbf{A}_{s,2} \end{bmatrix} \quad (8.1)$$

with

$$\mathbf{A}_{s,1} = \begin{bmatrix} \overline{\mathbf{M}}_1^{-1}\mathbf{K}_{11} & \overline{\mathbf{M}}_1^{-1}\mathbf{K}_{12} & \overline{\mathbf{M}}_1^{-1}\mathbf{K}_{13} \\ \mathbf{0}_{m\times g} & \mathbf{0}_{m\times m} & \mathbf{0}_{m\times n_3} \\ \mathbf{M}_3^{-1}\mathbf{K}_{31} & \mathbf{M}_3^{-1}\mathbf{K}_{32} & \mathbf{M}_3^{-1}\mathbf{K}_{33} \end{bmatrix}, \quad \mathbf{A}_{s,2} = \begin{bmatrix} \overline{\mathbf{M}}_1^{-1}\check{\mathbf{C}}_{11} & \overline{\mathbf{M}}_1^{-1}\check{\mathbf{C}}_{12} & \overline{\mathbf{M}}_1^{-1}\check{\mathbf{C}}_{13} \\ \mathbf{0}_{m\times g} & \mathbf{0}_{m\times m} & \mathbf{0}_{m\times n_3} \\ \mathbf{M}_3^{-1}\check{\mathbf{C}}_{31} & \mathbf{M}_3^{-1}\check{\mathbf{C}}_{32} & \mathbf{M}_3^{-1}\check{\mathbf{C}}_{33} \end{bmatrix}$$

Then, $\Pi = \left[\mathbf{D} \quad \mathbf{A}_s\mathbf{D} \quad \cdots \quad \mathbf{A}_s^{n-1}\mathbf{D}\right]$, with $\mathbf{A}_s$ given by (8.1) and $\mathbf{D}$ given by (7.7b), $\lambda = \text{rank}\Pi$ and the general solution for $\mathbf{F}$ is given by $\mathbf{F} = \mathbf{F}_s + \hat{\mathbf{R}}\Xi$, where $\hat{\mathbf{R}}$ is an $m \times (n - \lambda)$ arbitrary matrix and Ξ is a basis for $\text{Ker}\,\Pi$.

The above general results can be easily applied to shear-type frames. According to the shear-type frame concept, horizontal beams are assumed to be rigid. Thus a control at the ground story can produce a rigid body motion at the upper stories. Therefore, by eliminating displacements of the control floor (output), one achieves the elimination of the displacements of all upper stories as well. The two-story, single bay steel frame of Figure 5 is considered in the present illustrative example, for which all data are in compatible units. In particular, we assume that the masses are $m_1 = m_2 = m_3 = 16$. The additional

mass is set equal to $m_g = 5000\left(\cong 100 \times \sum_{i=1}^{3} m_i\right)$. It is also assumed that only one

control force is applied, i.e. m=1 and that $\mathbf{B}_0 = \begin{bmatrix} 0 & 1 & 0 \end{bmatrix}^T$. Note also that

$\mathbf{M}_0 = \begin{bmatrix} 5000 & 0 & 0 \end{bmatrix}^T$. Moreover, in this problem one has k=3 degrees of freedom and one horizontal base acceleration (g=1). The stiffness matrix is

$$\mathbf{K} = \begin{bmatrix} 1500 & -1500 & 0 \\ -1500 & 3000 & 0 \\ 0 & -1500 & 1500 \end{bmatrix}$$

under the assumption of equal storeys with equivalent shear stiffness equal to 1500 for each story. A simple damping $\breve{\mathbf{C}} = \mathrm{diag}\{1.6, 1.6, 1.6\}$ is assumed. Finally, the displacement of the first story is measured, i.e. $y = u_2$ and $\mathbf{C} = \begin{bmatrix} 0 & 1 & 0 & 0 & 0 & 0 \end{bmatrix}$.

With these data one has $\mathbf{x} = \begin{bmatrix} u_1 & u_2 & u_3 & \dot{u}_1 & \dot{u}_2 & \dot{u}_3 \end{bmatrix}^T$ and

$$\mathbf{A} = \begin{bmatrix} 0 & 0 & 0 & 1 & 0 & 0 \\ 0 & 0 & 0 & 0 & 1 & 0 \\ 0 & 0 & 0 & 0 & 0 & 1 \\ -0.2990 & 0.2990 & 0 & -0.0003 & 0 & 0 \\ 93.7500 & -187.5000 & 93.7500 & 0 & -0.1000 & 0 \\ 0 & 93.7500 & -93.7500 & 0 & 0 & -0.1000 \end{bmatrix}$$

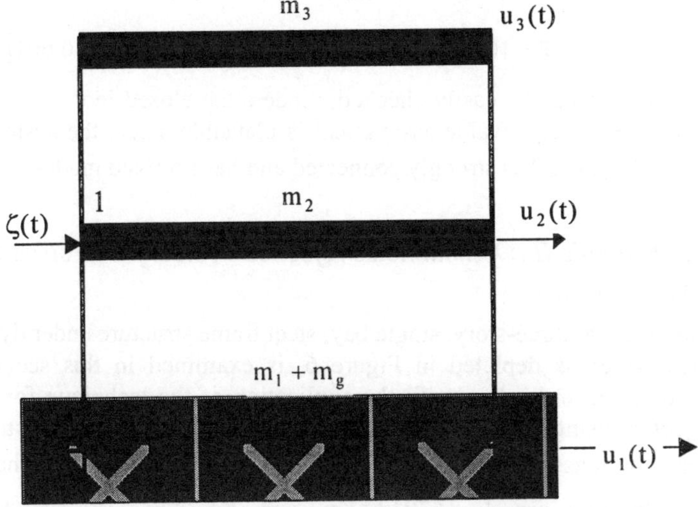

Figure 5. A shear-type frame with one control force and the additional mass concept for its simplified modelling

Application of the method presented in Section 3 yields the following special form $\mathbf{F}_s$ of the controller matrix $\mathbf{F}$

$$\mathbf{F}_s = 10^3 \times \begin{bmatrix} -1.5 & 3 & -1.5 & 0 & 0.0016 & 0 \end{bmatrix}$$

However, as it can be easily checked the closed-loop system becomes unstable, since matrix $\mathbf{A}_s = \mathbf{A} + \mathbf{B}\mathbf{F}_s$ has a double unstable eigenvalue at s=0. Thus, the calculated special feedback law is useless. To find a stabilizing feedback law, observe that

$$\Pi = \begin{bmatrix} 0.9968 & 0 & -0.2980 & 0.0001 & 0.0891 & -0.0001 \\ 0 & 0 & 0 & 0 & 0 & 0 \\ 0 & 0 & 0 & 0 & 0 & 0 \\ 0 & -0.2980 & 0.0001 & 0.0891 & -0.0001 & -0.0266 \\ 0 & 0 & 0 & 0 & 0 & 0 \\ 0 & 0 & 0 & 0 & 0 & 0 \end{bmatrix}$$

Clearly rankΠ=2, and basis for KerΠ is given by $\Xi = \begin{bmatrix} 0_{4\times2} & I_{4\times4} \end{bmatrix}$. Thus the general form of $\mathbf{F}$ is

$$\mathbf{F} = 10^3 \times \begin{bmatrix} -1.5 & 3 & -1.5 & 0 & 0.0016 & 0 \end{bmatrix} + \begin{bmatrix} \hat{r}_1 & \hat{r}_2 & \hat{r}_3 & \hat{r}_4 \end{bmatrix}\begin{bmatrix} 0_{4\times2} & I_{4\times4} \end{bmatrix}$$

Selecting $\hat{r}_1 = \hat{r}_3 = \hat{r}_4 = -1$ and $\hat{r}_2 = 0$, we obtain

$$\mathbf{F} = 10^3 \times \begin{bmatrix} -1.5 & 3 & -1.501 & 0 & 0.0006 & -0.001 \end{bmatrix}$$

which, as it can be easily checked, renders the closed-loop system stable. Note that arbitrary eigenvalue assignment is plausible since, the system with matrix triplet $(\mathbf{A}_s, \mathbf{B}, \Xi)$ is strongly connected and has no fixed modes.

8.2. Example II (Optimal noise rejection in steel frame structures using GSHF control)

The IPB 800 three-story, single bay, steel frame structure under dynamic loading $q(t)$, which is depicted in Figure 6, is examined in this second illustrative example, in order to clarify the application of the technique for optimal noise rejection using GSHF control presented in Section 4.7. For the structure studied here, the mass of each floor is $m_i = 16\,t$, i=1,2,3. The other characteristics of the structure are $\tilde{F}_i = 670.13\,t/m$ and $h_i = 3\,m$, i=1,2,3, $\mathbf{B}_0 = I_{3\times3}$, and

$$\mathbf{c} = \begin{bmatrix} 3.2 & 0 & 0 \\ 0 & 3.2 & 0 \\ 0 & 0 & 3.2 \end{bmatrix}.$$ With these values, the matrices of the state space description of the structure is obtained by (7.3), as

$$\mathbf{A} = \begin{bmatrix} 0 & 0 & 0 & 1 & 0 & 0 \\ 0 & 0 & 0 & 0 & 1 & 0 \\ 0 & 0 & 0 & 0 & 0 & 1 \\ -83.7663 & 83.7663 & 0 & -0.2 & 0 & 0 \\ 83.7663 & -167.5325 & 83.7663 & 0 & -0.2 & 0 \\ 0 & 83.7663 & -167.5325 & 0 & 0 & -0.2 \end{bmatrix}$$

$$\mathbf{B} = \mathbf{D} = \begin{bmatrix} 0 & 0 & 0 \\ 0 & 0 & 0 \\ 0 & 0 & 0 \\ 1/16 & 0 & 0 \\ 0 & 1/16 & 0 \\ 0 & 0 & 1/16 \end{bmatrix}$$

In order to apply the GSHF based optimal regulation technique to the above civil structure, we next focus our attention to the case where the outputs of the structure are the velocities of the three stories. In this case, we have $\mathbf{C} = \begin{bmatrix} \mathbf{0}_{3\times3} & \mathbf{I}_{3\times3} \end{bmatrix}$. It can be easily checked that the system with the above state space matrices is controllable and observable. Therefore, the technique presented in Section 4.7, is applicable in the present case. The covariance kernels of $\mathbf{w}(t)$ and $\mathbf{v}(kT_0)$ are

$$\mathbf{R}_{\mathbf{W}} = \begin{bmatrix} & \mathbf{0}_{3\times6} & \\ \mathbf{0}_{3\times3} & 0.1 \times \mathbf{I}_{3\times3} \end{bmatrix}, \quad \mathbf{R}_{\mathbf{v}} = 0.002 \times \mathbf{I}_{3\times3} > 0 \quad (8.2)$$

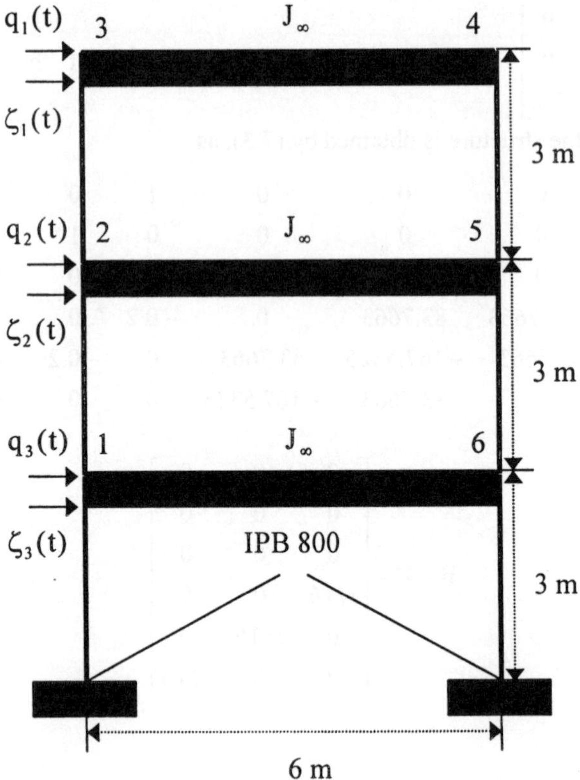

Figure 6. The IPB 800 three-story, single bay, steel frame structure with the loading
$q_i(t), i = 1,2,3$ and the control forces $\zeta_i(t), i = 1,2,3$

Selecting $\mathbf{Q} = \mathbf{I}_{6\times6}$, $T_0 = 2.5$ sec and applying the proposed technique in the case where $\mathbf{F}(t)$ does not have a prespecified structure, we obtain

$$\Phi = \begin{bmatrix} -0.6566 & 0.0683 & 0.0486 & -0.0792 & -0.0494 & -0.0276 \\ 0.0683 & 0.6763 & 0.0197 & -0.0494 & -0.0574 & -0.0218 \\ 0.0486 & 0.0197 & 0.7249 & -0.0276 & -0.0218 & -0.0298 \\ 2.4930 & -0.6666 & 0.4863 & -0.6407 & 0.0782 & 0.0541 \\ -0.6666 & 3.6459 & -1.1529 & 0.0782 & -0.6648 & 0.0241 \\ 0.4863 & -1.1529 & 3.1596 & 0.0541 & 0.0241 & -0.7189 \end{bmatrix}$$

$$\mathbf{K} = 10^{-4} \times \begin{bmatrix} 52 & 31 & 14 & -60 & -28 & -14 \\ & 35 & 17 & -28 & -46 & -14 \\ & & 20 & -14 & -14 & -32 \\ & \text{Symmetry} & & 1262 & 118 & 90 \\ & & & & 1234 & 28 \\ & & & & & 1144 \end{bmatrix}$$

$$\widetilde{\mathbf{F}}_{\phi} = 10^{-4} \times \begin{bmatrix} 505 & 400 & 238 \\ 400 & 343 & 162 \\ 238 & 162 & 105 \\ 7334 & -505 & -253 \\ -505 & 7586 & -252 \\ -253 & -252 & 7839 \end{bmatrix}$$

Matrix $\widetilde{\mathbf{F}}_{\phi}$ presents a certain kind of symmetry and the 3×3 regulator matrix $\mathbf{F}(t)$ is symmetric. In Figures 7a and 7b, the elements $f_{11}(t)$ and $f_{23}(t)$ of the admissible hold function $\mathbf{F}(t)$, are depicted.

In the case where our objective is to design hold functions with piecewise constant behavior, one must select N_I's and solve equation (2.6). For example, in the case where $N_1=N_2=N_3=15$, we obtain a symmetric multirate hold function $\mathbf{F}(t)$, whose elements $f_{11}(t)$ and $f_{23}(t)$ are depicted in Figures 8a and 8b. Similar results are obtained in the case where N_i's have different values. For example in the case where $N_1=30$, $N_2=45$ and $N_3=60$, the elements $f_{11}(t)$, $f_{23}(t)$ and $f_{32}(t)$ of the admissible multirate hold function are depicted in Figures 9a-9c. Obviously, in this case the matrix $\mathbf{F}(t)$ is not symmetric, due to different values of N_i's. Moreover, it is obvious that, as $N_i \to \infty$, the multirate GSHF obtained tends to the unconstrained GSHF depicted in Figures 7a and 7b. Note that, in all the three cases presented above, the minimum of the cost function is

$$J_{min} = \text{tr}\{\mathbf{KQ}\} = \text{tr}\{\mathbf{K}\} = 0.3747$$

With regard to the robustness properties of the above designed GSHF based

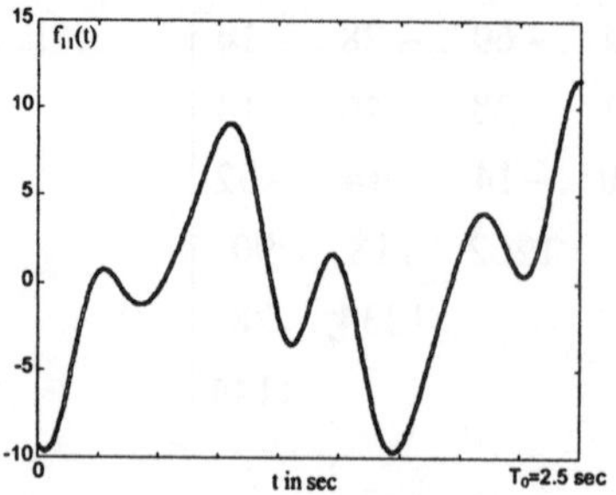

Figure 7a. The 1-1 entry of $\mathbf{F}(t)$ in the case of an unconstrained GSHF and $T_0 = 2.5$ sec.

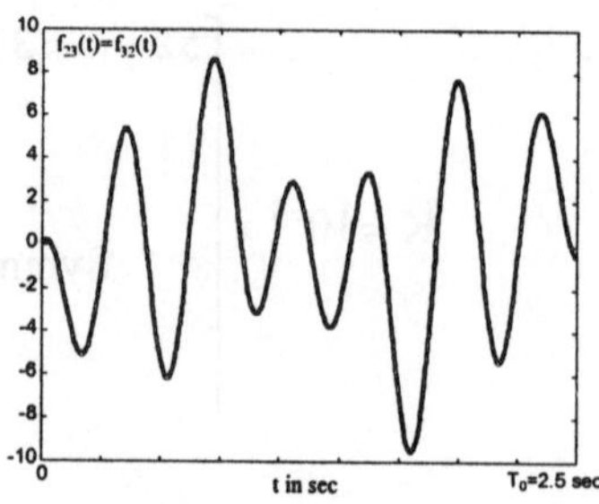

Figure 7b. Entries 2-3 and 3-2 of $\mathbf{F}(t)$ in the case of an unconstrained GSHF and $T_0 = 2.5$ sec.

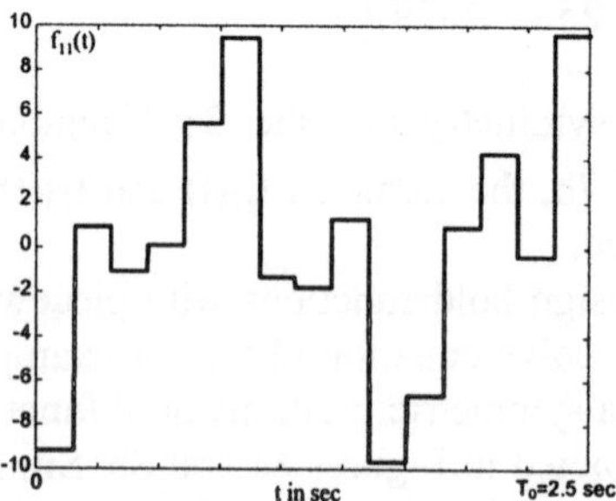

Figure 8a. The 1-1 entry of $\mathbf{F}(t)$ in the case of a multirate GSHF with $N_1=N_2=N_3=15$ and $T_0 = 2.5$ sec.

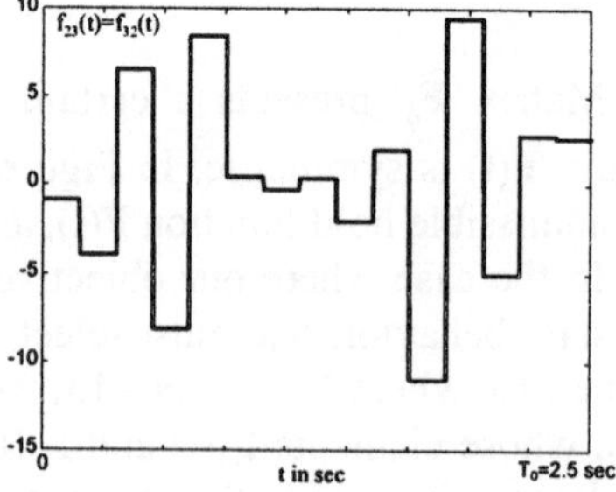

Figure 8b. Entries 2-3 and 3-2 of $\mathbf{F}(t)$ in the case of a multirate GSHF with $N_1=N_2=N_3=15$ and $T_0 = 2.5$ sec.

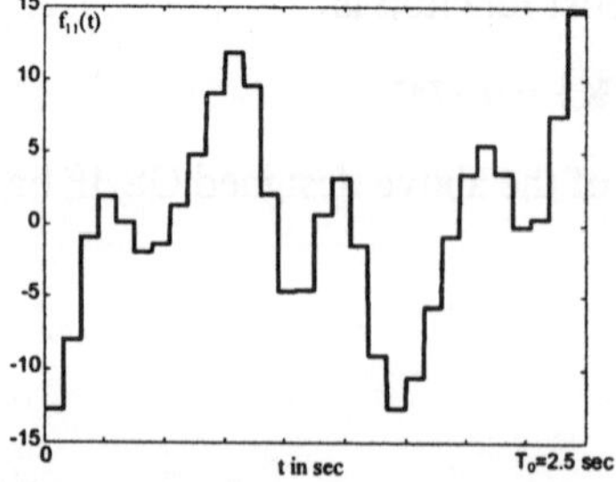

Figure 9a. The 1-1 entry of $\mathbf{F}(t)$ in the case of a multirate GSHF with $N_1=30$, $N_2=45$, $N_3=60$ and $T_0 = 2.5$ sec.

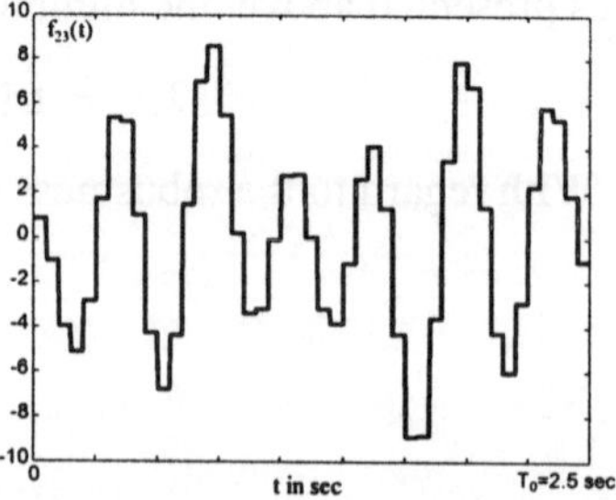

Figure 9b. The 2-3 entry of $\mathbf{F}(t)$ in the case of a multirate GSHF with $N_1=30$, $N_2=45$, $N_3=60$ and $T_0 = 2.5$ sec.

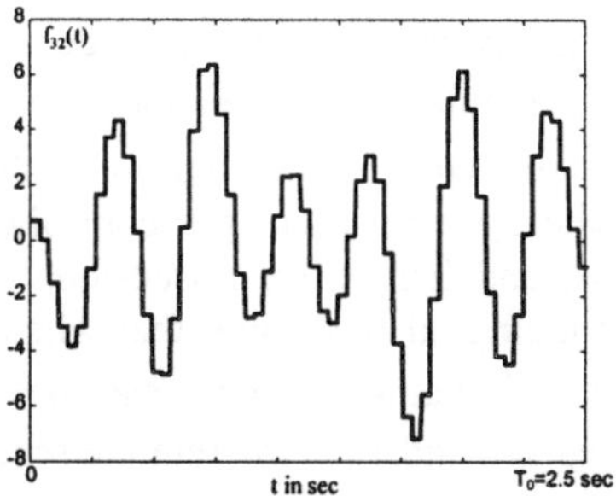

Figure 9c. The 3-2 entry of $\mathbf{F}(t)$ in the case of a multirate GSHF with $N_1=30$, $N_2=45$, $N_3=60$ and $T_0 = 2.5$ sec.

optimal regulator, following the results reported in [29], one can check that, in the present case, the estimated guaranteed stability margins of the GSHF based optimal regulator are quite tight and have the values

$$\text{GM}_{\text{ext}}^{\text{in}} = 0.9009 \text{ or } -0.9067 \text{ dB} \text{ , } \text{GM}_{\text{ext}}^{\text{up}} = 1.1236 \text{ or } 1.0125 \text{ dB}$$

$$\text{PM}_{\text{est}} = \pm 0.1101 \text{ rad} \text{ or } \pm 6.3072^{\circ}$$

Consider now the case where the covariance kernel

$$\mathbf{R}_W = \begin{bmatrix} & \mathbf{0}_{3\times6} \\ \mathbf{0}_{3\times3} & 10^{-2} \times \mathbf{I}_{3\times3} \end{bmatrix}$$

and all the other parameters remain the same as in the previous optimal design. In this case, the estimated guaranteed stability margins of the above designed GSHF based optimal regulator are

$$\text{GM}_{\text{ext}}^{\text{in}} = 0.7889 \text{ or } -2.0598 \text{ dB} \text{ , } \text{GM}_{\text{ext}}^{\text{up}} = 1.3654 \text{ or } 2.7053 \text{ dB}$$

$$\text{PM}_{\text{est}} = \pm 0.2684 \text{ rad} \text{ or } \pm 15.3797^{\circ}$$

Clearly, in the present case, the guaranteed stability margins of the GSHF based regulator are larger than the ones obtained in the previous case. In the case where the covariance kernel $\mathbf{R}_W = \begin{bmatrix} & \mathbf{0}_{3\times6} \\ \mathbf{0}_{3\times3} & 10^{-3} \times \mathbf{I}_{3\times3} \end{bmatrix}$, the guaranteed stability margins of the regulator, obtained by applying the results in [29] are

$$\text{GM}_{\text{ext}}^{\text{in}} = 0.6024 \text{ or } -4.4017 \text{ dB} \text{ , } \text{GM}_{\text{ext}}^{\text{up}} = 2.9405 \text{ or } 9.3684 \text{ dB}$$

$$\text{PM}_{\text{est}} = \pm 0.6725 \text{ rad} \text{ or } \pm 38.5327^{\circ}$$

In conclusion, when the covariance kernel $\mathbf{R}_W$ decreases, the robustness of the GSHF based optimal regulator is ameliorated. Of course, this fact is expected, since a smaller covariance kernel $\mathbf{R}_W$ corresponds to a smaller level of the noise disturbing the system. Then, the controller has to withstand to smaller system variations, and hence the closed-loop system becomes more robust. It is also worth noticing, at this point, that when the covariance kernel $\mathbf{R}_W$ decreases, the regulator's gains also decrease in magnitude. This can be easily identified by Figures 10a and 10b, wherein the regulator's gains $f_{11}(t)$ and

$f_{23}(t)$ are given for the case where $\mathbf{R}_W = \begin{bmatrix} \mathbf{0}_{3\times6} \\ \mathbf{0}_{3\times3} & 10^{-4} \times \mathbf{I}_{3\times3} \end{bmatrix}$. Clearly, this is due

to the fact that, the less the level of the disturbance acting on the system the smaller must be the control effort, for rejecting the disturbance.

We next consider the case where, the covariance kernel $\mathbf{R}_V = 2 \times \mathbf{I}_{3\times3}$, and all the other parameters remain the same as in the original optimal design. Then, the regulator gains $f_{11}(t)$ and $f_{23}(t)$ are identical to those of Figures 10a and 10b. We can conclude that, if the covariance kernel $\mathbf{R}_V$ is large, small gains of the hold function $\mathbf{F}(t)$ are expected, indicating a poorly regulated system, since, the measurement noise predominates.

We finally study the impact of the sampling period T_0 on the gain $\mathbf{F}(t)$. To this end, in what follows, the sampling period is decreased to the value $T_0 = 0.5$ sec and the covariance kernels maintain the values given by (8.2). In this case, we obtain

$$\mathbf{K} = 10^{-4} \times \begin{bmatrix} 12 & 8 & 3 & 29 & 13 & 4 \\ & 8 & 4 & 13 & 19 & 9 \\ & & 5 & 4 & 9 & 15 \\ & \text{Symmetry} & & 385 & -6 & 25 \\ & & & & 416 & -31 \\ & & & & & 391 \end{bmatrix}$$

$$\widetilde{\mathbf{F}}_\phi = \begin{bmatrix} -0.0877 & -0.0593 & -0.0510 \\ -0.0593 & -0.0794 & -0.0083 \\ -0.0510 & -0.0083 & -0.0284 \\ 0.1235 & 0.4023 & 0.7342 \\ 0.4023 & 0.4555 & -0.3320 \\ 0.7342 & -0.3320 & -0.2787 \end{bmatrix}$$

The minimum of the cost function is $J_{\min} = \text{tr}\{\mathbf{KQ}\} = \text{tr}\{\mathbf{K}\} = 0.1218$. The elements $f_{11}(t)$ and $f_{23}(t)$ of the admissible hold function $\mathbf{F}(t)$ are depicted in Fi-

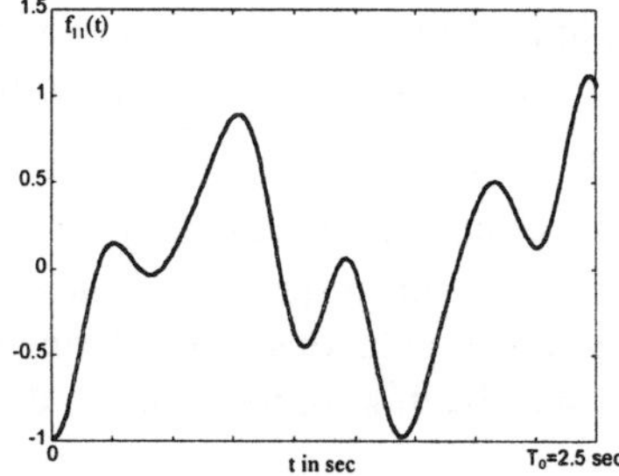

Figure 10a. Hold function $f_{11}(t)$ for T_0=2.5 sec and $\mathbf{R_W}$ decreased ($\mathbf{R_V}$ increased) 10^3 times compared to (8.2)

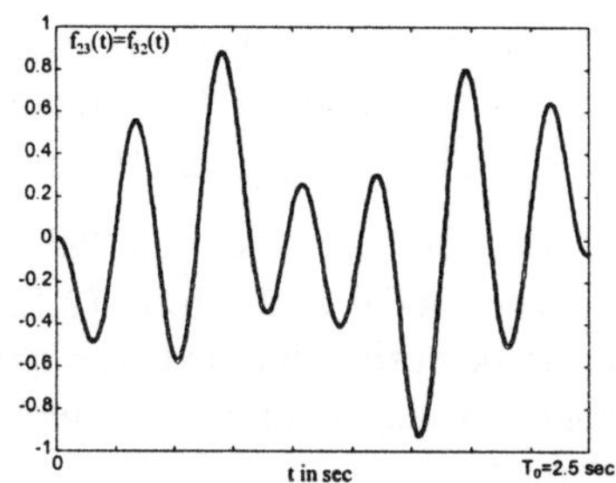

Figure 10b. Gains $f_{23}(t)$, $f_{32}(t)$ for T_0=2.5sec and $\mathbf{R_W}$ decreased ($\mathbf{R_V}$ increased) 10^3 times compared to (8.2)

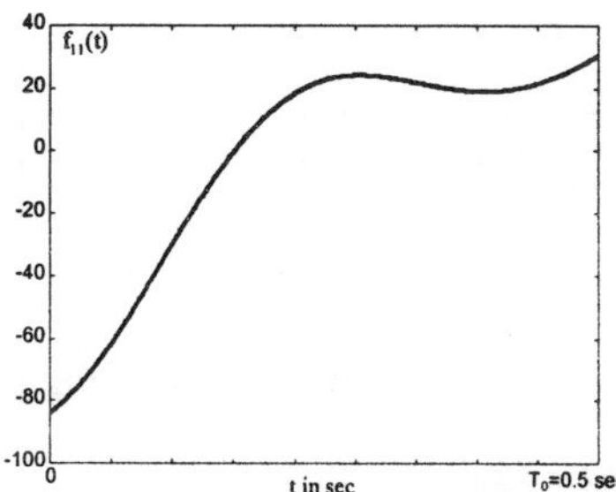

Figure 11a. The 1-1 entry of $\mathbf{F}(t)$ in the case of $T_0 = 0.5$ sec.

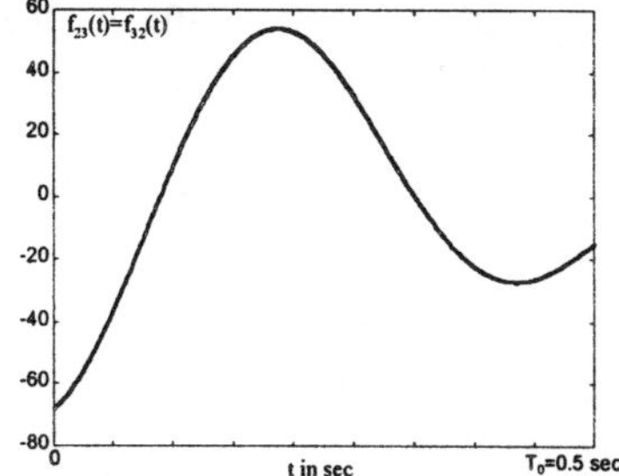

Figure 11b. Entries 2-3 and 3-2 of $\mathbf{F}(t)$ in the case of $T_0 = 0.5$ sec.

gures 11a and 11b, and they are larger than the gains obtained in the case where $T_0 = 2.5$ sec. This is due to the fact that, choosing a small T_0 yields a small $\mathbf{W(A, B, T_0)}$ or a small $\hat{\mathbf{B}}$ (in the case of multirate GSHF). Their inversion produces matrices with large entries, which are involved in the computation of the gain $\mathbf{F}(t)$.

The above conclusions, can be used to assess the detrimental effect of noise on the closed-loop system and the tradeoff involved in assuring good performance and sufficient robustness as well as in selecting the sampling period T_0.

8.3. Example III (LQR vs. H^∞ control for seismic excited buildings)

In this example, continuous-time state feedback LQ regulation is compared to the continuous-time state feedback H^∞ control technique. To this end, consider the six-story full-scale shear-beam-type building with identical floors reported in

[56]. In this structure, there are two actuators, one on the first floor and another on the third floor, both active brancing systems. The mass of each floor, and the stiffness and damping coefficients of each story unit are: $m_i = 345.6\,t$, $\kappa_i = 3.404 \times 10^5$ kN and $c_i = 2937$ kNs/m. According to the results in [56], a state–space model of the form (2.1a), (2.1c), (2.1d), with system matrices as in (7.3), can be easily obtained after straightforward manipulations. Considering as controlled system outputs the interstory drifts, we obtain k=6, n=12, m=2, d=1, $p_1=p_2=6$ and

$$
\mathbf{M}^{-1}\mathbf{K} =
\begin{bmatrix}
\dfrac{\kappa_1}{m_1} & -\dfrac{\kappa_2}{m_1} & 0 & 0 & 0 & 0 \\[2mm]
-\dfrac{\kappa_1}{m_1} & \dfrac{\kappa_2}{m_1}+\dfrac{\kappa_2}{m_2} & -\dfrac{\kappa_3}{m_2} & 0 & 0 & 0 \\[2mm]
0 & -\dfrac{\kappa_2}{m_2} & \dfrac{\kappa_3}{m_2}+\dfrac{\kappa_3}{m_3} & -\dfrac{\kappa_4}{m_3} & 0 & 0 \\[2mm]
0 & 0 & -\dfrac{\kappa_3}{m_3} & \dfrac{\kappa_4}{m_3}+\dfrac{\kappa_4}{m_4} & -\dfrac{\kappa_5}{m_4} & 0 \\[2mm]
0 & 0 & 0 & -\dfrac{\kappa_4}{m_4} & \dfrac{\kappa_5}{m_4}+\dfrac{\kappa_5}{m_5} & -\dfrac{\kappa_6}{m_5} \\[2mm]
0 & 0 & 0 & 0 & -\dfrac{\kappa_5}{m_5} & \dfrac{\kappa_6}{m_5}+\dfrac{\kappa_6}{m_6}
\end{bmatrix}
$$

$$
\mathbf{M}^{-1}\mathbf{B}_0 =
\begin{bmatrix}
\dfrac{1}{m_1} & 0 \\[2mm]
-\dfrac{1}{m_1} & 0 \\[2mm]
0 & \dfrac{1}{m_3} \\[2mm]
0 & -\dfrac{1}{m_3}-\dfrac{1}{m_4} \\[2mm]
0 & \dfrac{1}{m_4} \\[2mm]
0 & 0
\end{bmatrix}
, \quad
\mathbf{M}^{-1}\mathbf{M}_0 =
\begin{bmatrix}
-1 \\ 0 \\ 0 \\ 0 \\ 0 \\ 0
\end{bmatrix}
$$

$$
\mathbf{C} = \mathbf{E} = \begin{bmatrix} \mathbf{I}_6 & \mathbf{0}_{6\times 6} \end{bmatrix} , \quad \mathbf{J}_1 = \mathbf{J}_2 = \mathbf{0}_{6\times 2}
$$

The earthquake ground motion used for the simulation studies that follow is depicted in Figure 12.

In the sequel, our purpose is to design LQ and H^∞ regulators, in order to attenuate the effect of the earthquake ground motion to the interstory drifts.

We begin our analysis by first designing LQ optimal regulators. Applying the results of Section 4.1, and selecting matrices $\mathbf{Q} \in \mathbf{R}^{6\times 6}$ and $\mathbf{R} \in \mathbf{R}^{2\times 2}$ as

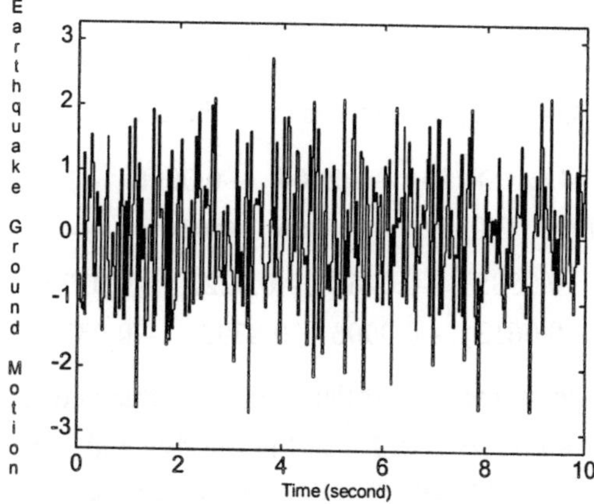

Figure 12. Earthquake ground motion for Example III

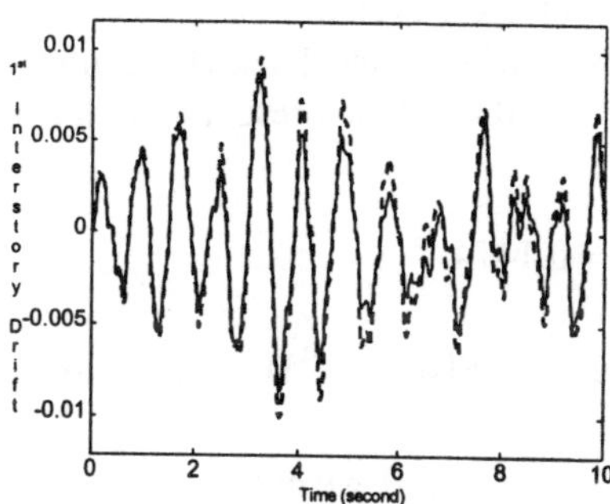

Figure 13a. First interstory drifts of the uncontrolled (dashed line) and the controlled system (solid line) with the controller (8.3a)

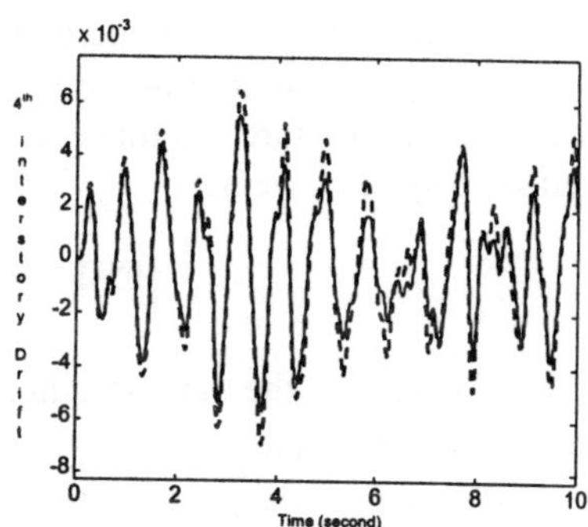

Figure 13b. Fourth interstory drifts of the uncontrolled (dashed line) and the controlled system (solid line) with the controller (8.3a)

$$\mathbf{Q} = \mathrm{diag}\{3 \times 10^7 \quad \text{and} \quad \mathbf{R} = \mathbf{I}_2, \text{ we obtain}$$

$$\mathbf{F} = 10^3 \times \begin{bmatrix} -0.0440 & 7.0053 & -3.7222 & -0.2609 & -2.2971 & -3.1243 \\ -0.2609 & 6.9059 & -4.5751 & 0.0440 & -8.9970 & 3.6387 \\ -3.9079 & 0.1188 & -0.1088 & -0.0386 & -0.0923 & -0.0725 \\ 0.0282 & 0.1522 & 0.0149 & 3.7224 & -0.0578 & 0.0448 \end{bmatrix} \tag{8.3a}$$

In Figures 13a and 13b, the first and fourth interstory drifts of both the uncontrolled and the controlled system are depicted, when the disturbance signal of Figure 12, excites the structure. Similar results can be obtained for the remaining interstory drifts. Clearly, the effect of the disturbance on the system outputs is attenuated in the controlled structure.

We next perform another LQ design with matrix $\mathbf{R}$ as above and matrix $\mathbf{Q}$ having the value $\mathbf{Q} = \text{diag}\{10^8$. In this case, the controller matrix is

$$\mathbf{F} = 10^4 \times \begin{bmatrix} -0.0147 & 3.8123 & -2.2004 & -0.0096 & -1.1796 & -1.2987 \\ -0.0096 & 3.0357 & -2.1866 & 0.0147 & -4.0439 & 1.4209 \\ -0.9498 & 0.0585 & -0.0462 & -0.0222 & -0.0379 & -0.0281 \\ 0.0215 & 0.0613 & -0.0006 & 0.8792 & -0.0327 & 0.0133 \end{bmatrix} \tag{8.3b}$$

Note that, the same controller can be obtained in the case where $\mathbf{Q} = \text{diag}\{3 \times 10^7$ and $\mathbf{R} = 0.3 \times \mathbf{I}_2$. In Figures 14a and 14b, the first and fourth interstory drifts of both the uncontrolled and the controlled system are depicted. As it can be seen by these figures, the effect of the disturbance on the system outputs is further attenuated. However, this improved performance is attained at the expense of additional control effort, since, obviously, the value of the control law is increased here, as compared to the previous LQ design.

We are now able to proceed with the design of state feedback H^∞ regulators for the above structure. To this end, observe first that, in the present case, for the open-loop system, we have

$$\left\| \mathbf{E}(s\mathbf{I} - \mathbf{A})^{-1}\mathbf{D} \right\|_\infty = 0.1474$$

In Figure 15, the largest singular value of $\mathbf{E}(s\mathbf{I} - \mathbf{A})^{-1}\mathbf{D}$ is depicted as function of the frequency ω.

Applying Corollary 6.1 for several values of γ and θ we obtain:

(i)   *For γ=0.1 and θ=10⁻⁹:*

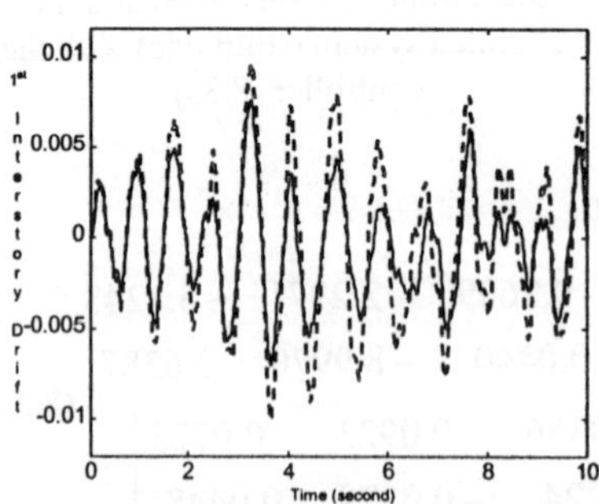

Figure 14a. First interstory drifts of the uncontrolled (dashed line) and the controlled system (solid line) with the controller (8.3b)

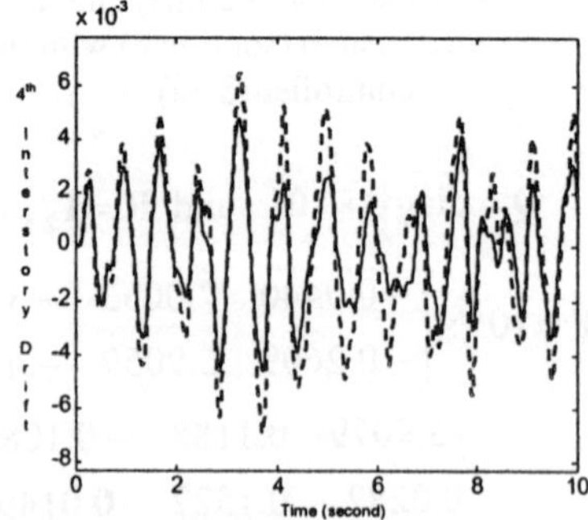

Figure 14b. Fourth interstory drifts of the uncontrolled (dashed line) and the controlled system (solid line) with the controller (8.3b)

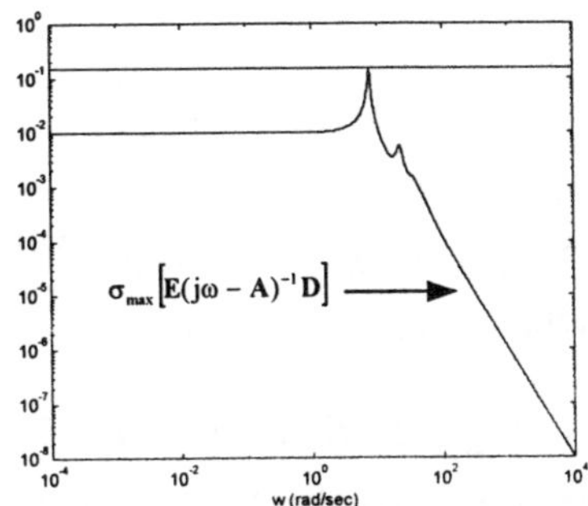

Figure 15. The largest singular value of the transfer function of the uncontrolled system

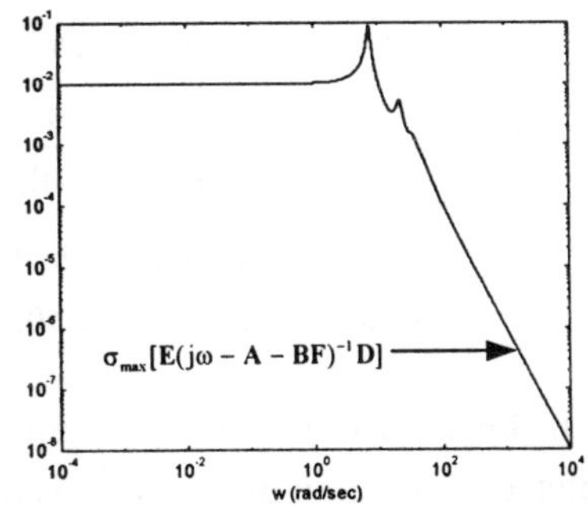

Figure 16. The largest singular value of the closed-loop transfer function in Case (i)

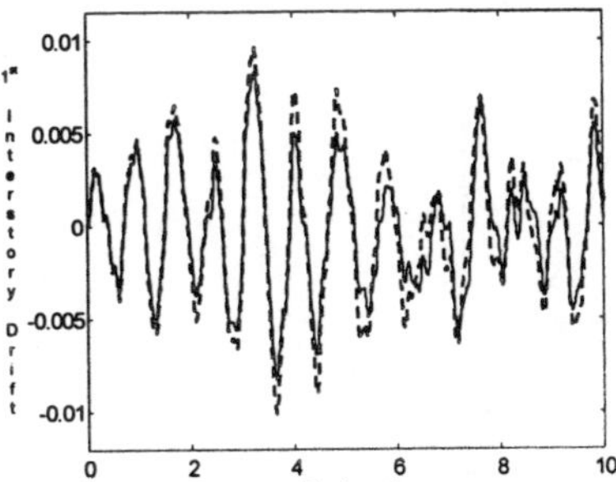

Figure 17a. First interstory drifts of the uncontrolled (dashed line) and the controlled system (solid line) with the H^{∞}-controller (8.4a)

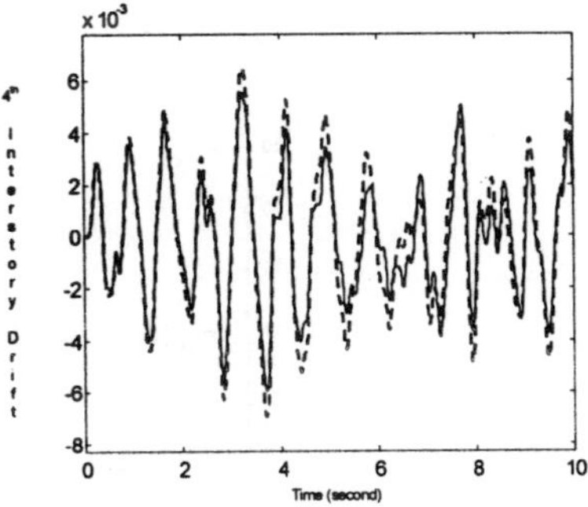

Figure 17b. Fourth interstory drifts of the uncontrolled (dashed line) and the controlled system (solid line) with the H^{∞}-controller (8.4a)

$$\mathbf{F} = 10^3 \times \begin{bmatrix} 4.3143 & 5.4671 & 4.8087 & 3.8602 & 2.6967 & 1.3846 \\ -3.8570 & -3.6359 & -3.2042 & -1.1102 & -1.8197 & -0.9397 \\ -1.5301 & -1.3306 & -1.1040 & -0.8526 & -0.5807 & -0.2942 \\ 0.8514 & 0.8309 & 0.7916 & 0.7364 & 0.4979 & 0.2511 \end{bmatrix} \tag{8.4}$$

In Figure 16, the largest singular value of $\mathbf{H}_{qy_c}(s) = \mathbf{E}(s\mathbf{I} - \mathbf{A} - \mathbf{BF})^{-1}\mathbf{D}$ is depicted as function of ω. Obviously the attenuation level is achieved. Moreover, in order to show the effect of the H^{∞} regulator, in Figures 17a and 17b, the first and fourth interstory drifts of both the controlled and the uncontrolled system are depicted, when the disturbance depicted in Figure 12, excites the structure.

(ii)   For $\gamma=0.075$ and $\theta=10^{-10}$:

$$F = 10^4 \times \begin{bmatrix} -1.4412 & 0.1218 & 0.0864 & 0.0299 & -0.0416 & -0.0523 \\ 0.0323 & 0.0405 & -0.0091 & 1.4390 & -0.1016 & -0.0650 \\ -0.6326 & -0.4984 & -0.3846 & -0.2823 & -0.1860 & -0.0925 \\ 0.2826 & 0.2886 & 0.3028 & 0.3296 & 0.2106 & 0.1027 \end{bmatrix} \quad (8.4b)$$

In Figure 18, the largest singular value of $H_{qy_c}(s) = E(sI - A - BF)^{-1}D$ is depicted as function of ω. The desired disturbance attenuation level is satisfied. Moreover, in Figures 19a and 19b, the first and fourth interstory drifts of both the controlled and the uncontrolled structure are depicted.

(iii)   For $\gamma=0.05$ and $\theta=10^{-11}$:

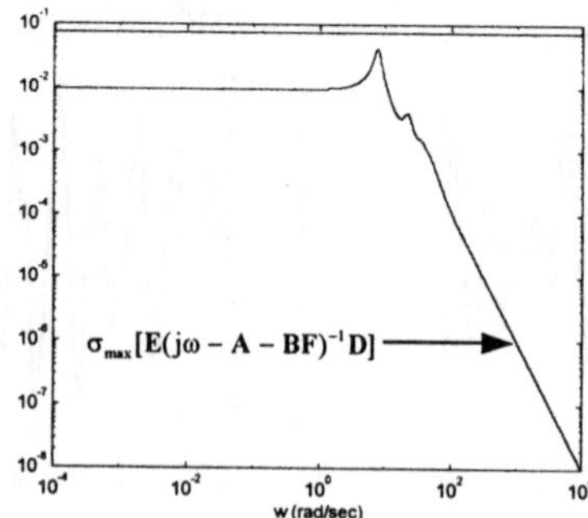

Figure 18. The largest singular value of the closed-loop transfer function in Case (ii)

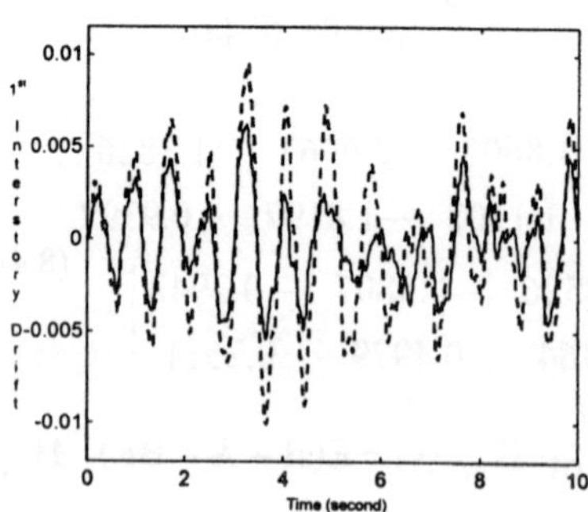

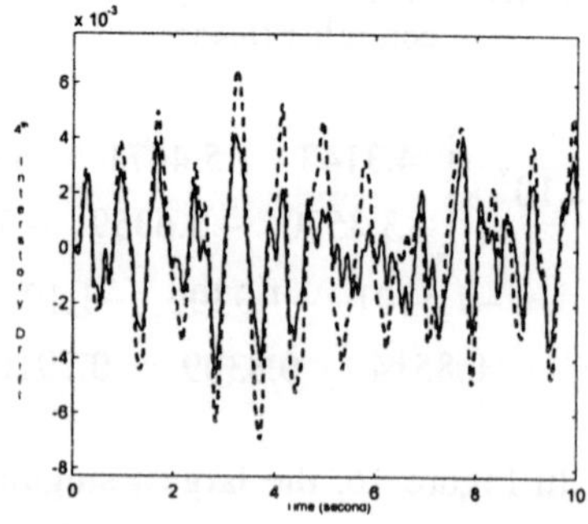

Figure 19a. First interstory drifts of the uncontrolled (dashed line) and the controlled system (solid line) with the H^∞-controller (8.4b)

Figure 19b. Fourth interstory drifts of the uncontrolled (dashed line) and the controlled system (solid line) with the H^∞-controller (8.4b)

$$
\mathbf{F} = 10^5 \times \begin{bmatrix} -1.2473 & 0.3170 & 0.1338 & 0.0367 & -0.1090 & -0.1203 \\ 0.0410 & 0.0945 & -0.0634 & 1.2429 & -0.2545 & -0.1197 \\ -0.2353 & -0.1640 & -0.1196 & -0.0852 & -0.0547 & -0.0266 \\ 0.0878 & 0.0917 & 0.1039 & 0.1336 & 0.0773 & 0.0360 \end{bmatrix} \quad (8.4c)
$$

In Figure 20, the largest singular value of $\mathbf{H}_{qy_c}(s) = \mathbf{E}(s\mathbf{I} - \mathbf{A} - \mathbf{BF})^{-1}\mathbf{D}$ is depicted as function of ω. Once again, the desired disturbance attenuation level is achieved. Moreover, in Figures 21a and 21b, the first and fourth interstory drifts of both the controlled and the uncontrolled structure are depicted.

Note that the minimum achievable disturbance attenuation level is $\gamma=0.011$ (for $\theta=10^{-13}$), since below this level, equation (6.5) has no positive semi-definite solutions.

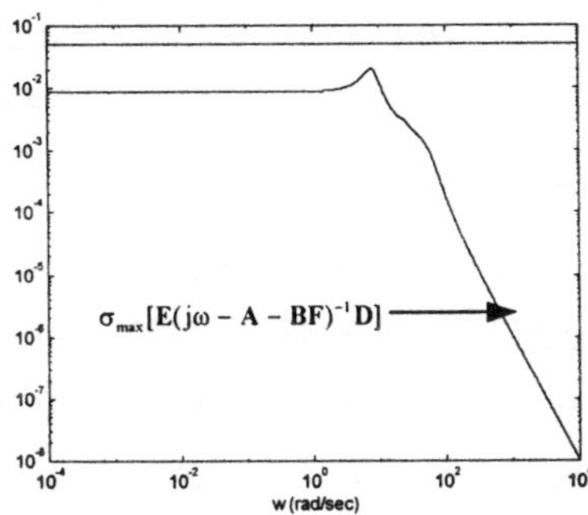

Figure 20. The largest singular value of the closed-loop transfer function in Case (iii)

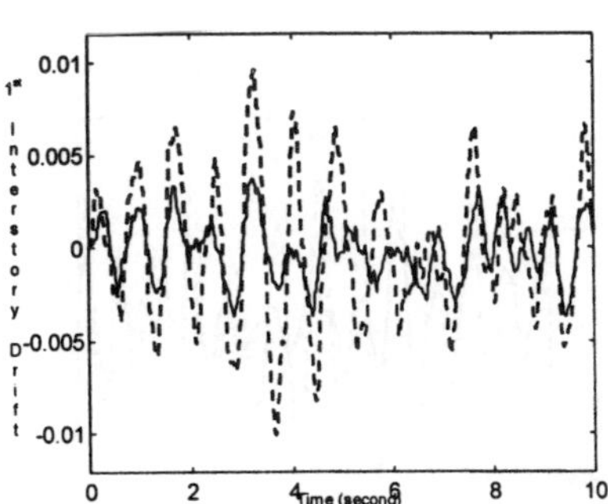

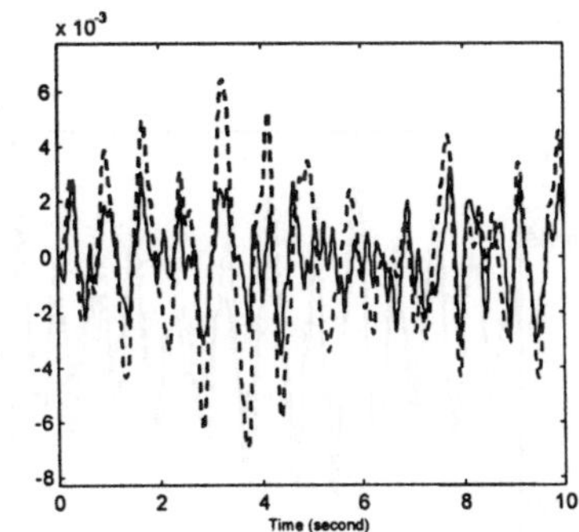

Figure 21a. First interstory drifts of the uncontrolled (dashed line) and the controlled system (solid line) with the H^∞-controller (8.4c)

Figure 21b. Fourth interstory drifts of the uncontrolled (dashed line) and the controlled system (solid line) with the H^∞-controller (8.4c)

Of course a state feedback H^∞ controller for the structure under study, can be easily designed on the basis of Proposition 6.1. Applying this result for $\widetilde{R} = I_2$ and $\widetilde{Q} = I_{12}$ we obtain:

(i)   *For γ=0.1 and ε=10⁻⁹:*

$$F = 10^3 \times \begin{bmatrix} -7.2000 & 0.5931 & 0.4171 & 0.1520 & -0.2106 & -0.2644 \\ 0.1588 & 0.2145 & -0.0344 & 7.1938 & -0.5061 & -0.3230 \\ -2.9242 & -2.2669 & -1.7252 & -1.2525 & -0.8188 & -0.4051 \\ 1.2563 & 1.2953 & 1.3838 & 1.5433 & 0.9792 & 0.4755 \end{bmatrix} \tag{8.5a}$$

(ii)   *For γ=0.075 and ε=10⁻¹⁰:*

$$F = 10^4 \times \begin{bmatrix} -7.9532 & 2.1971 & 0.8416 & 0.2324 & -0.7129 & -0.7823 \\ 0.2439 & 0.6232 & -0.4825 & 7.9415 & -1.7252 & -0.7704 \\ -1.3257 & -0.9045 & -0.6517 & -0.4605 & -0.2922 & -0.1406 \\ 0.4812 & 0.5041 & 0.5795 & 0.7654 & 0.4378 & 0.2026 \end{bmatrix} \tag{8.5b}$$

(iii)   *For γ=0.05 and ε=10⁻¹¹:*

$$F = 10^5 \times \begin{bmatrix} -5.5820 & 2.2003 & 0.4960 & 0.1157 & -0.3469 & -0.3773 \\ 0.1262 & 0.2777 & -0.7548 & 5.5723 & -1.3955 & -0.4719 \\ -0.5063 & -0.3601 & -0.2552 & -0.1788 & -0.1064 & -0.0472 \\ 0.1914 & 0.1953 & 0.2302 & 0.2983 & 0.1828 & 0.0836 \end{bmatrix} \tag{8.5c}$$

Figures 22a-22c show the effect of the disturbance on the first interstory drift of both the uncontrolled and the controlled structure, in each case examined above.

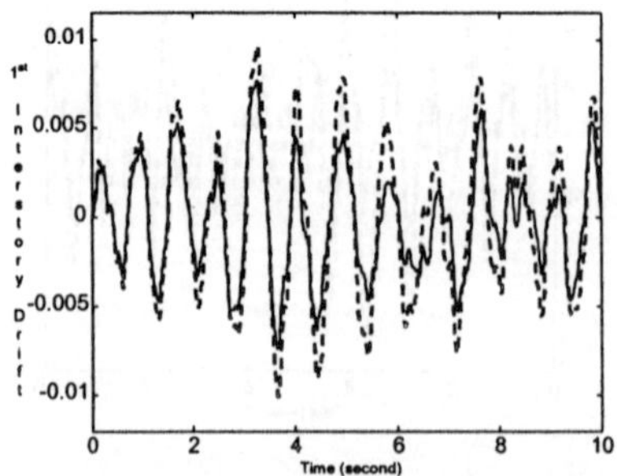

Figure 22a. First interstory drifts of the uncontrolled (dashed line) and the controlled system (solid line) with the H^∞-controller (8.5a)

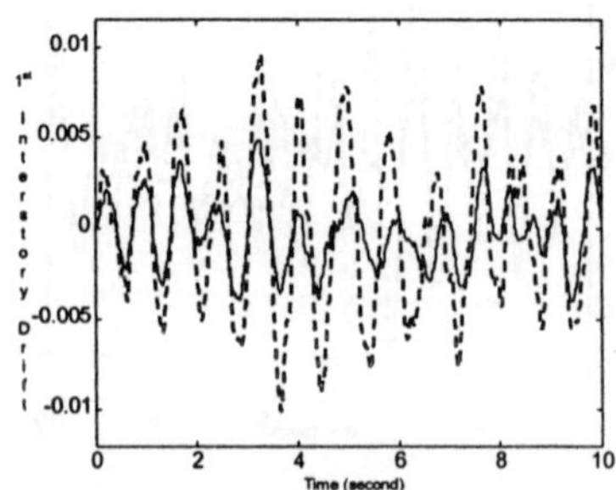

Figure 22b. First interstory drifts of the uncontrolled (dashed line) and the controlled system (solid line) with the H^∞-controller (8.5b)

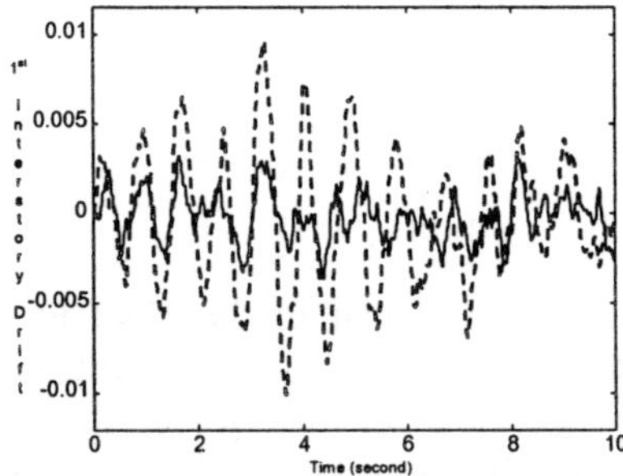

Figure 22c. First interstory drifts of the uncontrolled (dashed line) and the controlled system (solid line) with the H^∞-controller of (8.5c)

From the results obtained from the application of the H^∞ control technique, we arrive at the conclusion that, in the design of H^∞ controllers, similarly to that of LQ regulators, the less the effect of the disturbance on the system output we want to achieve, the largest must be the value of the control law.

It is also clear from a simple comparison of the results obtained from the application of a state feedback LQ regulator and a state feedback H^∞ controller, that an H^∞ controller attenuates system disturbances better than an LQ regulator, at the expense of almost the same (if not lower in some cases) control effort. For example, see, in comparison Figures 13a, 13b, 14a, 14b and 17a, 17b, 19a, 19b, respectively, and the values of the respective controllers. See also Figures 13a and 14a in comparison with Figures 22a and 22b. Therefore, an H^∞ controller is more suitable for disturbance attenuation than LQ regulators.

8.4. Example IV (Robust nonlinear controller design for an aseismic base isolated structure)

In this example we examine the approach presented in Section 5. In order to investigate the dynamic behavior of multi story buildings, prior to the implementation of an experimental program [57], a two story (N=2) prototype laboratory structure is considered. The floor masses are

$$m_1 = 20000\text{kg (base floor) and } m_2 = 10000\text{kg}.$$

The base isolation system coefficients are

$$\kappa_0 = 2000\text{N}/\text{m}, \, \breve{c}_0 = 200\text{Ns}/\text{m}.$$

The remaining stiffness and damping coefficients are

$$\kappa_1 = \kappa_2 = 100000\text{N}/\text{m}, \, \breve{c}_1 = \breve{c}_2 = 10000\text{Ns}/\text{m}$$

With these values we obtain $\mathbf{W} = \begin{bmatrix} 200 & 2000 \end{bmatrix}$ and

$$A = \begin{bmatrix} -0.51 & 0.5 & -5.1 & 5 \\ 1 & -2 & 10 & -20 \\ 1 & 0 & 0 & 0 \\ 0 & 1 & 0 & 0 \end{bmatrix}, \quad B = \begin{bmatrix} 5 \times 10^{-5} \\ 0 \\ 0 \\ 0 \end{bmatrix}, \quad D = \begin{bmatrix} 0.01 & 0.1 \\ 0 & 0 \\ 0 & 0 \\ 0 & 0 \end{bmatrix}$$

The simulated ground motion is that of the 1940 El Centro earthquake. Figure 23 shows the ground velocity and displacement. Figure 24 shows the uncontrolled velocity and displacement histories of the first floor under the assumed earthquake excitation. It is well known that the responses of the floors are essentially the same [57], [58]. This is not unexpected, in view of the relatively high stiffness of the building vis-a-vis that of the base isolation system. Therefore, in this case and in the results that follow we report only the first floor responses.

Our aim in the sequel is to design a full state feedback nonlinear robust controller along the lines reported in Section 5. To this end, the assumed maximum values of ground velocity and displacement are taken to be $\dot{y}_0^{max} = 0.35 \text{m/s}$ and $y_0^{max} = 0.11 \text{m}$ determining the maximum value of the control given by equation (5.6). As it can be easily checked matrix A is stable. Therefore, one may choose $F=0$ and $A_{cl} = A$ is an appropriate choice. Furthermore, the P matrix in the control is based on $Q = I_4$. With these observations, solving (5.7) we obtain

$$P = \begin{bmatrix} 6.2250 & 1.7613 & -0.5 & -0.1512 \\ & 6.8866 & 0.1512 & -0.5 \\ \text{symmetry} & & 2.6136 & 1.3547 \\ & & & 1.0641 \end{bmatrix} > 0$$

Note that in the present case, according to (5.8)

$$\rho(x(t)) = \max_{q \in Q} \|Wq(t)\| = \sqrt{(\kappa_0 y_0^{max})^2 + (\tilde{c}_0 \dot{y}_0^{max})^2} = 230.8679 \text{ N}$$

Simulation results are next presented for $\varepsilon = 9 \times 10^{-6}$. Figure 25 shows the first floor velocity and displacement time histories together with the control force. Obviously, there is considerable improvement over the uncontrolled situation in the amplitude of the responses. Simulation results show that, the smaller the value of ε is the greater the reduction in the responses will be, with a consequent, however, increase in control force requirements [57].

We next revisit the above design in order to apply partial state feedback based on the state of the first floor alone. In this case we have

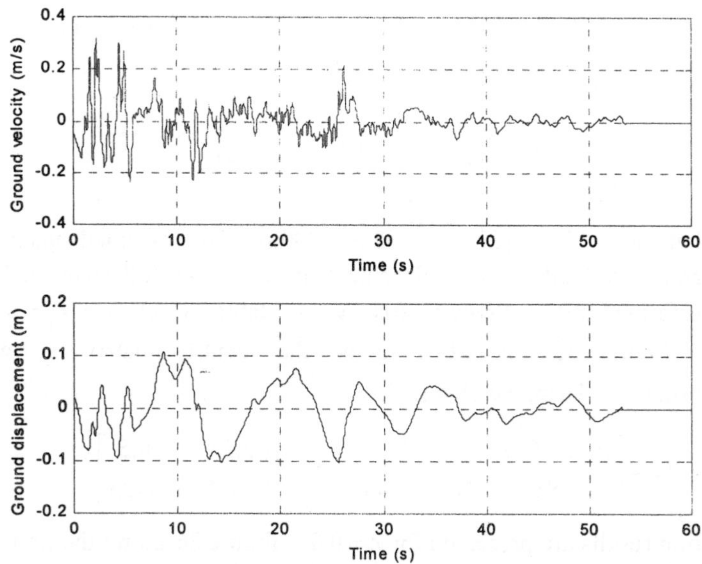

Figure 23. El Centro earthquake: Ground velocity and displacement.

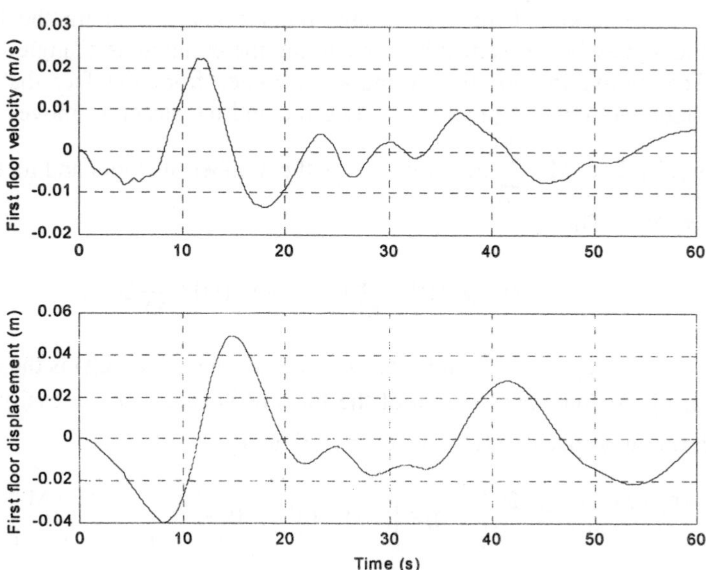

Figure 24. First floor uncontrolled velocity and displacement.

$$\mathbf{A}_{11} = \begin{bmatrix} -0.51 & -5.1 \\ 1 & 0 \end{bmatrix}, \quad \mathbf{B}_1 = \begin{bmatrix} 5 \times 10^{-5} \\ 0 \end{bmatrix}, \quad \mathbf{D}_1 = \begin{bmatrix} 0.01 & 0.1 \\ 0 & 0 \end{bmatrix}$$

$$\mathbf{A}_{12} = \begin{bmatrix} 0.5 & 5 \\ 0 & 0 \end{bmatrix}, \quad \mathbf{A}_{21} = \begin{bmatrix} 1 & 10 \\ 0 & 0 \end{bmatrix}, \quad \mathbf{A}_{22} = \begin{bmatrix} -2 & -20 \\ 1 & 0 \end{bmatrix}$$

Note that here, $\mathbf{W}_1 = \begin{bmatrix} \breve{c}_0 & \kappa_0 \end{bmatrix} = \begin{bmatrix} 200 & 2000 \end{bmatrix}$. The assumed maximum values of ground velocity and displacement are as in the full state feedback designed performed above. Since matrix $\mathbf{A}_{11}$ is stable, we choose $\mathbf{F}_1 = \mathbf{0}$ and $\mathbf{A}_{cl,1} = \mathbf{A}_{11}$. Matrix $\mathbf{Q}_2$ is taken to be the identity matrix $\mathbf{I}_2$ while $\mathbf{Q}_1 = \mathrm{diag}\{170, 170\}$. Then, we obtain

$$\mathbf{P}_1 = \begin{bmatrix} 1016.6667 & -85 \\ -85 & 207.8464 \end{bmatrix}, \quad \mathbf{P}_2 = \begin{bmatrix} 5.25 & -0.5 \\ -0.5 & 0.3125 \end{bmatrix}$$

Simulation results are presented for $\varepsilon = 0.01$. Figure 26 shows the first floor velocity and displacement time histories together with the control force. Again, there is considerable improvement over the uncontrolled responses of Figure 24. The control force is kept very similar to that obtained in full state feedback design. As expected, the velocity and displacement records are close to those of the full state feedback design.

It is worth remarking here that an approximate analysis along the lines presented above may be performed by considering the building as a single story structure. The motivation for this formulation comes from the fact that the individual floor motions are practically the same and the structure moves as a single mass [57]. Let, $M = \sum_{i=1}^{N} m_i$, where N is the number of floors and m_i their masses. Here, N=2. Then,

$$\ddot{y}(t) = -\frac{\breve{c}_0}{M}\left(\dot{y}_1(t) - \dot{y}_0(t)\right) - \frac{\kappa_0}{M}\left(y_1(t) - y_0(t)\right) + \frac{1}{M}\zeta(t),$$

where $y_1(t_0) = y_1^0$, $\dot{y}_1(t_0) = \dot{y}_1^0$ and t_0 is some initial time. Here, $\zeta(t)$ is the control force on M. A state-space model of the form (2.1) can, then, be easily obtained by defining, as usual, $\mathbf{x}(t) = \left(\dot{y}_1(t), y_1(t)\right)^{\mathrm{T}}$ and

$$\mathbf{A} = \begin{bmatrix} -\breve{c}_0/M & -\kappa_0/M \\ 1 & 0 \end{bmatrix}, \quad \mathbf{B} = \begin{bmatrix} 1/M & 0 \end{bmatrix}^{\mathrm{T}}, \quad \mathbf{D} = \begin{bmatrix} \breve{c}_0/M & \kappa_0/M \\ 0 & 0 \end{bmatrix}$$

It can be shown that this approximate analysis may be satisfactory in some typical situations. Let us clarify this point by reconsidering the previous full state feedback design, by taking a lumped mass model and by applying full state

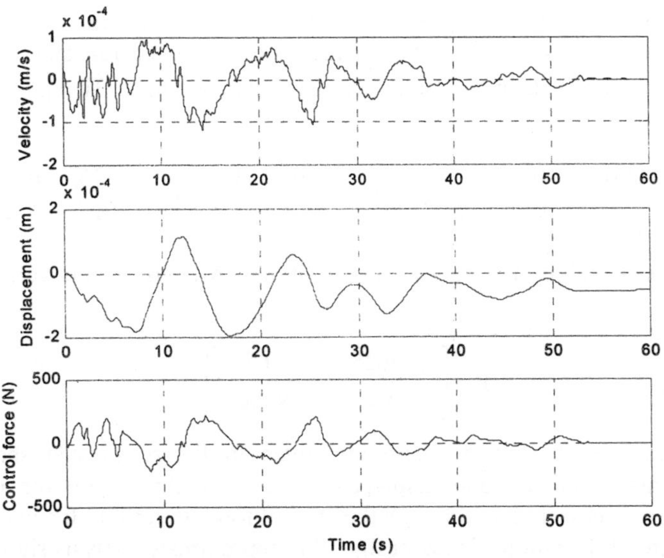

Figure 25. Full state feedback design: First floor controlled velocity, displacement and control force

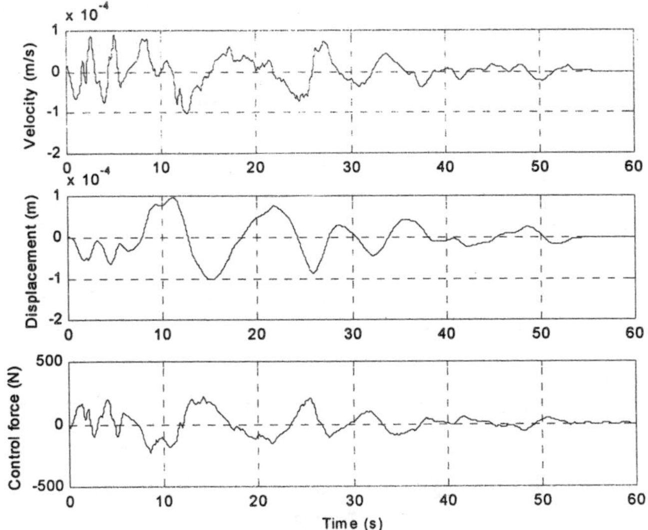

Figure 26. Partial state feedback design: First floor controlled velocity, displacement and control force

feedback which, in this case, coincides with first floor feedback. The procedure and the parameter values are those of the aforementioned full state feedback design based on a distributed model. In particular, we have $\mathbf{W} = \begin{bmatrix} 200 & 2000 \end{bmatrix}$ and

$$\mathbf{A} = \begin{bmatrix} -0.0067 & -0.0667 \\ 1 & 0 \end{bmatrix}, \ \mathbf{B} = 10^{-4} \times \begin{bmatrix} 0.3333 \\ 0 \end{bmatrix}, \ \mathbf{D} = \begin{bmatrix} 0.0067 & 0.0667 \\ 0 & 0 \end{bmatrix}$$

Since, $\mathbf{A}$ is stable, one may choose $\mathbf{F}{=}\mathbf{0}$ and $\mathbf{A}_{cl} = \mathbf{A}$. Moreover, $\mathbf{Q} = \mathbf{I}_2$. Then,

$$\mathbf{P} = \begin{bmatrix} 79.9625 & -0.5 \\ -0.5 & 1198.8884 \end{bmatrix}$$

Simulation results for this case are given in Figure 27 that shows the combined mass velocity and displacement records along with the required control force. It is seen that although the control force is close to the one used in the distributed full state feedback design, this approximate analysis shows some deterioration in the responses.

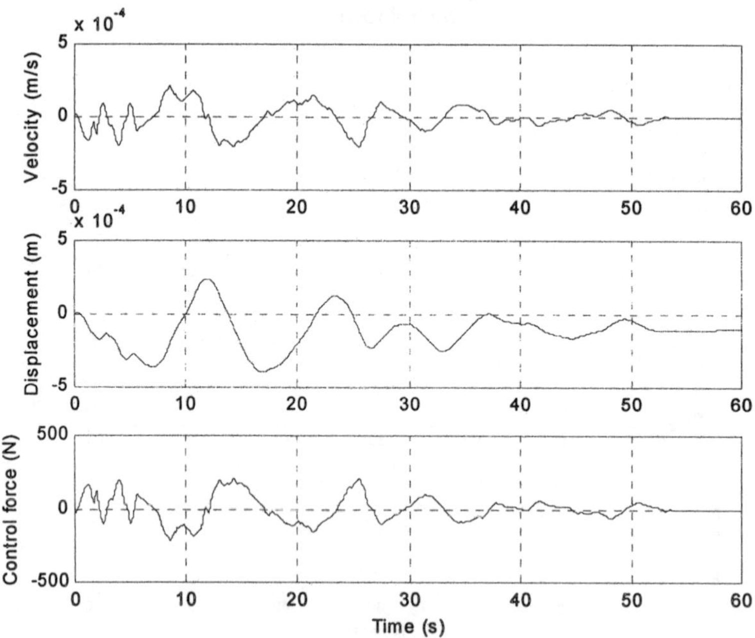

Figure 27. Mass velocity, displacement and control force in the case of a full state feedback design based on a lumped model

8.5. Example V (Discrete-time state-feedback H^∞ control of an operational self propelled sprayer)

In this example, we consider an operational self-propelled sprayer with the following parameter values [53]

$$m_1 = 4000 \text{ kg} \ , \ m_2 = 1015 \text{ kg} \ , \ I_1 = 1345 \text{ kgm}^2 \ , \ I_2 = 33560 \text{ kgm}^2$$

$$\lambda_1 = 0.8 \text{ m} \ , \ R = 0.95 \text{ m} \ , \ \lambda = 0.69 \text{ m} \ , \ \kappa_1 = 360000 \text{ N/m} \ , \ c_1 = 4500 \text{ Nsec/m}$$

$$g = 9.81 \text{ m/sec}^2 \ , \ \lambda_s = 13.75 \text{ m} \ , \ \kappa = 30000 \text{ N/m} \ , \ c = 1000 \text{ Nsec/m}.$$

We these parameter values we obtain

$$A = \begin{bmatrix} 0 & 0 & 0 & 1 & 0 & 0 \\ 0 & 0 & 0 & 0 & 1 & 0 \\ 0 & 0 & 0 & 0 & 0 & 1 \\ -143.5692 & 0 & 0 & -1.7946 & 0 & 0 \\ 0 & -133084.2721 & 13.0245 & 0 & -2.5622 & 0.4361 \\ 0 & 130483.1050 & -13.8530 & 0 & 2.5122 & -0.4570 \end{bmatrix}$$

$$B = \begin{bmatrix} 0 & 0 & 0 \\ 0 & 0 & 0 \\ 0 & 0 & 0 \\ 0.1994 & 0 & 0 \\ 0 & 0.4448 & -0.4361 \\ 0 & -0.4361 & 0.4570 \end{bmatrix} , \ C = E = \begin{bmatrix} 1 & 13.75 & 13.75 & 0 & 0 & 0 \\ 1 & -13.75 & -13.75 & 0 & 0 & 0 \end{bmatrix}$$

$$D = \begin{bmatrix} 0 & 0 & 0 & 0 \\ 0 & 0 & 0 & 0 \\ 0 & 0 & 0 & 0 \\ 71.7846 & 71.7846 & 0.8973 & 0.8973 \\ -128.1119 & 128.1119 & -1.6014 & 1.6014 \\ 125.6081 & -125.6081 & 1.5701 & -1.5701 \end{bmatrix}$$

Our purpose here is to apply the theoretical results of Section 6.2 in order to design state-feedback discrete-time H^∞ regulators for the above sprayer model. Note that a controller guaranteeing disturbance attenuation at a certain amount is indispensable in our case, since disturbances due to tractor movements have a detrimental effect on the vertical distances between the two boom tips and the soil, and hence on the application of the spray. A well known measure for evaluating the effect of the disturbances on the sprayer outputs is the H_∞-norm of the transfer function between disturbances and controlled outputs of the uncontrolled system, which in our case takes a large value, i.e. the value

$$\left\| E(sI - A)^{-1} D \right\|_\infty = 6.6951$$

We next design a state-feedback discrete-time H^∞, for several values of the sampling period T_0, of the disturbance attenuation level γ and of the parameter η. In particular, we obtain the following results:

(i)   *For $T_0=0.1$ sec, $\gamma=1$ and $\eta=10^{-5}$:*

$$F = 10^3 \times \begin{bmatrix} -0.4769 & 0 & 0 & -0.0852 & 0 & 0 \\ 0 & -0.3150 & -0.3163 & 0 & -0.0157 & -0.0160 \\ 0 & -6.6294 & -6.7425 & 0 & -0.6644 & -0.6780 \end{bmatrix}$$

The four disturbance signals used in the simulation are depicted in Figures 28a-28d. The deformations of the left two tip of both the uncontrolled and the controlled sprayer are depicted in Figures 29.

(ii)   *For $T_0=0.01$ sec, $\gamma=1$ and $\eta=10^{-3}$:*

$$F = 10^4 \times \begin{bmatrix} -0.0800 & 0 & 0 & -0.0097 & 0 & 0 \\ 0 & -0.0014 & -0.0029 & 0 & -0.000033 & -0.000035 \\ 0 & -1.6052 & -1.6391 & 0 & -0.1036 & -0.1056 \end{bmatrix}$$

The deformations of the left tip of both the uncontrolled and the controlled sprayer are in this case depicted in Figures 30.

(iii)   *For $T_0=0.01$ sec, $\gamma=0.1$ and $\eta=10^{-4}$:*

$$F = 10^5 \times \begin{bmatrix} -0.1089 & 0 & 0 & -0.0038 & 0 & 0 \\ 0 & -0.0207 & -0.0333 & 0 & -0.00027 & -0.00029 \\ 0 & -1.1215 & -1.1384 & 0 & -0.0274 & -0.0279 \end{bmatrix}$$

The deformations of the right tip of the controlled sprayer are in this case depicted in Figures 31.

(iv) *For $T_0=0.01$ sec, $\gamma=0.01$ and $\eta=10^{-6}$:*

$$F = 10^6 \times \begin{bmatrix} -0.0886 & 0 & 0 & -0.0009 & 0 & 0 \\ 0 & -1.6188 & -1.7242 & 0 & -0.0102 & -0.0106 \\ 0 & -0.5301 & -0.5468 & 0 & -0.0061 & -0.0062 \end{bmatrix}$$

The deformations of the right tip of the controlled sprayer are in this case depicted in Figures 32.

(v) *For $T_0=0.001$ sec, $\gamma=0.01$ and $\eta=10^{-5}$:*

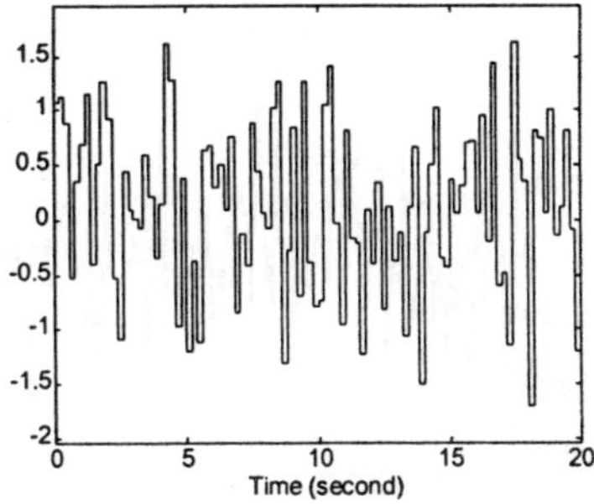

Figure 28a. Disturbance q₁(t) in Example V

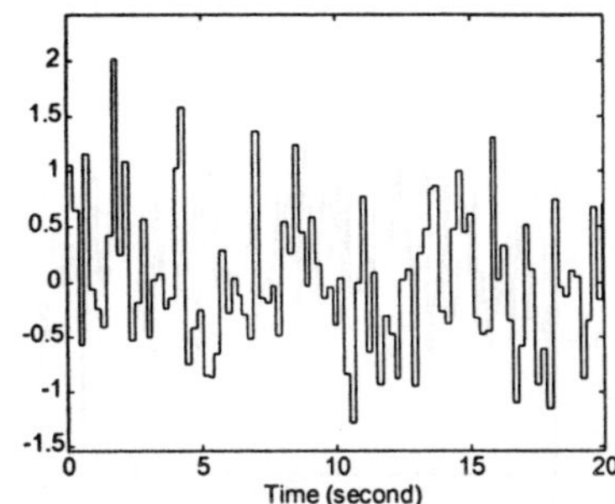

Figure 28c. Disturbance q₃(t) in Example V

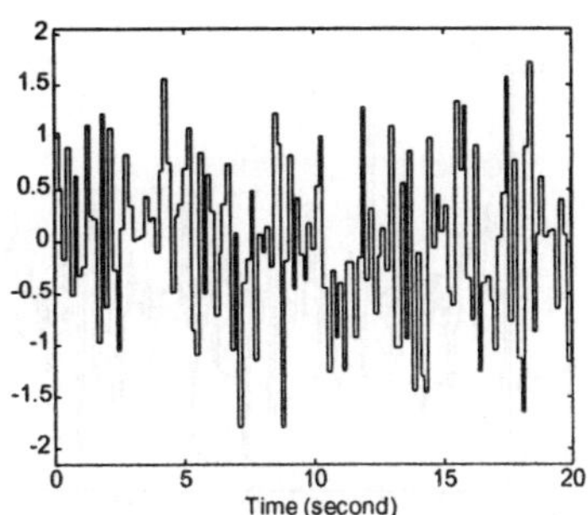

Figure 28b. Disturbance q₂(t) in Example V

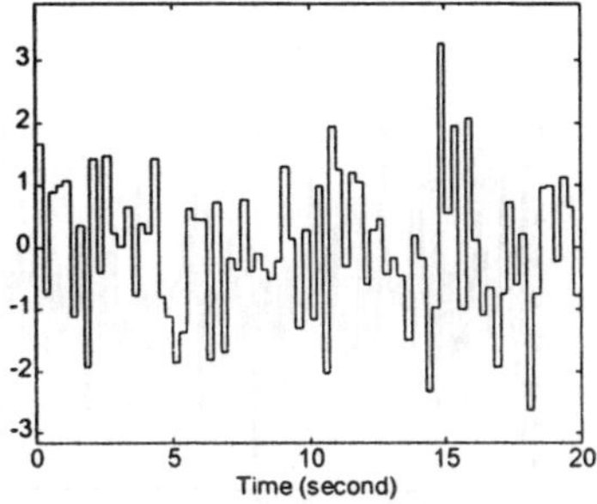

Figure 28d. Disturbance q₂(t) in Example V

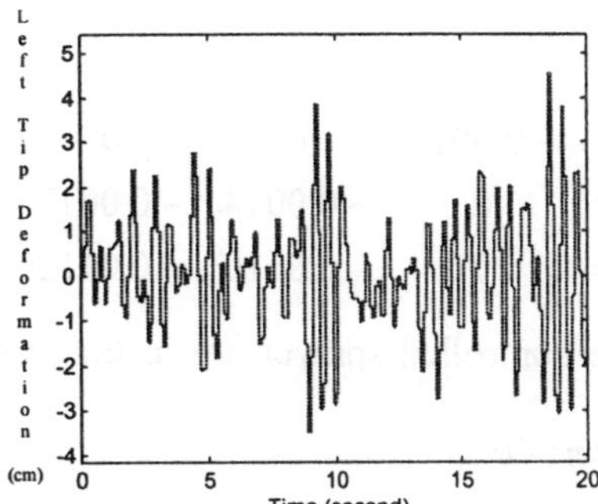

Figure 29. Left tip deformations of the uncontrolled (dotted line) and the controlled (black line) sprayer with the controller of Case (i)

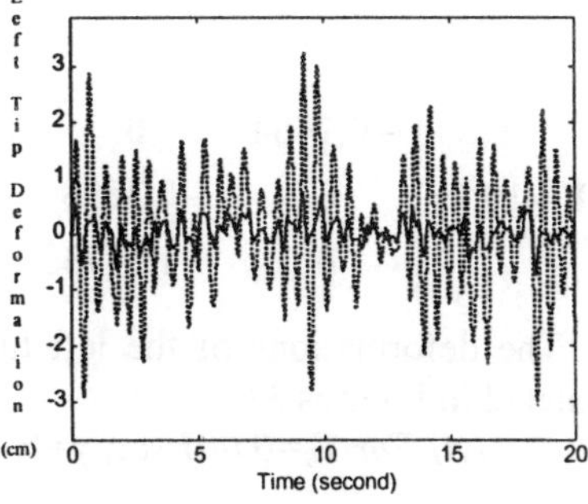

Figure 30. Left tip deformations of the uncontrolled (dotted line) and the controlled (black line) sprayer with the controller of Case (ii)

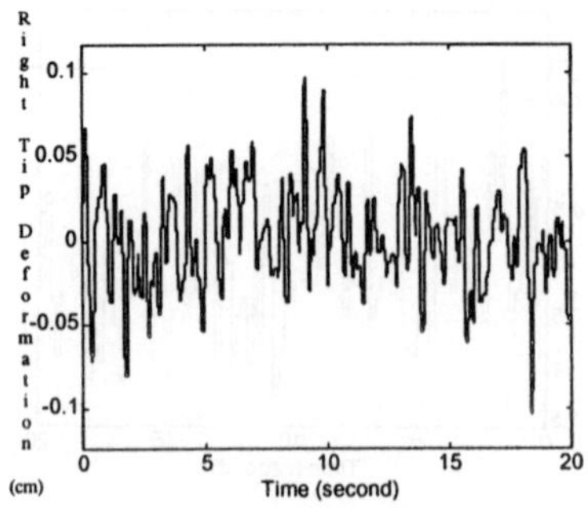

Figure 31. Right tip deformation of the controlled sprayer with the controller of Case (iii)

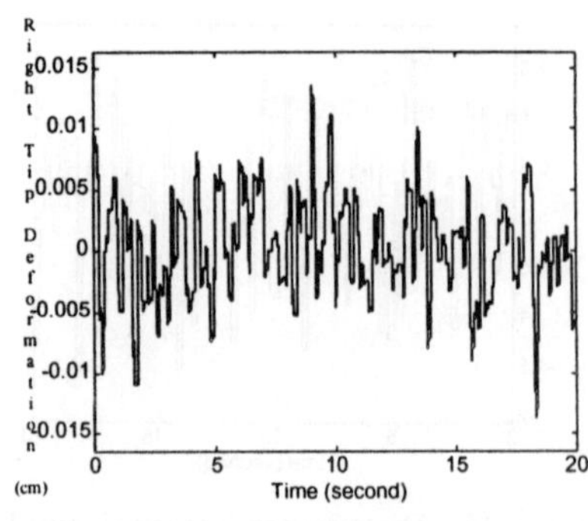

Figure 32. Right tip deformation of the controlled sprayer with the controller of Case (iv)

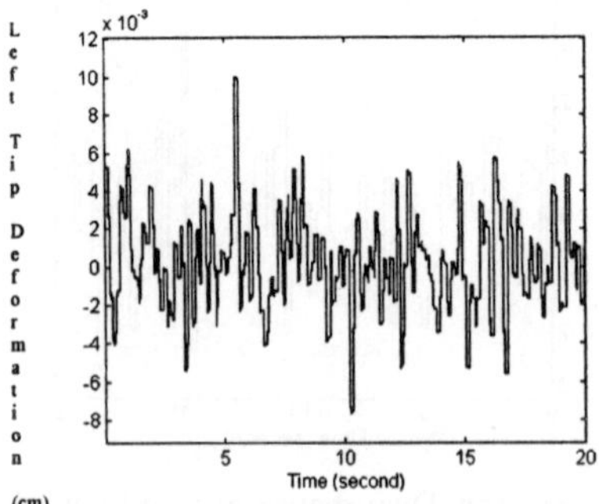

Figure 33. Left tip deformation of the controlled sprayer with the controller of Case (v)

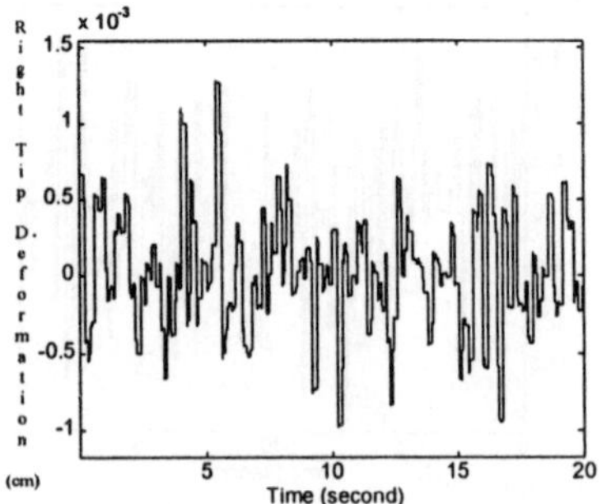

Figure 34. Right tip deformation of the controlled sprayer with the controller of Case (vi)

$$\mathbf{F} = 10^6 \times \begin{bmatrix} -0.1464 & 0 & 0 & -0.0014 & 0 & 0 \\ 0 & -0.3893 & -0.4009 & 0 & -0.0014 & -0.0013 \\ 0 & -1.5373 & -1.5894 & 0 & -0.0110 & -0.0112 \end{bmatrix}$$

The deformations of the left tip of the controlled sprayer are in this case depicted in Figures 33.

(vi) For $T_0=0.001$ sec, $\gamma=0.001$ and $\eta=10^{-6}$:

$$\mathbf{F} = 10^7 \times \begin{bmatrix} -0.1162 & 0 & 0 & -0.0004 & 0 & 0 \\ 0 & -0.6640 & -0.6693 & 0 & -0.0016 & -0.0016 \\ 0 & -1.0399 & -1.0520 & 0 & -0.0030 & -0.0030 \end{bmatrix}$$

The deformations of the right tip of the controlled sprayer are in this case depicted in Figures 34.

From the above analysis, it becomes clear that in the discrete-time case, similar to the continuous-time one, if the desired disturbance attenuation level is decreased, a higher gain feedback occurs. Moreover, in the discrete-time case, the H^∞ design is significantly affected by the sampling period T_0, which is a design factor of crucial importance, since almost all the matrices involved in the design algorithm depend on T_0.

For the above example, a continuous-time state feedback H^∞ controller can be designed on the basis of Corollary 6.1. For example, in the case where $\gamma=0.1$ and $\theta=10^{-9}$, we obtain the following continuous-time controller

$$\mathbf{F}_{\mathrm{CT}} = \begin{bmatrix} -44579.4251 & 0 & 0 \\ 0 & -35277.9307 & -35784.9519 \\ 0 & -596476.0529 & -613856.6001 \\ -668.3422 & 0 & 0 \\ 0 & -111.3749 & -92.7777 \\ 0 & -6501.7197 & -6645.4823 \end{bmatrix} \tag{8.6}$$

It is interesting to note that, if $\gamma=0.1$, $\eta=\sqrt{10^{-9}}$ and $T_0 =10^{-9}$, the discrete-time H^∞ controller is given by (8.6). Simulation results lead to a very interesting general conclusion: For the same disturbance attenuation level γ, in both the continuous-time and the discrete-time design, the continuous-time H^∞-controller, obtained for a certain value of the parameter θ, coincides with the discrete-time H^∞-controller, obtained for the value of the parameter $\eta = \sqrt{\theta}$ and for $T_0 \to 0$.

8.6. Example VI (Discrete H^∞ control of a hollow beam using MROCs)

In this example, we consider a flexible hollow beam of length 4 m with respective inside and outside cross-sections of 16 mm × 56 mm and 20 mm × 60 mm. A cradle on which the beam is clamped at a right angle to the direction of motion of the cradle is also available to represent (and replace) the tractor in experimental arrangements. The cradle is mounted on an inertial frame, fixed to the floor, with only a translational degree of freedom. The beam is mounted such that it has least stiffness in the horizontal plane. The cradle is forced into a prescribed motion, similar to longitudinal tractor vibrations, through a double-acting double-rod linear hydraulic cylinder driven by a four-way servo-valve. The beam can vibrate only in the horizontal plane. Simulations reveal a much larger effect of longitudinal accelerations than yawing angular accelerations. The two lowest horizontal natural frequencies of the beam, of 1.46 and 8.94 Hz, are in the range of the natural frequencies of common small spray booms. Elastic deformations of the beam are measured by an angle measuring device fixed onto the cradle. A needle in the sensor is connected with the beam tip through an inelastic rod. Relative displacements of the beam tip with respect to the cradle forces the needle to rotate over a measured angle that is directly proportional to the magnitude of the tip deformation. A torsional spring, with negligible stiffness with regard to the beam stiffness, tries to pull the needle back into its stationary position such that the needle follows perfectly all vibrations of the beam tip. Horizontal elastic deformations of the beam are attenuated by a control actuator, consisting of a second identical hydraulic cylinder with servo-valve. The control actuator is attached to the cradle and the beam at 1 m from the clamped end and parallel to the inertial frame.

An evaluation modal model of the structure can be obtained as described in Section 7. In particular, for the present case we have [54], $r = 4, k = 8, n = 16$

$$\mathbf{B}_{sm} = \begin{bmatrix} -0.064 & -0.276 & 0.486 & -0.471 & -0.230 & 0.063 & -0.269 & 0.099 \end{bmatrix}^{\mathrm{T}}$$

$$\mathbf{M}_{sm} = \begin{bmatrix} -2.375 & -1.316 & 0.769 & -0.526 & -0.377 & -0.306 & 0.215 & -0.046 \end{bmatrix}^{\mathrm{T}}$$

$$\Omega = \mathrm{diag}\{9.2, 58, 164, 323, 600, 964, 1528, 2506\}, \quad Z = \operatorname*{diag}_{i=1,2,\cdots,8}\{\zeta_i\}, \quad \zeta_i = 0.005$$

The beam tip y is related to the state modal variables through the relation [54], $y = \mathbf{C}\mathbf{x}$, where

$$\mathbf{C} = \begin{bmatrix} 0.659 & -0.661 & -0.668 & -0.656 & 0.696 & -0.751 & -0.758 & -1.628 & \mathbf{0}_{1\times 8} \end{bmatrix}$$

Design models of the hollow beam can be obtained from the above evaluation model using truncation. In this example we assume two different design models. The first design model only preserves the fundamental mode of the beam and is described by the matrices

$$\mathbf{A}_1 = \begin{bmatrix} 0 & 1 \\ -85.53 & -0.09 \end{bmatrix} \quad \mathbf{B}_1 = \begin{bmatrix} 0 \\ -0.064 \end{bmatrix} \quad \mathbf{D}_1 = \begin{bmatrix} 0 \\ -2.375 \end{bmatrix}, \quad \mathbf{C}_1 = \begin{bmatrix} 0.659 & 0 \end{bmatrix} \tag{8.7a}$$

and $\mathbf{E}_1 = \mathbf{C}_1$. The second design model retains the fundamental mode and the first harmonic of the beam. Its state space matrices are

$$\mathbf{A}_2 = \begin{bmatrix} 0 & 0 & 1 & 0 \\ 0 & 0 & 0 & 1 \\ -85.53 & 0 & -0.09 & 0 \\ 0 & -3366.88 & 0 & -0.58 \end{bmatrix} \quad \mathbf{B}_2 = \begin{bmatrix} 0 \\ 0 \\ -0.064 \\ -0.276 \end{bmatrix}, \quad \mathbf{D}_2 = \begin{bmatrix} 0 \\ 0 \\ -2.375 \\ -1.316 \end{bmatrix} \tag{8.7b}$$

$$\mathbf{C}_2 = \begin{bmatrix} 0.659 & -0.661 & 0 & 0 \end{bmatrix}, \quad \mathbf{E}_2 = \mathbf{C}_2$$

Disturbances have a significant effect in the output of both models. As it can be easily checked, for both models we have

$$\left\| \mathbf{E}_1 (s\mathbf{I} - \mathbf{A}_1)^{-1} \mathbf{D}_1 \right\|_\infty = \left\| \mathbf{E}_2 (s\mathbf{I} - \mathbf{A}_2)^{-1} \mathbf{D}_2 \right\|_\infty = 1.8165$$

Our aim in the sequel will be the design MROC based H^∞ optimal regulators for the hollow beam, in order to attenuate the disturbance effects. To this end, we first observe that for both models, condition (4.24) is satisfied. Then, for model (8.7a), we select $T_0 = 0.1$ sec. Then for several values of the disturbance attenuation level γ, of the parameter η and of the output multiplicity of the sampling N, we obtain:

(i)   *For $\gamma=0.5$, $\eta=10^{-2}$ and $N=3$:* In this case, the fictitious state-feedback controller is

$$F=[-30.6852 \quad 7.0076] \qquad (8.8)$$

Then, the MROC gains obtained by solving (4.26) are

$$\mathbf{L}_\gamma = [422.2751 \quad -1067.8690 \quad 598.555] \, , \, \mathbf{L}_\zeta = 0$$

(ii)   *For $\gamma=0.5$, $\eta=10^{-2}$ and $N=6$:* The fictitious state-feedback controller is given by (8.5) and the MROC gains are $\mathbf{L}_\zeta = 0$

$$\mathbf{L}_\gamma = [280.9385 \quad -110.7693 \quad -296.8914 \quad -273.3260 \quad -40.9457 \quad 394.4306]$$

(iii)   *For $\gamma=0.5$, $\eta=10^{-2}$ and $N=11$:* The fictitious state-feedback controller is given by (8.8) and the MROC gains are

$$\mathbf{L}_\gamma = [177.6762 \quad 55.0839 \quad -39.4878 \quad -105.3940 \quad -142.6478$$
$$-149.6478 \quad -127.7307 \quad -76.6203 \quad 3.2988 \quad 111.4386 \quad 247.0124] \, , \, \mathbf{L}_\zeta = 0$$

(iv)   *For $\gamma=0.1$, $\eta=10^{-3}$ and $N=3$:* In this case, the fictitious state-feedback controller is

$$F=[-146.6512 \quad 59.4598] \qquad (8.9)$$

Then, the MROC gains are

$$\mathbf{L}_\gamma = [3759.0942 \quad -9561.2901 \quad 5579.6601] \, , \, \mathbf{L}_\zeta = 0$$

(v)   *For $\gamma=0.1$, $\eta=10^{-3}$ and $N=7$:* The fictitious state-feedback controller is given by (8.9) and the MROC gains are

$$\mathbf{L}_\gamma = [2211.5651 \quad -362.2878 \quad -1853.4890$$
$$-2237.4552 \quad -1508.9209 \quad 317.9945 \quad 3210.0575] \, , \, \mathbf{L}_\zeta = 0$$

(vi)   *For $\gamma=0.03$, $\eta=10^{-4}$ and $N=3$:* In this case, the fictitious state-feedback controller is given by

$$F=[715.1938 \quad 204.3380] \qquad (8.10)$$

and the MROC gains are

$$\mathbf{L}_\gamma = \begin{bmatrix} 14762.9028 & -38222.9471 & 24545.3155 \end{bmatrix} , \quad \mathbf{L}_\zeta = 0$$

(vii)   *For $\gamma=0.03$, $\eta=10^{-4}$ and $N=6$:* The fictitious state-feedback controller is given by (8.10) and the MROC gains are

$$\mathbf{L}_\gamma = \begin{bmatrix} 9419.1599 & -3938.2979 & -10098.4000 \\ -8925.9795 & -459.8059 & 15088.5948 \end{bmatrix}, \quad \mathbf{L}_\zeta = 0$$

In all cases examined above, the obtained MROC controller is static since $\mathbf{L}_\zeta = 0$. It is worth noticing that, in the present case, there is no possibility to arbitrarily assign the value of $\mathbf{L}_\zeta$, since as it can be easily checked,

$$\mathrm{rank}\begin{bmatrix} \mathbf{A}_1 & \mathbf{B}_1 & \mathbf{D}_1 \\ \mathbf{C}_1 & 0 & 0 \end{bmatrix} = 3 \neq n + m + d = 4.$$

For the aforementioned Case (iv), the deformations of both the uncontrolled and the controlled hollow beam, when the disturbance of Figure 35 is applied are depicted in Figure 36.

We next consider the model (8.7b) and we select $T_0 = 0.1$ sec. Then for several values of the disturbance attenuation level γ, of the parameter ε and of the output multiplicity of the sampling N, we obtain:

(i)   *For $\gamma=0.5$, $\eta=10^{-2}$ and $N=5$:* In this case, the fictitious state-feedback controller is

$$\mathbf{F} = \begin{bmatrix} -30.6903 & 1.4544 & 7.0074 & 0.0019 \end{bmatrix} \qquad \text{(8.11)}$$

Then, the MROC gains are

$$\mathbf{L}_\gamma = \begin{bmatrix} 1453.0115 & -5027.2903 & 7181.8077 & -5522.1985 & 1868.0922 \end{bmatrix}$$

$$\mathbf{L}_\zeta = -2.09563 \times 10^{-5} \cong 0$$

(ii)   *For $\gamma=0.5$, $\eta=10^{-2}$ and $N=15$:* In this case, the fictitious state-feedback controller is given by (8.8) and the MROC gains are

$$\mathbf{L}_\gamma = \begin{bmatrix} 136.2438 & 63.1608 & 33.2488 & -51.1814 & -59.0050 \\ -87.9257 & -122.3558 & -87.9218 & -110.9530 & -77.0769 \\ -31.2269 & -20.6045 & 69.5142 & 114.4022 & 185.1039 \end{bmatrix}$$

$$\mathbf{L}_\zeta = -2.09563 \times 10^{-5} \cong 0$$

(iii)   *For $\gamma=0.1$, $\eta=10^{-3}$ and $N=5$:* In this case, the fictitious state-feedback controller is

$$\mathbf{F} = \begin{bmatrix} -147.0572 & 138.6909 & 59.4512 & 0.0746 \end{bmatrix} \text{ (8.12)}$$

Then, the MROC gains are

$$\mathbf{L}_\gamma = \begin{bmatrix} 12860.3220 & -45007.3331 & 64983.6560 & -50195.5429 & 17135.7441 \end{bmatrix}$$

$$\mathbf{L}_\zeta = \text{-}6.296778 \times 10^{-6} \cong 0$$

(iv)   *For $\gamma=0.1$, $\eta=10^{-3}$ and $N=15$:* In this case, the fictitious state-feedback controller is given by (8.12) and the MROC gains are

$$\mathbf{L}_\gamma = \begin{bmatrix} 1192.4864 & 556.8286 & 260.4837 & -427.6570 & -532.3749 \\ -775.0137 & -1047.0012 & -789.1502 & -942.5481 & -660.3528 \\ -269.8309 & -129.0458 & 618.1479 & 1045.9000 & 1675.8743 \end{bmatrix}$$

$$\mathbf{L}_\zeta = \text{-}6.296778 \times 10^{-6} \cong 0$$

(v)   *For $\gamma=0.03$, $\eta=10^{-4}$ and $N=5$:* In this case, the fictitious state-feedback controller is given by

$$\mathbf{F} = \begin{bmatrix} 714.0298 & 867.4733 & 204.6136 & -2.7820 \end{bmatrix}$$

and the MROC gains are

$$\mathbf{L}_\gamma = \begin{bmatrix} 51770.0425 & -18220.0209 & 26269.0767 & -20067.4775 & 69497.6361 \end{bmatrix}$$

$$\mathbf{L}_\zeta = 0.00113$$

Note that, in all cases examined above, the respective MROC is a stable controller, since $|\mathbf{L}_\zeta| < 1$. It is worth noticing that, in the present case, there is no possibility to arbitrarily assign the value of $\mathbf{L}_\zeta$, since as it can be easily checked, $\text{rank}\begin{bmatrix} \mathbf{A}_2 & \mathbf{B}_2 & \mathbf{D}_2 \\ \mathbf{C}_2 & 0 & 0 \end{bmatrix} = 5 \neq n + m + d = 6$.

For the aforementioned Case (iii), the deformations of both the uncontrolled and the controlled hollow beam, when the disturbance of Figure 35 is applied are depicted in Figure 37.

We next consider the design of a MROC based H^∞ controller for model (8.7b), in the case where $T_0 = 0.01$, $\gamma=0.5$, $\eta=10^{-2}$ and $N=5$. In this case, the fictitious state feedback controller is given by

$$\mathbf{F} = \begin{bmatrix} -1.7781 & 0.0890 & 7.9376 & -0.0003 \end{bmatrix}$$

and the MROC gains are

$$\mathbf{L}_\gamma = \begin{bmatrix} 12502.2546 & -39981.2820 & 53961.8821 & -44002.6461 & 17516.8010 \end{bmatrix}$$

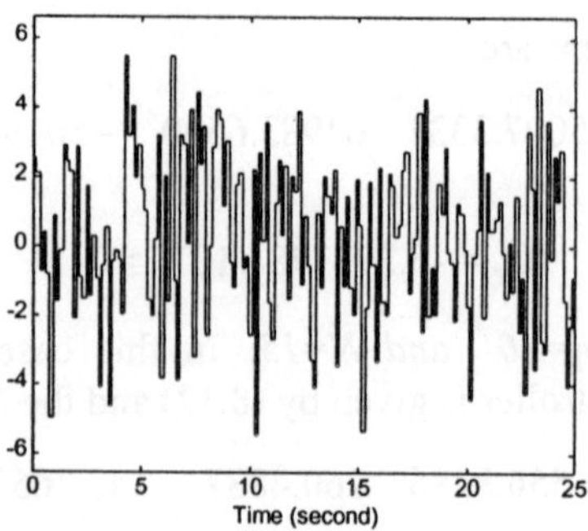

Figure 35. Disturbance signal for Examples
VI and VII

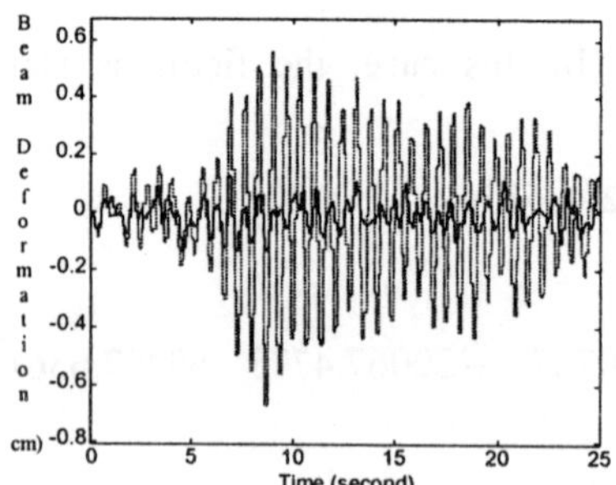

Figure 36. Deformations of the
uncontrolled (dashed line) and the
controller model (8.7a) of the hollow
beam with the controller of Case (iv)

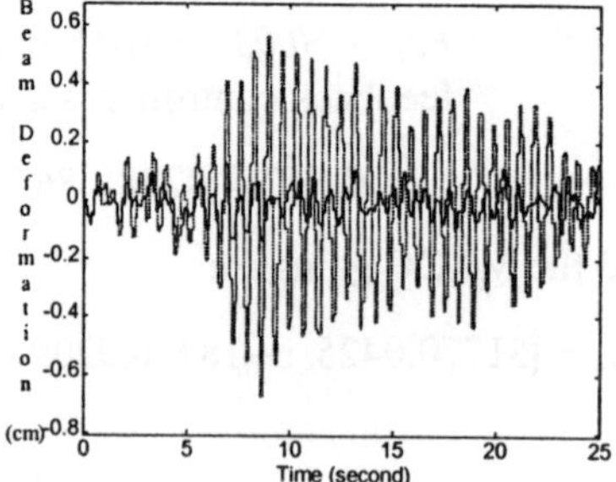

Figure 37. Deformations of the
uncontrolled (dashed line) and the
controller model (8.7b) of the hollow
beam with the controller of Case (iii)

$$\mathbf{L}_\zeta = -1.210762 \times 10^{-6} \cong 0$$

From the above analysis it becomes clear that, in the design of MROC based H^∞ controllers, both the fundamental sampling period T_0 and the output multiplicity of the sampling, are factors of crucial importance. In particular, the fundamental sampling period T_0 affects both the fictitious controller gain and matrices $\begin{bmatrix} \mathbf{H} & \Theta_q \end{bmatrix}$ and Θ_ζ, while the output multiplicity of the sampling affect matrices $\begin{bmatrix} \mathbf{H} & \Theta_q \end{bmatrix}$ and and Θ_ζ, and hence the MROC gains. In general, a deficiency of MROC performance may arise in cases where matrix $\begin{bmatrix} \mathbf{H} & \Theta_q \end{bmatrix}$ approaches a rank-deficient matrix. In these cases, the controller gain matrix $\mathbf{L}_\gamma$

will have extremely large entries. Large gains will have the effect of amplifying any inaccuracies in the output due, for example, to noise or nonlinearity. As a consequence, although the ideal plant input $\zeta(kT_0)$ will remain well behaved, taking the value $\zeta(kT_0) = -\mathbf{F}\mathbf{x}(kT_0)$, the actual plant input will not remain well behaved and the closed-loop design may become extremely sensitive to noise or to small parameter variations. In order to avoid such problems, one must follow the guidelines proposed in [59]. Taking into account the analysis presented in [59], for the proper operation of MROCs the output multiplicities of the sampling as well the sampling period T_0 should be chosen sufficiently large, but not too large in the case of open-loop unstable systems. Violation of these guidelines will lead to noise amplification in the controller and to deficiency of stability robustness (see also [22]-[24], for an analysis of this issue).

8.7. Example VII (Singlerate vs. Multirate LQ control of the hollow beam)

In this example, the techniques for singlerate and multirate sampled-data LQ control, presented in Sections 4.3 and 4.4, respectively, will be compared, through its application to the LQ control of a hollow beam. To this end, we consider again, the second design model of the hollow beam given by (8.7b) and retaining the fundamental mode and the first harmonic of the beam.

We first design a singlerate sampled-data LQ regulator for the hollow beam in the case where $\mathbf{Q} = 10^8 \times \mathbf{I}_4$ and $\mathbf{R} = 1$. In this case, selecting $T_0 = 0.01$ sec and applying the technique reported in Section 4.3, yields

$$
\hat{\mathbf{Q}} = \begin{bmatrix}
1240419.544 & 0 & -421064.302 & 0 \\
 & 179680395936.825 & 0 & -10628984.816 \\
\text{symmetry} & & 996290.866 & 0 \\
 & & & 463419.920
\end{bmatrix}
$$

$$
\hat{\mathbf{N}} = \begin{bmatrix}
180.966 \\
147293.076 \\
-318.810 \\
-8.713
\end{bmatrix}, \quad \hat{\mathbf{R}} = 0.267
$$

Then, (4.11), (4.12) give the admissible feedback gains having the value

$$
\hat{\mathbf{F}} = \begin{bmatrix} 638.8684 & -491733.9432 & -1228.5489 & 817.1851 \end{bmatrix} \quad (8.13)
$$

We next design a multirate sampled-data LQ regulator for the hollow beam, for the same values of $\mathbf{Q}$, $\mathbf{R}$ and T_0. In particular, we select N=5. In this case

$$\mathbf{B}_N = 10^{-5} \times \begin{bmatrix} 3.6722 & 0 & 0 & 0 \\ -1.3607 & 3.0192 & 0 & 0 \\ 555.6402 & 271.5038 & 160.0523 & 0 \\ -396.8767 & -88.6422 & 653.3990 & 1579.7320 \end{bmatrix}$$

$$\widetilde{\mathbf{Q}}_N = \begin{bmatrix} 1240419.544 & 0 & -421064.303 & 0 \\ & 179680395999.067 & 0 & -10628984.830 \\ \text{symmetry} & & 996290.867 & 0 \\ & & & 463419.920 \end{bmatrix}$$

$$\widetilde{\mathbf{G}}_N = \begin{bmatrix} -1950.6303 & -238.5570 & -41.9512 & 162.9773 \\ -2560842.5510 & 2047260.7467 & -446032.5261 & 583673.8258 \\ 3663.1184 & -1.7023 & -0.2779 & 0.9271 \\ 849.0622 & -1405.9564 & -228.7775 & 4417.8704 \end{bmatrix}$$

$$\widetilde{\mathbf{R}}_N = \begin{bmatrix} 56.7914 & -34.1027 & 6.3139 & -3.8913 \\ & 40.4842 & -8.2579 & -5.2772 \\ \text{symmetry} & & 8.8826 & -1.5220 \\ & & & 59.8567 \end{bmatrix}$$

Then, from (4.18), (4.19) we obtain the admissible controller having the value

$$\mathbf{F}_m = \begin{bmatrix} 29.1341 & -25658.7946 & 116.8442 & 13.5205 \\ -5.1973 & 26817.9179 & 80.4724 & -18.9664 \\ -29.5808 & -323.7046 & 0.3040 & -37.3689 \\ 7.0140 & 10395.8056 & 18.3229 & 72.2729 \end{bmatrix} \qquad (8.14)$$

In Figure 38, the deformation of the controlled hollow beam in the case of singlerate LQ control (dashed line) is compared with the its deformation in the case of multirate LQ control (solid line), when the disturbance signal is the one depicted in Figure 35. Clearly, the multirate LQ design provides a better disturbance attenuation, at the expense of smaller control action, as compared to the singlerate LQ design. Of course, this is due to the fact that the multirate design produces a less LQR cost than that of the singlerate one.

When state variables are not accessible, the above singlerate and multirate LQ designs can easily be realized on the basis of MROCs and TPMRCs, respectively.

In particular, the controller (8.13) can be equivalently realized by a MROC as follows: Since, as it can be easily checked (4.24) holds, we can select $N \geq 5$. Then, we choose N=6 and solving (4.26), we obtain the following MROC gains

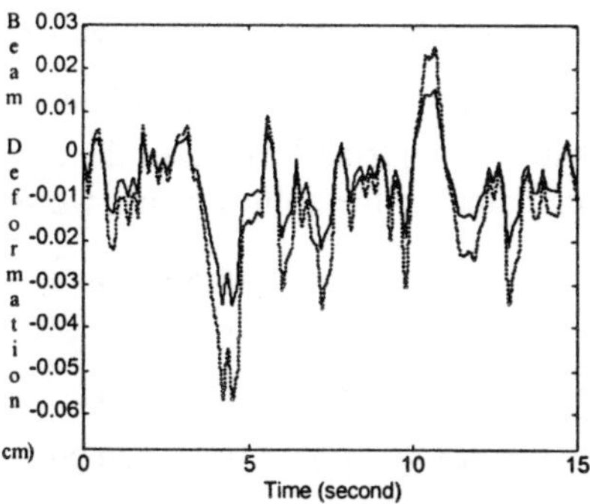

Figure 38. Deformations of model (8.7b) of
the hollow beam controlled by the
singlerate controller (8.13) (dashed line)
and the multirate controller (8.14) (solid
line)

$$\mathbf{L}_{\hat{\gamma}} = \begin{bmatrix} -2604126.3955 & 6908350.8948 & -4559079.3443 \\ -3851361.2201 & 7883546.5686 & -3776465.9636 \end{bmatrix}, \quad \mathbf{L}_{\zeta} = -0.3509$$

Note that, since $\left|\mathbf{L}_{\zeta}\right| < 1$, the designed MROC is stable. Therefore, in addition to LQ regulation, strong stabilization (i.e. stabilization with a stable controller) is achieved.

The controller (8.14) can be realized by a TPMRC, in a similar manner. For example, selecting M=6 and solving (4.32) we obtain

$$\mathbf{L}_{\hat{\gamma}} = \begin{bmatrix} 138657.9093 & -324688.9830 & 180962.2259 \\ 168750.4215 & -429279.3900 & 261619.7030 \\ -13754.4559 & 15859.0148 & 16613.9634 \\ 63222.0984 & -121247.3702 & 20337.4789 \end{bmatrix}$$

$$\begin{bmatrix} 155318.5618 & -354376.6737 & 204165.6940 \\ 237408.9786 & -459482.3637 & 220980.4919 \\ -10035.9516 & -24679.5042 & 15951.3482 \\ 68693.9171 & -46166.4644 & 15173.2060 \end{bmatrix}$$

$$\mathbf{L}_{\hat{\zeta}} = \begin{bmatrix} 0.3207 & -0.5074 & 0.4778 & -1.7484 \\ 0.2280 & -1.4201 & 0.8495 & -2.2442 \\ -0.0677 & -0.0323 & 0.3835 & 0.3262 \\ 0.1893 & -0.2089 & -0.6013 & -1.1256 \end{bmatrix}$$

However, matrix $\mathbf{L}_{\hat{\zeta}}$ is unstable, since its eigenvalues are $\lambda_1 = -1.0963$, $\lambda_2 = 0.0709$, $\lambda_3 = -0.3665$ and $\lambda_4 = 0.0002$. Therefore, the above TPMRC is useless, since it fails to ensure strong stabilization of the closed-loop system. To design a stable TPMRC regulator for the hollow beam, we select M=10 and

$$
\mathbf{L}_{\hat{\zeta},\mathrm{sp}} = \begin{bmatrix} 0 & 0 & -0.5 & 0 \\ 0 & -0.4 & 0 & -0.5 \\ -0.1 & 0 & 0 & 0 \\ 0.1 & -0.5 & 0 & -0.2 \end{bmatrix}
$$

which has stable eigenvalues $\lambda_1 = -0.8099$, $\lambda_2 = -0.2236$, $\lambda_3 = 0.2236$, $\lambda_4 = 0.2099$. In this case, matrix $\begin{bmatrix} \mathbf{H} & \Theta_{\hat{\zeta}} & \Theta_q \end{bmatrix}$ takes the form

$$
\begin{bmatrix} \mathbf{H} & \Theta_{\hat{\zeta}} & \Theta_q \end{bmatrix} =
$$

$$
\begin{bmatrix}
0.6562 & -0.5879 & -0.0066 & -0.0005 \\
0.6567 & -0.3237 & -0.0059 & -0.0010 \\
0.6572 & 0.0463 & -0.0053 & -0.0011 \\
0.6576 & 0.4009 & -0.0046 & -0.0009 \\
0.6580 & 0.6242 & -0.0040 & -0.0004 \\
0.6583 & 0.6430 & -0.0033 & 0.0003 \\
0.6585 & 0.4514 & -0.0026 & 0.0008 \\
0.6587 & 0.1122 & -0.0020 & 0.0011 \\
0.6589 & -0.2635 & -0.0013 & 0.0010 \\
0.6590 & -0.5528 & -0.0007 & 0.0006
\end{bmatrix}
$$

$$
\begin{bmatrix}
0.0239 \times 10^{-4} & 0.3516 \times 10^{-4} & 0.1399 \times 10^{-4} & 0.0835 \times 10^{-4} & -0.7794 \times 10^{-4} \\
-0.0061 \times 10^{-4} & 0.2538 \times 10^{-4} & 0.1631 \times 10^{-4} & 0.1483 \times 10^{-4} & -0.6205 \times 10^{-4} \\
-0.0253 \times 10^{-4} & 0.1332 \times 10^{-4} & 0.1697 \times 10^{-4} & 0.1473 \times 10^{-4} & -0.4731 \times 10^{-4} \\
-0.0339 \times 10^{-4} & 0.0226 \times 10^{-4} & 0.1446 \times 10^{-4} & 0.0999 \times 10^{-4} & -0.3419 \times 10^{-4} \\
-0.0336 \times 10^{-4} & -0.0496 \times 10^{-4} & 0.0832 \times 10^{-4} & 0.0410 \times 10^{-4} & -0.2315 \times 10^{-4} \\
-0.0273 \times 10^{-4} & -0.0778 \times 10^{-4} & 0.0102 \times 10^{-4} & -0.0076 \times 10^{-4} & -0.1447 \times 10^{-4} \\
-0.0184 \times 10^{-4} & -0.0709 \times 10^{-4} & -0.0455 \times 10^{-4} & -0.0290 \times 10^{-4} & -0.0817 \times 10^{-4} \\
-0.0101 \times 10^{-4} & -0.0452 \times 10^{-4} & -0.0659 \times 10^{-4} & -0.0279 \times 10^{-4} & -0.0402 \times 10^{-4} \\
-0.0042 \times 10^{-4} & -0.0216 \times 10^{-4} & -0.0453 \times 10^{-4} & -0.0168 \times 10^{-4} & -0.0158 \times 10^{-4} \\
-0.0012 \times 10^{-4} & -0.0060 \times 10^{-4} & -0.0127 \times 10^{-4} & -0.0047 \times 10^{-4} & -0.0036 \times 10^{-4}
\end{bmatrix}
$$

Matrix $\begin{bmatrix} \mathbf{H} & \Theta_{\hat{\zeta}} & \Theta_q \end{bmatrix}$ has full column rank, and consequently, equation (4.33) is solvable with respect to $\mathbf{L}_{\hat{\gamma}}$.

Similarly, for M=10, in the case of a static TPMRC (i.e. the case where $\mathbf{L}_{\hat{\zeta},\mathrm{sp}} = \mathbf{0}_{4\times4}$), solving (4.33) we obtain the admissible static TPMRC gain $\mathbf{L}_{\hat{\gamma}}$.

9. Conclusions

We have shown that robust control and especially the notions of H-infinity control lead to more powerfull optimal control schemes than can be used for engineering design. Two examples from aseismic design and agricultural engineering have been presented in some details. There are more reasons to choose these two areas of applications, besides our research interests and the interests of our co-workers. Both areas, namely civil engineering and agricultural engineering, are believed to be areas of low-technology, in comparison with, say, aerospace engineering. This short contribution and our recent research shows that application of more complicated modeling and design techniques is possible with the available theoretical results and computer programs. Real-life applications, although feasible, have not been discussed in this paper. Some relevant results can be found in the literature.

Additional material on optimal structural control can be found in the recent monographs [60]-[61]. Applications of smart structures in other areas of mechanics have been reported in other places of this book series. Some recent developments are described, among others, in the review article [62]. All these approaches can be combined with the methods discussed in this paper and extended for the study of civil and agricultural engineering structures.

For some additional information on H-infinity control and civil engineering the reader may also consult the references given in our previous publications [29], [35] and [63]-[65].

References

[1].Chen, C.T., *Introduction to Linear System Theory*. New York: Holt, Reinhart and Winston Inc., 1970.

[2].Kailath, T., *Linear Systems*. Englewood Cliffs, New Jersey: Prentice Hall, 1980.

[3].Sontag, E.D., *Mathematical Control Theory: Deterministic Finite Dimensional Systems*. New York: Springer-Verlag, 1990.

[4].Basile, G. and Marro, G., *Controlled and Conditioned Invariants in Linear System Theory*. Englewood Cliffs, New Jersey: Prentice Hall, 1992.

[5].Ogata, K., *Discrete-Time Control Systems*. Englewood Cliffs, New Jersey: Prentice Hall, 1987.

[6].Perdon, A.M., Conte, G. and Longhi, S., "Invertibility and inversion of linear periodic systems", *Automatica,* vol. 28, pp. 645-648, 1992.

[7].Arvanitis, K.G. and Paraskevopoulos, P.N., "Disturbance localization for left invertible linear periodic discrete-time systems", *Syst. Control Lett.,* vol. 31, pp. 185-198, 1997.

[8].Anderson, B.D.O. and Moore, J.B., *Linear Optimal Control*. Englewood-Cliffs, New Jersey: Prentice Hall, 1972.

[9].Kwakernaak, H.and Sivan, R., *Linear Optimal Control Systems*. New York: John Wiley, 1972.

[10].   Safonov, M.G. and Athans, M., "Gain and phase margins of multiloop LQG regulators", *IEEE Trans. Autom. Control,* vol. AC-22, pp. 173-179, 1977.

[11].   Safonov, M.G., Laub, A.J. and Hartmann, G.L., "Feedback properties of multivariable systems: The role and use of the return difference matrix", *IEEE Trans. Autom. Control,* vol. AC-26, pp. 47-65, 1981.

[12].   Chung, D., Kang, T. and Lee, J.G., "Stability robustness of LQ optimal regulators for the performance index with cross-product terms", *IEEE Trans. Autom. Control,* vol. AC-39, pp. 1698-1702, 1994.

[13].   Doyle, J.C., "Guaranteed margins for LQG regulators", *IEEE Trans. Autom. Control,* vol. AC-23, pp. 756-757, 1978.

[14].   Kondo, R. and Furuta, K., "Sampled-data optimal control of continuous systems for quadratic criterion function taking account of delayed control action", *Int. Journal Control,* vol. 41, pp. 1051-1060, 1985.

[15].   Pappas, T., Laub, A.J. and Sandell, N.R., "On the numerical solution of the discrete-time algebraic Riccati equation", *IEEE Trans. Autom. Control,* vol. AC-25, pp. 631-641, 1980.

[16].   Shaked, U., "Guaranteed stability margins for the discrete-time linear quadratic regulator", *IEEE Trans. Autom. Control,* vol. AC-31, pp. 162-165, 1986.

[17]. Arvanitis, K.G. and Kalogeropoulos, G., "Guaranteed stability margins for discrete-time LQ optimal regulators for the performance index with cross product terms", *Circuits Syst. Signal Process.,* vol. 16, pp. 663-701, 1997.

[18]. Al-Rahmani, H.M. and Franklin, G.F., "A new optimal multirate control of linear periodic and time-invariant systems", *IEEE Trans. Autom. Control,* vol. AC-35, pp. 406-415, 1990.

[19]. Al-Rahmani, H.M. and Franklin, G.F., "Multirate control: A new aproach", *Automatica,* vol. 28, pp. 35-44, 1992.

[20]. Van Loan, C.F., "Computing integrals involving the matrix exponential", *IEEE Trans. Autom. Control,* vol. AC-23, pp. 395-404, 1978.

[21]. Hagiwara, T. and Araki, M., "Design of a stable state feedback controller based on the multirate sampling of the plant output", *IEEE Trans. Autom. Control,* vol. AC-33, pp. 812-819, 1988.

[22]. Arvanitis, K.G., "An indirect model reference adaptive controller based on the multirate sampling of the plant output", *Int. J. Adaptive Contr. Signal Process.,* vol. 10, pp. 673-705, 1996.

[23]. Arvanitis, K.G. and Kalogeropoulos, G., "Stability robustness of LQ optimal regulators designed on the basis of the multirate sampling of the plant output", *J.Optimiz. Theory Appl.,* vol. 97, pp. 299-337, 1998.

[24]. Arvanitis, K.G. and Kalogeropoulos, G., "Stability robustness to unstructured uncertainties of linear systems controlled on the basis of the multirate sampling of the plant output", *IMA J. Math. Control Inform.,* vol. 15, pp. 241-268, 1998.

[25]. Arvanitis, K.G. and Paraskevopoulos, P.N., "Sampled-data minimum H∞-norm regulation of linear continuous-time systems using multirate-output controllers", *J. Optimiz. Theory Appl.,* vol. 87, pp. 235-267, 1995.

[26]. Arvanitis, K.G., "A new multirate LQ optimal regulator for linear time-invariant systems and its stability robustness properties", *Appl. Mathematics Comp. Science,* vol. 8, pp. 529-584, 1998.

[27]. Kabamba, P.T., "Control of linear systems using generalized sampled-data hold functions", *IEEE Trans. Autom. Control,* vol. AC-32, pp. 772-783, 1987.

[28]. Arvanitis, K.G., "Adaptive decoupling of linear systems using multirate generalized sampled-data hold functions", *IMA J. Math. Control Inform.,* vol. 12, pp. 157-177, 1995.

[29]. Arvanitis, K.G., Zacharenakis, E.C. and Soldatos, A.G., "Optimal noise rejection in structural analysis by means of generalized sampled-data hold functions", *J. Global Optimiz.,* vol. 17(1-4), pp. 19-42, 2000.

[30]. Weinmann, A., *Uncertain Models and Robust Control,* Wien: Springer-Verlag, 1991.

[31]. Chandrasekharan, P.C., *Robust Control of linear Dynamical Systems,* New York: Academic Press, 1996.

[32]. Soldatos, A.G. and Corless, M., "Stabilizing uncertain systems with bounded control", *Dynamics and Control,* vol. 1, pp. 227-238, 1991.

[33]. Leitmann, G., "Deterministic control of uncertain systems", *Acta Astronautica,* vol. 7, pp. 1457, 1980.

[34]. Leitmann, G., Ryan, E.P. and Steinberg, A., "Feedback control of uncertain systems: robustness with respect to neglected actuator and sensor dynamics", *Int. J. Control,* vol. 43, p. 1243, 1986.

[35]. Soldatos, A.G., Arvanitis, K.G. and Zacharenakis, E.C., "Active control schemes for aseismic base-isolated structures", *J. Applied Mechanics,* accepted for publication, 2000.

[36]. Zames, G., "Feedback and optimal sensitivity: model reference transformations, multiplicative seminorms and approximate inverses", *IEEE Trans. Autom. Control,* vol. AC-26, pp. 301-320, 1981.

[37]. Petersen, I.R., "Disturbance attenuation and H$^{\infty}$-optimization: A design method based on the algebraic Riccati equation", *IEEE Trans. Autom. Control,* vol. AC-32, pp. 427-429, 1987.

[38]. Doyle, J., Klover, K., Khargonekar, P. and Francis, B., "State-space solutions to standard H_∞ and H_2 control problems, *IEEE Trans. Autom. Control,* vol. AC-34, pp. 831-847, 1989.

[39]. Scherer, C., "H∞-control by state feedback: An iterative algorithm and characterization of high-gain occurrence", *Syst. Control Lett.,* vol. 12, pp. 383-391, 1989.

[40]. Stoorvogel, A., *The H_∞ Control Problem: A State Space Approach.* London: Prentice-Hall, 1992.

[41]. Green, M. and Limebeer, D.J.N., *Linear Robust Control,* Englewood Cliffs, New Jersey: Prentice Hall, 1995.

[42]. Yaesh, I. and Shaked, U., "Minimum H∞-norm regulation of linear discrete-time systems and its relation to linear quadratic discrete games", *IEEE Trans. Autom. Control,* vol. AC-35, pp. 1061-1064, 1990.

[43]. Grimble, M.J., *Robust Industrial Control: Optimal Design Approach for Polynomial Systems.* Hemel Hempstead, U.K.: Prentice Hall International Ltd., 1994.

[44]. Basar, T. and Bernhard, P., *H^∞-Optimal Control and Related Minimax Design Problems: A Dynamic Game Approach.* Boston, MA: Birkhauser, 1995.

[45]. Yaesh, I. and Shaked, U., "A transfer function approach to the problems of discrete-time systems: H∞-optimal linear control and filtering", *IEEE Trans. Autom. Control,* vol. AC-36, pp. 1264-1271, 1991.

[46]. Yaesh, I. and Shaked, U., "H∞-optimal one-step-ahead output feedback control of discrete-time systems", *IEEE Trans. Autom. Control,* vol. AC-37, pp. 1245-1250, 1992.

[47]. Geradin, M. and Rixen, D., *Mechanical Vibrations: Theory and Application to Structural Dynamics.* New York: John Wiley and Sons, 1997.

[48]. Nishimura, H. and Kojima, A., "Seismic isolation control for a buildinglike structure", *IEEE Control Syst. Magzine,* vol. 19, pp. 38-44, 1999.

[49]. Palazzo, B. and Petti, L., "Feedback control of base isolated systems", *Proc. 10th Europ. Conf. Earthquke Engineering,* Duma (ed.), Balkema, Rotterdam, pp. 2113-2119, 1995.

[50]. Chaplin, J. and Wu, C., "Dynamic modelling of field sprayers", *Trans. ASAE,* vol. 32, pp. 1857-1863, 1989.

[51]. Sinfort, C., Miralles, A., Sevila, F. and Maniere, G.M., "Study and development of a test method for spray boom suspensions", *J.Agr.Eng.Res.,*vol. 59, pp. 245-252, 1994.

[52]. Langenakens, J.J., Ramon, H. and De Baerdemaeker, J., "A model for measuring the effect of tire pressure and driving speed on horizontal sprayer boom movements and spray pattern", *Trans. ASAE,* vol. 38, pp. 65-72, 1995.

[53]. Ramon, H. and De Baerdemaeker, J., "Using principal gains for evaluating and optimizing the perfor-mance of sprayers", *Trans. ASAE,* vol. 38, pp. 1327-1333, 1995.

[54]. Ramon, H., Anthonis, J., Moschou, D. and De Baerdemaeker, J., "Evaluation of a cascade compensator for horizontal vibrations of a flexible spray boom", *J.Agr.Eng.Res.,*vol. 71, pp. 81-92-252, 1998.

[55]. Weaver, W. and Johnson, P., *Structural Dynamics by Finite Elements.* Englewood Cliffs, New Jersey: Prentice Hall, 1987.

[56]. Jabbari, F., Schmitendorf, W.E. and Yang, J.N. "H∞ control for seismic-excited buildings with acceleration feedback", *J.Enging. Mechanics,* vol. 121, pp. 994-1002, 1995.

[57]. Kelly, J.M., Leitmann, G. and Soldatos, A.G., "Robust control of base-isolated structures under earthquake excitation", *J. Optimiz. Theory Appl.*, vol.53, pp. 159-180, 1987.

[58]. Kelly, J.M., Leitmann, G. and Soldatos, A.G., "Seismic Protection of Structures Using Base Isolation and Active Control", *Proc. American Control Conference*, Minneapolis, Minnesota, pp. 1885-1889, 1987.

[59]. Er, M.J. and Anderson, B.D.O., "Practical issues in multirate-output controllers", *Int.J.Control,* vol. 53, pp. 1005-1020, 1991.

[60]. Preumont, A., *Vibration Control of Active Structures*. Kluwer Academic Publishers, 1997.

[61]. Beards, C., Engineering Vibration Analysis with Application to Control Systems. Edward Arnold Press, 1995.

[62]. Gaul, L., Ströbener, U. Active Control of Structures. In: *Modal Analysis and Testing,* J.M. M. Silva and N.M.M. Maia (editors), pp. 453-486, Kluwer Academic Publishers, 1999.

[63]. Zacharenakis, E.C. "On the input-output decoupling with simultaneous disturbance attenuation and H∞ optimization in structural analysis", *Computers and Structures,* vol. 60, pp. 627-633, 1996.

[64]. Zacharenakis, E.C. . "On the disturbance attenuation and H∞ optimization in structural analysis", *ZAMM,* vol. 77, pp. 189-195, 1997.

[65]. Zacharenakis, E.C. and Stavroulakis, G.E., "On the seismic disturbance rejection of structures", *J. Global Optimiz.,* vol. 17(1-4), pp. 403-410, 2000.

IONIC POLYMER-CONDUCTOR COMPOSITES (IPCC) AS BIOMIMETIC SENSORS, ACTUATORS AND ARTIFICIAL MUSCLES

MOHSEN SHAHINPOOR

Artificial Muscle Research Institute, School of Engineering and School of Medicine
The University of New Mexico, Albuquerque, NM 87131
ARDESHIR GURAN

Institute of Structronics, 275 Slater Street, 9th Floor, Ottawa, Canada K1P-5H9

The fundamentals of Ionic Polymer-Conductor Composites (IPCC's) as biomimetic sensors/transducers/actuators and artificial muscles are briefly reviewed in this short paper. Briefly, the current state-of-the art IPCC manufacturing techniques, governing phenomenological laws, and mechanical and electrical characteristics are described.

1 Introduction

It is now well documented that ionic polymers (such as a perfluorinated sulfonic acid polymer, i.e. Nafion™) in a composite form with a conductive metallic medium (herein, called Ionic Polymer-Conductor Composites, IPCC's) can exhibit large dynamic deformation if placed in a time-varying electric field (see Figure 1) [1-6]. Conversely, dynamic deformation of such ionic polymers produces dynamic electric fields (see Figure 2). A recently presented model by de Gennes, Okumura, Shahinpoor, and Kim [7] describes the underlying principle of electro-thermodynamics in ionic polymers based upon internal ionic transport and electrophoresis. It is obvious that IPCC's show a great potential as soft robotic actuators, artificial muscles, and dynamic/static sensors and transducers in micro-to-macro size scales.

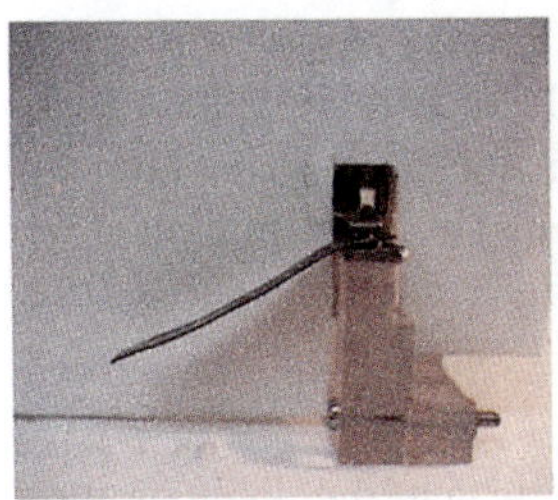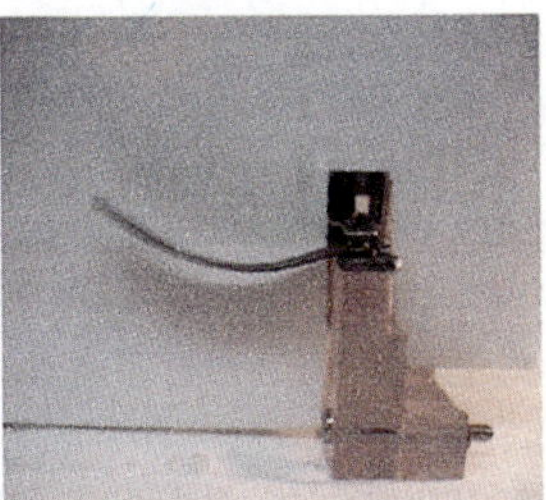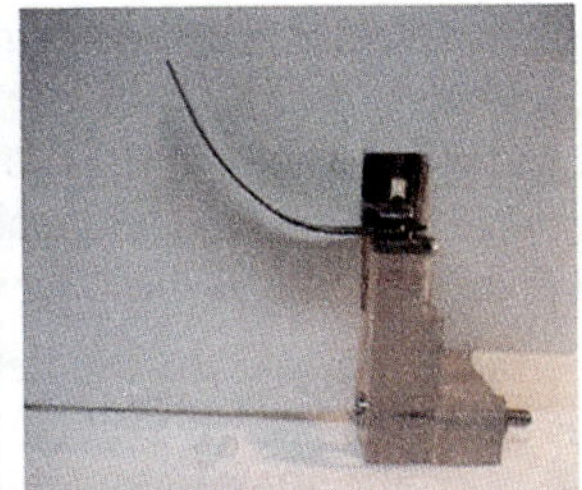

Figure 1. Successive photographs of an IPCC strip that shows very large deformation (up to 4 cm) in the presence of low voltage. Note that Δt=0.5 sec, 2 volts applied. The sample is 1 cm wide, 4 cm long, and 0.2 mm thick.

Manufacturing of an IPCC starts with Ion Exchange Membranes (IEM) via. metal compositing by means of chemical reduction processes. The term *Ion Exchange Membranes* refers to materials designed to selectively pass through ions of a single charge (either cations or anions). They are often manufactured from polymers that consist of fixed covalent ionic groups. The currently available IEM are:

(i) Perfluorinated alkenes with short side-chains terminated by ionic groups [typically sulfonate or carboxylate (SO_3^- or COO^-) for cation exchange or ammonium cations for anion exchange (see Figure 3)]. The large polymer backbones determine their mechanical strength. Short side-chains provide ionic groups to interact with waters being served as the passage of appropriate ions. A popular product is Nafion^{TM} of Du Pont Co.

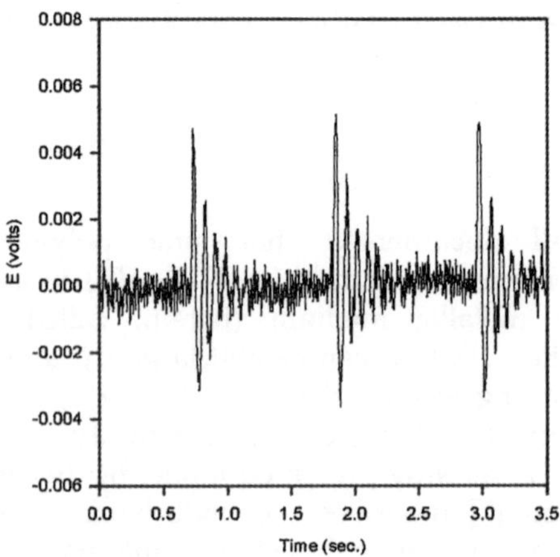

Figure 2. A typical sensing response of an IPCC.

It shows dynamic sensing response of a strip of an IPCC (a thickness of 0.2 mm) subject to a dynamic impact loading in a cantilever configuration. A damped oscillatory electric response is observed which is highly repeatable with a high bandwidth of up to 1000's of Hz. Such direct mechanoelectric behaviors are related to the endo-ionic mobility due to imposed stresses. This implies that, if we impose a finite solvent (=water) flux, $|Q|$, not allowing a current flux, $J=0$, a certain conjugate electric field, $\vec{E}$, is produced that can be dynamically monitored as discussed later in Section 3.

(ii) Styrene/divinylbenzene-based polymers in which the ionic groups have been substituted from the phenyl rings where the nitrogen atom to fix an ionic group. These polymers are highly crosslinked and are rigid. Ionic groups are high and analogous to gels. A popular product is CR/AR series of Ionics Inc.

In Nafion^{TM} there are relatively few fixed ionic groups. They are located at the end of side-chains so as to position themselves in their preferred orientation at some extent. Therefore, it can create hydrophilic nano-channels, so called *cluster networks.* Such a fact is completely different in nature relative to styrene/divinylbenzene-based ones that are primarily limited by cross-linking, ability of the IEM to expand (due to hydrophilic nature).

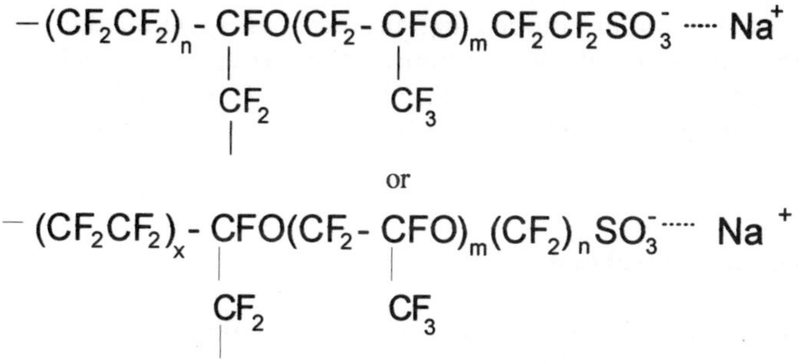

$$-(CF_2CF_2)_n - \underset{\underset{CF_2}{|}}{C}FO(CF_2 - \underset{\underset{CF_3}{|}}{C}FO)_m CF_2CF_2SO_3^- \cdots Na^+$$

or

$$-(CF_2CF_2)_x - \underset{\underset{CF_2}{|}}{C}FO(CF_2 - \underset{\underset{CF_3}{|}}{C}FO)_m (CF_2)_n SO_3^- \cdots Na^+$$

Figure 3. Perfluorinated sulfonic acid IEMs. The counter ion, Na^+ in this case, can simply replaced by other ions.

2 Manufacturing Techniques

The preparation of ionic polymer-metal composites (IPCC's) requires intense laboratory work. The current state-of-the-art IPCC manufacturing technique [4, 5] incorporates two distinct preparation processes: *initial compositing process* and *surface electroding process.* Due to different preparation processes, morphologies of precipitated platinum are significantly different. The initial compositing process requires an appropriate platinum salt such as $Pt(NH_3)_4HCl$ in the context of chemical reduction processes similar to the processes evaluated by a number of investigators including Takenaka [8] and Millet [9]. The principle of the compositing process is to metalize the inner surface of the membrane by a chemical-reduction means such as $LiBH_4$ or $NaBH_4$. The IEM is soaked in a salt solution to allow platinum-containing cations to diffuse through the thin membrane *via* the ion-exchange process. Later, a proper reducing agent such as $LiBH_4$ or $NaBH_4$ is introduced to platinize the membrane.

The platinum layer is buried submicron deep (typically 1-20 μm) within the IPCC surface and is porous. The fabricated muscles can be optimized for producing a maximum force by changing multiple process parameters including time-dependent concentrations of the salt and the reducing agents (applying the Taguchi technique to identify the optimum process parameters seems quite attractive [10]). The primary reaction is,

$$LiBH_4 + 4[Pt(NH_3)_4]^{2+} + 8OH^- \Rightarrow 4Pt^o + 16NH_3 + LiBO_2 + 6H_2O \tag{1}$$

In the subsequent surface electroding process, multiple reducing agents are introduced (under optimized concentrations) to carry out the reducing reaction similar to Equation (1), in addition to the initial platinum layer formed by the initial compositing process. In general, the majority of platinum salts stays in the solution and precedes the reducing reactions and production of platinum metal. Other metals (or conducing mediums) also successfully used include palladium, silver, gold, carbon, graphite, and nanotubes.

3   Phenomenological Law

A recent study by de Gennes, Okumura, Shahinpoor, and Kim [4] has presented the standard Onsager formulation on the underlining principle of IPCC actuation/sensing phenomena using linear irreversible thermodynamics: when static conditions are imposed, a simple description of *mechanoelectric effect* is possible based upon two forms of transport: *ion transport* (with a current density, *J*, normal to the material) and *electrophoretic solvent transport* (with a flux, *Q*, *we can assume that this term is water flux*). The conjugate forces include the electric field, $\vec{E}$, and the pressure gradient, $-\nabla p$. The resulting equation has the concise form of,

$$J = \sigma\vec{E} - L_{12}\nabla p \tag{2}$$

$$Q = L_{21}\vec{E} - K\nabla p \tag{3}$$

where σ and K are the membrane conductance and the Darcy permeability, respectively.

A cross coefficient is usually $L_{12} = L_{21} = L$, believed to be on the order of 10^{-8} (ms^{-1})/(volts-m^{-1}). The simplicity of the above equations provides a compact view of underlining principles of both actuation and sensing of IPCC. So, we can illustrate it simply in Figure 4.

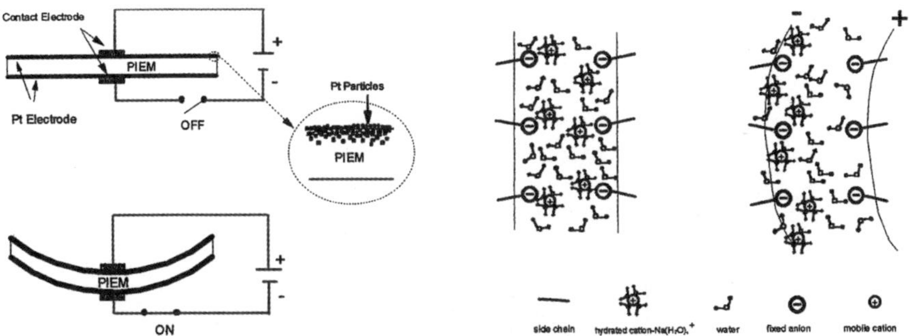

Figure 4. A schematic of the typical IPCC artificial muscle and its actuation principle.

Several models to describe the richness variation along a disturbance gradient have been proposed [11,13,15]; however, these models are either phenomenological [15] or very general [12], and they fail to predict, with the same mathematical and-biological framework, a variety of data sets [1]. Models like those proposed by Dayton and Hessler [8], Connell [6], Huston [12] and Barradas [2] predict highest species richness under intermediate perturbations, but, in some cases [6,13], their predictions are qualitative; in the case of [2] the model only consider competitive interactions. There are no testable mathematical models considering simultaneously competition and stress. Moreover, the classical succession theory [5] states the existence of immature communities in fluctuating environments and mature or climacic communities in stable environments; on the light of this theory, the possibility of a low richness at intermediate disturbance regimes becomes plausible.

4   Characteristics

Mechanical

Figure 5 shows tensile testing results, in terms of normal stress versus normal strain, on a typical IPMC (H^+ form) relative to Nafion^{TM}-117 (H^+ form). Recognizing that Nafion^{TM}-117 is the adopted starting material for this IPMC, this comparison is useful. There is a little increase in mechanical strength of IPMC (both the stiffness and the modulus of elasticity), but it still follows the intrinsic nature of Nafion itself. This means that, in the tensile (positive) strain, the stress/strain behavior is predominated by the polymer material rather than metallic powders (composited electrode materials).

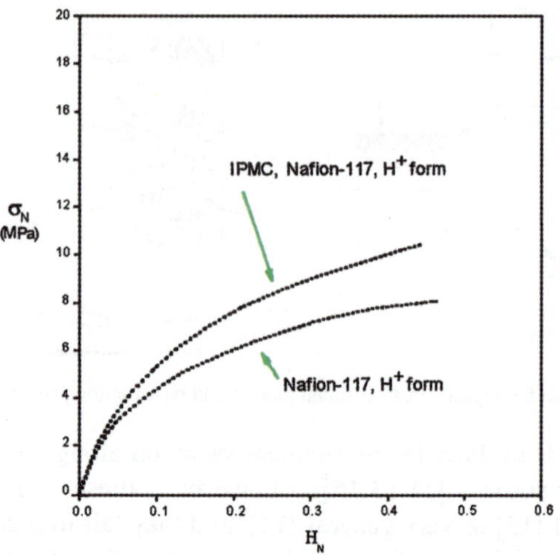

Figure 5. Tensile testing results.

Figure 5 shows normal stress, σ_N, vs. normal strain, ε_N; IPMC and Nafion-117™. Note that both samples were fully hydrated when they were tested.

Although the tensile testing results show the intrinsic nature of the IPMC, a problem arises when the IPMC operates in a bending mode. Dissimilar mechanical properties of the metal particles (the electrode) and polymer network seem to affect each other. Therefore, in order to construct the effective stress-strain curves for IPMCs, strips of IPMCs are suitably cut and tested in a cantilever configuration. In a cantilever configuration, the end deflection δ due to a distributed load $w(s,t)$, where s is the arc length of a beam of length L and t is the time, can be related approximately to the radius of the curvature ρ of the cantilever beam, i.e.,

$$\rho \cong \frac{L^2 + \delta^2}{2\delta} \tag{4}$$

Note that the radius of curvature ρ is in turn related to the maximum tensile (positive) or compressive (negative) strains in the beam as,

$$\varepsilon \cong \frac{h}{2\rho} \tag{5}$$

where h is the thickness of the beam at the built-in end. Note that in the actuation mode of the IPMC, the tensile strain can be simply realized, but difficult to isolate. In the negative strain (material compression) the metal particles become predominant so as to experience much higher stiffness and modulus of elasticity than the ones in the positive strain regime. Thus, the mathematical description regarding the physics of the cantilever beam of the IPMC is somewhat challenging and should be addressed carefully. Obviously, experimental approaches are available and should be pursued. The stress σ can be related to the strain ε by simply using Hooke's law, assuming linear elasticity or non-linear Neo-Hookean type constitutive equations. This leads to,

$$\sigma = \frac{Mh}{2I} \tag{6}$$

where σ is the stress tensor, M is the maximum moment at the built-in end and I is the moment of inertia of the cross-section of the beam. Thus, the moment M can be calculated based on the distributed load on the beam or the applied electrical activation of the IPMC beam. Having also calculated the moment of inertia I, which for a rectangular cross-section of width b will be $I = bh^3/12$, the stress σ can be related to the strain ε and the representative results are plotted in Figure 6-(i) and (ii). These figures include the effect of swelling (i) and stiffening behavior under electric activation (ii). Here, electric activation refers the IPMC in the electro-mechanical mode exhibiting increased stiffness due to redistributed hydrated ions or non-linear characteristics of electromechanical properties of the IPMC. A further investigation of such observed findings is currently underway.

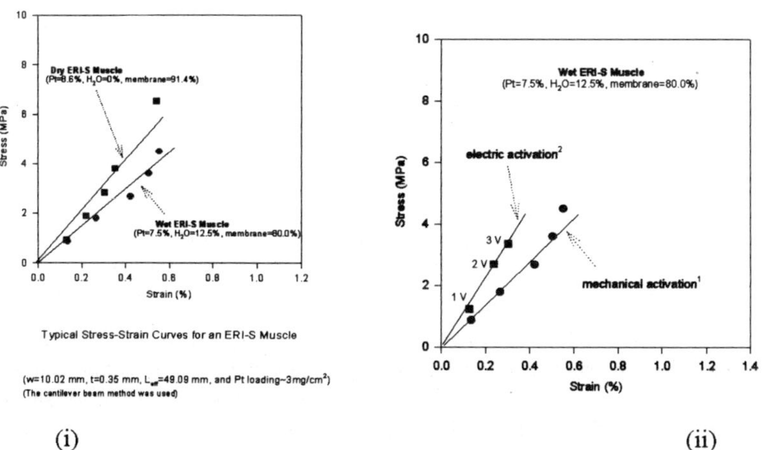

(i)   (ii)

Figure 6. A cantilever configuration of the IPMC (i) and an illustration of positive/negative strains experienced in the operation mode of the IPMC.

Note in Figure 6 that the bottom graphs show the effect of swelling (i) and stiffening behavior under electric activation (ii).   Swelling is also an important parameter to affect the mechanical property.   Indicatively speaking, swelling causes mechanical weakening.   Electric activation has a tendency to stiffen the material due to redistributed ions within the IPMC.

Electrical

In order to assess the electrical properties of the IPMC, the standard AC impedance method, that can reveal the equivalent electric circuit, has been adopted.   A typical measured impedance plot, provided in Figure 7, shows the frequency dependency of impedance of the IPMC.   Overall, it is interesting to note that the IPMC is nearly resistive (> 50Ω) in the high frequency range and fairly capacitive (> 100 μF) in the low frequency range.

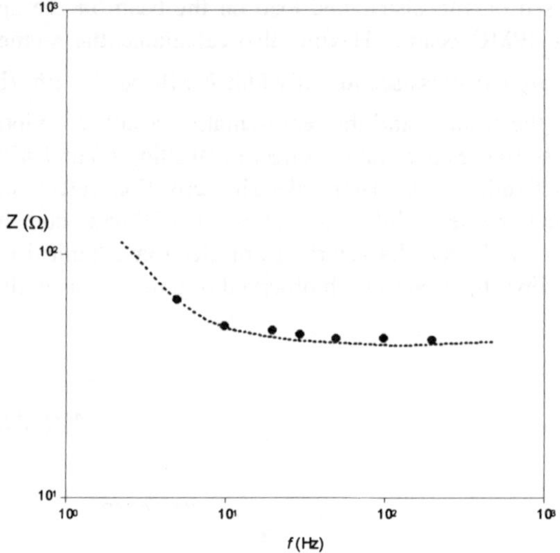

Figure 7. The measured AC impedance characteristics of an IPMC sample (the wet IPMC sample has a dimension of 5 mm width, 20 mm length, and 0.2 mm thickness).

Based upon the above findings, we consider a simplified equivalent electric circuit of the typical IPMC such as the one shown in Figure 8 [6]. In this approach, each single unit-circuit (i) is assumed to be connected in a series of arbitrary surface-resistance (R_{ss}) in the surface.   This approach is based upon the experimental observation of the considerable surface-electrode resistance (see Figure 8).   We assume that there are four components to each single unit-circuit: the surface-electrode resistance (R_s), the polymer resistance (R_p), the capacitance related to the ionic polymer and the double layer at the surface-electrode/electrolyte interface (C_d) and an intricate impedance (Z_w) due to a charge transfer resistance near the surface electrode. For the typical IPMC, the importance of R_{ss} relative to R_s may be interpreted from $\Sigma R_{ss} / R_s \approx L/t >> 1$, where notations L and t are the length and thickness of the electrode (therefore, it becomes two dimensional, considering that the typical values of t is ~ 1-10 μm, so this statement is valid). So, a significant overpotential is required to maintain the effective voltage condition along the surface of the typical IPMC. An effective technique to solve this problem is to overlay a thin layer of a highly conductive metal (such as gold) on top of the platinum surface-electrode [6].

Realizing that water residing in the perfluorinated IPMC is the sole solvent that can create useful strains in the actuation mode, another issue to be dealt with is the so-called "decomposition voltage." As can be clearly seen in Figure 9, the decomposition voltage is the minimum voltage at which significant electrolysis occurs. This figure contains the graph of steady-state current, I, versus applied DC voltage, E_{app}, showing that as the voltage increases, there is little change in current (obeying Faraday's Law). However, a remarkable increase in DC current is observed with a small change of voltage. Even though the intrinsic voltage causing water electrolysis is about 1.23 V, a small overpotential (*ca.* 0.3-0.5 V) was observed. It should be noted that such water electrolysis leads to lower thermodynamic efficiency of the IPMC.

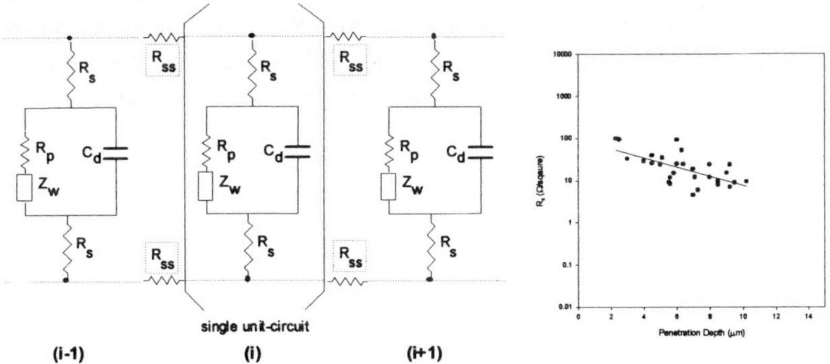

Figure 8. A possible electric circuit equivalent to the typical IPMC (left) and measured surface resistance, R_s, as a function of platinum penetration depth (right).

Note that SEM was used to estimate the penetration depth of platinum into the membrane. The four-probe method was used to measure the surface resistance, R_s, of the IPMCs. Obviously, the deeper the penetration, the lower the surface resistance.

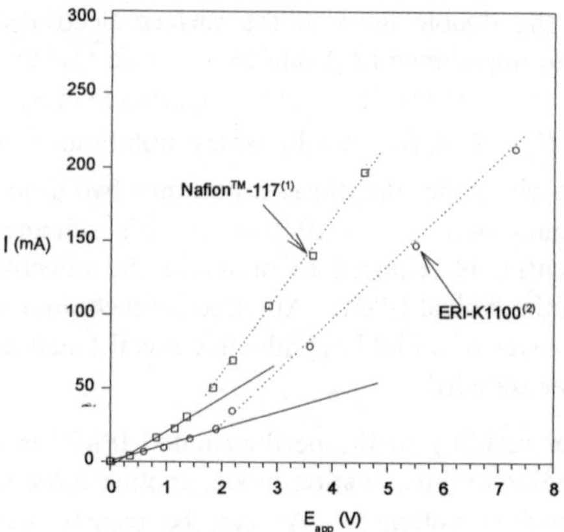

Figure 9. Steady state current, I, versus, applied voltages, E_{app}, on the typical IPMCs.

ERI-K1100 stands for a propriety IPMC fabricated by Environmental Robots, Inc – it has a thickness of 2.9 mm and is suitably platinum/gold electroded. Figure 10 depicts measured cyclic current/voltage responses of a typical IPMC (the scan rate of 100 mV/sec). As can be seen, a rather simple behavior with a small hysteresis is obtained. Note that the reactivity of the IPMC is mild such that it does not show any distinct reduction or re-oxidation peaks within +/- 4 volts, except for a decomposition behavior at $\sim \pm 1.5\,\mathrm{V}$ where the extra current consumption is apparent due to electrolysis. Overall behavior of the IPMC shows a simple behavior of ionic motions caused under an imposed electric field.

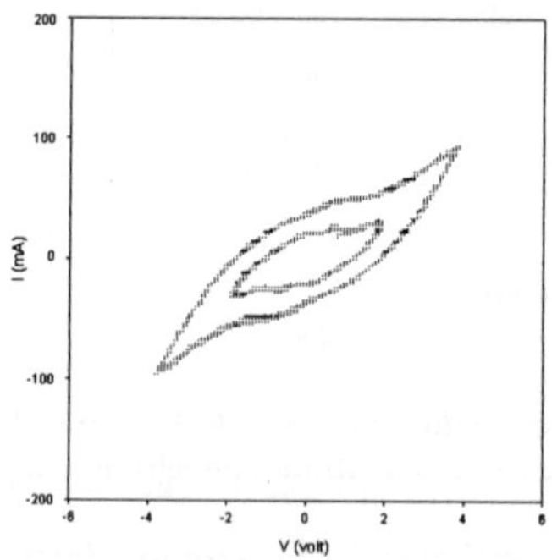

Figure 10. I/V curves for a typical IPMC. (Nafion™-117 based IPMC)

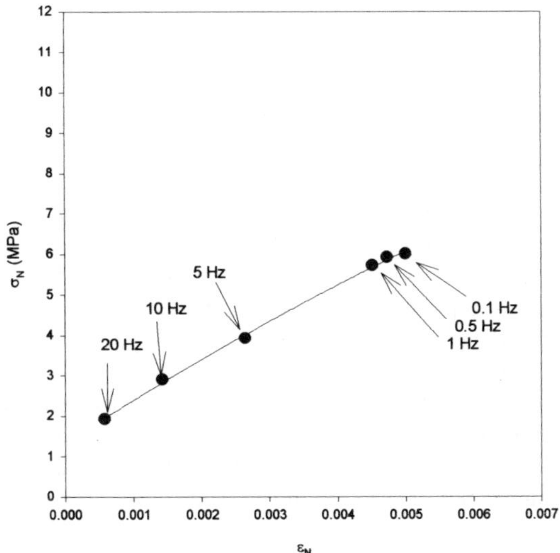

Figure 11. Frequency dependency of the IPMC in terms of the normal stress, σ_N, vs. the normal strain, ε_N, under an imposed step voltage of 1 V.
(This Nafion^{TM}-117 IPMC has a cation of Li$^+$ and a size of 5 x 20 mm)

In Figure 11, frequency dependency of the IPMC is expressed in terms of the normal stress versus the normal strain. Its frequency dependency follows the fact that as frequency increases the beam displacement decreases. However, a careful observation must be given to realize that, at low frequencies (0.1-1 Hz), the effective modulus by the IPMC under an imposed voltage seems to be rather small meaning that water transport is high such that it gushes out of the porous surface electrode, mainly platinum particles, (one can normally and routinely observe this effect in the laboratory) and does not return to the polymer network (this causes large irreversibility). On the other hand, at high frequencies (5-20 Hz) such an effect is mild so that most of water transport phenomena occur within the polymer network (since it does not have enough time to gush out of the surface electrode). Therefore, water transport within the IPMC can experience larger modulus than that at low frequencies. This is of interest in a similar analogy to hydraulic actuators. Obviously, water leakage is a definite disadvantage in achieving high efficiency for the IPMC. This issue should be resolved to obtain much higher efficiency and specific power of the IPMC.

A recent work by Shahinpoor and Kim [11] was an attempt to identify effective dispersion agents (additives). As a successful outcome, the use of the effective dispersing agent during the platinum metalization process has given dramatically improved force density characteristics. The results are shown in Figure 12 where reported is the measured force of the improved IPMC relative to conventional one. As clearly seen, the additive treaded IPMC has shown:
  i)   a much sharper response to the input electric field and
  ii) a dramatically increased force density generation by as much as 100%.
One key observation is that the virtual disappearance of the delay response, which has been often observed in the conventional IPMC, prevents the water leakage from the surface electrode (an analogy to *sand bagging*). Such an effect can be translated into a higher power IPMC than any other IPMCs reported so far [11].

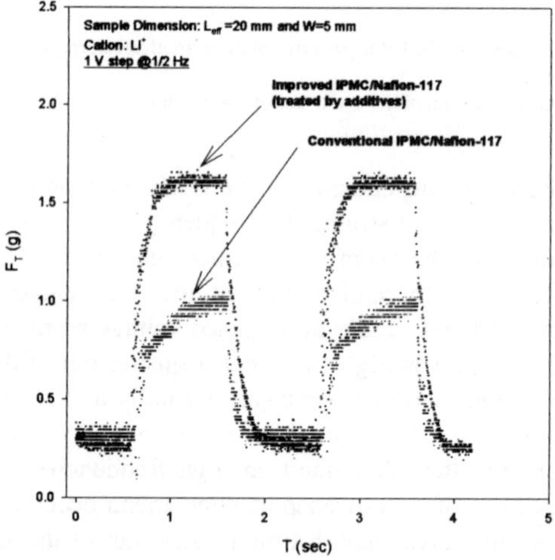

Figure 12. Force response characteristics of the improved IPMC versus the conventional IPMC.

In Figure 13, a SEM micrograph along with its X-ray line-scan is provided. As can be seen, a good platinum penetration is achieved meaning that an effective additive enhances platinum dispersion that leads to better penetration into the

IEM. A convenient way to handle this situation (free diffusion into finite porous slab or membrane) is to use an effective diffusivity, D_{eff}, and, then, to consider it as one-dimensional. Assuming that fast kinetics for the metal precipitation reaction of $\left[Pt(NH_3)_4\right]^{2+} + 2e^- \Rightarrow Pt^o + 4NH_3$, the precipitated platinum concentration, N_x, can be expressed as,

$$N_x = \frac{C_{PT}(\delta_i)}{C_{Pt,i}} = 1 - erf\left(\frac{\delta_i}{\sqrt{4D_{eff}t}}\right) \tag{7}$$

where notations $C_{Pt}(\delta_i)$, $C_{Pt,i}$, and δ_i are the platinum concentration, the platinum concentration at the interface, and the particle penetration depth, respectively. For a typical reduction time of $t=15$ minutes (in Figures 12 and 13), Equation (7) is plotted for values of $D_{eff}=1\text{x}10^{-10}$, $1\text{x}10^{-9}$, and $1\text{x}10^{-8}$ cm^2/sec, respectively. The effective diffusivity, D_{eff}, could be estimated on the order of $1\text{x}10^{-8}$ cm^2/sec for the improved IPMC. Although this situation is somewhat complicated due to simultaneous effect of a mass transfer and significant kinetics involved, nevertheless, the estimated value of $D_{eff}\sim1\text{x}10^{-8}$ cm^2/sec, would be a convenient value for engineering the platinum metalization process described here for the improved IPMC.

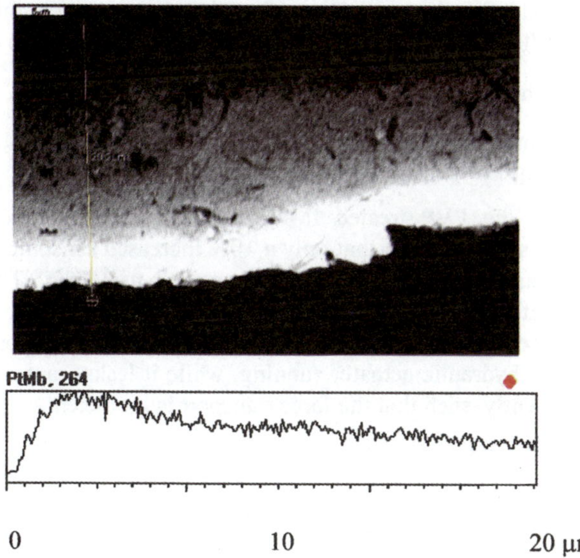

0   10   20 μm

Figure 13. A SEM micrograph of an IPMC treated with a dispersing agent (top) and its X-ray line-scan (bottom).

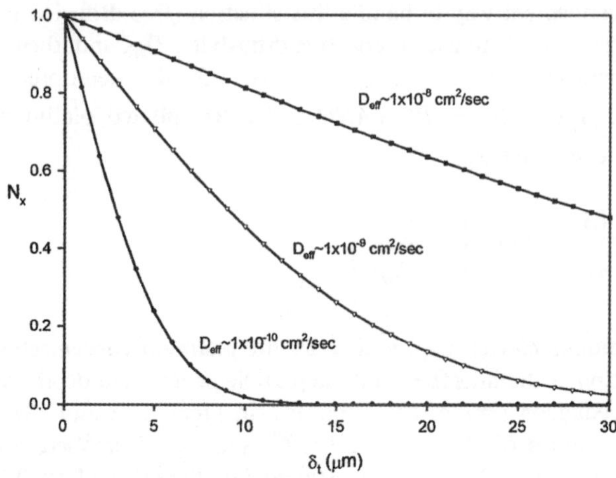

Figure 14. Platinum penetration profiles in an IPMC sample.

In Figure 15, the results of potentiostatic analysis are presented. Assuming that there are no side reactions (other than electrolysis) that consume current. The current passed following the application of an electric potential to the IPMCs (both the PVP treated IPMC and the conventional IPMC) is shown. The current decays exponentially. The charge passed after time t (Q_t) is, $Q_t = \int_0^t I_t dt$. It is useful to make a direct comparison between $Q_{t,PVP}$ (for the PVP treated IPMC) and Q_t (for the conventional IPMC). The data shown in Figure 19 gives $Q_{t,PVP}/Q_t \cong 1.1$. This means that the PVP treated IMPC consumes approximately 10% more charges. This raises a question that only a 10% increased consumption of charges is not the only reason to increase the force by as much as 100%. Therefore, it can be interpreted that "the sand-bagging effect" that prevents the water leakage out of the porous surface electrode, when the IPMC strip bends, is important. A good analogy may be a hydraulic actuator running, while it leaks, causes the solvent to drain out permanently, such that the force transport is inefficient.

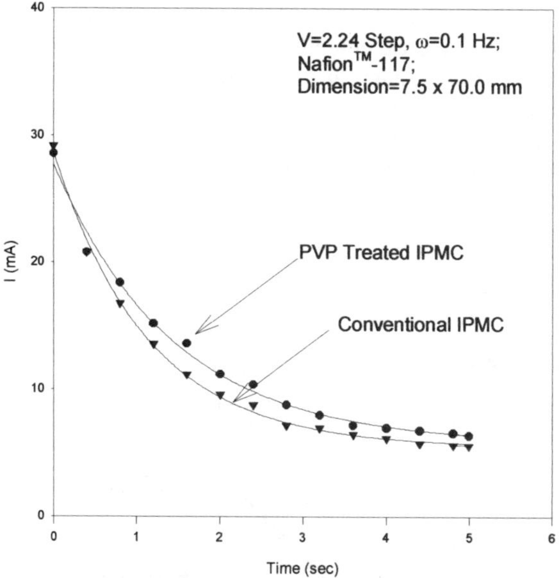

Figure 15. Potentiostatic Coulometric analysis for the additive treated IPMC and the conventional IPMC.

High Force Density IPMC – A View from Linear Irreversible Thermodynamics

In connection with the phenomenological laws and irreversible thermodynamics considerations previously discussed in Section 3, when one considers the actuation with ideal impermeable electrodes, which results in $Q=0$ from Equation (3), one has

$$\nabla p = \frac{L}{K}\vec{E} \tag{8}$$

Also, the pressure gradient can be estimated from,

$$\nabla p \cong \frac{2\sigma_{max}}{h} \tag{9}$$

where σ_{max} and h are the maximum stress generated under an imposed electric field and the thickness of the membrane, respectively. The values of σ_{max} can be obtained when the maximum force (=blocking force) was measured at the tip of the IPMC per a given electric potential. In Figure 16, we present the maximum stresses generated, σ_{max}, under an imposed electric potentials, E_o, for both calculated values and experimental values of the conventional IPMC and the improved IPMC. It should be noted that the improved IPMC (by the method of using additives) is superior to the conventional IPMC approaching the theoretically obtained values.

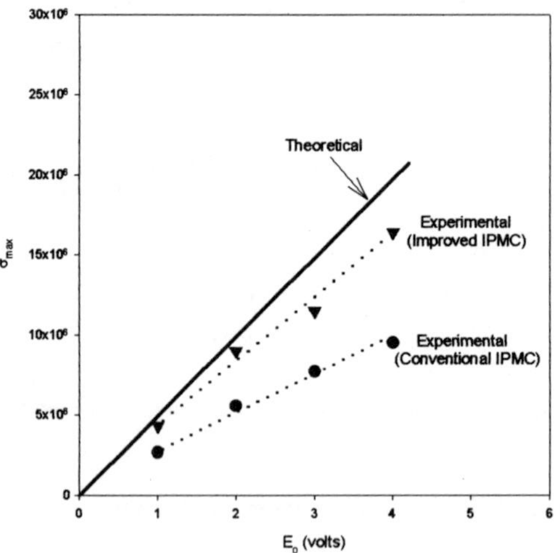

Figure 16. Maximum stresses generated by the IPMC at given voltages.

For theoretical calculation the following values were used; i) $L_{12}=L_{21}=2\times10^{-8}$ {cross coefficient, (m/s)/(V/m)}, ii) $k=1.8\times10^{-18}$ {hydraulic permeability, m^2 [12]}, and iii) $\vec{E} = E_0 / h$ where h=200 μm {membrane thickness}.

High Force Density IPMC – Thermodynamic Efficiency

The bending force of the IPMC is generated by the effectively strained IPMC due to water transport due to ionic redistribution. This is an ion-induced hydraulic actuation phenomenon. Typically, such a bending force is field-dependently distributed along the length of the IPMC strip. Further, a surface voltage drop occurs which can be minimized [6].   The IPMC strip bends due to these ion migration-induced hydraulic actuation and redistribution. The total bending force, F_t, can be approximated as,

$$F_t \cong \int_0^L f \, dL \tag{10}$$

where f is the force density per unit length. Assuming a uniformly distributed load over the length of the IPMC, then, the mechanical power produced by the IPMC can be obtained from,

$$P_{out} = F_t \int_{-S}^{S} v \, dS \tag{11}$$

Notations S and v are the arc length and tip velocity of the IPMC in action. Finally, the thermodynamic efficiency, $E_{ff,em}$, can be obtained as,

$$E_{ff,em} \ (\%) = \frac{P_{out}}{P_{in}} \times 100 \tag{12}$$

where P_{in} is the electrical power input to the IPMC. Based on Equation (12), one can construct a graph (see Figure 17) which shows the thermodynamic efficiency of the IPMC as a function of frequency.

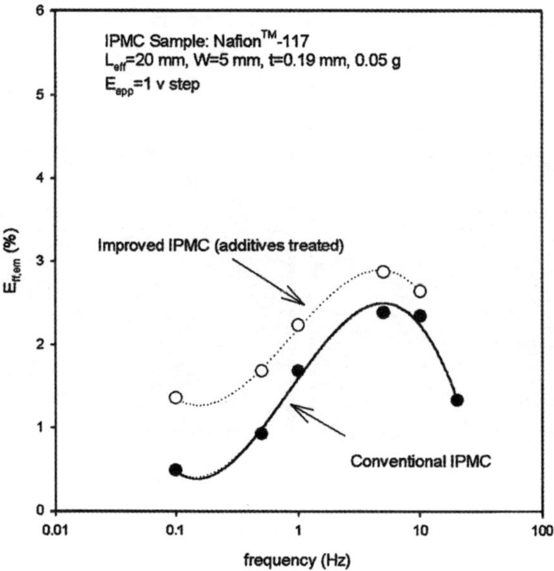

Figure 17. Thermodynamic efficiency of the IPMC as a function of frequency.

Note that this graph presents the experimental results for the conventional IPMC and the additive-treated improved IPMC. It is of note that the optimun efficiencies occur at near 5-10 Hz for these IPMCs. The optimum values of these IPMCs are approximately 2.5-3.0 %. At low frequencies, the water leakage out of the surface electrode seems to cost the efficiency significantly. However, the additive-treated IPMC shows a dramatic improvement in efficiency since water transports within the IPMC rather than gushing out. This is a definite benefit. The important sources of energy consumption for the IPMC actuation could be from

i) the necessary mechanical energy needed to cause the positive/negative strains for the IPMC strip,

ii) the I/V hysteresis due to the diffusional water transport within the IPMC,

iii) the thermal losses-Joule heating (see Figure 18),

iv) the decomposition due to water electrolysis, and v) the water leakage out of the electrode.

Despite our effort to improve the performance of the IPMC by blocking water leakage out of the porous surface electrode, the overall thermodynamic efficiencies of all IPMC samples tested in a frequency range of 0.1-10 Hz remain somewhat low. However, it should be noted that the obtained values are favorable realizing that other types of bending actuators, i.e. conducting polymers and piezoelectric materials at similar conditions, exhibit considerablely lower efficiencies [13, 14].

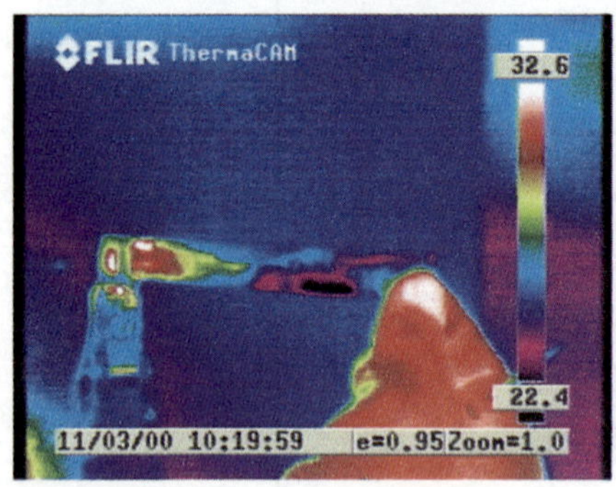
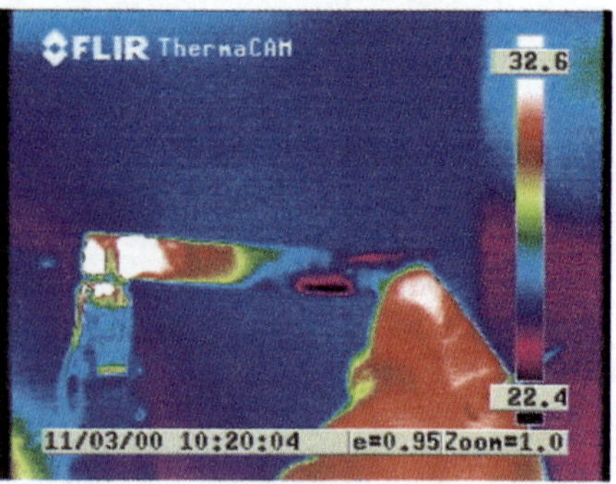

t= 0 sec;   t= 5 sec;

Figure 18. IR thermographs taken for an IPMC in action.

The hot stop starts from the electrode and propagates toward the tip of the IPMC strip. The temperature difference is more than 10 degrees C when a DC voltage of 3 was applied for the IPMC sample size of 1.2 x 7.0 cm.

5   CLOSURE AND FUTURE CHALLENGES

In this paper, the fundamental properties and characteristics of Ionic Polymeric-Metal Composites (IPMCs) as biomimetic sensors, actuators and artificial muscles are briefly presented in a summary form. The present paper presented a summary of recent findings and its current state-of-the art manufacturing techniques, phenomenological law, and mechanical and electrical characteristics.

It is obvious that the successful commercialization of the IPMC is highly dependent upon further improvement of its thermodynamic efficiency and force density generation. In particular, the dimensional scale-up of the IPMC is of a major factor enhancing its useful forces. The ultimate goal of using the IPMC as soft biomimetic sensors and actuators can be clearly envisioned upon the successful development of three-dimensional systems of IPMC's to be used as electrically controllable/deformable three-dimensional smart structures.   In future communications, we will discuss our effort to achieve these challenges.

6   Acknowledgements

This work was partially supported by U.S. NRL/DARPA. The authors thank the laboratory work done by Environmental Robots, Inc. and the Artificial Muscle Research Institute of the University of New Mexico. Discussions with Prof. P. G. De Gennes at the College de France were helpful.

7   References

1. K. Oguro, K. Asaka, and H. Takenaka, "Actuator Element," *U.S. Patent* #5,268,082 (1993).
2. A. Asada, K. Oguro, Y. Nishimura, M. Misuhata, and H. Takenaka, "Bending of Polyelectrolyte Membrane-Platinum Composites by Electric Stimuli, I. Response Characteristics to Various Wave Forms," *Polymer Journal*, **27**, pp. 436-440 (1995).
3. M. Shahinpoor, D. Adolf, D. Segalman, and W. Witkowski, "Electrically Controlled Polymeric Gel Actuators," *U.S. Patent* #5,250,167 (1993).
4. M. Shahinpoor and M. Mojarrad, "Soft Actuators and Artificial Muscles," *U.S. Patent*, #6,109,852 (2000).
5. M. Shahinpoor, Y. Bar-Cohen, J. O. Simpson, and J. Smith, "Ionic Polymer-Metal Composites (IPMC) as Biomimetic Sensors and Structures-A Review," *Smart Materials and Structures*, **7**, pp. 15-30 (1998).

6. M. Shahinpoor and K. J. Kim, "The Effect of Surface-Electrode Resistance on the Performance of Ionic Polymer-Metal Composites (IPMC) Artificial Muscles, *Smart Materials and Structures*, **9**, pp. 543-551 (2000).

7. P. G. de Gennes, K. Okumura, M. Shahinpoor, and K. J. Kim, "Mechanoelectric Effects in Ionic Gels," *Europhysics Letters*, **50(4)**, pp. 513-518 (2000).

8. H. Takenaka, E. Torikai, Y. Kwami, and N. Wakabayshi, "Solid Polymer Electrolyte Water Electrolysis," *International Journal of Hydrogen Energy*, 7, pp. 397-403 (1982).

9. P. Millet, M. Pineri, and R. Durand, "New Solid Polymer Electrolyte Composites for Water Electrolysis," *Journal of Applied Electrochemistry*, **19**, pp. 162-166 (1989).

10. K. J. Kim, M. Shahinpoor, and A. Razani, "Preparation of IPMCs for Use in Fuel Cells, Electrolysis, and Hydrogen Sensors," *Proceedings of SPIE 7th International Symposium on Smart Structures and Materials*, **3687**, pp. 110-120 (March 2000).

11. D. M. Bernardi and M. W. Verbrugge, "A Mathematical Model of the Solid-Polymer-Electrolyte Fuel Cell," *Journal of Electrochemical Society*, **139(9)**, pp. 2477-2491 (1992).

12. Q. Wang, X. Du, B. Xu, and L. E. Cross, "Electromechanical Coupling and Output Efficiency of Piezoelectric Bending Actuators," *IEEE Transactions on Ultrasonic, Ferroelectrics, and Frequency Control*, **46(3)**, pp. 638-646 (1999).

13. Q. Pei, O. Inganas, and I. Lundstrom, "Bending Bilayer Strips Built from Polyaniline for Artificial Electrochemical Muscles," *Smart Materials and Structures*, **2(1)**, pp. 1-5 (1993).

Index